Basic Biomechanics

Basic Biomechanics

THIRD EDITION

Susan J. Hall, Ph.D.

Department of Kinesiology
The University of Texas at El Paso

WCB
McGraw-Hill

Boston   Burr Ridge, IL   Dubuque, IA   Madison, WI
New York   San Francisco   St. Louis
Bangkok   Bogotá   Caracas   Lisbon   London   Madrid   Mexico City
Milan   New Delhi   Seoul   Singapore   Sydney   Taipei   Toronto

WCB/McGraw-Hill

*A Division of The **McGraw·Hill** Companies*

BASIC BIOMECHANICS, THIRD EDITION

This book is printed on acid-free paper.

1 2 3 4 5 6 7 8 9 0 QPF/QPF 9 3 2 1 0 9 8

ISBN 0–07–092118–0

Vice president and editorial director: *Kevin T. Kane*
Publisher: *Edward E. Bartell*
Executive editor: *Vicki Malinee*
Developmental editor: *Patricia A. Schissel*
Senior marketing manager: *Pamela S. Cooper*
Project manager: *Sheila M. Frank*
Senior production supervisor: *Mary E. Haas*
Design manager: *Stuart D. Paterson*
Senior photo research coordinator: *Lori Hancock*
Supplement coordinator: *Stacy A. Patch*
Compositor: *York Graphic Services, Inc.*
Typeface: *10/12 New Baskerville*
Printer: *Quebecor Printing Book Group/Fairfield, PA*

Cover/interior design: *The Publishing Services Group*
Cover image: *Uniphoto, Inc.*

Library of Congress Cataloging-in-Publication Data

Hall, Susan J. (Susan Jean), 1953– .
   Basic biomechanics / Susan J. Hall. — 3rd ed.
     p.   cm.
   Includes index.
   ISBN 0–07–092118–0
   1. Biomechanics. 2. Biomechanics—Problems, exercises, etc.
I. Title.
QP303.H35   1999
612.7'6—dc21
                               98–34960
                                  CIP

Contents in Brief

1   What is Biomechanics?   2

2   Kinematic Concepts for Analyzing Human Motion   28

3   Kinetic Concepts for Analyzing Human Motion   62

4   The Biomechanics of Human Bone Growth and Development   90

5   The Biomechanics of Human Skeletal Articulations   118

6   The Biomechanics of Human Skeletal Muscle   146

7   The Biomechanics of the Human Upper Extremity   186

8   The Biomechanics of the Human Lower Extremity   234

9   The Biomechanics of the Human Spine   282

10   Linear Kinematics of Human Movement   326

11   Angular Kinematics of Human Movement   368

12   Linear Kinetics of Human Movement   396

13   Equilibrium and Human Movement   436

14   Angular Kinetics of Human Movement   476

15   Human Movement in a Fluid Medium   506

Appendices

A   Basic Mathematics and Related Skills   542

B   Trigonometric Functions   548

C   Common Units of Measurement   552

D   Anthropometric Parameters for the Human Body   554

Glossary   556

Index   566

Contents

Preface xvii

1 What is Biomechanics? 2

Introduction 2
Biomechanics: Definition and Perspective 3
    Problems Studied by Biomechanists 5
    Why Study Biomechanics? 12
Problem-Solving Approach 13
    Quantitative versus Qualitative Problems 13
    Solving Qualitative Problems 14
    Formal versus Informal Problems 15
    Solving Formal Quantitative Problems 16
    Units of Measurement 17
Summary 21

2 Kinematic Concepts for Analyzing Human Motion 28

Introduction 28
Standard Reference Terminology 29
    Anatomical Reference Position 29
    Directional Terms 29
    Anatomical Reference Planes 30
    Anatomical Reference Axes 32
Forms of Motion 32
    Linear Motion 33
    Angular Motion 34
    General Motion 36
    Mechanical Systems 37
Joint Movement Terminology 37
    Sagittal Plane Movements 37
    Frontal Plane Movements 38
    Transverse Plane Movements 40
    Other Movements 43

Qualitative Analysis of Human Movement   44
   Prerequisite Knowledge for a Qualitative Analysis   45
   Planning a Qualitative Analysis   46
   Conducting a Qualitative Analysis   49
Tools for Measuring Kinematic Quantities   52
   Cinematography and Videography   53
   Other Movement Monitoring Systems   54
   Other Assessment Tools   55
Summary   56

3   **Kinetic Concepts for Analyzing Human Motion   62**

Introduction   62
Basic Concepts Related to Kinetics   63
   Mass   63
   Inertia   63
   Force   63
   Center of Gravity   65
   Weight   66
   Pressure   67
   Volume   67
   Density   70
   Torque   71
   Impulse   72
Mechanical Loads on the Human Body   73
   Compression, Tension, and Shear   73
   Mechanical Stress   74
   Torsion, Bending, and Combined Loads   75
   The Effects of Loading   77
   Repetitive versus Acute Loads   78
Tools for Measuring Kinetic Quantities   79
   Electromyography   79
   Dynamography   80
Vector Algebra   81
   Vector Composition   81
   Vector Resolution   82
   Graphic Solution of Vector Problems   84
   Trigonometric Solution of Vector Problems   84
Summary   86

4   **The Biomechanics of Human Bone Growth and Development   90**

Introduction   90
Composition and Structure of Bone Tissue   91
   Material Constituents   91

Structural Organization   91
Types of Bones   94
Bone Growth and Development   94
Longitudinal Growth   97
Circumferential Growth   97
Bone Response to Stress   97
Bone Modeling and Remodeling   98
Bone Hypertrophy   99
Bone Atrophy   100
Osteoporosis   102
Adult Bone Development   102
Postmenopausal and Age-Associated Osteoporosis   103
Female Athlete Triad   104
Preventing and Treating Osteoporosis   105
Common Bone Injuries   107
Fractures   107
Epiphyseal Injuries   110
Summary   110

5   **The Biomechanics of Human Skeletal Articulations   118**

Introduction   118
Joint Architecture   118
Classification of Joints   118
Articular Cartilage   122
Articular Fibrocartilage   122
Articular Connective Tissue   123
Joint Stability   124
Shape of the Articulating Bone Surfaces   124
Arrangement of Ligaments and Muscles   125
Other Connective Tissues   126
Joint Flexibility   127
Measuring Joint Range of Motion   128
Factors Influencing Joint Flexibility   130
Flexibility and Injury   130
Techniques for Increasing Joint Flexibility   132
Neuromuscular Response to Stretch   132
Active and Passive Stretching   134
Ballistic and Static Stretching   136
Proprioceptive Neuromuscular Facilitation   136
Common Joint Injuries and Pathologies   138
Sprains   138
Dislocations   138
Bursitis   138
Arthritis   139

Rheumatoid Arthritis   139
Osteoarthritis   139
Summary   139

6   The Biomechanics of Human Skeletal Muscle   146

Introduction   146
Behavioral Properties of the Musculotendinous Unit   147
Extensibility and Elasticity   147
Irritability and the Ability to Develop Tension   149
Structural Organization of Skeletal Muscle   149
Muscle Fibers   149
Motor Units   154
Fiber Types   156
Fiber Architecture   158
Skeletal Muscle Function   162
Recruitment of Motor Units   162
Change in Muscle Length with Tension Development   162
Roles Assumed by Muscles   164
Two-Joint and Multijoint Muscles   166
Factors Affecting Muscular Force Generation   167
Force-Velocity Relationship   167
Length-Tension Relationship   169
Electromechanical Delay   170
Muscular Strength, Power, and Endurance   170
Muscular Strength   171
Muscular Power   175
Muscular Endurance   176
Muscle Fatigue   176
Effect of Muscle Temperature   177
Summary   178

7   The Biomechanics of the Human Upper Extremity   186

Introduction   186
Structure of the Shoulder   186
Sternoclavicular Joint   187
Acromioclavicular Joint   188
Coracoclavicular Joint   189
Glenohumeral Joint   189
Scapulothoracic Joint   190
Bursae   191
Movements of the Shoulder Complex   192
Muscles of the Scapula   193
Muscles of the Glenohumeral Joint   194

Flexion at the Glenohumeral Joint 194
Extension at the Glenohumeral Joint 197
Abduction at the Glenohumeral Joint 197
Adduction at the Glenohumeral Joint 198
Medial and Lateral Rotation of the Humerus 199
Horizontal Adduction and Abduction at the Glenohumeral
Joint 199
Loads on the Shoulder 199
Common Injuries of the Shoulder 203
Dislocations 203
Rotator Cuff Damage 204
Rotational Injuries 205
Subscapular Neuropathy 206
Structure of the Elbow 206
Humeroulnar Joint 206
Humeroradial Joint 206
Proximal Radioulnar Joint 206
Movements at the Elbow 207
Muscles Crossing the Elbow 207
Flexion and Extension 208
Pronation and Supination 210
Loads on the Elbow 211
Common Injuries of the Elbow 213
Sprains and Dislocations 213
Overuse Injuries 215
Structure of the Wrist 216
Movements of the Wrist 217
Flexion 217
Extension and Hyperextension 218
Radial and Ulnar Deviation 219
Structure of the Joints of the Hand 219
Carpometacarpal and Intermetacarpal Joints 220
Metacarpophalangeal Joints 220
Interphalangeal Joints 221
Movements of the Hand 221
Common Injuries of the Wrist and Hand 226
Summary 227

8 **The Biomechanics of the Human Lower Extremity 234**

Introduction 234
Structure of the Hip 234
Movements at the Hip 235
Muscles of the Hip 237
Flexion 237

Extension   237
Abduction   239
Adduction   239
Medial and Lateral Rotation of the Femur   242
Horizontal Abduction and Adduction   243
Loads on the Hip   243
Common Injuries of the Hip   245
Fractures   245
Contusions   245
Strains   245
Structure of the Knee   247
Tibiofemoral Joint   247
Menisci   248
Ligaments   248
Patellofemoral Joint   250
Joint Capsule and Bursae   250
Movements at the Knee   251
Muscles Crossing the Knee   251
Flexion and Extension   251
Rotation and Passive Abduction and Adduction   251
Patellofemoral Joint Motion   251
Loads on the Knee   253
Forces at the Tibiofemoral Joint   253
Forces at the Patellofemoral Joint   255
Common Injuries of the Knee and Lower Leg   256
Ligament Injuries   256
Meniscus Injuries   258
Iliotibial Band Friction Syndrome   258
Breaststroker's Knee   259
Chondromalacia   259
Shin Splints   259
Structure of the Ankle   260
Movements at the Ankle   261
Structure of the Foot   265
Subtalar Joint   265
Tarsometatarsal and Intermetatarsal Joints   265
Metatarsophalangeal and Interphalangeal Joints   265
Plantar Arches   265
Movements of the Foot   266
Muscles of the Foot   266
Toe Flexion and Extension   266
Inversion and Eversion   267
Pronation and Supination   269
Loads on the Foot   269
Common Injuries of the Ankle and Foot   270
Ankle Injuries   270

Overuse Injuries   271
Alignment Anomalies of the Foot   272
Summary   274

9   **The Biomechanics of the Human Spine   282**

Introduction   282
Structure of the Spine   282
Vertebral Column   282
Vertebrae   284
Intervertebral Discs   286
Ligaments   290
Spinal Curves   292
Movements of the Spine   294
Flexion, Extension, and Hyperextension   294
Lateral Flexion and Rotation   294
Muscles of the Spine   296
Anterior Aspect   298
Posterior Aspect   298
Lateral Aspect   304
Loads on the Spine   304
Common Injuries of the Back and Neck   313
Low Back Pain   313
Soft Tissue Injuries   315
Acute Fractures   315
Stress Fractures   316
Disc Herniations   317
Summary   318

10   **Linear Kinematics of Human Movement   326**

Introduction   326
Linear Kinematic Quantities   326
Distance and Displacement   328
Speed and Velocity   330
Acceleration   335
Average and Instantaneous Quantities   340
Kinematics of Projectile Motion   341
Horizontal and Vertical Components   341
Influence of Gravity   342
Influence of Air Resistance   343
Factors Influencing Projectile Trajectory   344
Projection Angle   345
Projection Speed   347
Relative Projection Height   348

Optimum Projection Conditions   350
Analyzing Projectile Motion   352
Equations of Constant Acceleration   352
Summary   360

11   **Angular Kinematics of Human Movement   368**

Introduction   368
Observing the Angular Kinematics of Human Movement   368
Measuring Angles   369
Relative versus Absolute Angles   370
Tools for Measuring Body Angles   372
Instant Center of Rotation   373
Angular Kinematic Relationships   374
Angular Distance and Displacement   375
Angular Speed and Velocity   378
Angular Acceleration   380
Angular Motion Vectors   382
Average versus Instantaneous Angular Quantities   382
Relationships Between Linear and Angular Motion   382
Linear and Angular Displacement   382
Linear and Angular Velocity   384
Linear and Angular Acceleration   387
Summary   390

12   **Linear Kinetics of Human Movement   396**

Introduction   396
Newton's Laws   396
Law of Inertia   397
Law of Acceleration   397
Law of Reaction   398
Law of Gravitation   402
Mechanical Behavior of Bodies in Contact   402
Friction   403
Momentum   411
Impulse   412
Impact   417
Work, Power, and Energy Relationships   420
Work   420
Power   421
Energy   422
Conservation of Mechanical Energy   424
Principle of Work and Energy   425
Summary   429

13   **Equilibrium and Human Movement   436**

Introduction   436
Equilibrium   436
   Torque   436
   Resultant Joint Torques   441
   Levers   445
   Anatomical Levers   450
   Equations of Static Equilibrium   452
   Equations of Dynamic Equilibrium   454
Center of Gravity   458
   Locating the Center of Gravity   459
   Locating the Human Body Center of Gravity   461
Stability and Balance   465
Summary   470

14   **Angular Kinetics of Human Movement   476**

Introduction   476
Resistance to Angular Acceleration   476
   Moment of Inertia   476
   Determining Moment of Inertia   479
   Human Body Moment of Inertia   481
Angular Momentum   482
   Conservation of Angular Momentum   485
   Transfer of Angular Momentum   488
   Change in Angular Momentum   490
Angular Analogues of Newton's Laws of Motion   494
   Newton's First Law   495
   Newton's Second Law   495
   Newton's Third Law   496
Centripetal Force   497
Summary   498

15   **Human Movement in a Fluid Medium   506**

Introduction   506
The Nature of Fluids   507
   Relative Motion   507
   Laminar versus Turbulent Flow   507
   Fluid Properties   509
Buoyancy   510
   Characteristics of the Buoyant Force   510
   Flotation   512
   Flotation of the Human Body   512

Drag   514
    Skin Friction   516
    Form Drag   518
    Wave Drag   522
Lift Force   523
    Foil Shape   524
    Magnus Effect   529
Propulsion in a Fluid Medium   530
    Propulsive Drag Theory   532
    Propulsive Lift Theory   533
    Vortex Generation   533
    Stroke Technique   533
Summary   534

Appendices

A   **Basic Mathematics and Related Skills**   **542**

B   **Trigonometric Functions**   **548**

C   **Common Units of Measurement**   **552**

D   **Anthropometric Parameters for the Human Body**   **554**

Glossary   **556**

Index   **566**

Preface

The third edition of *Basic Biomechanics* constitutes a significant update of the previous edition. As the interdisciplinary field of biomechanics grows in both breadth and depth, it is important that even introductory textbooks reflect the nature of the science. Accordingly, the text material has been revised, expanded, and updated, with the objectives being to present relevant information from recent research findings as well as to prepare students to *analyze* human biomechanics.

The approach taken remains an integrated balance of qualitative and quantitative examples, applications, and problems designed to illustrate the principles discussed. The third edition also maintains the important sensitivity to the fact that some beginning students of biomechanics possess weak backgrounds in mathematics and includes numerous sample problems and applications along with practical advice on approaching quantitative problems.

ORGANIZATION

The new edition incorporates minor organizational changes within several of the chapters to address the revision and expansion of material from the previous edition. Students will still find that each chapter follows a logical and readable format, with the introduction of new concepts consistently accompanied by practical human movement examples and applications from across the lifespan and across sport, ergonomic, and daily living activities.

NEW CONTENT HIGHLIGHTS

New content has been added to provide updated scientific information on relevant topics. *All chapters* contain recently documented examples of the concepts presented from the scientific literature accompanied by *new, improved illustrations*. The chapters on the biomechanics of human bone and muscle, in particular, have been significantly revised and expanded to reflect recent scientific developments in these topical areas. The introductory chapter and chapters on the upper and lower extremities and the spine also contain additional, updated coverage.

Balanced Coverage

The Biomechanics Academy of AAHPERD recommends that preparation for undergraduate students in the area of biomechanics be devoted approximately one-third to anatomical considerations, approximately one-third to mechanical considerations, and the remainder to applications. The integrated approach to coverage of these areas taken in the previous editions of the book is continued in this third edition.

Applications Oriented

All chapters in this new edition contain discussion of a broad range of updated human movement applications, many of which are taken from the recent biomechanics research literature. Special emphasis has been placed on the inclusion of examples that span all ages and address clinical and daily living issues, as well as sport applications.

The Use of Problems

The Introductory and Additional Problems sections have been revised, with greater emphasis on critical thinking added. The sample problems with solutions provided throughout the text have also been expanded.

Laboratory Exercises

The section on suggested laboratory experiences has also been revised, with more practical suggestions for a variety of laboratory situations included. While many of the exercises can be used "as is," students and instructors are also encouraged to view these suggestions as seed concepts that can be modified and expanded.

Related Web Sites

A new section on related web sites has been added at the end of each chapter. The sections provide selected links to the wealth of related scientific and medical information now available on the World Wide Web.

PEDAGOGICAL FEATURES

Aside from the sample problems, problem sets, and laboratory experiences, and web sites, the book retains other pedagogical features from previous editions. These include **chapter objectives,** which serve as focal points for understanding chapter content; **marginal definitions and notes,** which highlight important terms and concepts; **chapter sum-**

maries; **suggested readings,** for enrichment beyond the material presented in each chapter; a comprehensive **glossary;** and several **appendices,** containing useful reference or review information. Newly added to the third edition are numerous revised and updated illustrations, in a two-color format for superior clarity.

ANCILLARY

A new Instructor's Manual to accompany the book has been developed by Darla Smith, University of Texas at El Paso. The new Instructor's Manual provides updated pedagogical materials, problem solutions, and a bank of test questions corresponding to each chapter. Also included are 100 overhead transparencies displaying selected illustrations and sample problems.

ACKNOWLEDGMENTS

I would like to thank Pat Schissel, the Developmental Editor assigned to this project by WCB/McGraw-Hill, as well as Executive Editor Vicki Malinee for assistance and encouragement. I also greatly appreciate the capable work of Project Manager Sheila Frank and her staff. I also wish to extend appreciation for their suggestions to the following reviewers arranged through WCB/McGraw-Hill:

Nancy L. Meyer, Calvin College
M. Christopher Washam, Northeast Louisiana University
Mike Lester, Idaho State University
Nena Amundson, California Lutheran University
Jim Carr, University of Tulsa
Marilyn Miller, University of Wisconsin—LaCrosse

I also very much appreciate and wish to acknowledge the excellent critical suggestions I have received from numerous students and colleagues.

Susan J. Hall

Department of Kinesiology
The University of Texas at El Paso

Basic
Biomechanics

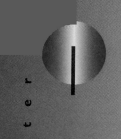

WHAT IS BIOMECHANICS?

After completing this chapter, the reader will be able to:

Define the terms biomechanics, statics, dynamics, kinematics, and kinetics and explain the ways in which they are related.

Describe the scope of scientific inquiry addressed by biomechanists.

Distinguish between qualitative and quantitative approaches for analyzing human movement.

Explain how to formulate questions for qualitative analysis of human movement.

Use the 11 steps identified in the chapter in solving formal problems.

Learning to walk is an ambitious task from a biomechanical perspective.

Courses in anatomy, physiology, mathematics, physics, and engineering provide background knowledge for biomechanists.

Why do some golfers slice the ball? Under what circumstances can orthotic shoe implants relieve low back pain? What cues can a physical education teacher provide to help students learn the underhand volleyball serve? Why do some elderly individuals tend to fall? We have all admired the fluid, graceful movements of highly skilled performers in various sports. We have also observed the awkward first steps taken by a young child, the slow progress of an injured person with a walking cast, and the hesitant, uneven gait of an elderly person using a cane. Virtually every activity class includes a student who seems to acquire new skills with utmost ease and a student who trips when executing a jump or misses the ball when attempting to catch, strike, or serve. What enables some individuals to execute complex movements so easily, while others appear to have difficulty with relatively simple movement skills?

Although the answers to these questions may be rooted in physiological, psychological, or sociological issues, the problems identified are all biomechanical in nature. This book is written to provide a foundation for identifying, analyzing, and solving problems related to the biomechanics of human movement.

BIOMECHANICS: DEFINITION AND PERSPECTIVE

During the early 1970s, the international community adopted the term **biomechanics** to describe the science involving the study of biological systems from a mechanical perspective (32). Biomechanists use the tools of **mechanics,** the branch of physics involving analysis of the actions of forces, to study the anatomical and functional aspects of living organisms (Figure 1-1). **Statics** and **dynamics** are two major subbranches of mechanics. Statics is the study of systems that are in a state of constant motion, that is, either at rest (with no motion) or moving with a constant velocity. Dynamics is the study of systems in which acceleration is present.

Kinematics and kinetics are further subdivisions of biomechanical study. Kinematics is the description of motion, including the pattern and speed of movement sequencing by the body segments that often translates to the degree of coordination an individual displays. Whereas kinematics describes the appearance of motion, kinetics is the study of the forces associated with motion. The study of human biomechanics may include questions such as whether the amount of force the muscles are producing is optimal for the intended purpose of the movement. **Anthropometric** factors including the size, shape, and weight of the body segments are other important considerations in a kinetic analysis.

Although biomechanics is relatively young as a recognized field of scientific inquiry, biomechanical considerations are of interest in several different scientific disciplines and professional fields. Biomechanists may have academic backgrounds in zoology; orthopedic, cardiac, or sports medicine; biomedical or biomechanical engineering; physical therapy; or kinesiology, with the commonality being an interest in the biomechanical aspects of the structure and function of living things.

The biomechanics of human movement is one of the subdisciplines of **kinesiology,** the study of human movement (Figure 1-2). Although some biomechanists study topics such as ostrich locomotion, blood flow through constricted arteries, or micromapping of dental cavities, this book focuses primarily on the biomechanics of human movement from the perspective of the movement analyst.

biomechanics
application of mechanical principles in the study of living organisms

mechanics
branch of physics that analyzes the actions of forces on particles and mechanical systems

statics
branch of mechanics dealing with systems in a constant state of motion

dynamics
branch of mechanics dealing with systems subject to acceleration

kinematics
study of the description of motion, including considerations of space and time

kinetics
study of the action of forces

anthropometric
related to the dimensions and weights of body segments

kinesiology
study of human movement

FIGURE 1-1
Biomechanics uses the principles of mechanics for solving problems related to the structure and function of living organisms.

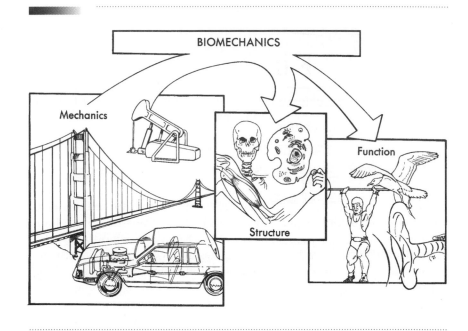

Anthropometric characteristics may predispose an athlete to success in one sport and yet be disadvantageous for participation in another.

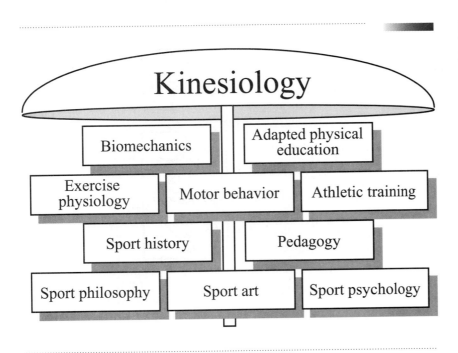

FIGURE 1-2
The subdisciplines of kinesiology.

As shown in Figure 1-3, biomechanics is also a scientific branch of **sports medicine.** Sports medicine has been defined by Lamb as "an umbrella term that encompasses both clinical and scientific aspects of exercise and sport" (28). The American College of Sports Medicine is an example of an organization that promotes interaction between scientists and clinicians with interests in sports medicine–related topics.

sports medicine
clinical and scientific aspects of sports and exercise

Problems Studied by Biomechanists

As expected given the different scientific and professional fields represented, biomechanists study questions or problems that are topically diverse. For example, zoologists have examined the locomotion patterns of dozens of species of animals walking, running, trotting, and galloping at controlled speeds on a treadmill to determine why animals choose a particular stride length and a particular stride rate at a given speed (36). They concluded that most vertebrates, including humans, select a gait that optimizes economy, or metabolic energy consumption, at a given speed. Although researchers have identified a variety of anthropometric, kinematic, and kinetic factors that are believed to influence the energy cost of locomotion, there is no clear set of criteria for identifying economic human gait (2). Interestingly, however, research has shown that as gait speed increases, both humans and horses go from a walk to a run or from a trot to a gallop before the transition is

In research, each new study, investigation, or experiment is usually designed to address a particular question or problem.

FIGURE 1-3
The branches of sports medicine.

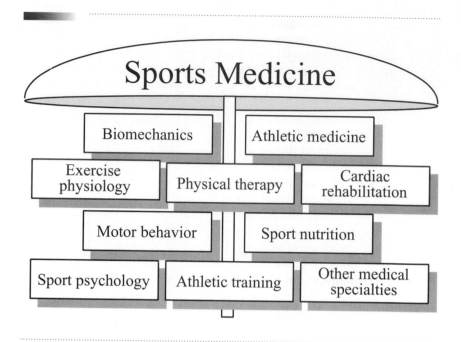

FIGURE 1-4
Research on the biomechanics of animal gait poses some interesting problems.

energetically optimal, suggesting that comfort does not strictly parallel economy (16, 23). A recent hypothesis is that the walk-run transition occurs when the centripetal and gravitational forces acting on the body during ground contact reach a certain critical ratio (25). One of the challenges of this type of research is determining how to persuade a cat, a pig, or a lion to run on a treadmill (Figure 1-4).

The U.S. National Aeronautics and Space Administration (NASA) sponsors another multidisciplinary line of biomechanics research to promote understanding of the effects of microgravity on the human musculoskeletal system. Of concern is the fact that astronauts who have been out of the earth's gravitational field for just a few days have returned with reduced bone density, mineralization, and strength, especially in the lower extremities. Since those early days of space flight, biomechanists have designed and built a number of exercise devices for use in space to take the place of normal bone-maintaining activities on earth. Recent research has focused on the design of treadmills for use in space that load the bones of the lower extremity with deformations and strain rates that are optimal for stimulating new bone formation (11).

Maintaining sufficient bone mineral density is also a topic of concern here on earth. Osteoporosis is a condition in which bone mineral mass and strength are so severely compromised that daily activities can cause bone pain and fracturing (18). This condition is found in most elderly individuals, with earlier onset in women, and is becoming increasingly prevalent with the increasing mean age of the population (42). Today, approximately 40% of women experience one or more osteoporotic fractures after the age of 50, and after age 60, about 90% of all fractures in both men and women are osteoporosis-related (29, 38). Osteoporosis is neither a disease with acute onset, nor an inevitable accompaniment of aging, but the result of a lifetime of habits that are erosive to the skeletal system (5). Known risk factors for developing osteoporosis include physical inactivity, cigarette smoking, deficiencies in estrogen, calcium, and vitamin D, and excessive consumption of protein, caffeine, and alcohol (12). Treatment of osteoporosis involves addressing these risk factors. Importantly, studies show that a regular program of weight-bearing exercise, such as walking, among individuals with osteoporosis can increase bone health and strength (4).

Another problem area challenging biomechanists who study the elderly is mobility impairment. Age is associated with decreased ability to balance, and older adults both sway more and fall more than young adults, although the reasons for these changes are not well understood (35). Falls, and particularly fall-related hip fractures, are extremely serious, common, and costly medical problems among the elderly (22). Biomechanists have studied strategies employed by the elderly in rising from a chair and determined that difficulty with balance is more likely to be a limiting factor than inadequate strength (17). Promising work in the development of intervention strategies has shown that exercise

Exercise in space is critically important for preventing loss of bone mass among astronauts as demonstrated by Astronaut Bob Crippen on Shuttle flight STS-7. Photo courtesy of NASA.

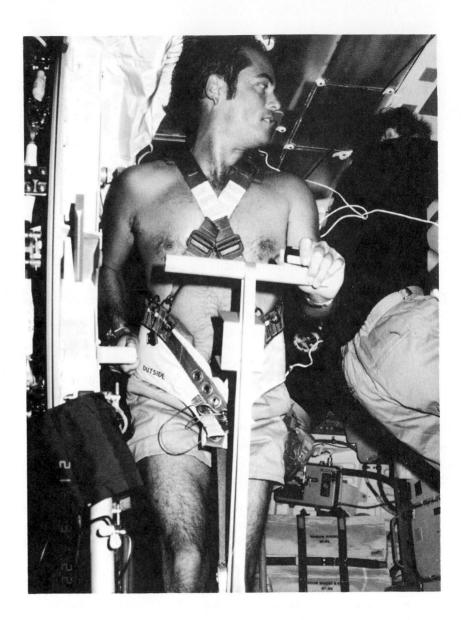

walking can be effective in improving balance and reducing the likelihood of falling among sedentary older adults (6).

Research by clinical biomechanists has resulted in improved gait among children with cerebral palsy, a condition involving high levels of muscle tension and spasticity. The gait of the cerebral palsy individual is characterized by excessive knee flexion during stance. This problem is treated by surgical lengthening of the hamstring tendons, which is done to improve knee extension during stance. In some patients, however, the procedure also diminishes knee flexion during the swing phase of gait, resulting in dragging of the foot. After research showed that patients with this problem exhibited significant co-contraction of the rectus femoris with the hamstrings during the swing phase, orthopedists began treating the problem by surgically attaching the rectus femoris to the sartorius insertion (19). This creative, biomechanics research–based approach has enabled a major step toward gait normalization for children with cerebral palsy.

Research by biomedical engineers has also resulted in improved gait for children and adults with below-knee amputations. Ambulation with a prosthesis creates an added metabolic demand, which can be particularly significant for elderly amputees and for young active amputees who participate in sports requiring aerobic conditioning. In response to this problem, researchers have developed an array of lower limb and foot prostheses that store and return mechanical energy during gait, thereby reducing the metabolic cost of locomotion (3, 15, 31, 43). Studies have shown that the more compliant prostheses are better suited for active and fast walkers, whereas prosthetics that provide a more stable base of support are generally preferred for the elderly population (7).

Occupational biomechanics is a field that focuses on the prevention of work-related injuries and the improvement of working conditions and worker performance (8). Researchers in this field have learned that work-related low back pain can be derived not only from the handling of heavy materials, but from unnatural postures, sudden and unexpected motions, and the characteristics of the individual worker (46). Occupational biomechanists are recognizing the importance of workers being both physically and mentally prepared for jobs in industry in order to prevent low back pain (46). In recent years, although the number of overall workplace injuries has decreased, an overuse injury of the wrist called **carpal tunnel syndrome** has steadily increased in frequency (20). Because carpal tunnel syndrome is particularly associated with repetitive keyboard use, studies are being conducted to explore novel keyboard designs that may be more biomechanically optimal than the traditional keyboard. Interesting new designs being tested include a keyboard split into left and right halves with each half positioned directly in front of a shoulder, and split, vertically aligned keyboards that allow maintaining the wrists in neutral position (30).

Biomechanists have also contributed to performance improvements in selected sports. The recent innovations in wheel and helmet designs for cycle racing have resulted from findings of experiments conducted in experimental chambers called wind tunnels that involved controlled simulation of the air resistance actually encountered during cycling (27). Wind tunnel experiments have been conducted to identify optimal body configuration during ski jumping (39) and the effects of wind resistance and altitude on track events such as sprinting (10, 26). Figure 1-5 demonstrates one way equipment can be tested in a wind tunnel.

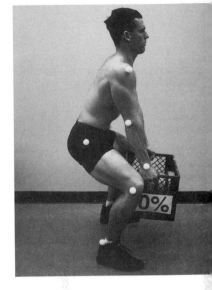

Occupational biomechanics involves study of safety factors in activities such as lifting.

carpal tunnel syndrome overuse condition caused by compression of the median nerve in the carpal tunnel and involving numbness, tingling, and pain in the hand

The aerodynamic cycling equipment introduced in the 1984 Olympic Games contributed to the new world records set.

FIGURE 1-5
A wind tunnel is a chamber designed to measure aerodynamic forces. New designs in sport apparel and equipment have resulted from wind tunnel testing.

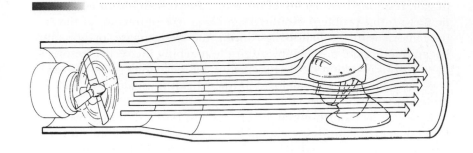

Sport biomechanists have generated other efforts at improving athletic performance. They have learned, for example, that even at the elite level, running style is highly individual (45). This makes it more difficult to coach runners of all levels, since there is no precise model for correct running technique. Factors found to contribute to superior performance in the long jump, high jump, and pole vault include large horizontal velocity going into takeoff and a shortened last step that facilitates continued elevation of the total body center of mass (9, 21). Study of freestyle swimmers has shown that the rolling of the trunk is as important as the motion of the hand relative to the trunk in determining the overall mediolateral motion of the hand through the water during the stroke cycle (34).

One rather dramatic example of performance improvement partially attributable to biomechanical analysis is the case of four-time Olympic discus champion Al Oerter. Mechanical analysis of the discus throw requires precise evaluation of the major mechanical factors affecting the flight of the discus. These factors are

1. the speed of the discus when it is released by the thrower,
2. the projection angle at which the discus is released,
3. the height above the ground at which the discus is released, and
4. the angle of attack (the orientation of the discus relative to the prevailing air current).

By using computer simulation techniques, researchers can predict the needed combination of values for these four variables that will result in a throw of maximum distance for a given athlete. High-speed cameras can record performances in great detail, and when the film or videotape is analyzed, the actual projection height, velocity, and angle of attack can be compared to the computer-generated values required for optimal performance. At the age of 43, Oerter bettered his best Olympic performance by 8.2 meters. Although it is difficult to determine the contributions of motivation and training to such an improvement, some part of Oerter's success was a result of enhanced technique following biomechanical analysis (40). Most adjustments to skilled athletes' techniques produce relatively modest results because their performances are already characterized by above-average technique.

Some of the research produced by sport biomechanists has been done in conjunction with the Sports Medicine Division of the United States Olympic Committee (USOC). The general goal of USOC research is to examine the ways in which mechanical factors limit the performances of elite American athletes training for Olympic and other international competition (14). Typically, this work is done in direct cooperation with the national coach of the sport to ensure the practicality of results. USOC-sponsored research has yielded much new information about the mechanical characteristics of elite performance in various sports. Because of continuing advances in scientific analysis equipment, the role of sport biomechanists in contributing to performance improvements is likely to be increasingly important in the future.

The USOC began funding sports medicine research in 1981. Other countries began sponsoring research to boost the performance of elite athletes in the early 1970s.

Other concerns of sport biomechanists relate to minimizing sport injuries through both identifying dangerous practices and designing safe equipment and apparel. In recreational runners, for example, research shows that the most serious risk factors for overuse injuries are training errors such as a sudden increase in running distance or intensity, excess cumulative mileage, running on cambered surfaces, and improper footwear (33). The complexity of safety-related issues increases when the sport is equipment-intensive. Evaluation of protective helmets involves not only ensuring that the impact characteristics offer reliable protection, but also that the helmet does not overly restrict wearers' peripheral vision.

Impact testing of protective sport helmets is carried out scientifically in engineering laboratories.

An added complication is that equipment designed to protect one part of the human body may actually contribute to the likelihood of injury in another part of the musculoskeletal system. Modern ski boots and bindings, while effective in protecting the ankle and lower leg against injury, unfortunately contribute to severe bending moments at the knee when the skier loses balance. This factor has contributed to the nearly threefold increase in the incidence of knee injuries in skiing since 1972 (24). Injuries in snowboarding are also more frequent with rigid, as compared to pliable, boots, although nearly half of all snowboarding injuries are to the upper extremity (37).

Another challenging area of research for biomechanists in the realm of sport safety is investigation of the efficacy of prophylactic knee braces designed to provide extra stability for athletes' knees. The wearing of such braces by healthy individuals has been a contentious issue since the American Academy of Orthopaedics issued a position statement against their use in 1987. Recent work on this topic shows that knee braces can contribute 20% to 30% added resistance against lateral blows to the knee, with custom-fitted braces providing the best protection (1). A possible concern, however, is that knee braces act to change the pattern of lower extremity muscle activity during gait, with less work performed at the knee and more at the hip (13). Other documented problems that appear to affect some athletes more than others and may be brace-specific include reduced sprinting speed and earlier onset of fatigue (1).

An area of biomechanics research with implications for both safety and performance is sport shoe design. Today sport shoes are designed both to prevent excessive loading and related injuries and to enhance performance. Because the ground or playing surface, the shoe, and the human body compose an interactive system, athletic shoes are specifically designed for particular sports, surfaces, and anatomical considerations. Aerobic dance shoes are constructed to cushion the metatarsal arch. Football shoes to be used on artificial turf are designed to minimize the risk of knee injury. Running shoes are available for training, racing, and running on snow and ice. In fact, sport shoes today are so specifically designed for designated activities that wearing an inappropriate shoe can contribute to the likelihood of injury (41).

These examples illustrate the diversity of topics addressed in biomechanics research, including some examples of success as well as areas of continuing challenge. Clearly, biomechanists are contributing to the knowledge base on the full gamut of human movement, from the gait of the physically challenged child to the technique of the elite athlete. Although varied, all of the research described is based on applications of mechanical principles in solving specific problems in living organisms. This book is designed to provide an introduction to many of those principles and to focus on some of the ways in which biomechanical principles may be applied in the analysis of human movement.

Why Study Biomechanics?

As is evident from the preceding section, biomechanical principles are applied by scientists and professionals in a number of fields in addressing problems related to human health and performance. Knowledge of basic biomechanical concepts is also essential for the competent physical education teacher, physical therapist, physician, coach, personal trainer, or exercise instructor.

An introductory course in biomechanics provides foundational understanding of mechanical principles and how they can be applied in analyzing movements of the human body. The knowledgeable human movement analyst should be able to answer the following questions related to biomechanics: Why is swimming *not* the best form of exercise for individuals with osteoporosis? What is the biomechanical principle behind variable resistance exercise machines? What is the safest way to lift a heavy object? Is it possible to judge what movements are more/less economical from visual observation? At what angle should a ball be thrown for maximum distance? From what distance and angle is it best to observe a patient walk down a ramp or a volleyball player execute a serve? What strategies can an elderly person or a football lineman employ to maximize stability? Why are some individuals unable to float?

Perusing the objectives at the beginning of each chapter of this book is a good way to further examine the scope of biomechanical topics to be covered at the introductory level. For those planning careers that involve visual observation and analysis of human movement, knowledge of these topics will be invaluable.

PROBLEM-SOLVING APPROACH

Scientific research is usually aimed at providing a solution for a particular problem or answering a specific question. Even for the nonresearcher, however, the ability to solve problems is a practical necessity for functioning in modern society. The use of specific problems is also an effective approach for illustrating basic biomechanical concepts.

Quantitative versus Qualitative Problems

Analysis of human movement may be either **quantitative** or **qualitative.** The word *quantitative* implies that numbers are involved, and *qualitative* refers to a description of quality without the use of numbers. After watching the performance of a standing long jump, an observer might qualitatively state, "That was a very good jump." Another observer might quantitatively announce that the same jump was 2.1 meters in length. Other examples of qualitative and quantitative descriptors are displayed in Figures 1-6 and 1-7.

quantitative
involving the use of numbers

qualitative
involving nonnumeric description of quality

FIGURE 1-6

It is important to recognize that the term *qualitative* does not mean *general*. Qualitative descriptions may be general, but they may also be extremely detailed. It can be stated qualitatively and generally, for example, that a man is walking down the street. It might also be stated that the same man is walking very slowly, appears to be leaning to the left, and is bearing weight on his right leg for as short a time as possible. The second description is entirely qualitative but provides a more detailed picture of the movement.

Both qualitative and quantitative descriptions play important roles in the biomechanical analysis of human movement. Biomechanical researchers rely heavily on quantitative techniques in attempting to answer specific questions related to the mechanics of living organisms. Clinicians, coaches, and teachers of physical activities regularly employ qualitative observations of their patients, athletes, or students to formulate opinions or give advice.

Solving Qualitative Problems

Qualitative problems commonly arise during daily activities. Questions such as what clothes to wear, whether to major in botany or English, and whether to study or watch television are all problems in the sense that they are uncertainties that may require resolution. Thus, a large portion of our daily lives is devoted to the solution of problems.

Analyzing human movement, whether to identify a gait anomaly or to refine a student's technique, is essentially a process of problem solving. Whether the analysis is to be qualitative or quantitative, this includes identifying, then studying or analyzing, and finally answering a question or problem of interest.

To effectively analyze a movement, it is first essential to formulate one or more questions regarding the movement. Depending on the specific purpose of the analysis, the questions to be framed may be general or specific. General questions, for example, might include the following:

1. Is the movement being performed with adequate (or optimal) force?
2. Is the movement being performed through an appropriate range of motion?
3. Is the sequencing of body movements appropriate (or optimal) for execution of the skill?
4. Why does this elderly woman have a tendency to fall?
5. Why is this shot putter not getting more distance?

More specific questions might include these:

1. Is there excessive pronation taking place during the stance phase of gait?
2. Is release of the ball taking place at the instant of full elbow extension?
3. Does selective strengthening of the vastus medialis obliquus alleviate mistracking of the patella for this person?

Once one or more questions have been identified, the next step in analyzing a human movement is to collect data. The form of data most commonly collected by teachers, therapists, and coaches is qualitative visual observation data. That is to say, the movement analyst carefully observes the movement being performed and makes either written or mental notes. To acquire the best observational data possible, it is useful to plan ahead as to the optimal distance(s) and perspective(s) from which to make the observations. These and other important considerations for qualitatively analyzing human movement are discussed in detail in Chapter 2.

Coaches rely heavily on qualitative observations of athletes' performances in formulating advice about technique.

Formal versus Informal Problems

When confronted with a stated problem taken from an area of mathematics or science, many individuals believe they are not capable of finding a solution. Clearly, a stated math problem is different from a problem such as what to wear to a particular social gathering. In some ways, however, the informal type of problem is the more difficult one to solve. According to Wickelgren (44), a formal problem (such as a stated math problem) is characterized by three discrete components:

1. a set of given information;
2. a particular goal, answer, or desired finding; and
3. a set of operations or processes that can be used to arrive at the answer from the given information.

In dealing with informal problems, however, individuals may find the given information, the processes to be used, and even the goal itself to be unclear or not readily identifiable.

Solving Formal Quantitative Problems

Formal problems are effective vehicles for translating nebulous concepts into well-defined, specific principles that can be readily understood and applied in the analysis of human motion. People who believe themselves incapable of solving formal stated problems do not recognize that, to a large extent, problem-solving skills can be learned. Entire books on problem-solving approaches and techniques are available. However, most students are not exposed to coursework involving general strategies of the problem-solving process. A simple procedure for approaching and solving problems involves 11 sequential steps, which are as follows:

1. Read the problem *carefully*. It may be necessary to read the problem several times before proceeding to the next step. Only when the information given and the question(s) to be answered are clearly understood should step 2 be undertaken.
2. Write down the given information in list form. It is acceptable to use symbols (such as v for velocity) to represent physical quantities if the symbols are meaningful.
3. Write down what is wanted or what is to be determined, using list form if more than one quantity is to be solved for.
4. Draw a diagram representing the problem situation, clearly indicating all known quantities and representing those to be identified with question marks. (Although certain types of problems may not easily be represented diagrammatically, it is critically important to carry out this step whenever possible to accurately visualize the problem situation.)
5. Identify and write down the relationships or formulas that might be useful in solving the problem. (More than one formula may be useful and/or necessary.)
6. From the formulas that were written down in step 5, select the formula(s) containing both given variables (from step 2) and the

variables that are desired unknowns (from step 3). If a formula contains only one unknown variable that is the variable to be determined, skip step 7 and proceed directly to step 8.

7. If a workable formula cannot be identified (in more difficult problems), certain essential information was probably not specifically stated but can be determined by using **inference** and further thought and analysis of the given information. If this occurs, it may be necessary to repeat step 1 and review the pertinent information relating to the problem presented in the text.

8. Once the appropriate formula(s) have been identified, write the formula(s) and carefully substitute the known quantities given in the problem for the variable symbols.

9. Using the simple algebraic techniques reviewed in Appendix A, solve for the unknown variable by a) rewriting the equation so that the unknown variable is isolated on one side of the equals sign and b) reducing the numbers on the other side of the equation to a single quantity.

10. Do a commonsense check of the answer derived. Does it seem too small or too large? If so, recheck the calculations. Also check to ensure that *all* questions originally posed in the statement of the problem have been answered.

11. Clearly box in the answer and include the correct units of measurement.

inference
the process of forming deductions from available information

Figure 1-8 provides a summary of this procedure for solving formal quantitative problems. These steps should be carefully studied, referred to, and used in working the quantitative problems included at the end of each chapter. Figures 1-9 and 1-10 illustrate the use of this procedure.

Units of Measurement

Providing the correct units of measurement associated with the answer to a quantitative problem is important. Clearly, an answer of 2 centimeters is quite different from an answer of 2 kilometers. It is also important to recognize the units of measurement associated with particular physical quantities. Ordering 10 kilometers of gasoline for a car when traveling in a foreign country would clearly not be appropriate.

The predominant system of measurement still used in the United States is the **English system.** The English system of weights and measures arose over the course of several centuries primarily for purposes of commerce and land parceling. Specific units came largely from royal decrees. For example, a yard was originally defined as the distance from the end of the nose of King Henry I to the thumb of his extended arm. The English system of measurement displays little logic. There are 12 inches to the foot, 3 feet to the yard, 5280 feet to the mile, 16 ounces to the pound, and 2000 pounds to the ton.

English system
system of weights and measures originally developed in England and used in the United States today

FIGURE 1-8

Summary of Steps for Solving Formal Problems

1. Read the problem carefully.

2. List the given information.

3. List the desired (unknown) information for which you are to solve.

4. Draw a diagram of the problem situation showing the known and unknown information.

5. Write down formulas that may be of use.

6. Identify the formula to use.

7. If necessary, reread the problem statement to determine whether any additional needed information can be inferred.

8. Carefully substitute the given information into the formula.

9. Solve the equation to identify the unknown variable (the desired information).

10. Check that the answer is both reasonable and complete.

11. Clearly box in the answer.

FIGURE 1-9

S A M P L E   P R O B L E M   1

A baseball player hits a triple to deep left field. As he is approaching third base, he notices that the incoming throw to the catcher is wild and he decides to run for home plate. The catcher retrieves the ball 10 m from the plate and runs back toward the plate at a speed of 5 m/s. As the catcher starts running, the base runner, who is traveling at a speed of 9 m/s, is 15 m from the plate. Given that time = distance/speed, who will reach the plate first?

Solution
STEP 1   Read the problem carefully.
STEP 2   Write down the given information:

$$\text{Base runner's speed} = 9 \text{ m/s}$$
$$\text{Catcher's speed} = 5 \text{ m/s}$$
$$\text{Distance of base runner from plate} = 15 \text{ m}$$
$$\text{Distance of catcher from plate} = 10 \text{ m}$$

STEP 3   Write down the variable to be identified: Find which player reaches home plate in the shortest time.

STEP 4   Draw a diagram of the problem situation.

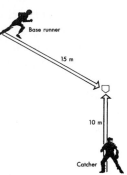

STEP 5   Write down formulas of use:

Time = distance/speed

STEP 6   Identify the formula to be used: It may be assumed that the formula provided is appropriate because no other information relevant to the solution has been presented.

STEP 7   Reread the problem if all necessary information is not available: It may be determined that all information appears to be available.

STEP 8   Substitute the given information into the formula:

$$time = \frac{distance}{speed}$$

Catcher:

$$time = \frac{10 \text{ m}}{5 \text{ m/s}}$$

Base runner:

$$time = \frac{15 \text{ m}}{9 \text{ m/s}}$$

STEP 9   Solve the equations:

Catcher:

$$time = \frac{10 \text{ m}}{5 \text{ m/s}}$$

$$time = 2 \text{ s}$$

Base runner:

$$time = \frac{15 \text{ m}}{9 \text{ m/s}}$$

$$time = 1.67 \text{ s}$$

STEP 10 Check that the answer is both reasonable and complete.

STEP 11 Box in the answer:

The base runner arrives at home plate first, by 0.33 seconds.

FIGURE 1-10

━━━━━━

S A M P L E   P R O B L E M   2

A man sits in a 20 kg wheelchair at the top of a short ramp. When the brakes on the wheelchair are suddenly released, the wheelchair begins to roll down the ramp, accelerating at a rate of 0.5 meters per second[2]. If the net force causing the wheelchair to roll down the ramp is 45 Newtons, what is the mass of the man in the wheelchair? (*Hint:* The relationship to be used in solving this problem is force = mass × acceleration. For more information on the metric units of kilograms, meters, and Newtons, refer to Appendix C.) *This time, work through the 11 steps on your own.*

STEP 2

$$\text{Net force} = 45 \text{ Newtons}$$
$$\text{Wheelchair mass} = 20 \text{ kg}$$
$$\text{Acceleration} = 0.5 \text{ m/s}^2$$

STEP 3   Find: mass of man

STEP 4 $\text{mass}_{total} = \text{mass}_{chair} + \text{mass}_{man}$

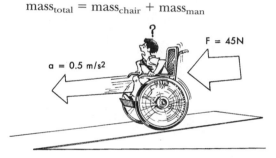

Mass total = mass chair + mass man

STEPS 5 and 6
STEPS 8 and 9

$$\text{Force} = \text{mass} \times \text{acceleration}$$
$$\text{Force} = (\text{mass}_{chair} + \text{mass}_{man}) \times \text{acceleration}$$
$$45 \text{ N} = (20 \text{ kg} + \text{mass}_{man}) \times 0.5 \text{ m/s}^2$$
$$\frac{45 \text{ N}}{0.5 \text{ m/s}^2} = (20 \text{ kg} + \text{mass}_{man})$$
$$\frac{45 \text{ N}}{0.5 \text{ m/s}^2} - 20 \text{ kg} = \text{mass}_{man}$$
$$\text{mass}_{man} = 70 \text{ kg}$$

STEP 11   70 kg

....

metric system
system of weights and measures used internationally in scientific applications and adopted for daily use by every major country except the United States

The system of measurement that is presently used by every major country in the world except the United States is Le Système International d'Unites (the International System of Units), which is commonly known as the S.I. or the **metric system.** The metric system originated as the result of a request of King Louis XVI to the French

Academy of Sciences during the early 1790s. Although the system fell briefly from favor in France, it was readopted in 1837. In 1875 the Treaty of the Meter was signed by 17 countries agreeing to adopt the use of the metric system.

Since that time the metric system has enjoyed worldwide popularity for several reasons. First, it entails only four base units—the meter of length, the kilogram of mass, the second of time, and the degree Kelvin of temperature. Second, the base units are precisely defined, reproducible quantities that are independent of factors such as gravitational force. Third, all units excepting those for time relate by factors of 10, in contrast to the numerous conversion factors necessary in converting English units of measurement. Last, the system is used internationally.

For these reasons as well as the fact that the metric system is used almost exclusively by the scientific community, it is the system that is used in this book. For those who are not familiar with the metric system, it is useful to be able to recognize the approximate English system equivalent of metric quantities. Two conversion factors that are particularly valuable are 2.54 centimeters for every inch and approximately 4.45 Newtons for every pound. All of the relevant units of measurement in both systems and common English-metric conversion factors are presented in Appendix C.

SUMMARY

Biomechanics is a multidisciplinary science involving the application of mechanical principles in the study of the structure and function of living organisms. Because biomechanists come from different academic backgrounds and professional fields, biomechanical research addresses a spectrum of problems and questions.

Basic knowledge of biomechanics is essential for competent professional analysts of human movement, including physical education teachers, physical therapists, physicians, coaches, personal trainers, and exercise instructors. The structured approach presented in this book is designed to facilitate the identification, analysis, and solution of problems or questions related to human movement.

INTRODUCTORY PROBLEMS

1. Locate and read three articles from the scientific literature that report the results of biomechanical investigations. (The *Journal of Biomechanics, Journal of Applied Biomechanics,* and *Medicine and Science in Sports and Exercise* are possible sources.) Write a one-page summary of each article and identify whether the investigation involved statics or dynamics and kinetics or kinematics.
2. List 8 to 10 World Wide Web sites that are related to biomechanics and write a paragraph describing each site.

3. Write a brief discussion about how knowledge of biomechanics may be useful in your intended profession or career.

4. Choose three jobs or professions and write a discussion about the ways in which each involves quantitative and qualitative work.

5. Write a summary list of the problem-solving steps identified in the chapter using your own words.

6. Write a description of one informal problem and one formal problem.

7. Step by step, show how to arrive at a solution to one of the problems you described in Problem 6.

8. Solve for x in each of the equations below. Refer to Appendix A for help if necessary.
   a. $x = 5^3$
   b. $7 + 8 = x/3$
   c. $4 \times 3^2 = x \times 8$
   d. $-15/3 = x + 1$
   e. $x^2 = 27 + 35$
   f. $x = \sqrt{79}$
   g. $x + 3 = \sqrt{38}$
   h. $7 \times 5 = -40 + x$
   i. $3^3 = x/2$
   j. $15 - 28 = x \times 2$
   (Answer: a. 125; b. 45; c. 4.5; d. −6; e. 7.9; f. 8.9; g. 3.2; h. 75; i. 54; j. −6.5)

9. Two schoolchildren race across a playground for a ball. Tim starts running at a distance of 15 meters from the ball, and Jan starts running at a distance of 12 meters from the ball. If Tim's average speed is 4.2 m/s and Jan's average speed is 4.0 m/s, which child will reach the ball first? Show how you arrived at your answer. (See Figure 1-9.) (Answer: Jan reaches the ball first.)

10. A 0.5 kg ball is kicked with a force of 40 Newtons. What is the resulting acceleration of the ball? (See Figure 1-10.) (Answer: 80 m/s^2)

ADDITIONAL PROBLEMS

1. Select a specific movement or sport skill of interest and read two or three articles from the scientific literature that report the results of biomechanical investigations related to the topic. Write a short paper that integrates the information from your sources into a scientifically based description of your chosen movement.

2. When attempting to balance your checkbook, you discover that your figures show a different balance in your account than was calculated by the bank. List an ordered, logical set of procedures that may be used to discover the error. You may use list, outline, or block diagram format.

3. Sarah goes to the grocery store and spends half of her money. On the way home, she stops for an ice cream cone that costs $0.78. Then she stops and spends a quarter of the remaining money on a $5.50 bill at the dry cleaners. How much money did Sarah have originally? (Answer: $45.56)

4. Wendell invests $10,000 in a stock portfolio made up of Petroleum Special at $30 per share, Newshoe at $12 per share, and Beans & Sprouts at $2.50 per share. He places 60% of the money in P.S., 30% in N, and 10% in B & S. With market values changing (P.S. down $3.12, N up 80%, and B & S up $.20), what is his portfolio worth 6 months later? (Answer: $11,856)

5. The hypotenuse of right triangle ABC (shown at right) is 4 cm long. What are the lengths of the other two sides? (Answer: A = 2 cm; B = 3.5 cm)

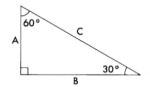

6. In triangle ABC, side B is 4 cm long and side C is 7 cm long. If the angle between sides B and C is 50 degrees, how long is side A? (Answer: 5.4 cm)

7. An orienteer runs 300 m north and then 400 m to the southeast (at a 45-degree angle to north). If he has run at a constant speed, how far away is he from the starting position? (Answer: 283.4 m)

8. John is out for his daily noontime run. He runs 2 km west, then 2 km south, and then runs on a path that takes him directly back to the place he started at. a. How far did John run? b. If he has run at an average speed of 4 m/s, how long did the entire run take? (Answer: a. 6.83 km; b. 28.5 minutes)

9. John and Al are in a 15 km race. John averages 4.4 m/s during the first half of the race and then runs at a speed of 4.2 m/s until the last 200 m, which he covers at 4.5 m/s. At what average speed must Al run to beat John? (Answer: > 4.3 m/s)

10. A sailboat heads north at 3 m/s for 1 hour and then tacks back to the southeast (at 45 degrees to north) at 2 m/s for 45 minutes. a. How far has the boat sailed? b. How far is it from its starting location? (Answer: a. 16.2 km; b. 8.0 km)

LABORATORY EXPERIENCES

1. Working within a group of 3–5 students, choose three human movements or motor skills with which you are all familiar. (A vertical jump is an example.) For each movement make a list of at least three general questions and at least three specific questions that an analyst might choose to answer.

2. Working within a group of 3–5 students, choose a human movement or motor skill with which you are all familiar and have two members of the group simultaneously perform the movement several times as the group observes. Based on your comparative observations, make a list of differences and similarities that you can

detect. Which of these are of potential importance and which are more a matter of personal style?

3. Working within a group of 3–5 students, view a previously taken videotape or film of a human movement or motor skill performance. After viewing the movement several times, make a list of at least three general questions and at least three specific questions that an analyst might choose to answer regarding the movement.

4. Having completed laboratory exercises 1–3, discuss within your group the relative advantages and disadvantages of each exercise in terms of your ability to formulate meaningful questions.

5. Have one member of your group perform several trials of walking as the group observes from front, side, and rear views. The subject may walk either on a treadmill or across the floor. What observations can be made about the subject's gait from each view that are not visible or apparent from the other views?

REFERENCES

1. Albright JP, Saterbak A, and Stokes J: Use of knee braces in sport. Current recommendations, Sports Med 20:281, 1995.
2. Anderson T: Biomechanics and running economy, Sports Med 22:76, 1996.
3. Arya AP, Lees A, Nirula HC, and Klenerman L: A biomechanical comparison of the SACH, Seattle and Jaipur feet using ground reaction forces, Prosthet Orthot Int 19:37, 1995.
4. Barlet JP, Coxam V, and Davicco MJ: Physical exercise and the skeleton, Arch Physiol Biochem 103:681, 1995.
5. Bockman RS: Osteoporosis and its management in older women, J Am Med Women's Assoc 52:121, 1997.
6. Bucher DM, Cres ME, deLateur BJ, Esselman PC, Margherita AJ, Price R, and Wagner EH: A comparison of the effects of three types of endurance training on balance and other risk factors in older adults, Aging (Milano) 9:112, 1997.
7. Casillas JM, Dulieu V, Cohen M, Marcer I, and Didier JP: Bioenergetic comparison of a new energy-storing foot and SACH foot in traumatic below-knee vascular amputations, Arch Phys Med Rehabil 76:39, 1995.
8. Chaffin DB and Andersson GBJ: *Occupational Biomechanics* (2nd ed.) New York, 1991, John Wiley & Sons.
9. Dapena J and Chung CS: Vertical and radial motions of the body during the take-off phase of high jumping, Med Sci Sports Exerc 20:290, 1988.
10. Davies CTM: Effects of wind resistance and assistance on the forward motion of a runner, J Appl Physiol 48:702, 1980.
11. Davis BL, Cavanagh PR, Sommer HJ III, and Wu G: Ground reaction forces during locomotion in simulated microgravity, Aviat Space Environ Med 67:235, 1996.
12. Deal CL: Am J Med 102:35S, 1997.
13. DeVita P, Torry M, Glover KL, and Speroni DL: A functional knee brace alters joint torque and power patterns during walking and running, J Biomech 29:583, 1996.

14. Dillman CJ: Overview of the United States Olympic Committee sports medicine biomechanics program. In Butts NK, Gushiken TT, and Zarins B, eds: *The elite athlete,* New York, 1985, Spectrum Publications.

15. Farber BS and Jacobson JS: An above-knee prosthesis with a system of energy recovery: a technical note, J Rehabil Res Dev 32:337, 1995.

16. Farley CT and Taylor CR: A mechanical trigger for the trot-gallop transition in horses, Science 253:306, 1991.

17. Ferrucci L, Guralnik JM, Buchner D, Kasper J, Lamb SE, Simonsick EM, Corti MC, Bandeen-Roche K, and Fried LP: Departures from linearity in the relationship between measures of muscular strength and physical performance of the lower extremities: the Women's Health and Aging Study, J Gerontol A Biol Sci Med Sci 52:M275, 1997.

18. Frost HM: Defining osteopenias and osteoporoses: another view (with insights from a new paradigm), Bone 20:385, 1997.

19. Gage JR, Perry J, Hicks RR, Koop S, and Werntz JR: Rectus femoris transfer to improve knee function of children with cerebral palsy. Develop Med Child Neur 29:159, 1987.

20. Gassett RS, Hearne B, and Keelan B: Ergonomics and body mechanics in the work place, Orthop Clin North Am 27:861, 1996.

21. Hay JG and Nohara H: The techniques used by elite long jumpers in preparation for take-off, J Biomech 23:229, 1990.

22. Hayes WC and Myers ER: Biomechanical considerations of hip and spine fractures in osteoporotic bone, Instr Course Lect 46:431, 1997.

23. Hreljac A: Preferred and energetically optimal gait transition speeds in human locomotion, Med Sci Sports Exerc 25:1158, 1993.

24. Johnson SC: Anterior cruciate ligament injury in elite Alpine competitors, Med Sci Sports Exerc 27:323, 1995.

25. Kram R, Domingo A, and Ferris DP: Effect of reduced gravity on the preferred walk-run transition speed, J Exp Biol 200:821, 1997.

26. Kyle CR: Reduction of wind resistance and power output of racing cyclists and runners traveling in groups, Ergon 22:387, 1979.

27. Kyle CR: Athletic clothing, Sci Am 254:104, 1986.

28. Lamb DR: The sports medicine umbrella, Sports Med Bull 5:8, 1984.

29. Lips P: Epidemiology and predictors of fractures associated with osteoporosis, Am J Med 103:3S, 1997.

30. Marklin RW, Simoneau GG, and Monroe JF: The effect of split and vertically-inclined computer keyboards on wrist and forearm posture, Proceedings of the Human Factors and Ergonomics Society 41st Annual Meeting, Albuquerque, NM, 642–646, 1997.

31. Miller LA and Childress DS: Analysis of a vertical compliance prosthetic foot, J Rehabil Res Dev 34:52, 1997.

32. Nelson RC: Biomechanics: past and present. In Cooper JM and Haven B, eds: Proceedings of the Biomechanics Symposium, Bloomington, Ind, 1980.

33. O'Toole ML: Prevention and treatment of injuries to runners, Med Sci Sports Exerc 24:S360, 1992.

34. Payton CJ, Hay JG, and Mullineaux DR: The effect of body roll on hand velocity in the front crawl, Comm XVth Congr of the Int Soc Biomechanics, Jyväskylä, Finland, 1995.

35. Perrin PP, Jeandel C, Perrin CA, Béné MC: Influence of visual control, conduction, and central integration on static and dynamic balance in healthy older adults, Gerontology 43:223, 1997.

36. Perry AK, Blickhan R, Biewener AA, Heglund NC, and Taylor CR: Preferred speeds in terrestrial vertebrates: are they equivalent? J Exp Biol 137:207, 1988.
37. Pigozzi F, Santori N, Di Salvo V, Parisi A, and Di-uigi L: Snowboard traumatology: an epidemiological study, Orthopedics 20:505, 1997.
38. Recker RR: Osteoporosis, Contemp Nutr 8:1, 1983.
39. Remizov LP: Biomechanics of optimal flight in ski jumping, J Biomech 17:167, 1984.
40. Ruby D: Biomechanics—how computers extend athletic performance to the body's far limits, Popular Science p 58, Jan 1982.
41. Senatore JR: Functional components of a sport shoe, Orthop Nurs 15:19, 1996.
42. Snow-Harter C and Marcus R: Exercise, bone mineral density, and osteoporosis, Exerc Sport Sci Rev 19:351, 1991.
43. Torburn L, Powers CM, Guiterrez, and Perry J: Energy expenditure during ambulation in dysvascular and traumatic below-knee amputees: a comparison of five prosthetic feet, J Rehabil Res Dev 32:111, 1995.
44. Wickelgren WA: *How to solve problems,* San Francisco, 1974, WH Freeman & Co.
45. Williams KR: Biomechanics of distance running. In Grabiner MD, ed: *Current issues in biomechanics,* Champaign, 1993, Human Kinetics.
46. Yamamoto S: A new trend in the study of low back pain in workplaces, Ind Health 35:173, 1997.

ANNOTATED READINGS

Bartlett RM: Current issues in the mechanics of athletic activities. A position paper, J Biomech 30:477, 1997.
Reviews important or contentious current issues in the study of the biomechanics of athletic activities.

Bransford JD and Stein BS: The ideal problem solver, New York, 1984, WH Freeman & Co.
Discusses techniques for improving problem-solving skills.

Cavanagh PR: Biomechanics: a bridge builder among the sport sciences, Med Sci Sports Exerc 22:546, 1990.
Discusses a variety of sport biomechanics applications and illustrates potential linkages between biomechanics and other branches of sports medicine.

Chaffin DB and Andersson GBJ: Occupational biomechanics (2nd ed.), New York, 1991, John Wiley & Sons.
Serves as a comprehensive text on the field of occupational biomechanics.

Grabiner MD (ed.): Current issues in biomechanics, Champaign, 1993, Human Kinetics.
Includes chapters on selected topics in sport, occupational, tissue, and neuromuscular aspects of biomechanics.

RELATED WEB SITES

American College of Sports Medicine—Biomechanics Interest Group
http://www.pepperdine.edu/seaver/natsci/spme/acsmbiomech
This is the home page of the Biomechanics Interest Group in the American College of Sports Medicine.

American Society of Biomechanics
http://osu.orst.edu/dept/HHP/ASB

> *This home page of the American Society of Biomechanics provides information about the organization, conference abstracts, and a list of graduate programs in biomechanics.*

The Biomch-L Newsgroup
http://www.lri.ccf.org./isb/biomch-l

> *Provides information about an e-mail discussion group for biomechanics and human/animal movement science.*

Biomechanics Books
http://www.makura.com/bookstore/biomechanics.html

> *Lists a large number of biomechanics-related books available through Amazon.com.*

Biomechanics, Inc.
http://www.biomechanics-inc.com

> *Provides information on technology used in biomechanics-related work and includes a number of downloadable video clips.*

Biomechanics World Wide
http://www.per.ualberta.ca/biomechanics/bwwnofrm.htm

> *A comprehensive site with links to other web sites for a wide spectrum of topics related to biomechanics.*

International Society of Biomechanics
http://www.bme.ccf.org/isb

> *This home page of the International Society of Biomechanics (ISB) provides information on ISB, biomechanical software and data, and pointers to other sources of biomechanics-related information.*

International Society of Biomechanics in Sports
http://www.uni-stuttgart.de/External/isbs

> *This home page of the International Society of Biomechanics in Sports (ISBS) provides information about the organization.*

2

KINEMATIC CONCEPTS FOR ANALYZING HUMAN MOTION

After completing this chapter, the reader will be able to:

Identify and describe the reference positions, planes, and axes associated with the human body.

Provide examples of linear, angular, and general forms of motion.

Define and appropriately use directional terms and joint movement terminology.

Explain how to plan and conduct an effective qualitative human movement analysis.

Identify and describe the uses of available instrumentation for measuring kinematic quantities.

Is it best to observe walking gait from a side view, front view, or back view? From what distance can a coach best observe an athlete's pitching style? What are the advantages and disadvantages of analyzing a movement captured on videotape? To the untrained observer, there may be no differences in the forms displayed by an elite hurdler and a novice hurdler or in the functioning of a normal knee and an injured, partially rehabilitated knee. What skills are necessary and what procedures are used for effective analysis of human movement kinematics?

One of the most important steps in learning a new subject is mastering the associated terminology. Likewise, learning a general analysis protocol that can be adapted to specific questions or problems within a field of study is invaluable. In this chapter, human movement terminology is introduced and the problem-solving approach is adapted to provide a template for qualitative solving of human movement analysis problems.

STANDARD REFERENCE TERMINOLOGY

Communicating specific information about human movement requires specialized terminology that precisely identifies body positions and directions.

Anatomical Reference Position

Anatomical reference position is an erect standing position with the feet just slightly separated and the arms hanging relaxed at the sides, with the palms of the hands facing forward. It is not a natural standing position, but is the body orientation conventionally used as the reference position or starting place when movement terms are defined.

anatomical reference position
erect standing position with all body parts, including the palms of the hands, facing forward; considered the starting position for body segment movements

Directional Terms

In describing the relationship of body parts or the location of an external object with respect to the body, the use of directional terms is necessary. The following are commonly used directional terms:

Superior: Closer to the head (In zoology, the synonymous term is *cranial.*)

Inferior: Farther away from the head (In zoology, the synonymous term is *caudal.*)

Anterior: Toward the front of the body (In zoology, the synonymous term is *ventral.*)

Posterior: Toward the back of the body (In zoology, the synonymous term is *dorsal.*)

Medial: Toward the midline of the body

Lateral: Away from the midline of the body

Proximal: Closer to the trunk (For example, the knee is proximal to the ankle.)

Distal: Away from the trunk (For example, the wrist is distal to the elbow.)

Superficial: Toward the surface of the body

Deep: Inside the body and away from the body surface

Anatomical reference position.

Reference planes and axes are useful in describing gross body movements and in defining more specific movement terminology.

All of these directional terms can be paired as antonyms—words having opposite meanings. Saying that the elbow is proximal to the wrist is as correct as saying that the wrist is distal to the elbow. Similarly, the nose is superior to the mouth and the mouth is inferior to the nose.

Anatomical Reference Planes

cardinal planes
three imaginary perpendicular reference planes that divide the body in half by mass

sagittal plane
plane in which forward and backward movements of the body and body segments occur

frontal plane
plane in which lateral movements of the body and body segments occur

transverse plane
plane in which horizontal body and body segment movements occur when the body is in an erect standing position

Although most human movements are not strictly planar, the cardinal planes provide a useful way to describe movements that are primarily planar.

The three imaginary **cardinal planes** bisect the mass of the body in three dimensions. A *plane* is a two-dimensional surface with an orientation defined by the spatial coordinates of three discrete points not all contained in the same line. It may be thought of as an imaginary flat surface. The **sagittal plane,** also known as the anteroposterior (AP) plane, divides the body vertically into left and right halves, with each half containing the same mass. The **frontal plane,** also referred to as the coronal plane, splits the body vertically into front and back halves of equal mass. The horizontal or **transverse plane** separates the body into top and bottom halves of equal mass. For an individual standing in anatomical reference position, the three cardinal planes all intersect at a single point known as the body's center of mass or center of gravity (Figure 2-1). These imaginary reference planes exist only with respect to the human body. If a person turns at a 45° angle to the right, the reference planes also turn at a 45° angle to the right.

Although the entire body may move along or parallel to a cardinal plane, the movements of individual body segments may also be described as sagittal plane movements, frontal plane movements, and transverse plane movements. When this occurs, the movements being described are usually in a plane that is parallel to one of the cardinal planes. For example, movements that involve forward and backward motion are referred to as sagittal plane movements. When a forward roll is executed, the entire body moves parallel to the sagittal plane. During running in place, the motion of the arms and legs is generally forward and backward, although the planes of motion pass through the shoulder and hip joints rather than the center of the body. Marching, bowling, and cycling are all largely sagittal plane movements (Figure 2-2). Frontal plane movement is lateral (side-to-side) movement, and an example of total body frontal plane movement is the cartwheel. Jumping jacks, side stepping, and side kicks in soccer require frontal plane movement at certain body joints. Examples of total body transverse plane movement include a twist executed by a diver, trampolinist, or airborne gymnast or a dancer's pirouette.

Although many of the movements conducted by the human body are not oriented sagittally, frontally, or transversely or are not planar at all, the three major reference planes are still useful. Gross body movements and specifically named movements that occur at joints are often described as primarily frontal, sagittal, or transverse plane.

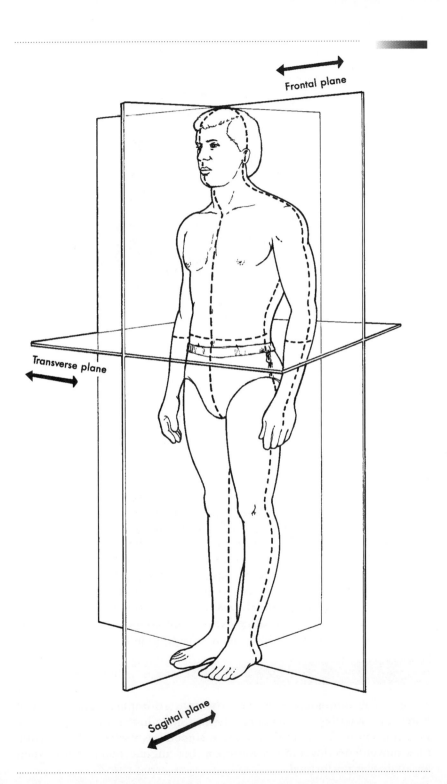

FIGURE 2-1
The three cardinal
reference planes.

FIGURE 2-2
Cycling requires sagittal plane movement of the legs.

Sagittal plane

Anatomical Reference Axes

frontal axis
imaginary line around which sagittal plane rotations occur
sagittal axis
imaginary line around which frontal plane rotations occur
longitudinal axis
imaginary line around which transverse plane rotations occur
general motion
involving translation and rotation simultaneously
linear
along a line that may be straight or curved, with all parts of the body moving in the same direction at the same speed
angular
involving rotation around a central line or point

When a segment of the human body moves, it rotates around an imaginary axis of rotation that passes through a joint to which it is attached. There are three reference axes for describing human motion, and each is oriented perpendicular to one of the three planes of motion. The **frontal axis,** also known as the transverse axis, is perpendicular to the sagittal plane. Rotation in the frontal plane occurs around the **sagittal axis** or anteroposterior (AP) axis (Figure 2-3). Transverse plane rotation is around the **longitudinal axis** or vertical axis. It is important to recognize that each of these three axes is always associated with the same single plane—the one to which the axis is perpendicular.

FORMS OF MOTION

Most human movement is **general motion,** a complex combination of **linear** and **angular** motion components. Since linear and angular motion are "pure" forms of motion, it is sometimes useful to break complex movements down into their linear and angular components when an analysis is performed.

Rotation of a body segment at a joint occurs around an imaginary line known as the axis of rotation that passes through the joint center.

Linear Motion

Pure linear motion involves uniform motion of the system of interest, with all system parts moving in the same direction at the same speed. Linear motion is also referred to as translatory motion, or **translation.** When a body experiences translation it moves as a unit and portions of the body do not move relative to each other. For example, a sleeping passenger on a smooth airplane flight is being translated through the air. If the passenger lifts an arm to reach for a magazine, however, pure translation is no longer occurring because the position of the arm respective to the body has changed.

Linear motion may also be thought of as motion along a line. If the line is straight, the motion is **rectilinear.** If the line is curved, the motion is **curvilinear.** A motorcyclist maintaining a motionless posture as the bike moves along a straight path is moving rectilinearly. If the motorcyclist jumps the bike and the frame of the bike does not rotate, both rider and bike (with the exception of the spinning wheels) are moving curvilinearly while airborne. Likewise, a Nordic skier coasting in a locked static position down a short hill is in rectilinear motion. If the skier jumps over a gully with all body parts moving in the same direction at

translation
linear motion

rectilinear
along a straight line

curvilinear
along a curved line

FIGURE 2-3
For a jumping jack, the major axes of rotation are sagittal axes passing through the shoulders and hips.

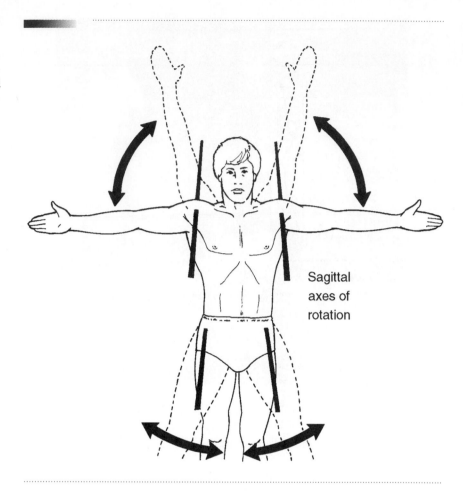

Sagittal axes of rotation

the same speed along a curved path, the motion is curvilinear. When a motorcyclist or skier goes over the crest of a hill, the motion is *not* linear because the top of the body is moving at a greater speed than lower body parts. Figure 2-4 displays a gymnast in rectilinear, curvilinear, and rotational motion.

Angular Motion

Angular motion is rotation around a central imaginary line known as the **axis of rotation,** which is oriented perpendicular to the plane in which the rotation occurs. When a gymnast performs a giant circle on a bar, the entire body rotates, with the axis of rotation passing through the center of the bar. When a springboard diver executes a somersault in midair, the entire body is again rotating, this time around an imaginary axis of rotation that moves along with the body. Almost all volitional human move-

axis of rotation
imaginary line perpendicular to the plane of rotation and passing through the center of rotation

FIGURE 2-4

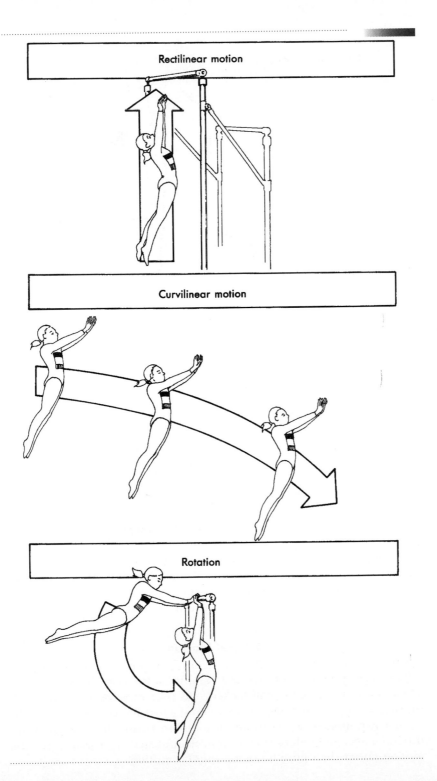

FIGURE 2-5
General motion is a
combination of linear and
angular motion.

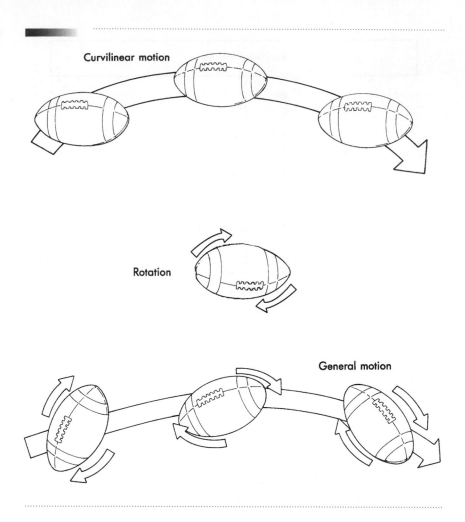

Curvilinear motion

Rotation

General motion

ment involves rotation of a body segment around an imaginary axis of rotation that passes through the center of the joint to which the segment
attaches. When angular motion or rotation occurs, portions of the body
in motion are constantly moving relative to other portions of the body.

General Motion

When translation and rotation are combined, the resulting movement
is general motion. A football kicked end over end translates through
the air as it simultaneously rotates around a central axis (Figure 2-5). A
runner is translated along by angular movements of body segments at
the hip, knee, and ankle. Human movement usually consists of general
motion rather than pure linear or angular motion.

*Most human movement
activities are categorized as
general motion.*

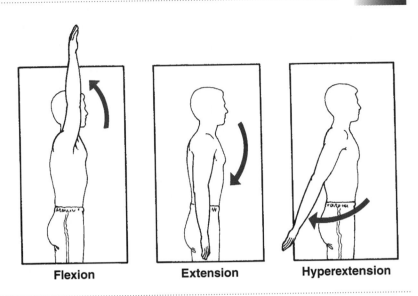

| Flexion | Extension | Hyperextension |

FIGURE 2-6
Sagittal plane movements
at the shoulder.

Mechanical Systems

Before determining the nature of a movement, the mechanical **system** of interest must be defined. In many circumstances the entire human body is chosen as the system to be analyzed. In other circumstances, however, the system might be defined as the right arm or perhaps even a ball being projected by the right arm. When an overhand throw is executed the body as a whole displays general motion, the motion of the throwing arm is primarily angular, and the motion of the released ball is linear. The mechanical system to be analyzed is chosen by the movement analyst according to the focus of interest.

system
mechanical system chosen
by the analyst for study

JOINT MOVEMENT TERMINOLOGY

When the human body is in anatomical reference position, all body segments are considered to be positioned at zero degrees. Rotation of a body segment away from anatomical position is named according to the direction of motion and is measured as the angle between the body segment's position and anatomical position.

Sagittal Plane Movements

From anatomical position, the three primary movements occurring in the sagittal plane are *flexion, extension,* and *hyperextension* (Figure 2-6). Flexion includes anteriorly directed sagittal plane rotations of the head,

*Sagittal plane movements
include flexion, extension, and
hyperextension; and dorsiflexion
and plantar flexion.*

Figure 2-7
Sagittal plane movements
of the foot.

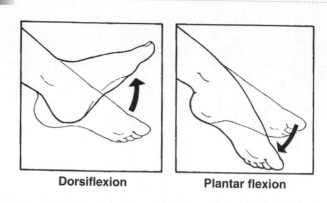

Dorsiflexion	**Plantar flexion**

trunk, upper arm, forearm, hand, and hip, and posteriorly directed sagittal plane rotation of the lower leg. Extension is defined as the movement that returns a body segment to anatomical position from a position of flexion, and hyperextension is the rotation beyond anatomical position in the direction opposite the direction of flexion. If the arms or legs are internally or externally rotated from anatomical position, flexion, extension, and hyperextension at the knee and elbow may occur in a plane other than the sagittal.

Sagittal plane rotation at the ankle occurs both when the foot is moved relative to the lower leg and when the lower leg is moved relative to the foot. Motion bringing the top of the foot toward the lower leg is known as *dorsiflexion,* and the opposite motion, which can be visualized as "planting" the ball of the foot, is termed *plantar flexion* (Figure 2-7).

Frontal Plane Movements

The major frontal plane rotational movements are *abduction* and *adduction.* Abduction moves a body segment away from the midline of the body and adduction moves a body segment closer to the midline of the body (Figure 2-8).

Frontal plane movements include abduction and adduction, lateral flexion, elevation and depression, inversion and eversion, and radial and ulnar deviation.

Other frontal plane movements include sideways rotation of the trunk, which is termed right or left *lateral flexion* (Figure 2-9). *Elevation* and *depression* of the shoulder girdle refer to movement of the shoulder girdle in superior and inferior directions respectively (Figure 2-10). Rotation of the hand at the wrist in the frontal plane toward the radius (thumb side) is referred to as *radial deviation,* and *ulnar deviation* is hand rotation toward the ulna (little finger side) (Figure 2-11).

Movements of the foot that occur largely in the frontal plane are eversion and inversion. Outward rotation of the sole of the foot is termed

A cartwheel is primarily frontal plane movement.

eversion, and inward rotation of the sole of the foot is called *inversion* (Figure 2-12). Abduction and adduction are also used to describe outward and inward rotation of the entire foot. *Pronation* and *supination* are often used to describe motion occurring at the subtalar joint. Pronation at the subtalar joint consists of a combination of eversion, abduction, and dorsiflexion, and supination involves inversion, adduction, and plantar flexion.

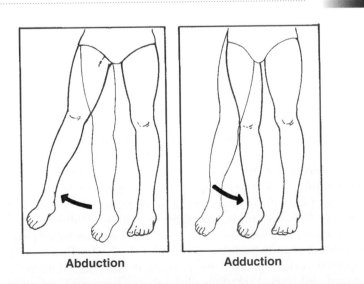

Abduction   **Adduction**

FIGURE 2-8
Frontal plane movements at the hip.

FIGURE 2-9
Frontal plane movements
of the spinal column.

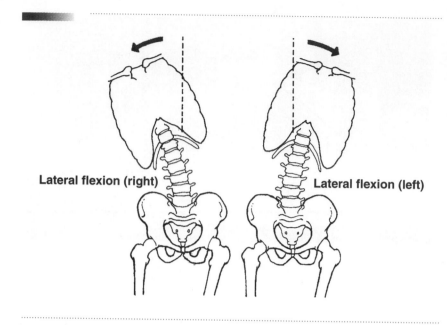

FIGURE 2-10
Frontal plane movements
of the shoulder girdle.

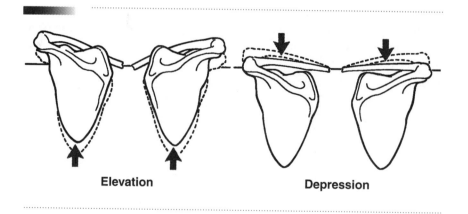

Transverse Plane Movements

*Transverse plane movements
include left and right rotation,
medial and lateral rotation,
supination and pronation, and
horizontal abduction and
adduction.*

Body movements in the transverse plane are rotational movements about
a longitudinal axis. *Left rotation* and *right rotation* are used to describe
transverse plane movements of the head, neck, and trunk. Rotation of
an arm or leg as a unit in the transverse plane is called *medial rotation*
when rotation is toward the midline of the body and *lateral rotation* when
the rotation is away from the midline of the body (Figure 2-13).

Specific terms are used for rotational movements of the forearm.
Outward and inward rotations of the forearm are respectively known as

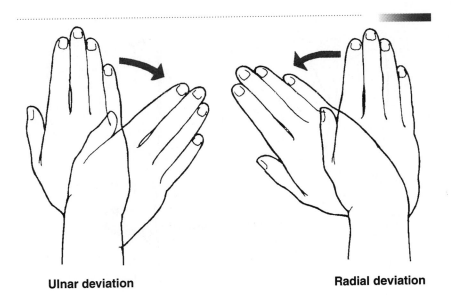

FIGURE 2-11
Frontal plane movements of the hand.

Ulnar deviation

Radial deviation

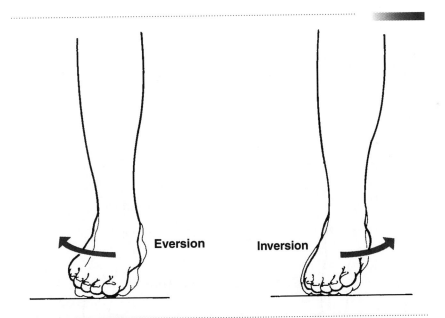

FIGURE 2-12
Frontal plane movements of the foot.

Eversion

Inversion

supination and *pronation* (Figure 2-14). In anatomical position the forearm is in a supinated position.

Although abduction and adduction are frontal plane movements, when the arm or thigh is flexed to a 90° position, movement of these segments in the transverse plane from an anterior position to a lateral

FIGURE 2-13
Transverse plane
movements of the leg.

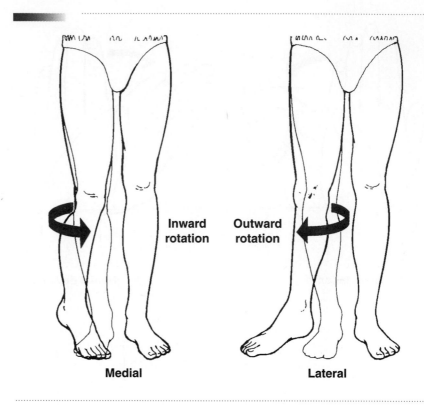

Inward
rotation

Outward
rotation

Medial

Lateral

FIGURE 2-14
Transverse plane
movements of the
forearm.

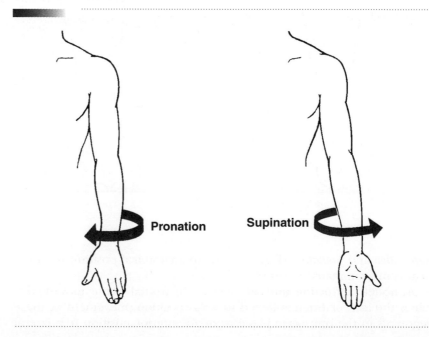

Pronation

Supination

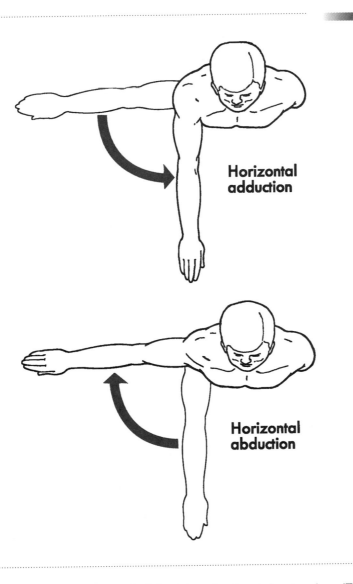

FIGURE 2-15
Transverse plane movements at the shoulder.

Horizontal adduction

Horizontal abduction

position is termed *horizontal abduction* or horizontal extension (Figure 2-15). Movement in the transverse plane from a lateral to an anterior position is called *horizontal adduction* or horizontal flexion.

Other Movements

Many movements of the body limbs take place in planes that are oriented diagonally to the three traditionally recognized cardinal planes. Because human movements are so complex, however, nominal identification of every plane of human movement is impractical.

A tennis serve requires arm movement in a diagonal plane.

FIGURE 2-16
Circumduction of the
index finger at the
metacarpophalangeal
joint.

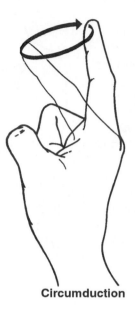

Circumduction

One special case of general motion involving circular movement of a body segment is designated as *circumduction*. Tracing an imaginary circle in the air with a fingertip while the rest of the hand is stationary requires circumduction at the metacarpophalangeal joint (Figure 2-16). Circumduction combines flexion, extension, abduction, and adduction, resulting in a conical trajectory of the moving body segment.

QUALITATIVE ANALYSIS OF HUMAN MOVEMENT

Qualitative analysis requires knowledge of the specific biomechanical purpose of the movement and the ability to detect the causes of errors.

Qualitative assessment of human movement requires both knowledge of the movement characteristics desired and the ability to observe and analyze whether a given performance incorporates these characteristics (6). As introduced in Chapter 1, the word *qualitative* refers to a description of quality without the use of numbers. Visual observation is the most commonly used approach for qualitatively analyzing the mechanics of human movement. Based on information gained from watching an athlete perform a skill, a patient walk down a ramp, or a student attempt a novel task, coaches, clinicians, and teachers make judgments and recommendations on a daily basis. To be effective, however, a qualitative analysis cannot be conducted haphazardly, but must be carefully planned and conducted by an analyst with knowledge of the biomechanics of the movement.

Prerequisite Knowledge for a Qualitative Analysis

There are two main sources of information for the analyst diagnosing a motor skill. The first is the kinematics or technique exhibited by the performer, and the second is the performance outcome (13). Evaluating performance outcome is of limited value, since the root of optimal performance outcome is appropriate biomechanics.

To effectively analyze a motor skill, the analyst must first understand the specific purpose of the skill from a biomechanical perspective (12). The general goal of a volleyball player serving a ball is to legally project the ball over the net and into the opposite court. Specifically, this requires a coordinated summation of forces produced by trunk rotation, shoulder flexion, elbow extension, and forward translation of the total body center of gravity, as well as contacting the ball at an appropriate height and angle. Whereas the ultimate purpose of a competitive sprint cyclist is to maximize speed while maintaining balance in order to cross the finish line first, biomechanically this requires factors such as maximizing perpendicular force production against the pedals and maintaining a low body profile to minimize air resistance.

Without knowledge of relevant biomechanical principles, analysts may have difficulty in identifying the factors that contribute to (or hinder) performance and may misinterpret the observations noted. More specifically, to effectively analyze a motor skill, the analyst must be able to identify the *cause* of a technique error, as opposed to a symptom of the error, or a performance idiosyncrasy (12). Inexperienced coaches of tennis or golf may focus on getting the performer to display an appropriate follow-through after hitting the ball. Inadequate follow-through, however, is merely a symptom of the underlying performance error, which may be failure to begin the stroke or swing with sufficient trunk rotation and backswing, or failure to swing the racquet or club with sufficient velocity. The ability to identify the cause of a performance error is dependent on understanding the biomechanics of the motor skill.

One potential source of knowledge about the biomechanics of a motor skill is experience in performing the skill. A person who performs a skill proficiently is usually better equipped to qualitatively analyze that skill than a person less familiar with the skill. For example, research shows that experienced tennis and softball players are more proficient than novices at detecting errors in the forehand tennis stroke and in batting (2, 7). This is at least partially due to the fact that expert performers often focus their attention more appropriately than novices when analyzing performance (3). Research on physical education student teachers also shows that when they analyze a movement, the focus of attention differs with personal background experiences (1). In most cases a high level of familiarity with the skill or movement being performed improves the analyst's ability to focus attention on the critical aspects of the event.

Many jobs require conducting qualitative analyses of human movement daily.

Analysts should be able to distinguish the cause of a problem from symptoms of the problem or an unrelated movement idiosyncrasy.

Experience in performing a motor skill does not necessarily translate to proficiency in analyzing the skill.

Direct experience in performing a motor skill, however, is not the only or necessarily the best way to acquire expertise in analyzing the skill. Skilled athletes often achieve success not because of the form or technique they display, but in spite of it! Furthermore, highly accomplished athletes do not always become the best coaches, and highly successful coaches may have had little to no participatory experience in the sports that they coach.

The conscientious coach, teacher, and clinician typically use several avenues to develop a knowledge base from which to evaluate a motor skill. One is to read available materials from textbooks, scientific journals, and lay (coaching) journals, despite the fact that not all movement patterns and skills have been researched and some biomechanics literature is so esoteric that advanced training in biomechanics is required to understand it. However, when selecting reading material it is important to distinguish between articles supported by research and those based primarily on opinion, as "commonsense" approaches to skill analyses may be flawed. There are also opportunities to interact directly with individuals who have expert knowledge of particular skills at conferences and symposia.

Planning a Qualitative Analysis

Even the simplest qualitative analysis may yield inadequate or faulty information if approached haphazardly. As the complexity of the skill and/or the level of desired analytical detail increases, the level of required planning increases.

The first step in any analysis is to identify the major question or questions of interest to be answered. Often these questions have already been formulated by the analyst, or they serve as the original purpose for the observation. For example, has a post–knee surgery patient's gait returned to normal? Why is a volleyball player having difficulty hitting cross-court? What might be causing a secretary's wrist pain? Or simply, is a given skill being performed as effectively as possible? Having one or more particular questions or problems in mind helps to focus the analysis. The ability to identify appropriate analysis questions is dependent on the analyst's knowledge of the biomechanics of the movement.

Analysts should next determine the optimal perspective(s) from which to view the movement. If the major movements are primarily planar, such as with the legs during cycling or the pitching arm during an underhand softball pitch, a single viewing perspective such as a side view or a rear view may be sufficient. If the movement occurs in more than one plane, as with the motions of the arms and legs during the breaststroke or the arm motion during a baseball batter's swing, the observer may need to view the movement from more than one perspective to see all critical aspects of interest. For example, a rear view, a side view, and

Primarily planar skills

Multiplanar skills

FIGURE 2-17
Whereas skills that are primarily planar may require only one viewing perspective, the movement analyst should view multiplanar skills from more than one direction.

A tennis player's eyes should follow the oncoming ball long enough to enable contacting the ball with the racket.

a top view of a martial artist's kick all yield different information about the movement (Figure 2-17).

Even when a movement is primarily planar, using more than a single viewing perspective may be useful. Sometimes, for example, observing the performer's eye movements yields valuable information. A common error among beginning tennis players is looking across the net to where they expect to hit the ball rather than visually tracking enough of the

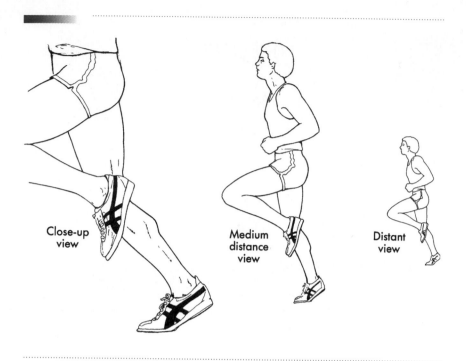

Close-up
view

Medium
distance
view

Distant
view

oncoming ball's trajectory to establish racket contact with it. In fencing,
focusing the eyes on the rapidly moving tip of an opponent's blade can
be a drastic performance error, facilitating the effectiveness of the opponent's feints (16).

The analyst's viewing distance from the performer should also be selected thoughtfully (Figure 2-18). If the analyst wishes to observe subtalar pronation and supination in a patient walking on a treadmill, a
close-up rear view of the lower legs and feet is necessary. Analyzing where
a particular volleyball player moves on the court during a series of plays
under rapidly changing game conditions is best accomplished from a
reasonably distant, elevated position.

Another consideration is the number of trials or executions of the
movement that should be observed in the course of formulating an
analysis. A skilled athlete may display movement kinematics that deviate
only slightly across performances, but a child learning to run may take
no two steps alike. Basing an analysis on observation of a single performance is usually unwise. The greater the inconsistency in the performer's kinematics, the larger the number of observations that should
be made.

Other factors that potentially influence the quality of observations of
human movement are the performer's attire and the nature of the sur-

*Use of a video camera
provides both advantages and
disadvantages to the
movement analyst.*

*Repeated observation of a
motor skill is useful in helping
the analyst to distinguish
consistent performance errors
from random errors.*

rounding environment. When biomechanic researchers study the kinematics of a particular movement, the subjects typically wear minimal attire so that movements of body segments will not be obscured. Although there are many situations, such as instructional classes, competitive events, and team practices, for which this may not be practical, analysts should be aware that loose clothing can obscure subtle motions. Adequate lighting and a nondistracting background of contrasting color also improve the visibility of the observed movement.

A final consideration is whether to rely on visual observation alone or to use a video camera. As the speed of the movement of interest increases, it becomes progressively less practical to rely on visual observation. The human eye cannot resolve events that occur in less than approximately one-fourth of a second (4). Consequently, even the most careful observer may miss important aspects of a rapidly executed movement. Videotape also enables the performer to view the movement, as well as repeated viewing of the movement by analyst and performer. Better-quality playback units also enable slow-motion viewing and single-picture advance that facilitate isolation of the critical aspects of a movement.

The analyst should be aware, however, that there is a potential drawback to the use of videotape. The subject's awareness of the presence of a camera sometimes results in changes in performance. Researchers found performance decrements among unskilled subjects and performance increments among skilled subjects performing basketball free throws when a camera was introduced (9). Movement analysts should be aware that subjects may be distracted or unconsciously modify their techniques when a recording device is used.

When performers are minimally attired, analysts are better able to observe the joint motions taking place.

Conducting a Qualitative Analysis

Despite careful planning of a qualitative analysis, occasionally new questions emerge during the course of collecting observations. Movement modifications may be taking place with each performance as learning occurs, especially when the performer is unskilled. Even when this is not the case, the observations made may suggest new questions of interest. For example, what is causing the inconsistencies in a golfer's swing? What technique changes are occurring over the 30 to 40 m range in a 100 m sprint? A careful analysis is not strictly preprogrammed, but often involves identifying new questions or problems to solve. The teacher, clinician, or coach is often involved in a continuous process of formulating an analysis, collecting additional observations, and formulating an updated analysis (Figure 2-19).

Answering questions that have been identified requires that the analyst be able to focus on the critical aspects of the movement. Once a biomechanical error has been generally identified, it is often useful for

A video camera can be a useful tool for human movement analysts.

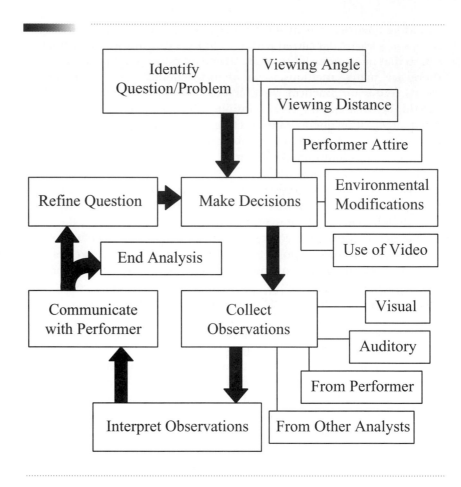

the analyst to watch the performer over several trials and to progressively zero in on the specific problem (12). Evaluating a softball pitcher's technique might begin with observation of insufficient ball speed, progress to an evaluation of upper-extremity kinematics, and end with an identification of insufficient wrist snap at ball release.

The analyst should also be aware that every performance of a motor skill is affected by the characteristics of the performer. These include the performer's age, gender, anthropometry, the developmental and skill levels at which the performer is operating, and any special physical or personality traits that may impact performance. Providing a novice, preschool-aged performer with cues for a skilled, mature performance may be counterproductive, since children are not simply scaled-down adults (8). Likewise, although training can ameliorate loss of muscular strength and joint range of motion once thought to be inevitably associated with aging, human movement analysts need increased knowledge of and sensitivity to the special needs of older adults who wish to develop specialized motor skills. Analysts should also be aware that although gender has traditionally been regarded as a basis for performance differences, research has shown that before puberty most gender-associated performance differences are probably culturally derived rather than biologically determined (14). Young girls are usually not expected to be as skilled or even as active as young boys. Unfortunately, in many settings these expectations extend beyond childhood into adolescence and adulthood. Analysts of female performers should not reinforce this cultural misunderstanding by lowering their expectations of girls or women based on gender. Analysts should also be sensitive to other factors that can influence performance. Has the performer experienced a recent emotional upset? Is the sun in his eyes? Is she tired?

To supplement visual observation, the analyst should be aware that nonvisual forms of information can also sometimes be useful during a movement analysis. For example, auditory information can provide clues about the way in which a movement was executed. Proper contact of a golf club with a ball sounds distinctly different from when a golfer "tops" the ball. Similarly, the crack of a baseball bat hitting a ball indicates that the contact was direct rather than glancing. The sound of a double contact of a volleyball player's arms with the ball may identify an illegal hit. The sound of a patient's gait usually reveals whether an asymmetry is present.

Auditory information is often a valuable source in the analysis of human motor skills.

Another potential source of information is feedback from the performer (Figure 2-20). A performer who is experienced enough to recognize the way a particular movement feels as compared to the way a slight modification of the same movement feels is a useful source of information. However, not all performers are sufficiently kinesthetically attuned to provide meaningful subjective feedback of this nature. The performer being analyzed may also assist in other ways. As Hoffman (5) has pointed out, performance deficiencies may result from errors in technique, perception, or decision making. Identification of perceptual and decision-making errors by the performer often requires more than visual observation of the performance. In these cases, asking meaningful questions of the performer may be useful. However, the analyst should consider subjective input from the performer in conjunction with more objective observations.

Another potential way to enhance the thoroughness of an analysis is to involve more than one analyst. This reduces the likelihood of oversight. Students in the process of learning a new motor skill may also benefit from teaming up to analyze each other's performances under appropriate teacher direction.

Finally, analysts must remember that the skill of observation improves with practice. Research indicates that training in both general analysis protocol and visual discrimination of critical features of a specific motor skill can dramatically improve an analyst's ability (11, 15).

The ability to effectively analyze human movement improves with practice.

FIGURE 2-20
Auditory information,
such as the sound of a
double contact, can be
useful to the movement
analyst.

SAMPLE APPLICATION

Problem: Sally, a powerful outside hitter on a high school volleyball team, has been out for two weeks with mild shoulder bursitis, but has recently received her physician's clearance to return to practice. Joan, Sally's coach, notices that Sally's spikes are traveling at a slow speed and are being easily handled by the defensive players.

Playing the Analysis:
1. What specific problems need to be solved or questions need to be answered regarding the movement? Joan first questions Sally to make sure that the shoulder is not painful. She then reasons that a technique error is present.
2. From what angle(s) and distance(s) can problematic aspects of the movement best be observed? Is more than one view needed? Although a volleyball spike involves transverse plane rotation of the trunk, the arm movement is primarily in the sagittal plane. Joan therefore decides to begin by observing a sagittal view from the side of Sally's hitting arm.
3. How many movement performances should be observed? Since Sally is a skilled player and her spikes are consistently being executed at reduced velocity, Joan reasons that only a few observations may be needed.
4. Is special subject attire, lighting, or background environment needed to facilitate observation? The gym where the team works out is well lit and the players wear sleeveless tops. Therefore, no special accommodations for the analysis seem necessary.
5. Will a video recording of the movement be necessary or useful? A volleyball spike is a relatively fast movement, but there are definite check points that the knowledgeable observer can watch in real time. Is the jump primarily vertical and is it high enough for the player to contact the ball above

As analysts gain experience, the analysis process becomes more natural and the analyses conducted are likely to become more effective and informative. The expert analyst is better able to both identify and diagnose errors than the novice (13). Novice analysts should take every opportunity to practice movement analysis in carefully planned and structured settings.

TOOLS FOR MEASURING KINEMATIC QUANTITIES

Biomechanic researchers have available a wide array of equipment for studying human movement kinematics. Knowledge gained through the use of this apparatus is often published in professional journals for teachers, clinicians, coaches, and others interested in human movement.

the net? Is the hitting arm positioned with the upper arm in maximal horizontal abduction prior to arm swing to allow a full range of arm motion? Is the hitting movement initiated by trunk rotation followed by shoulder flexion, then elbow extension, then snap-like wrist flexion? Is the movement being executed in a coordinated fashion to enable imparting a large force to the ball?

Conducting the Analysis:
1. Review, and sometimes reformulate, specific questions of focus. After watching Sally execute two spikes, Joan observes that her arm range of motion appears to be relatively small.
2. Repeated viewing of a movement enables gradual zeroing in on causes of performance errors. After watching Sally spike three more times, Joan suspects that Sally is not positioning her upper arm in maximal horizontal abduction in preparation for the hit.
3. Be aware of the influence of performer characteristics. Joan talks to Sally on the sideline and asks her to put her arm in the preparatory position for a hit. She asks Sally if this position is painful and Sally responds that it is not.
4. Pay attention to nonvisual cues. (None apparent in this situation.)
5. When appropriate, ask the performer to self-analyze. Joan tells Sally that she suspects Sally has been protecting the shoulder by not rotating her arm back far enough in preparation for spikes. She can correct the problem. Sally's next few spikes are executed at much faster velocity.
6. Consider involving other analysts to assist. Joan asks her assistant coach to watch Sally for the remainder of practice to determine whether the problem has been corrected.

Cinematography and Videography

Photographers began employing cameras in the study of human and animal movement during the late nineteenth century. One famous early photographer was Eadweard Muybridge, a British landscape photographer and a rather colorful character who frequently published essays praising his own work (10). Muybridge used electronically controlled still cameras aligned in sequence with an electromagnetic tripping device to capture serial shots of trotting and galloping horses, thereby resolving the controversy about whether all four hooves are ever airborne simultaneously (they are). More important, however, he amassed three volumes of photographic work on human and animal motions that provided scientific documentation of some of the subtle differences between normal and pathological gait.

A digitizer is an instrument that identifies x,y position coordinates for joint centers and other points of interest on a photograph, film projection, or video display.

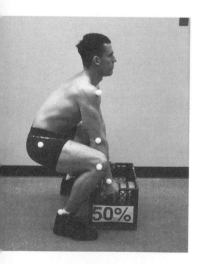

Reflective joint markers can be tracked by a camera for automatic digitizing of a movement.

The modern-day movement analyst has many camera types from which to choose. The type of movement and the requirements of the analysis largely determine the camera and analysis system of choice. Those most commonly used for documenting human movement sequences are video cameras and 8- and 16-mm movie cameras.

Because of the widespread availability, durability, and ease of use of modern video cameras and playback units, video is the most common motion picture medium used for qualitative analysis of human movement today. Standard video provides 30 resolvable pictures per second, which is perfectly adequate for most qualitative human movement applications. Scientists and clinicians performing detailed quantitative study of the kinematics of human motion typically require a more sophisticated video camera and playback unit, often with higher rates of picture capture. For both qualitative and quantitative analysis, however, a consideration often of greater importance than camera speed is the clarity of the captured images. A shuttered video camera allows user control of the exposure time, or length of time that the shutter is open when each picture in the video record is taken. This type of camera with the shutter speed appropriately set is necessary to provide crisp, unblurred images of rapid movements, irrespective of the rate of picture capture.

Another important consideration when analyzing human movement with video is the number of cameras required for adequately capturing the aspects of interest. Because most human movement is not constrained to a single plane, it is often necessary to use two or more cameras to ensure that all of the movements can be viewed and recorded accurately for a detailed analysis. When practicality dictates that a single camera be used, thoughtful consideration should be given to camera positioning relative to the movement of interest. Only when human motion is occurring perpendicular to the optical axis of a camera are the angles present at joints viewed without distortion.

Analysts usually perform a quantitative film or video analysis with computer-linked equipment that enables the calculation of estimates of kinematic quantities of interest for each picture. The traditional procedure for analyzing a film or video picture involves a process called *digitizing*. This involves the activation of a hand-held pen, cursor, or mouse over the image of the individual's joint centers or other points of interest, with the position coordinates of each point stored in a computer data file. Such files are subsequently accessed by software that calculates kinematic quantities of interest. Some more advanced video-based systems enable automated tracking and digitizing of high-contrast markers on the film or video by computer software.

Other Movement Monitoring Systems

Other technologies for scientific and clinical applications involving kinematic analysis of human movement are also available today. These systems provide real-time tracking of targets consisting of tiny electric lights known as light-emitting diodes (LEDs) or electromagnetic markers, which are attached to the skin over joint centers or other points of interest. Special computer-linked cameras track these targets, enabling on-line calculation of kinematic quantities of interest. A drawback to these systems, however, is that the precise camera array necessary for three-dimensional tracking of the joint markers typically restricts the analysis site to a constrained laboratory setting.

Other Assessment Tools

Other tools used by biomechanic researchers are available in commercial and homemade versions. The hand-operated goniometer used for assessing the angle present at a joint is also available in an electronic version known as an *electrogoniometer* or elgon (see Chapter 5). The center of the elgon is positioned over the center of rotation of the joint to be monitored, with the arms of the elgon aligned and firmly attached over the longitudinal axes of the adjacent body segments. When motion occurs at the joint, the electrical output provides a continuous record of the angle present at the joint.

Systems combining photocells, light beams, and timers can be used to directly measure movement velocity. The system is usually configured so that light beams intercept photocells at two or more carefully measured positions (Figure 2-21). The photocells are electrically connected to a timer so that the time interval between interruption of the light beams by a moving body segment or an object such as a thrown ball can be precisely recorded. The velocity of the moving body is

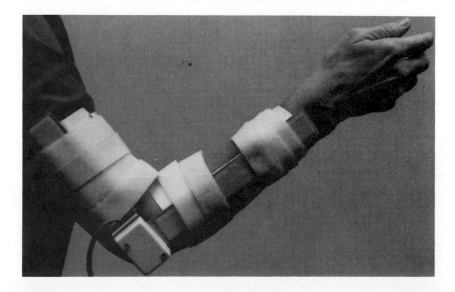

An electrogoniometer is a device that can be interfaced to a recorder to provide a graphical record of the angle present at a joint.

FIGURE 2-21

A light, photocell, and timer set up for measuring movement velocity. As a ball travels through the apparatus zone, the light beams focused on the photocells are interrupted, sending a signal to the timer. The ball's velocity is calculated as the measured distance between photocells divided by the measured time interval.

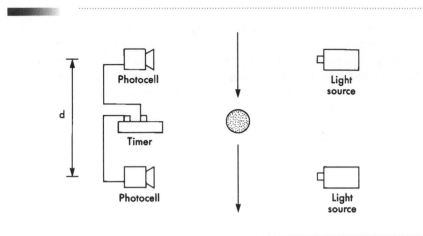

calculated as the measured distance between the photocells divided by the recorded time.

An *accelerometer* is a transducer used for the direct measurement of acceleration. The accelerometer is attached as rigidly as possible to the body segment or other object of interest, with electrical output channeled to a recording device. Two or more accelerometers must be used in combination to monitor acceleration during a nonlinear movement.

SUMMARY

Movements of the human body are referenced to the sagittal, frontal, and transverse planes, with their respectively associated frontal, sagittal, and longitudinal axes. Most human motion is general, with both linear and angular components. A set of specialized terminology is used to describe segment motions and joint actions of the human body.

Teachers of physical activities, clinicians, and coaches all routinely perform qualitative analyses to assess, correct, or improve human movements. Both knowledge of the specific biomechanical purpose of the movement and careful preplanning are necessary for an effective qualitative analysis. A number of special tools are available to assist researchers in collecting kinematic observations of human movement.

INTRODUCTORY PROBLEMS

1. Using appropriate movement terminology, write a qualitative description of the performance of a single jumping jack. Your description should be sufficiently detailed that the reader can completely and accurately visualize the movement.

2. Using appropriate movement terminology, write a qualitative description of the performance of a maximal vertical jump. Your description should be sufficiently detailed that the reader can completely and accurately visualize the movement.

3. Select a movement that occurs primarily in one of the three major reference planes. Qualitatively describe this movement in enough detail that the reader of your description can visualize the movement.

4. List five movements that occur primarily in each of the three cardinal planes. The movements may be either sport skills or activities of daily living.

5. Select a familiar animal. Does the animal move in the same major reference planes in which humans move? What are the major differences in the movement patterns of this animal and the movement patterns of humans?

6. Select a familiar movement and list the factors that contribute to skilled versus unskilled performance of that movement.

7. Test your observation skills by carefully observing the two drawings shown below. List the differences that you are able to identify between these two drawings.

8. Choose a familiar movement and list aspects of that movement that are best observed from close up, from 2 to 3 meters away, and from reasonably far away. Write a brief explanation of your choices.

9. Choose a familiar movement and list aspects of that movement that are best observed from the side view, front view, rear view, and top view. Write a brief explanation of your choices.

10. Choose one of the instrumentation systems described and write a short paragraph explaining the way in which it might be used to study a question related to analysis of a human movement of interest to you.

ADDITIONAL PROBLEMS

1. Select a familiar movement and identify the ways in which performance of that movement is affected by strength, flexibility, and coordination.

2. List three human movement patterns or skills that are best observed from a side view, from a front or rear view, and from a top view.

3. Select a movement that is nonplanar and write a qualitative description of that movement sufficiently detailed to enable the reader of your description to picture the movement.

4. Select a nonplanar movement of interest and list the protocol you would employ in analyzing that movement.

5. What special expectations, if any, should the analyst have of movement performances if the performer is an older adult? An elementary school–aged girl? A novice? An obese high school–aged boy?

6. What are the advantages and disadvantages of collecting observational data on a sport skill during a competitive event as opposed to a practice session?

7. Select a movement with which you are familiar and list at least five questions that you, as a movement analyst, might ask the performer of the movement to gain additional knowledge about a performance.

8. List the auditory characteristics of five movements and explain in each case how these characteristics provide information about the nature of the movement performance.

9. List the advantages and disadvantages of using a video camera as compared to the human eye for collecting observational data.

10. Locate an article in a professional or research journal that involves kinematic description of a movement of interest to you. What instrumentation was used by the researchers? What viewing distances and perspectives were used? How might the analysis described have been improved?

1. Observe and analyze a single performer executing two similar but different versions of a particular movement (for example, two pitching styles or two gait styles). Explain what viewing perspectives and distances you selected for collecting observational data on each movement. Write a paragraph comparing the kinematics of the two movements.

2. Observe a single sport skill as performed by a highly skilled individual, a moderately skilled individual, and an unskilled individual. Qualitatively describe the differences observed.

3. Select a movement at which you are reasonably skilled. Plan and carry out observations of a less-skilled individual performing the movement and provide verbal learning cues for that individual if appropriate. Write a short description of the cues provided with a rationale for each cue.

4. Select a partner and plan and carry out an observational analysis of a movement of interest. Write a composite summary analysis of the movement performance. Write a paragraph identifying in what ways the analysis process was changed by the inclusion of a partner.

5. Plan and carry out a videotape session of a slow movement of interest as performed by two different subjects. Write a comparative analysis of the subjects' performances.

REFERENCES

1. Allison PC: What and how preservice physical education teachers observe during an early field experience, Res Q Exerc Sport 58:242, 1987.

2. Armstrong CW and Hoffman SJ: Effect of teaching experience, knowledge of performer competence, and knowledge of performance outcome on performance error identification, Res Q 50:318, 1979.

3. Bard C et al: Analysis of gymnastics judges' visual search, Res Q Exerc Sport 51:267, 1980.

4. Eastman Kodak Company: High speed photography, Standard Book No 0-87985-165-1, 1979.

5. Hoffman SJ: The contributions of biomechanics to clinical competence: a view from the gymnasium. In Shapiro R and Marett JR, eds: *Proceedings of the second national symposium on teaching kinesiology and biomechanics in sports,* Colorado Springs, Colo, 1984, US Olympic Committee.

6. Hoffman SJ: Toward a pedagogical kinesiology, Quest 28:38, 1977.

7. Hoffman SJ and Sembiante JL: Experience and imagery in movement analysis. In Alderson GJK and Tyldesley DA, eds: *British proceedings of sport psychology,* Salford, England, 1975, British Society of Sports Psychology.

8. Hudson JL: The value of visual variables in biomechanical analysis. In Kreighbaum E and McNeill A, eds: *Proceedings of the 1988 symposium of the International Society of Biomechanics in Sports,* Bozeman, Mont, 1990, Montana State University.

9. Hudson JL, Lee EJ, and Disch JG: The influence of biomechanical measurement systems on performance. In Adrian M and Deutsch H, eds: *Biomechanics: the 1984 Olympic Scientific Congress proceedings,* Eugene, Ore, 1986, Norman Ross-Microform Academic Publishers.
10. Mozley AM: Introduction to the Dover edition. In Muybridge's complete human and animal locomotion, New York, 1979, Dover Publications, Inc.
11. Nielsen AB and Beauchamp L: The effect of training in conceptual kinesiology on feedback provision patterns, J Teach Phys Educ 11:126, 1992.
12. Norman RWK: How to use biomechanical knowledge. In Taylor J, ed: *How to be an effective coach,* Don Mills, Ont, 1975, Manufacturer's Life Insurance and The Coaching Association of Canada.
13. Pinheiro VED and Simon HA: An operational model of motor skill diagnosis, J Teach Phys Educ 11:288, 1992.
14. Thomas JR and Thomas KT: Development of gender differences in physical activity, Quest 40:219, 1988.
15. Wilkinson S: A training program for improving undergraduates' analytic skill in volleyball, J Teach Phys Educ 11:177, 1991.
16. Williams D and Bradford B: Fighting eyes, Strategies 2:21, 1989.

ANNOTATED READINGS

Dainty DA and Norman RW: Standardizing biomechanical testing in sport, Champaign, Ill, 1987, Human Kinetics Publishers, Inc.
   Includes expanded discussion of the characteristics, advantages, and disadvantages of tools commonly employed in biomechanics research for collecting kinematic data.
Hay JG: The development of deterministic models for qualitative analysis, Proceedings of the second national symposium on teaching kinesiology and biomechanics in sports, Colorado Springs, Colo, 1984, US Olympic Committee.
   Presents a procedure for constructing and using models of variables affecting performance in analyzing human movements.
Hudson JL: Core Concepts of Kinesiology, JOPERD 66:54, 1995.
   Describes 10 biomechanical concepts related to skillful movement using the analogy of a volume knob for music.
Muybridge E: Muybridge's complete human and animal locomotion, New York, 1979, Dover Publications.
   Includes a collection of photographic plates done by early cinematographer Eadweard Muybridge during the late 1800s, including the running stills used by Muybridge to prove the then controversial claim that all four hooves of a horse are simultaneously elevated from the ground during part of the gallop stride.

RELATED WEB SITES

Ariel Dynamics Worldwide
http://www.arielnet.com
   Provides product information on the Ariel Performance Analysis System as well as related links, including Cyber Sport E-zine.
Human Performance Technologies, Inc.
http://www.hpt-biolink.com
   Free papers and other information on golf swing biomechanics are available, along with product information for 3-D video assessment of human performance and injury potential.

Innovative Sports Training, Inc.
http://www.innsport.com
Information on virtual reality–based systems for use in biomechanics research and clinical applications.

Mikromak
http://www.mikromak.com
Advertises video hardware and software for sports, medicine, and product research.

Motion Analysis Corporation
http://www.motionanalysis.com
Offers an optical motion capture system utilizing reflective markers for entertainment, biomechanics, character animation, and motion analysis.

Musculographics, Inc.
http://www.musculographics.com
Includes descriptions of biomechanics CAD software for interactive musculoskeletal modeling, surgery simulation products, and computer-assisted surgery products, and also includes an anatomy image gallery.

The Neat System
http://www.neatsys.com
Describes inexpensive video motion analysis software for Windows and includes demo video frames from a variety of sports.

Northern Digital, Inc.
http://www.ndigital.com
Presents optoelectronic 3-D motion measurement systems that track light-emitting diodes for real-time analysis.

Peak Performance Technologies
http://www.peakperform.com
Describes video 3-D applications in sport science, medicine, industry, biology, and animation and includes demo files of animations.

Qualisys, Inc.
http://www.qualisys.com
Presents a system in which infrared cameras track reflective markers, enabling real-time calculations; applications described for research, clinical use, industry, and animation.

Redlake Imaging
http://www.redlake.com/imaging
Advertises high-speed video products for scientific and clinical applications.

SIMI Reality Motion Systems
http://www.simi.net
Describes computer-based video analysis for the human body and cellular applications; includes demo of gait analysis, among others.

Skill Technologies, Inc.
http://www.primenet.com/~skilltec
Advertises a motion capture and analysis system that provides 3-D graphical models and parameter graphs in real time based on 120 Hz tracking of 16 electromagnetic markers.

Chapter 3

KINETIC CONCEPTS FOR ANALYZING HUMAN MOTION

After completing this chapter, the reader will be able to:

Define and identify common units of measurement for mass, force, weight, pressure, volume, density, specific weight, torque, and impulse.

Identify and describe the different types of mechanical loads that act on the human body.

Identify and describe the uses of available instrumentation for measuring kinetic quantities.

Distinguish between vector and scalar quantities.

Solve quantitative problems involving vector quantities using both graphic and trigonometric procedures.

When muscles on opposite sides of a joint develop tension, what determines the direction of joint motion? In which direction will a swimmer swimming perpendicular to a river current actually travel? What determines whether a push can move a heavy piece of furniture? The answers to these questions are rooted in kinetics, the study of forces.

The human body both generates and resists forces during the course of daily activities. The forces of gravity and friction enable walking and manipulation of objects in predictable ways when internal forces are produced by muscles. Sport participation involves application of forces to balls, bats, racquets, and clubs and absorption of forces from impacts with balls, the ground or floor, and opponents in contact sports. This chapter introduces basic kinetic concepts that form the basis for understanding these activities.

BASIC CONCEPTS RELATED TO KINETICS

Understanding the concepts of mass, inertia, weight, pressure, volume, density, specific weight, torque, and impulse provides a useful foundation for understanding the effects of forces.

Mass

Mass is the quantity of matter composing a body. The conventional symbol for mass is *m*. The common unit of mass in the metric system is the kilogram (kg), with the English unit of mass being the *slug*, which is much larger than a kg.

Inertia

In common usage, **inertia** means resistance to action or to change (Figure 3-1). Similarly, the mechanical definition is resistance to acceleration. Inertia is the tendency of a body to maintain its current state of motion, whether motionless or moving with a constant velocity. For example, a 150 kg weight bar lying motionless on the floor has a tendency to remain motionless. A skater gliding on a smooth surface of ice has a tendency to continue gliding in a straight line with a constant speed.

Although inertia has no units of measurement, the amount of inertia a body possesses is directly proportional to its mass. The more massive an object is, the more it tends to maintain its current state of motion and the more difficult it is to disrupt that state.

Force

A **force** can be thought of as a push or a pull acting on a body. Each force is characterized by its magnitude, direction, and point of application to a given body. Body weight, friction, and air or water resistance are all forces that commonly act on the human body. Force may also be defined as the product of a body's mass and the acceleration of that body resulting from the application of the force:

$$F = ma$$

mass
quantity of matter contained in an object

inertia
tendency of a body to resist a change in its state of motion

force
push or pull; the product of mass and acceleration

A skater has a tendency to continue gliding with constant speed and direction due to inertia.

F is the conventional symbol for force. Units of force are units of mass multiplied by units of acceleration. In the metric system, the most common unit of force is the Newton (N), which is defined as the product of 1 kg of mass and 1 m/s^2 of acceleration:

$$1 \text{ N} = (1 \text{ kg})(1 \text{ m/s}^2)$$

In the English system, the most common unit of force is the pound (lb). A lb of force is the amount of force necessary to accelerate a mass of 1 slug at 1 ft/s^2, and 1 lb is equal to 4.45 N:

$$1 \text{ lb} = (1 \text{ slug})(1 \text{ ft/s}^2)$$

free body diagram
a sketch that shows a
defined system in isolation
with all of the force vectors
acting on the system

Because a number of forces act simultaneously in most situations, constructing a **free body diagram** is usually the first step when analyzing the effects of forces on a body or system of interest. A *free body* is any object, body, or body part that is being focused upon for analysis. A free body diagram consists of a sketch of the system being analyzed and vector representations of the acting forces (Figure 3-2). Even though a hand must be applying force to a tennis racket in order for the racket to forcefully contact a ball, if the racket is the free body of interest, the hand is represented in the free body diagram of the racket only as a force vec-

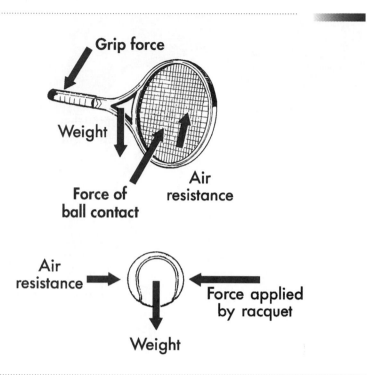

FIGURE 3-2
Two free body diagrams
showing the acting forces.

tor. Similarly, if the tennis ball constitutes the free body being studied, the force of racket acting on the ball is displayed as a vector.

Since a force rarely acts in isolation, it is important to recognize that the overall effect of many forces acting on a system or free body is a function of the **net force,** which is the vector sum of all of the acting forces. When all acting forces are balanced, or cancel each other out, the net force is zero and the body remains in its original state of motion, either motionless or moving with a constant velocity. When a net force is present the body moves in the direction of the net force and with an acceleration that is proportional to the magnitude of the net force.

net force
resultant force derived from the composition of two or more forces

Center of Gravity

A body's **center of gravity** or **center of mass** is the point around which a body's weight and mass are equally balanced in all directions (see Chapter 13). In motion analyses the motion of the center of gravity serves as an index of total body motion. From a kinetic perspective, the location of the center of mass determines the way in which the body responds to external forces.

center of gravity
point around which a body's weight and mass are equally balanced in all directions

Although a body's mass remains unchanged on the moon, its weight is less due to smaller gravitational acceleration. Photo courtesy of NASA.

Weight

weight
attractive force that the earth exerts on a body

Weight is defined as the amount of gravitational force exerted on a body. Algebraically, its definition is a modification of the general definition of a force, with weight (wt) being equal to mass (m) multiplied by the acceleration of gravity (a_g).

$$wt = ma_g$$

Since weight is a force, units of weight are units of force—either N or lb.

As the mass of a body increases, its weight increases proportionally. The factor of proportionality is the acceleration of gravity, which is -9.81 m/s^2. The negative sign indicates that the acceleration of gravity is directed downward, or toward the center of the earth. On the moon or another planet with a different gravitational acceleration, a body's weight would be different, although its mass would remain the same.

Because weight is a force, it is also characterized by magnitude, direction, and point of application. The direction in which weight acts is always toward the center of the earth. Because the point at which weight is assumed to act on a body is the body's center of gravity, it is at the center of gravity that the weight vector is shown to act in free body diagrams.

Although body weights are often reported in kilograms, the kilogram is actually a unit of mass. To be technically correct, weights should be identified in Newtons and masses reported in kilograms. The sample problem presented in Figure 3-3 illustrates the relationship between mass and weight.

Pressure

Pressure (p) is defined as force (F) distributed over a given area (A).

$$p = \frac{F}{A}$$

pressure
force per unit of area over which the force acts

Units of pressure are units of force divided by units of area. Common units of pressure in the metric system are N per square centimeter (N/cm^2) and Pascals (Pa). One Pascal represents one Newton per square meter ($Pa = N/m^2$). In the English system, the most common unit of pressure is pounds per square inch (psi or lb/in^2).

The pressure exerted by the sole of a shoe on the floor beneath it is the body weight resting on the shoe divided by the surface area between the sole of the shoe and the floor. As illustrated in the sample problem in Figure 3-4, the smaller amount of surface area on the bottom of a spike heel as compared to a flat sole results in a much larger pressure being exerted.

Volume

A body's **volume** is the amount of space that it occupies. Because space is considered to have three dimensions (width, height, and depth), a unit of volume is a unit of length multiplied by a unit of length multiplied by a unit of length. In mathematical shorthand, this is a unit of length raised to the exponential power of three, or a unit of length *cubed*. In the metric system, common units of volume are cubic centimeters (cm^3), cubic meters (m^3), and liters (l):

volume
space occupied by a body

$$1\ l = 1000\ cm^3$$

In the English system of measurement, common units of volume are cubic inches (in^3) and cubic feet (ft^3). Another unit of volume in the English system is the quart (qt):

$$1\ qt = 57.75\ in^3$$

Volume should not be confused with weight or mass. An 8 kg shot and a softball occupy approximately the same volume of space, but the weight of the shot is much greater than that of the softball.

FIGURE 3-3

SAMPLE PROBLEM I

1. If a scale shows that an individual has a mass of 68 kg, what is that individual's weight?

Known

m = 68 kg

Solution

Wanted: weight

Formulas: $wt = ma_g$

$1 \text{ kg} = 2.2 \text{ lb}$

(Mass may be multiplied by the acceleration of gravity to convert to weight within either the English or the metric system.)

$$wt = ma_g$$
$$wt = (68 \text{ kg})(9.81 \text{ m/s}^2)$$
$$wt = 667 \text{ N}$$

Mass in kg may be multiplied by the conversion factor 2.2 lb/kg to convert to weight in pounds:

$$(68 \text{ kg})(2.2 \text{ lb/kg}) = 150 \text{ lb}$$

2. What is the mass of an object weighing 1200 N?

Known

wt = 1200 N

Solution

Wanted: mass

Formula: $wt = ma_g$

(Weight may be divided by the acceleration of gravity within a given system of measurement to convert to mass.)

$$wt = ma_g$$
$$1200 \text{ N} = m\,(9.81 \text{ m/s}^2)$$
$$\frac{1200 \text{ N}}{9.81 \text{ m/s}^2} = m$$
$$m = 122.32 \text{ kg}$$

FIGURE 3-4

S A M P L E   P R O B L E M   2

Is it better to be stepped on by a woman wearing a spike heel or by a woman wearing a smooth-soled court shoe? If a woman's weight is 556 N, the surface area of the spike heel is 4 cm^2, and the surface area of the court shoe is 175 cm^2, how much pressure is exerted by each shoe?

Known

wt = 556 N
A_s = 4 cm^2
A_c = 175 cm^2

Solution

Wanted: Pressure exerted by the spike heel
        Pressure exerted by the court shoe
        Formula: p = F/A
Deduction: It is necessary to recall that weight is a force.

For the spike heel: $p = \dfrac{556}{4 \text{ cm}^2}$

$p = 139 \text{ N/cm}^2$

For the court shoe: $p = \dfrac{556 \text{ N}}{318 \text{ cm}^2}$

$p = 3.18 \text{ N/cm}^2$

Comparison of the amounts of pressure exerted by the two shoes:

$$\frac{P_{\text{spike heel}}}{P_{\text{court shoe}}} = \frac{139}{3.18} = 43.75$$

Therefore, 43.75 times more pressure is exerted by the spike heel than by the court shoe worn by the same woman.

Pairs of balls that are similar in volume but markedly different in weight.

Density

density
mass per unit of volume

The concept of **density** combines the mass of a body with the body volume. Density is defined as mass per unit of volume. The conventional symbol for density is the small Greek letter rho (ρ).

$$\text{density} \ (\rho) = \text{mass/volume}$$

Units of density are units of mass divided by units of volume. In the metric system, a common unit of density is the kilogram per cubic meter (kg/m^3). In the English system of measurement, units of density are not commonly used. Instead, units of specific weight (weight density) are employed.

specific weight
weight per unit of volume

Specific weight is defined as weight per unit of volume. Because weight is proportional to mass, specific weight is proportional to density. Units of specific weight are units of weight divided by units of volume. The metric unit for specific weight is Newtons per cubic meter (N/m^3), and the English system uses pounds per cubic foot (lb/ft^3).

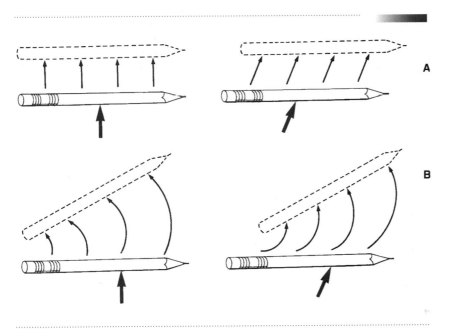

FIGURE 3-5
A, Centric forces produce translation. **B,** Eccentric forces produce translation and rotation.

Although a golf ball and a table tennis ball occupy approximately the same volume, the golf ball has a greater density and specific weight than the table tennis ball because the golf ball has more mass and more weight. Similarly, a lean person with the same body volume as an obese person has a higher total body density because muscle is denser than fat. Thus, percent body fat is inversely related to body density.

Torque

When a force is applied to an object such as a pencil lying on a desk, either translation or general motion may result. If the applied force is directed parallel to the desk top and through the center of the pencil (a *centric force*), the pencil will be translated in the direction of the applied force. If the force is applied parallel to the desk top but directed through a point other than the center of the pencil (an *eccentric force*), the pencil will undergo both translation and rotation (Figure 3-5).

The rotary effect created by an eccentric force is known as **torque** (T) or moment of force. Torque, which may be thought of as *rotary force,* is the angular equivalent of linear force. Algebraically, torque is the product of force (F) and the perpendicular distance (d) from the force's line of action to the axis of rotation.

$$T = Fd_\perp$$

torque
rotary effect of a force

The greater the amount of torque acting at the axis of rotation, the greater the tendency for rotation to occur. Units of torque in both the metric and the English systems follow the algebraic definition. They are units of force multiplied by units of distance: Newton-meters (N-m) or foot-pounds (ft-lb).

Impulse

When a force is applied to a body, the resulting motion of the body is dependent not only on the magnitude of the applied force, but also on the duration of force application. The product of force and time is known as **impulse:**

impulse
product of force and the time over which the force acts

$$\text{Impulse} = \text{Ft}$$

A large change in an object's state of motion may result from a small force acting for a relatively long time or from a large force acting for a relatively short time. A golf ball rolling across a green gradually loses speed because of the small force of rolling friction. The speed of a baseball struck vigorously by a bat changes because of the large force exerted by the bat during the fraction of a second it is in contact with the ball. When a vertical jump is executed, the larger the impulse generated against the floor, the greater the jumper's takeoff velocity and the higher the resulting jump. Units of physical quantities commonly used in biomechanics are shown in Table 3-1.

TABLE 3-1

COMMON UNITS FOR KINETIC QUANTITIES			
QUANTITY	SYMBOL	METRIC UNIT	ENGLISH UNIT
mass	m	kg	slug
force	F	N	lb
pressure	p	Pa	psi
volume (solids) (liquids)	V	m^3 liter	ft^3 gallon
density	ρ	kg/m^3	
specific weight	γ	N/m^3	lb/ft^3
torque	T	N-m	ft-lb
impulse		$N \cdot s$	$lb \cdot s$

FIGURE 3-6

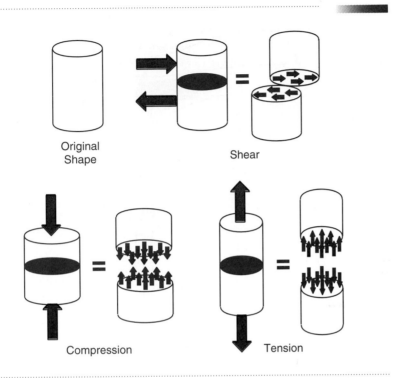

MECHANICAL LOADS ON THE HUMAN BODY

Muscle forces, gravitational force, and the bone-breaking force encountered in a skiing accident all affect the human body differently. The effect of a given force depends on its direction and duration as well as its magnitude, as described in the following section.

Compression, Tension, and Shear

Compressive force or **compression** can be thought of as a squeezing force (Figure 3-6). An effective way to press wildflowers is to place them inside the pages of a book and to stack other books on top of that book. The weight of the books creates a compressive force on the flowers. Similarly, the weight of the body acts as a compressive force on the bones that support it. When the trunk is erect, each vertebra in the spinal column must support the weight of that portion of the body above it.

The opposite of compressive force is tensile force or **tension** (Figure 3-6). Tensile force is a pulling force that creates tension in the object to which it is applied. When a child sits in a playground swing, the child's

compression
pressing or squeezing force directed axially through a body

tension
pulling or stretching force directed axially through a body

FIGURE 3-7

During performance of
squat or knee bend
exercises, shear force
acting at the knee is
maximal when flexion at
the knee is maximal. The
shear at the joint is
produced by the axial
force in the femur.

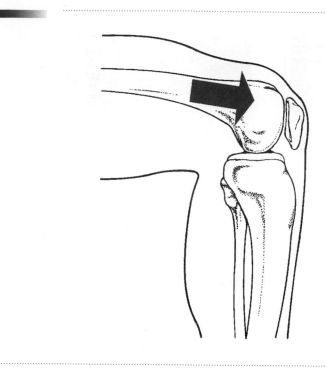

shear
force directed parallel to a
surface

stress
distribution of force within a
body, quantified as force
divided by the area over
which the force acts

weight creates tension in the chains supporting the swing. A heavier
child creates even more tension in the supports of the swing. Muscles
produce tensile force that pulls on the attached bones.

A third category of force is termed **shear.** Whereas compressive and
tensile forces act along the longitudinal axis of a bone or other body to
which they are applied, shear force acts parallel or tangent to a surface.
Shear force tends to cause one portion of the object to slide, displace,
or shear with respect to another portion of the object (Figure 3-6). For
example, a force acting at the knee joint in a direction parallel to the
tibial plateau is a shearing force at the knee (2). During the performance of a squat exercise, joint shear at the knee is greatest at the full
squat position (2). This position places a large amount of stress on the
ligaments and muscle tendons that prevent the femur from sliding off
the tibial plateau (Figure 3-7).

Mechanical Stress

Another factor affecting the outcome of the action of forces on the human body is the way in which the force is distributed. Whereas pressure
represents the distribution of force external to a solid body, **stress** rep-

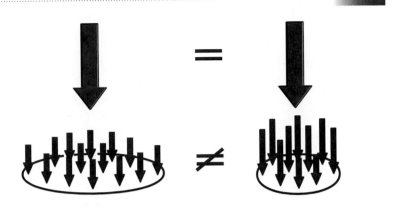

FIGURE 3-8
The amount of
mechanical stress created
by a force is inversely
related to the size of the
area over which the force
is spread.

resents the resulting force distribution inside a solid body when an external force acts. Stress is quantified in the same way as pressure: force per unit of area over which the force acts. As shown in Figure 3-8, a given force acting on a small surface produces greater stress than the same force acting over a larger surface. When a blow is sustained by the human body, the likelihood of injury to body tissue is related to the magnitude and direction of the stress created by the blow. Compressive stress, tensile stress, and shear stress are terms that indicate the direction of the acting stress.

Because the lumbar vertebrae bear more of the weight of the body than the thoracic vertebrae when a person is in an upright position, the compressive stress in the lumbar region should logically be greater. However, the amount of stress present is not directly proportional to the amount of weight borne, because the load-bearing surface areas of the lumbar vertebrae are greater than those of the vertebrae higher in the spinal column (Figure 3-9). This increased surface area reduces the amount of compressive stress present. Nevertheless, the L5-S1 intervertebral disc (at the bottom of the lumbar spine) is the most common site of disc herniations, although other factors also play a role (see Chapter 9). Quantification of mechanical stress is demonstrated in the sample problem shown in Figure 3-10.

Torsion, Bending, and Combined Loads

A somewhat more complicated type of loading is called **bending.** Pure compression and tension are both **axial** forces; that is, they are directed along the longitudinal axis of the affected structure. When an eccentric (or nonaxial) force is applied to a structure, the structure bends, creating compressive stress on one side and tensile stress on the opposite side (Figure 3-11).

bending
asymmetric loading that
produces tension on one
side of a body's longitudinal
axis and compression on the
other side

axial
directed along the
longitudinal axis of a body

FIGURE 3-9

The surfaces of the vertebral bodies increase in surface area as more weight is supported.

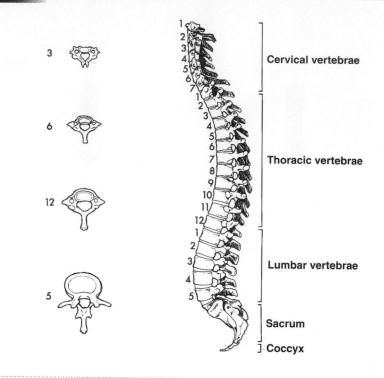

FIGURE 3-10

SAMPLE PROBLEM 3

How much compressive stress is present on the L1, L2 vertebral disk of a 625 N woman, given that approximately 45% of body weight is supported by the disk a) when she stands in anatomical position? b) when she stands erect holding a 222 N suitcase? (Assume that the disk is oriented horizontally and that its surface area is 20 cm^2.)

Solution

a. Given: $\quad$ F = (625 N)(0.45)

$\qquad\qquad$ A = 20 cm^2

Formula: $\quad$ Stress = F/A

$$\text{Stress} = \frac{(625 \text{ N})(0.45)}{20 \text{ cm}^2}$$

$\boxed{\text{Stress} = 14 \text{ N/cm}^2}$

b. Given: $\quad$ F = (625 N)(0.45) + 222 N

Formula: $\quad$ Stress = F/A

$$\text{Stress} = \frac{(625 \text{ N})(0.45) + 222 \text{ N}}{20 \text{ cm}^2}$$

$\boxed{\text{Stress} = 25.2 \text{ N/cm}^2}$

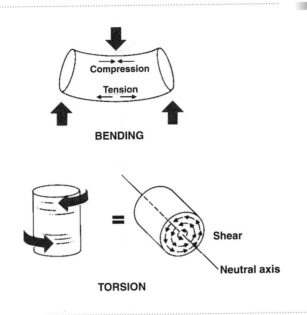

FIGURE 3-11
Objects loaded in bending are subject to compression on one side and tension on the other. Objects loaded in torsion develop internal shear stress, with maximal stress at the periphery and no stress at the neutral axis.

Torsion occurs when a structure is caused to twist about its longitudinal axis, typically when one end of the structure is fixed. Torsional fractures of the tibia are not uncommon in football and skiing accidents in which the foot is held in a fixed position while the rest of the body undergoes a twist.

The presence of more than one form of loading is known as **combined loading.** Because the human body is subjected to a myriad of simultaneously acting forces during daily activities, this is the most common type of loading on the human body.

torsion
load producing twisting of a body around its longitudinal axis

combined loading
simultaneous action of more than one of the pure forms of loading

The Effects of Loading

When a force acts on an object, there are two potential effects. The first is acceleration and the second is **deformation,** or change in shape. When a diver applies force to the end of a springboard, the board both accelerates and deforms. The amount of deformation that occurs in response to a given force depends on the stiffness of the object acted upon.

When an external force is applied to the human body, several factors influence whether an injury occurs. Among these are the magnitude and direction of the force, as well as the area over which the force is distributed. Also important, however, are the material properties of the loaded body tissues.

deformation
change in shape

FIGURE 3-12
When a structure is loaded it deforms, or changes shape. The deformation is temporary within the elastic region and permanent in the plastic region. Structural integrity is lost at the ultimate failure point.

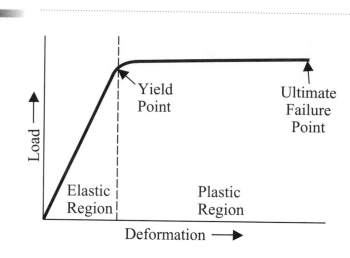

The relationship between the amount of force applied to a structure and the structure's response is illustrated by a load-deformation curve (Figure 3-12). With relatively small loads, deformation occurs but the response is elastic, meaning that when the force is removed the structure returns to its original size and shape. Since stiffer materials display less deformation in response to a given load, greater stiffness translates to a steeper slope of the load-deformation curve in the elastic region. If the force applied causes the deformation to exceed the structure's **yield point** or **elastic limit,** however, the response is plastic, meaning that some amount of deformation is permanent. Deformations exceeding the ultimate failure point produce mechanical **failure** of the structure, which in the human body means fracturing of bone or rupturing of soft tissues.

yield point (elastic limit)
point on the load-deformation curve past which deformation is permanent

failure
loss of mechanical continuity

Repetitive versus Acute Loads

repetitive loading
repeated application of a subacute load that is usually of relatively low magnitude

acute loading
application of a single force of sufficient magnitude to cause injury to a biological tissue

The distinction between **repetitive** and **acute loading** is also important. When a single force large enough to cause injury acts on biological tissues, the injury is termed acute and the causative force is termed macrotrauma. The force produced by a fall, a rugby tackle, or an automobile accident may be sufficient to fracture a bone.

Injury can also result from the repeated sustenance of relatively small forces. For example, each time a foot hits the pavement during running, a force of approximately 2 to 3 times body weight is sustained. Although a single force of this magnitude is not likely to result in a fracture of healthy bone, numerous repetitions of such a force may cause a fracture of an otherwise healthy bone somewhere in the lower extrem-

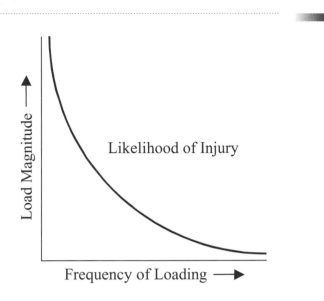

Likelihood of Injury

Load Magnitude ⟶

Frequency of Loading ⟶

FIGURE 3-13
The general pattern of injury likelihood as a function of load magnitude and repetition. Injury can be sustained, but is less likely, with a single large load and with a repeated small load.

ity. When repeated or chronic loading over a period of time produces an injury, the injury is called a chronic injury or a stress injury, and the causative mechanism is termed microtrauma. The relationship between the magnitude of the load sustained, the frequency of loading, and the likelihood of injury is shown in Figure 3-13.

TOOLS FOR MEASURING KINETIC QUANTITIES

Biomechanic researchers use equipment for studying both muscle forces and forces generated by the feet against the ground during gait and other activities. Knowledge gained through the use of this apparatus is often published in professional journals for teachers, clinicians, coaches, and others interested in human movement.

Electromyography

Eighteenth-century Italian scientist Galvani made two interesting discoveries about skeletal muscle: 1. it develops tension when electrically stimulated, and 2. it produces a detectable current or voltage when developing tension, even when the stimulus is a nerve impulse (1). The latter discovery was of little practical value until the twentieth century when technology became available for the detection and recording of extremely small electrical charges. The technique of recording electrical activity produced by muscle, or **myoelectric activity,** is known today as *electromyography* (EMG).

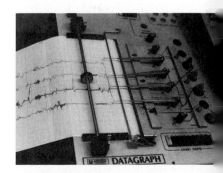

An electromyography system.

myoelectric activity
electric current or voltage produced by a muscle developing tension

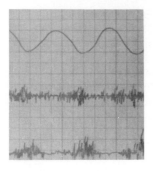

A graphic printout showing changes in joint range of motion (top trace) and myoelectric activity recorded from two muscle groups.

transducers
devices that detect a signal

Electromyography is used to study neuromuscular function, including identification of which muscles develop tension throughout a movement and which movements elicit more or less tension from a particular muscle or muscle group. It is also used clinically to assess nerve conduction velocities and muscle response in conjunction with the diagnosis and tracking of pathological conditions of the neuromuscular system. Scientists also employ electromyographic techniques to study the ways in which individual motor units respond to central nervous system commands.

The process of electromyography involves the use of **transducers** known as *electrodes* that sense the level of myoelectric activity present at a particular site over time. Depending on the questions of interest, either surface electrodes or fine wire electrodes are used. Surface electrodes, consisting of small discs of conductive material, are positioned on the surface of the skin over a muscle or muscle group to pick up global myoelectric activity. When more localized pickup is desired, indwelling, fine wire electrodes are injected directly into a muscle. Output from the electrodes is amplified and graphically displayed or mathematically processed and stored by a computer.

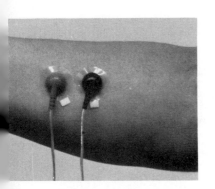

Surface EMG electrodes are small discs that attach directly to the skin over a muscle or muscle group of interest to transduce electrical activity in the underlying tissue.

Dynamography

Scientists have devised several types of platforms and portable systems for the measurement of forces and pressure on the plantar surface of the foot. These systems have been employed primarily in gait research but have also been used to study phenomena such as starts, takeoffs, landings, baseball and golf swings, and balance. Those systems that provide a graphical time-history of recorded force also allow the calculation of impulse as the area under the force-time curve.

Both commercially available and homemade *force platforms* and *pressure platforms* are typically built rigidly into a floor flush with the surface and are interfaced to a computer that calculates kinetic quantities of interest. Force platforms are usually designed to transduce ground reaction forces in vertical, lateral, and anteroposterior directions with respect to the platform itself; pressure platforms provide graphical or digital maps of pressures across the plantar surfaces of the feet. The force platform is a relatively sophisticated instrument, but its limitations include the restrictions of a laboratory setting and potential difficulties associated with the subject's consciously targeting the platform.

Portable systems for measuring plantar forces and pressures are also available in commercial and homemade models as instrumented shoes, shoe inserts, and thin transducers that adhere to the plantar surfaces of the feet. These systems provide the advantage of data collection outside the laboratory but lack the precision of the built-in platforms.

VECTOR ALGEBRA

A **vector** is a quantity that has both magnitude and direction. Vectors are represented by arrow-shaped symbols. The magnitude of a vector is its size; for example, the number 12 is of greater magnitude than the number 10. A vector symbol's orientation on paper represents direction, and its length represents magnitude. Force, weight, pressure, specific weight, and torque are kinetic vector quantities; displacement, velocity, and acceleration (see Chapter 10) are kinematic vector quantities. None is fully defined without the identification of both its magnitude and its direction. **Scalar** quantities possess magnitude but have no particular direction associated with them. Mass, volume, length, and speed are examples of scalar quantities.

vector
physical quantity that possesses both magnitude and direction

scalar
physical quantity that is completely described by its magnitude

Vector Composition

When vectors are added together, the operation is called **vector composition.** The composition of two or more vectors that have exactly the same direction results in a single vector that has a magnitude equal to the sum of the magnitudes of the vectors being added (Figure 3-14). The single vector resulting from a composition of two or more vectors is known as the resultant vector or the **resultant.** If two vectors that are oriented in exactly opposite directions are composed, the resultant has the direction of the longer vector and a magnitude that is equal to the difference in the magnitudes of the two original vectors (Figure 3-15).

It is also possible to add vectors that are not oriented in the same or opposite directions. When the vectors are coplanar, that is, contained in the same plane, a procedure that may be used is the *"tip-to-tail"* method, in which the tail of the second vector is placed on the tip of the first vector, and the resultant is then drawn with its tail on the tail of the first vector and its tip on the tip of the second vector. This procedure may be used for combining any number of vectors if each successive vector is positioned with its tail on the tip of the immediately preceding vector and the resultant connects the tail of the first vector to the tip of the last vector (Figure 3-16).

Through the laws of vector combination, we often can calculate or better visualize the resultant effect of combined vector quantities. For example, a canoe floating down a river is subject to both the force of the current and the force of the wind. If the magnitudes and directions of these two forces are known, the single resultant or *net force* can be derived through the process of vector composition (Figure 3-17). The canoe travels in the direction of the net force.

vector composition
process of determining a single vector from two or more vectors by vector addition

resultant
single vector that results from vector composition

FIGURE 3-14
The composition of vectors with the same direction requires adding their magnitudes.

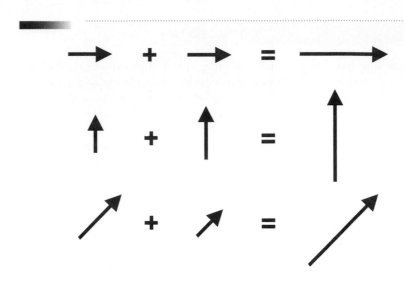

FIGURE 3-15
Composition of vectors with opposite directions requires subtracting their magnitudes.

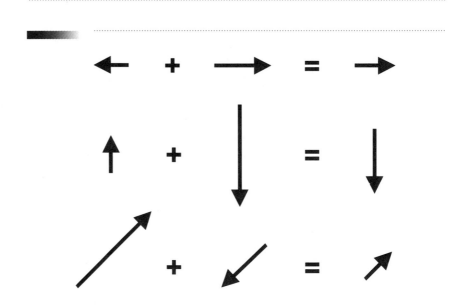

Vector Resolution

Determining the perpendicular components of a vector quantity relative to a particular plane or structure is often useful. For example, when a ball is thrown into the air, the horizontal component of its velocity determines the distance it travels, and the vertical component of its velocity determines

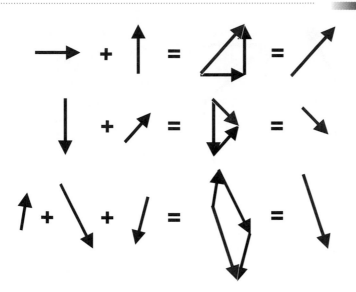

FIGURE 3-16
The "tip-to-tail" method of vector composition.

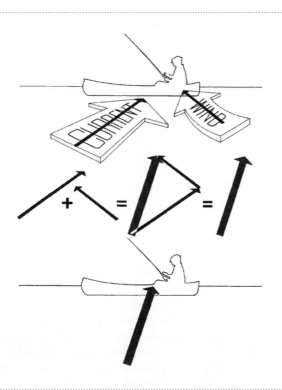

FIGURE 3-17
The net force is the resultant of all acting forces.

FIGURE 3-18
Vectors may be resolved
into perpendicular
components. The vector
composition of each
perpendicular pair of
components yields the
original vector.

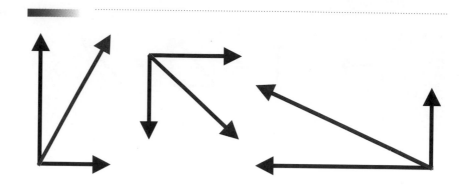

vector resolution
operation that replaces a
single vector with two
perpendicular vectors such
that the vector composition
of the two perpendicular
vectors yields the original
vector

the height it reaches (see Chapter 10). When a vector is resolved into
perpendicular components—a process known as **vector resolution**—
the vector sum of the components always yields a resultant that is
equal to the original vector (Figure 3-18). The two perpendicular components therefore are a different but equal representation of the original vector.

Graphic Solution of Vector Problems

When vector quantities are uniplanar (contained in a single plane), vector manipulations may be done graphically to yield approximate results.
Graphic solution of vector problems requires the careful measurement
of vector orientations and lengths to minimize error. Vector lengths,
which represent the magnitudes of vector quantities, must be drawn
to scale. For example, 1 cm of vector length could represent 10 N of
force. A force of 30 N would then be represented by a vector 3 cm
in length, and a force of 45 N would be represented by a vector of
4.5 cm length.

Trigonometric Solution of Vector Problems

A more accurate procedure for quantitatively dealing with vector problems involves the application of trigonometric principles. Through the
use of trigonometric relationships, the tedious process of measuring and
drawing vectors to scale can be eliminated (see Appendix B). Figure
3-19 provides an example of the processes of both graphic and trigonometric solutions using vector quantities.

FIGURE 3-19

S A M P L E   P R O B L E M   4

Terry and Charlie must move a refrigerator to a new location. They both push parallel to the floor, Terry with a force of 350 N and Charlie with a force of 400 N, as shown in the diagram below. a. What is the magnitude of the resultant of the forces produced by Terry and Charlie? b. If the amount of friction force that directly opposes the direction of motion of the refrigerator is 700 N, will they be able to move the refrigerator?

Graphic Solution

a. Use the scale 1 cm = 100 N to measure the length of the resultant.

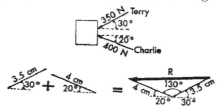

The length of the resultant is approximately 6.75 cm or   675 N.
b. Since 675 N < 700 N, they will not be able to move the refrigerator.

Trigonometric Solution

Given: F_T = 350 N
     F_C = 400 N
Wanted: The magnitude of the resultant force
Horizontal plane free body diagram:

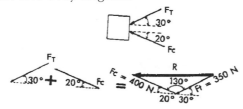

Formula:

$$C^2 = A^2 + B^2 - 2(A)(B) \cos \Theta \text{ (the Law of Cosines)}$$
$$R^2 = 400^2 + 350^2 - 2(400)(350) \cos 130$$
$$R = 680 \text{ N}$$

c. Since 680 N < 700 N, they will not be able to move the refrigerator unless they exert more collective force while pushing at these particular angles. (If both Terry and Charlie pushed at a 90° angle to the refrigerator, their combined force would be sufficient to move it.)

SUMMARY

This chapter introduces basic concepts related to kinetics, including mass, the quantity of matter composing an object; inertia, the tendency of a body to maintain its current state of motion; force, a push or pull that alters or tends to alter a body's state of motion; center of gravity, the point around which a body's weight is balanced; weight, the gravitational force exerted on a body; pressure, the amount of force distributed over a given area; volume, the space occupied by a body; density, the mass or weight per unit of body volume; and torque, the rotational effect of a force.

Several types of mechanical loads act on the human body. These include compression, tension, shear, bending, and torsion. Generally, some combination of these loading modes is present. The distribution of force within a body structure is termed mechanical stress. The nature and magnitude of stress determines the likelihood of injury to biological tissues.

Vector quantities have magnitude and direction; scalar quantities possess magnitude only. Problems with vector quantities can be solved using either a graphic or a trigonometric approach. Of the two procedures, the use of trigonometric relationships is more accurate and less tedious.

INTRODUCTORY PROBLEMS

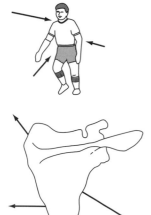

1. William Perry, defensive tackle and part-time running back better known as *The Refrigerator*, weighed in at 1352 N during his 1985 rookie season with the Chicago Bears. What was Perry's mass? (Answer: 138 kg)
2. How much force must be applied to a 0.5 kg ice hockey puck to give it an acceleration of 30 m/s^2? (Answer: 15 N)
3. A rugby player is contacted simultaneously by three opponents who exert forces of the magnitudes and directions shown in the diagram at left. Using a graphic solution, show the magnitude and direction of the resultant force.
4. Using a graphic solution, compose the muscle force vectors to find the net force acting on the scapula.
5. Draw the horizontal and vertical components of the vectors shown below.

6. A gymnastics floor mat weighing 220 N has dimensions of 3 m × 4 m × 0.04 m. How much pressure is exerted by the mat against the floor? (Answer: 18.33 Pa)
7. What is the volume of a milk crate with sides of 25 cm, 40 cm, and 30 cm? (Answer: 30,000 cm^3 or 30 l)

8. Choose three objects that are within your field of view and estimate the volume of each. List the approximate dimensions you used in formulating your estimates.
9. If the contents of the crate described in Problem 7 weigh 120 N, what are the average density and specific weight of the box and contents? (Answer: 0.0004 kg/cm^3; 0.004 N/cm^3)
10. Two children sit on opposite sides of a playground seesaw. Joey, who weighs 220 N, sits 1.5 m from the axis of the seesaw, and Suzy, who weighs 200 N, sits 1.7 m from the axis of the seesaw. How much torque is created at the axis by each child? In which direction will the seesaw tip? (Answer: Joey: 330 N-m; Suzy: 340 N-m; Suzy's end)

ADDITIONAL PROBLEMS

1. What is your own body mass in kg?
2. Gravitational force on planet X is 40% of that found on the earth. If a person weighs 667.5 N on earth, what is the person's weight on planet X? What is the person's mass on the earth and on planet X? (Answer: Weight on planet X = 267 N; mass = 68 kg on either planet.)
3. A football player is contacted by two tacklers simultaneously. Tackler A exerts a force of 400 N, and Tackler B exerts a force of 375 N. If the forces are coplanar and directed perpendicular to each other, what is the magnitude and direction of the resultant force acting on the player? (Answer: 548 N at an angle of 43° to the line of action of Tackler A.)
4. A 75 kg skydiver in free fall is subjected to a crosswind exerting a force of 60 N and to a vertical air resistance force of 100 N. Describe the resultant force acting on the skydiver. (Answer: 638.6 N at an angle of 5.4° to vertical)
5. Use a trigonometric solution to find the magnitude of the resultant of the following coplanar forces: 60 N at 90°, 80 N at 120°, and 100 N at 270°. (Answer: 49.57 N)
6. If 37% of body weight is distributed above the superior surface of the L5 intervertebral disc and the area of the superior surface of the disc is 25 cm^2, how much pressure exerted on the disc is attributable to body weight for a 930 N man? (Answer: 13.8 N/cm^2)
7. In the nucleus pulposus of an intervertebral disc, the compressive load is 1.5 times the externally applied load. In the annulus fibrosus the compressive force is 0.5 times the external load (2). What are the compressive loads on the nucleus pulposus and annulus fibrosus of the L5-S1 intervertebral disc of a 930 N man holding a 445 N weight bar across his shoulders, given that 37% of body weight is distributed above the disc? (Answer: 1183.7 N acts on the nucleus pulposus; 394.5 N acts on the annulus fibrosus.)
8. Estimate the volume of your own body. Construct a table that shows the approximate body dimensions you used in formulating your estimate.

9. Given the mass or weight and the volume of each of the following objects, rank them in the order of their densities.

OBJECT	WEIGHT OR MASS	VOLUME
A	50 kg	15.00 in^3
B	90 lb	12.00 cm^3
C	3 slugs	1.50 ft^3
D	450 N	0.14 m^3
E	45 kg	30.00 cm^3

10. Two muscles develop tension simultaneously on opposite sides of a joint. Muscle A, attaching 3 cm from the axis of rotation at the joint, exerts 250 N of force. Muscle B, attaching 2.5 cm from the joint axis, exerts 260 N of force. How much torque is created at the joint by each muscle? What is the net torque created at the joint? In which direction will motion at the joint occur? (Answer: A: 7.5 N-m; B: 6.5 N-m; net torque equals 1 N-m in the direction of A)

LABORATORY EXPERIENCES

1. Use a ruler to measure the dimensions of the sole of one of your shoes in centimeters. Being as accurate as possible, calculate an estimate of the surface area of the sole. (If a planimeter is available, use it to more accurately assess surface area by tracing around the perimeter of the sole.) Knowing your own body weight, calculate the amount of pressure exerted over the sole of one shoe. How much change in pressure would result if your body weight changed by 22 N (5 lb)?

2. Place a large container filled 3/4 full of water on a scale and record its weight. To assess the volume of an object of interest, completely submerge the object in the container, holding it just below the surface of the water. Record the *change* in weight on the scale. Remove the object from the container. Carefully pour water from the container into a measuring cup until the container weighs its original weight less the change in weight recorded. The volume of water in the measuring cup is the volume of the submerged object.

3. Secure one end of a pencil by firmly clamping it in a vise. Grip the other end of the pencil with an adjustable wrench and slowly apply a bending load to the pencil until it begins to break. Observe the nature of the break. On which side of the pencil did it begin? Is the pencil stronger in resisting compression or tension? Repeat the exercise using another pencil and applying a torsional (twisting) load. What does the nature of the initial break indicate about the distribution of shear stress within the pencil?

4. Experiment with pushing open a door by applying force with one finger. Apply force at distances of 10 cm, 20 cm, 30 cm, and 40 cm

from the hinges. Write a brief paragraph explaining at which force application distance it is easiest/hardest to open the door.

5. Stand on a bathroom scale and perform a vertical jump as a partner carefully observes the pattern of change in weight registered on the scale. Repeat the jump several times, as needed for your partner to determine the pattern. Trade positions and observe the pattern of weight change as your partner performs a jump. In consultation with your partner, sketch a graph of the change in exerted force (vertical axis) across time (horizontal axis) during the performance of a vertical jump. What does the area under the curve represent?

REFERENCES

Basmajian JV: *Muscles alive*, Baltimore, 1978, Williams & Wilkins.

ANNOTATED READINGS

Barham JN: Introduction to kinetics. In Barham JN: *Mechanical kinesiology*, St Louis, 1978, The CV Mosby Co.
Introduces basic kinetic concepts from a mechanical perspective.
Dainty DA and Norman RW: *Standardizing biomechanical testing in sport*, Champaign, Ill, 1987, Human Kinetics Publishers, Inc.
Includes expanded discussion of the characteristics, advantages, and disadvantages of tools commonly employed in biomechanics research for collecting kinetic data.
Wiktorin CV and Nordin M: *Introduction to problem solving in biomechanics*, Philadelphia, 1986, Lea & Febiger.
Provides an introduction to kinetic concepts in the context of clinical applications for physical therapists.

RELATED WEB SITES

Advanced Medical Technology, Inc.
http://www.amtiweb.com
Provides information on the AMTI force platforms, with reference to force and torque sensors, gait analysis, balance and posture, and other topics.
Bioengineering Technology Systems
http://www.bts.it/bts
Advertises telemetered electromyography and optoelectronic equipment.
Bortec Electronics, Inc.
http://www.cadvision.com/bortec/bortec.html
Describes a multichannel telemetered electromyography system.
Delsys
http://www.delsys.com
Provides description of surface electromyography equipment.
Kistler
http://www.kistler.com
Describes a series of force platforms.
Vicon
http://www.vicon.com
Advertises instrumentation systems for the study of biomechanics, animation, and gait analysis.

4

THE BIOMECHANICS OF HUMAN BONE GROWTH AND DEVELOPMENT

After completing this chapter, the reader will be able to:

Explain how the material constituents and structural organization of bone affect its ability to withstand mechanical loads.

Describe the processes involved in the normal growth and maturation of bone.

Describe the effects of exercise and of weightlessness on bone mineralization.

Explain the significance of osteoporosis and discuss current theories on its prevention.

Explain the relationship between different forms of mechanical loading and common bone injuries.

lever
a relatively rigid object that may be made to rotate about an axis by the application of force

What determines when a bone stops growing? How are stress fractures caused? Why does space travel cause reduced bone mineral density in astronauts? What is osteoporosis and how can it be prevented?

The word *bone* typically conjures up a mental image of a dead bone—a dry, brittle chunk of mineral that a dog would enjoy chewing. Given this picture, it is difficult to realize that living bone is an extremely dynamic tissue that is continually modeled and remodeled by the forces acting on it. Bone fulfills two important mechanical functions for human beings. First, it provides a rigid skeletal framework that supports and protects other body tissues. Second, it forms a system of rigid **levers** that can be moved by forces from the attaching muscles (see Chapter 12). This chapter discusses the biomechanical aspects of bone composition and structure, bone growth and development, bone response to stress, osteoporosis, and common bone injuries.

COMPOSITION AND STRUCTURE OF BONE TISSUE

The material constituents and structural organization of bone influence the ways in which bone responds to mechanical loading. The composition and structure of bone yield a material that is strong for its relatively light weight.

Material Constituents

The major building blocks of bone are calcium carbonate, calcium phosphate, collagen, and water. The relative percentages of these materials vary with the age and health of the bone. Calcium carbonate and calcium phosphate generally constitute approximately 60% to 70% of bone weight. These minerals give bone its **stiffness** and are the primary determiners of its **compressive strength.** Other minerals, including magnesium, sodium, and fluoride, also have vital structural and metabolic roles in bone growth and development (34). Collagen is a protein that provides bone with flexibility and contributes to its **tensile strength.** There is a progressive loss of collagen and increase in bone brittleness with aging. Thus, the bones of children are more pliable than the bones of adults.

The water content of bone makes up approximately 25% to 30% of the total bone weight. The water present in bone tissue is an important contributor to bone strength. For this reason, scientists and engineers studying the material properties of different types of bone tissue must ensure that the bone specimens they are testing do not become dehydrated.

stiffness
the ratio of stress to strain in a loaded material; that is, the stress divided by the relative amount of change in the structure's shape

compressive strength
ability to resist pressing or squeezing force

tensile strength
ability to resist pulling or stretching force

Collagen resists tension and provides flexibility to bone.

Structural Organization

The relative percentage of bone mineralization varies not only with the age of the individual but also with the specific bone in the body. Some bones are more **porous** than others. The more porous the bone, the smaller the proportion of calcium phosphate and calcium carbonate and the greater the proportion of nonmineralized tissue. Bone tissue has been classified into two categories based on porosity (Figure 4-1). If the porosity is low, with 5% to 30% of bone volume occupied by nonmineralized tissue, the tissue is termed **cortical bone.** Bone tissue with a relatively high porosity, with 30% to greater than 90% of bone volume

porous
containing pores or cavities

cortical bone
compact mineralized connective tissue with low porosity that is found in the shafts of long bones

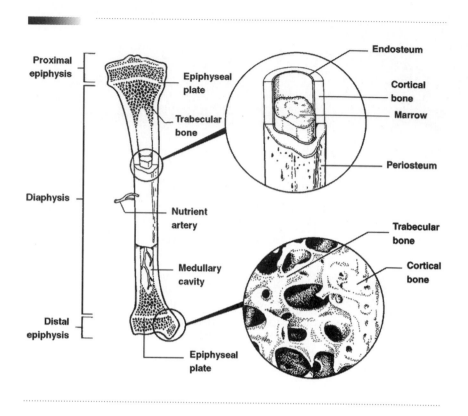

trabecular bone
less compact mineralized
connective tissue with high
porosity that is found in the
ends of long bones and in
the vertebrae

strain
amount of deformation
divided by the original length
of the structure or by the
original angular orientation
of the structure

*Because cortical bone is stiffer
than trabecular bone, it can
withstand greater stress but
less strain.*

occupied by nonmineralized tissue, is known as spongy, cancellous, or **trabecular bone.** Trabecular bone has a honeycomb structure with mineralized vertical and horizontal bars, called trabeculae, forming cells filled with marrow and fat.

The porosity of bone is of interest because it directly affects the mechanical characteristics of the tissue. With its higher mineral content, cortical bone is stiffer, so that it can withstand greater stress but less **strain** or relative deformation than trabecular bone. Because trabecular bone is spongier than cortical bone, it can undergo more strain before fracturing.

The function of a given bone determines its structure. The shafts of the long bones are composed of strong cortical bone. The relatively high trabecular bone content of the vertebrae contributes to their shock-absorbing capability. Trabecular bone develops four types of structure, depending on whether it must withstand relatively high or relatively low forces and whether the primary loading is axial (tension or compression) or asymmetric (bending) (22). The direction in which new bone tissue is formed is in line with the loads most habitually encountered, particularly in regions of high stress, such as the femoral neck.

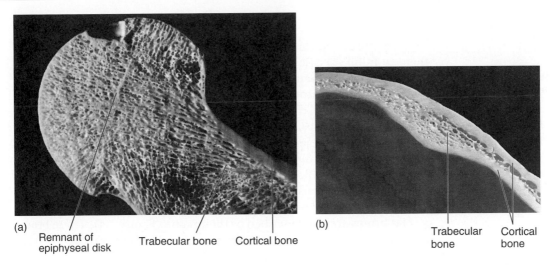

(a) Remnant of epiphyseal disk   Trabecular bone   Cortical bone

(b) Trabecular bone   Cortical bone

(a) In the femur, trabecular bone is encased by a thin layer of cortical bone. (b) In the skull, trabecular bone is sandwiched between plates of cortical bone. From Shier, Butler, and Lewis. *Hole's Human Anatomy and Physiology,* © 1996. Reprinted by permission of The McGraw-Hill Companies, Inc.

Both cortical and trabecular bone are **anisotropic;** that is, they exhibit different strength and stiffness in response to forces applied from different directions. Bone is strongest in resisting compressive stress and weakest in resisting shear stress (Figure 4-2).

anisotropic
exhibiting different mechanical properties in response to loads from different directions

Bone is strongest in resisting compression and weakest in resisting shear.

FIGURE 4-2
Relative bone strength in resisting compression, tension, and shear.

Stress to Fracture

Compression

Tension

Shear

Types of Bones

The structures and shapes of the 206 bones of the human body enable them to fulfill specific functions. The skeletal system is nominally subdivided into the central or **axial skeleton** and the peripheral or **appendicular skeleton** (Figure 4-3). The axial skeleton includes the bones that form the axis of the body, which are the skull, vertebrae, the sternum, and the ribs. The other bones form the body appendages or the appendicular skeleton. Bones are also categorized according to their general shapes and functions.

Short bones, which are approximately cubical, include only the carpals and the tarsals (Figure 4-4). These bones provide limited gliding motions and serve as shock absorbers.

Flat bones are also described by their name (Figure 4-4). These bones protect underlying organs and soft tissues and also provide large areas for muscle and ligament attachments. The flat bones include the scapulae, sternum, ribs, patellae, and some of the bones of the skull.

Irregular bones have different shapes to fulfill special functions in the human body (Figure 4-4). For example, the vertebrae provide a bony, protective tunnel for the spinal cord, offer several processes for muscle and ligament attachments, and support the weight of the superior body parts while enabling movement of the trunk in all three cardinal planes. The sacrum, coccyx, and maxilla are other examples of irregular bones.

Long bones form the framework of the appendicular skeleton (Figure 4-4). They consist of a long, roughly cylindrical shaft (also called the body or diaphysis) of cortical bone, with bulbous ends known as condyles, tubercles, or tuberosities. A self-lubricating **articular cartilage** protects the ends of long bones from wear at points of contact with other bones. Long bones also contain a central hollow area known as the medullary cavity or canal.

The long bones are adapted in size and weight for specific biomechanical functions. The tibia and femur are large and massive to support the weight of the body. The long bones of the upper extremity, including the humerus, radius, and ulna, are smaller and lighter to promote ease of movement. Other long bones include the clavicle, fibula, metatarsals, metacarpals, and phalanges.

BONE GROWTH AND DEVELOPMENT

Bone growth begins early in fetal development, and living bone is continually changing in composition and structure during the lifespan. Many of these changes represent normal growth and maturation of bone. Recent research shows that heredity, lifestyle, and race can also influence the quality of bone throughout life (56).

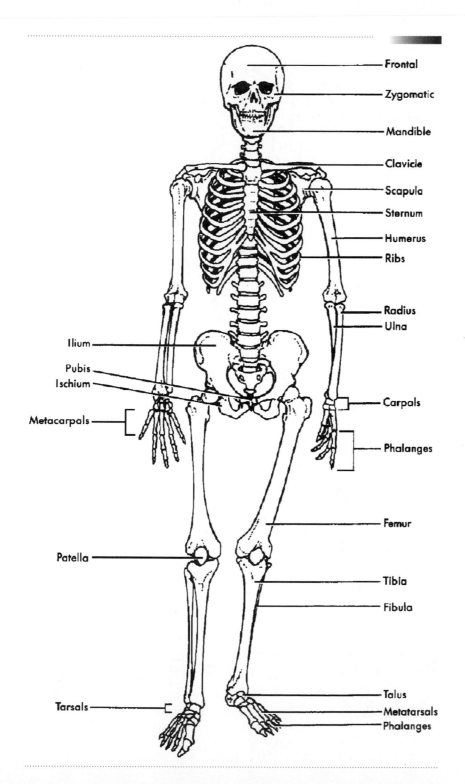

FIGURE 4-3
The human skeleton.

FIGURE 4-4
A, The carpals are categorized as short bones. **B**, The scapula is categorized as a flat bone. **C**, The vertebrae are examples of irregular bones. **D**, The femur represents the long bones.

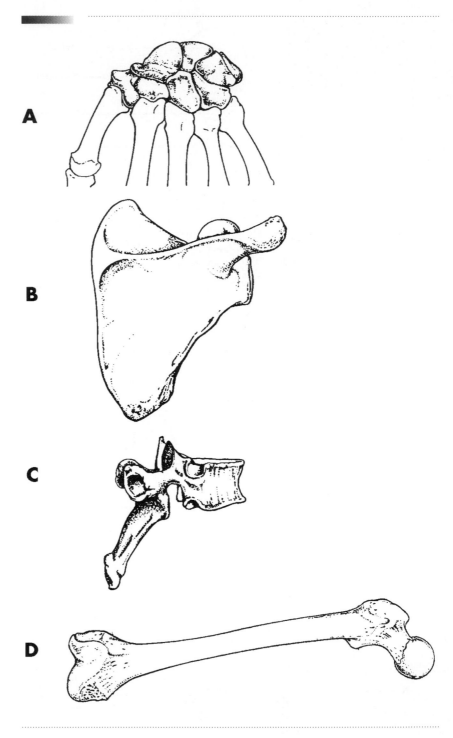

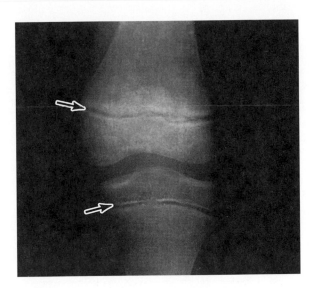

Epiphyseal plates are the sites of longitudinal growth in immature bone. From Shier, Butler, and Lewis. *Hole's Human Anatomy and Physiology,* © 1996. Reprinted by permission of The McGraw-Hill Companies, Inc.

Longitudinal Growth

Longitudinal growth of a bone occurs at the **epiphyses,** or epiphyseal plates (Figure 4-5). The epiphyses are cartilaginous discs found near the ends of the long bones. The diaphysis (central) side of each epiphysis continually produces new bone cells. During or shortly following adolescence the plate disappears and the bone fuses, terminating longitudinal growth. Most epiphyses close around age 18, although some may be present until about age 25.

Most epiphyses close around age 18, although some may be present until about age 25.

epiphysis
growth center of a bone that produces new bone tissue as part of the normal growth process until it closes during adolescence or early adulthood

Circumferential Growth

Long bones grow in diameter throughout most of the lifespan, although the most rapid bone growth occurs before adulthood. The internal layer of the **periosteum** builds concentric layers of new bone tissue on top of the existing ones. At the same time, bone is resorbed or eliminated around the circumference of the medullary cavity, so that the diameter of the cavity is continually enlarged. Specialized cells called **osteoblasts** and **osteoclasts** respectively form and resorb bone tissue. In healthy adult bone the activity of osteoblasts and osteoclasts is largely balanced.

periosteum
double-layered membrane covering bone; muscle tendons attach to the outside layer, and the internal layer is a site of osteoblast activity

osteoblasts
specialized bone cells that build new bone tissue

osteoclasts
specialized bone cells that resorb bone tissue

BONE RESPONSE TO STRESS

Other changes that occur in living bone throughout the lifespan are unrelated to normal growth and development. Bone responds dynamically to the presence or absence of different forces with changes in size, shape,

FIGURE 4-5

The structure of a long bone. From Shier, Butler, and Lewis. *Hole's Human Anatomy and Physiology,* © 1996. Reprinted by permission of The McGraw-Hill Companies, Inc.

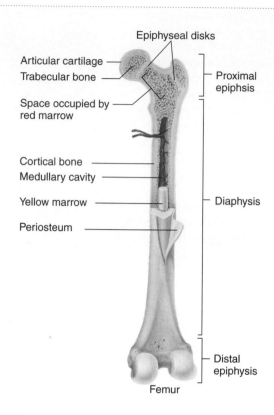

Epiphyseal disks

Articular cartilage
Trabecular bone
Proximal epiphsis

Space occupied by red marrow

Cortical bone
Medullary cavity
Yellow marrow
Diaphysis
Periosteum

Distal epiphysis

Femur

and density. This phenomenon was originally described by the German scientist Wolff in 1892:

> The form of a bone being given, the bone elements place or displace themselves in the direction of functional forces and increase or decrease their mass to reflect the amount of the functional forces (63).

Bone Modeling and Remodeling

Wolff's law indicates that bone strength increases and decreases as the functional forces on the bone increase and decrease.

According to Wolff's law, the densities and, to a much lesser extent, the shapes and sizes of the bones of a given human being are a function of the magnitude and direction of the mechanical stresses that act on the bones. Thus, bone mineralization and bone strength in children and adults are a function of the stresses on the skeleton. Since body weight provides the most constant mechanical stress to bones, bone mineral density generally parallels body weight, with heavier individuals having more massive bones. However, a given individual's physical activity profile, diet, lifestyle, and genetics can also dramatically influence bone density. Race is also a factor, with blacks tending to have greater skeletal density than whites, presumably because of a tendency for blacks to have greater muscle mass than whites (56).

Wolff's law is carried out through the actions of osteoblasts and osteoclasts, which continually act to increase, decrease, or reshape bone. A predominance of osteoblast activity produces bone modeling, with a net gain in bone mass. Bone remodeling involves a balance of osteoblast and osteoclast action or a predominance of osteoclast activity, with associated maintenance or loss of bone mass.

The processes causing bone remodeling are not fully understood and continue to be researched by scientists.

The malleability of bone is dramatically exemplified by the case of an infant who was born in normal physical condition but missing one tibia, the major weightbearing bone of the lower extremity. After the child was walking for a time, X rays revealed that modeling of the fibula in the abnormal leg had occurred to the extent that it could not be distinguished from the tibia of the other leg (2).

Another interesting case is that of a construction worker who had lost all but the fifth finger of one hand in a war injury. After 32 years the metacarpal and phalanx of the remaining finger had been modeled to resemble the third finger of the other hand (50).

Bone Hypertrophy

Although cases of complete changes in bone shape and size are unusual, there are many examples of bone modeling, or **bone hypertrophy,** in response to regular physical activity. The bones of physically active individuals tend to be denser and therefore more mineralized than those of sedentary individuals of the same age and gender. Moreover, the results of several studies indicate that occupations and sports particularly stressing a certain limb or region of the body produce accentuated bone hypertrophy in the stressed area. For example, professional tennis players display not only muscular hypertrophy in the tennis arm but also hypertrophy of that arm's radius (44). Similar bone hypertrophy has been observed in the arms of baseball pitchers (59).

bone hypertrophy
increase in bone mass resulting from a predominance of osteoblast activity

It also appears that the greater the forces or loads habitually encountered, the more dramatic the increased mineralization of the bone. In one interesting study, researchers measured the density of the femur among 64 nationally ranked athletes representing different sports (40). The femurs displaying the greatest density were those of the weight lifters, followed by throwers, runners, soccer players, and swimmers. As might be deduced from this research, it is the magnitude of skeletal loading rather than the frequency of loading that is related to bone mass (61).

Regular exercise seems to increase bone density, not only in the regions that are particularly stressed, but throughout the skeletal system. Investigations of men and women runners reveal bone densities higher than average not only in the lower extremity, but in the upper extremity as well (15, 17). Non-weightbearing exercises such as bicycle ergometry and swimming can also contribute positively to bone mineral density (12, 43). These findings suggest that factors other than mechanical stress to the bone tissue may contribute to bone hypertrophy. Improved systemic circulation, for example, may also contribute to bone health. There is some evidence, however, that swimmers who spend a lot of time in the water, where the buoyant force counteracts gravity, may have bone mineral density less than that of sedentary individuals (49). This may be primarily a problem for swimmers of low-average to below- average body weight, since increased body weight builds bone.

Bone Atrophy

bone atrophy
decrease in bone mass resulting from a predominance of osteoclast activity

Whereas bone hypertrophies in response to increased mechanical stress, it displays the opposite response to reduced stress. When the normal stresses exerted on bone by muscle contractions or weight bearing are reduced, bone tissue atrophies through remodeling. When **bone atrophy** occurs, the amount of calcium contained in the bone diminishes and both the weight and strength of the bone decrease. Loss of bone mass due to reduced mechanical stress has been found in bedridden patients, sedentary senior citizens, and astronauts. Four to six weeks of bed rest can result in significant decrements in bone mineral density that are not fully reversed after six months of normal weightbearing activity (11).

Bone demineralization is a potentially serious problem. From a biomechanical standpoint, as bone mass diminishes, strength and thus resistance to fracture also decrease, particularly in trabecular bone. When calcium compounds dissolve from bone tissue, they enter the bloodstream, where the kidneys filter them. Bone demineralization therefore also contributes to the likelihood of developing kidney stones.

The results of calcium loss studies conducted during the Skylab flights indicate that urinary calcium loss is related to time spent out of the earth's gravitational field. The pattern of bone loss observed is highly similar to that documented among patients during periods of bed rest (31). Encouraging reports on Russian cosmonauts, however, indicate that vigorous exercise, primarily in the form of treadmill walking and running, can reasonably maintain bone health in space for periods of up to a year (60).

It is not yet clear what specific mechanism or mechanisms are responsible for bone loss outside of the gravitational field. Studies of rats indicate that after seven days in space, osteoclast activity remains fairly constant but osteoblast activity diminishes, with the magnitude of the problem varying with the location of the bone in the skeleton (57). More research on this topic is needed.

Loss of bone mass during periods of time spent outside of the earth's gravitational field is a problem for astronauts.

A logical approach to combating the problem of bone loss in space is simulating gravity inside space vehicles. One technique that has been used to create artificial gravity is centrifuging. Centrifuging is a laboratory procedure that creates force on specimens spun at a high speed. Rats that were centrifuged aboard the Soviet Cosmos-936 biosatellite showed increased bone density and calcium content at the end of the experiment (46).

It remains to be seen if measures other than the artificial creation of gravity can effectively prevent bone loss during space travel. Astronauts' current exercise programs during flights in space are designed to prevent bone loss by increasing the mechanical stress placed on bones using muscular force. However, the muscles of the body exert mainly tensile forces on bone, whereas gravity provides a compressive force.

Therefore, it may be that no amount of physical exercise alone can completely compensate for the absence of gravitational force.

In the future, astronauts visiting other planets might maintain bone mass by carrying extra loads while walking on the planet surface. Researchers have calculated that individuals in average physical condition could walk for eight hours on the moon while carrying as much as 170% of body mass and on Mars while carrying 50% of body mass without undue fatigue (62). According to scientific projections, however, the daily walking times on these planets necessary for maintaining bone mass are unrealistically long (62).

Another important related question is to what extent bone mass can be recovered when astronauts reenter the earth's gravitational field. Research shows that some recovery of bone mass can be expected, but that the rate and extent of recovery vary with the individual and with the location of the bone (30, 31). Bones that are directly involved in weight bearing during ambulation can be expected to recover most quickly; one study of subjects who underwent 17 weeks of continuous bed rest showed an almost complete reversal of a 10.4% loss of bone density in the calcaneus after six months (31). An example of loading on a weightbearing bone is shown in Figure 4-6.

OSTEOPOROSIS

osteoporosis
a disorder involving decreased bone mass and strength with one or more resulting fractures

osteopenia
condition of reduced bone mineral density that predisposes the individual to fractures

Bone atrophy is a problem not only for astronauts and bedridden patients but also for a growing number of senior citizens and female athletes. **Osteoporosis** is found in most elderly individuals, with earlier onset in women, and is becoming increasingly prevalent with the increasing mean age of the population. The condition begins as **osteopenia,** reduced bone mass without the presence of a fracture, but often progresses to osteoporosis, a condition in which bone mineral mass and strength are so severely compromised that daily activities can cause bone pain and fracturing (21).

Adult Bone Development

Bone mineral normally accumulates throughout childhood and adolescence, reaching a peak at about age 25 to 28 in women and around age 30 to 35 in men (48, 56). Researchers disagree as to the length of time following this peak that bone density remains constant (54). However, an age-related, progressive decline in bone density and bone strength in both men and women may begin as soon as the early twenties (10, 39). Trabecular bone is particularly affected, with progressive disconnection and disintegration of trabeculae compromising the integrity of the bone's structure and seriously diminishing bone strength (54). Approximately 0.5–1.0% of bone mass is lost each year, until

FIGURE 4-6

S A M P L E   P R O B L E M   1

The tibia is the major weightbearing bone in the lower extremity. If 88% of body mass is proximal to the knee joint, how much compressive force acts on each tibia when a 600 N person stands in anatomical position? How much compressive force acts on each tibia if the person holds a 20 N sack of groceries?

Solution

Given: wt = 600 N

(It may be deduced that weight = compressive force, F_c.)

Formula: F_c on knees = (600 N)(0.88)

$$F_c \text{ on one knee} \quad = \frac{(600 \text{ N})(0.88)}{2}$$

$$F_c \text{ on one knee} \quad = 264 \text{ N}$$

$$F_c \text{ with groceries} = \frac{(600 \text{ N} + 20 \text{ N})(0.88)}{2}$$

$$F_c \text{ with groceries} = 272.8 \text{ N}$$

women reach about age 50 or menopause (10, 54). Following menopause there appears to be an increased rate of bone loss, with values as high as 6.5% per year reported during the first five to eight years (29).

Postmenopausal and Age-Associated Osteoporosis

The majority of those affected by osteoporosis are postmenopausal and elderly women, although elderly men are also susceptible. In women, the risk of experiencing an osteoporotic fracture once in life is 30–40%, or about twice the risk for men (14). The risk of sustaining additional fractures after the first osteoporotic fracture is significantly increased (18).

Type I osteoporosis, also known as postmenopausal osteoporosis, affects approximately 40% of women after the age of 50 (32). The first osteoporotic fractures usually begin to occur about 15 years postmenopause, with women suffering approximately three times as many femoral neck fractures, three times as many vertebral fractures, and six times as many wrist fractures as men of the same age (32). Type II osteoporosis, also called age-associated osteoporosis, affects most women and men past age 70 (5). After age 60, about 90% of all fractures in both men and women are osteoporosis-related, and these fractures are one of the leading causes of death in the elderly population (3, 47).

Osteoporosis is a serious health problem for most elderly individuals, with women affected more dramatically than men.

Although the radius and ulna, femoral neck, and spine are all common sites of osteoporotic fractures, the most common symptom of osteoporosis is back pain derived from fractures of the weakened trabecular bone of the vertebral bodies. Crush fractures of the lumbar vertebrae resulting from compressive loads created by weight bearing during activities of daily living frequently cause reduction of body height. Because most body weight is anterior to the spine, the resulting fractures often leave the vertebral bodies wedge-shaped, accentuating thoracic kyphosis (see Chapter 9). This disabling deformity is known as *dowager's hump*. Vertebral compression fractures are extremely painful and debilitating, and affect physical, functional, and psychosocial aspects of the person's life (51). As spinal height is lost there is added discomfort from the rib cage pressing on the pelvis. An estimated 26% of women aged 50 and over have vertebral compression fractures (49).

Painful, deforming, and debilitating crush fractures of the vertebrae are the most common symptom of osteoporosis.

As the skeleton ages in men, there is an increase in vertebral diameter that serves to reduce compressive stress during weight bearing (38). Thus, although osteoporotic changes may be taking place, the structural strength of the vertebrae is not reduced. Why the same compensatory change does not occur in women is unknown.

Female Athlete Triad

The desire to excel at competitive sports causes some young female athletes to strive to achieve an undesirably low body weight. This dangerous practice commonly involves a combination of disordered eating, **amenorrhea,** and osteoporosis, a combination that has come to be known as the "female athlete triad" (53). Because the triad can result in irreversible bone loss and death, friends, parents, coaches, and physicians need to be alert to the signs and symptoms.

amenorrhea
cessation of menses

Disordered eating, amenorrhea, and osteoporosis comprise a dangerous and potentially lethal triad for young female athletes.

As many as 62% of female athletes in certain sports display disordered eating behaviors, with those participating in endurance or appearance-related sports most likely to be involved (53). Prolonged disordered eating can lead to anorexia nervosa or bulimia nervosa, illnesses that affect from 1–10% of all adolescent and college-age women (24). Symptoms of anorexia nervosa in girls and women include body weight 15% or more below minimal normal weight for age and height, an intense fear of gaining weight, a disturbed body image, and amenorrhea. Symptoms of bulimia nervosa are a minimum of two eating binges a week for at least three months, a feeling of lack of control during binges, regular use of self-induced vomiting, laxatives, diuretics, strict dieting, or exercise to prevent weight gain, and excessive concern with body image and weight (24).

Anorexia nervosa and bulimia nervosa are life-threatening eating disorders.

The relationship between disordered eating and amenorrhea is not well understood. Amenorrhea is often associated with low estrogen levels, and it may be related to low body fat and/or excessive training. A reported 2–5% of premenopausal women in the United States have amenorrhea, but the prevalence in female athletes is higher. Studies of competitive women athletes in different sports indicate that from 3.4–66% have either primary amenorrhea, with menarche delayed beyond 16 years of age, or secondary amenorrhea, the absence of three to six consecutive menstrual cycles (53).

The link between cessation of menses and osteoporosis is a low level of endogenous estrogen (9). Although the incidence of osteoporosis among female athletes is unknown, the consequences of this disorder in young women are potentially tragic. Among one group of over 200 premenopausal female runners, those with amenorrhea had 10% less lumbar bone density than those with normal menses (25). In other research, young women with anorexia nervosa were found to have marked trabecular and cortical bone loss on the order of 4–10% per year (37). This pattern is of particular concern for adolescent athletes, because roughly 50% of bone mineralization and 15% of adult height are normally established during the teenage years (8, 58). Not surprisingly, amenorrheic premenopausal female athletes have a high rate of stress fractures, with more fractures related to later onset of menarche (58). Moreover, the loss of bone that occurs may be irreversible, and osteoporotic wedge fractures can ruin posture for life (53).

Competitive female athletes in endurance and appearance-related sports are particularly at risk for developing the dangerous female athlete triad (see text).

Regular exercise has been shown to be effective to some extent in mediating age-related bone loss.

Preventing and Treating Osteoporosis

Osteoporosis is neither a disease with acute onset, nor an inevitable accompaniment of aging, but the result of a lifetime of habits that are erosive to the skeletal system (13). Although proper diet, hormone levels, and exercise can work to increase bone mass at any stage in life, evidence suggests that it is easier to prevent osteoporosis than it is to treat it (6).

Weightbearing physical activity is necessary for maintaining skeletal integrity in both humans and animals. Importantly, studies show that a regular program of weightbearing exercise, such as walking, can increase bone health and strength even among individuals with osteoporosis (9). The American College of Sports Medicine Position Statement (7) makes five important points relative to the role of exercise in preventing and treating osteoporosis:

1. weightbearing physical activity is essential for developing and maintaining a healthy skeleton,
2. strength exercises may also be beneficial, particularly for non-weightbearing bones,
3. an increase in physical activity for sedentary women can prevent further inactivity-related bone loss and can even improve bone mass,

4. exercise is not an adequate substitute for postmenopausal hormone replacement, and
5. an appropriate exercise program for older women should include activities for improving strength, flexibility, and coordination, to lessen the likelihood of falls.

Estrogen and testosterone deficiencies promote the development of osteoporosis.

Although regular physical exercise promotes bone mineralization and bone health, it is clear that hormonal factors exert a more powerful influence, given the prevalence of osteoporosis in amenorrheic female athletes. High bone mineral density has been associated with early menarche, late menopause, and a long reproductive period, all of which support the effectiveness of estrogen in maintaining bone (28). Conversely, low levels of estrogen in females and low levels of testosterone in males promote bone loss in both children and adults. There are two ways in which estrogen deficiency potentially damages bone (54). First, it diminishes the efficiency with which calcium is absorbed in the body. Second, it facilitates the resorption of bone by osteoclasts. Because of its effectiveness and relatively low cost, estrogen replacement therapy is considered the first-line pharmacologic therapy for preventing osteoporosis in postmenopausal and amenorrheic women (1). Although the mechanism by which testosterone influences bone mass is not understood, testosterone replacement therapy also stimulates bone building (54).

Increased dietary calcium intake exerts a positive influence on bone mass for women with a dietary deficiency, with the amount of calcium absorbed influenced positively by calcitriol (the active form of vitamin D), and negatively by dietary fiber (45, 55). Although adequate dietary calcium is particularly important during the teenage years, unfortunately the median American girl falls below the recommended daily intake by age 11 (54). A modified diet or calcium supplementation can be critically important for the development of peak bone mass among adolescent females at a dietary deficiency. Clinicians are now recognizing that a predisposition for osteoporosis can begin in childhood and adolescence when a poor diet interferes with bone mass development (16).

Other lifestyle factors also affect bone mineralization. Known risk factors for developing osteoporosis include physical inactivity, tobacco smoking, deficiencies in estrogen, calcium, and vitamin D, and excessive consumption of protein, caffeine, and alcohol (18). A study of female twins, one of whom smoked more heavily than the other, showed that women who smoke one pack of cigarettes a day through adulthood will have a reduction in bone density of 5–10% by the time of menopause, which is sufficient to increase the risk of fracture (26). Genetic factors also influence both the peak bone mass achieved and the rate of change in bone mass (20).

Research indicates that in the future the use of pharmacologic agents that stimulate bone formation may enable the prevention and eventually the elimination of osteoporosis (4, 42). Recent clinical trials suggest that low doses of growth factors can stimulate osteoblast recruitment and promote bone formation (35). However, until this becomes a reality, young women in particular are encouraged to maximize peak bone mass and to minimize its loss by engaging in regular physical activity and avoiding the lifestyle factors that negatively affect bone health (19).

COMMON BONE INJURIES

Considering the important mechanical functions performed by bone, bone health is an important part of general health. Bone health can be impaired by injuries and pathologies.

Fractures

A **fracture** is a disruption in the continuity of a bone. The nature of a fracture depends on the direction, magnitude, loading rate, and duration of the mechanical load sustained, as well as the health and maturity of the bone at the time of injury. Fractures are classified as simple when the bone ends remain within the surrounding soft tissues and compound when one or both bone ends protrude from the skin. When the loading rate is rapid, a fracture is more likely to be comminuted, containing multiple fragments (Figure 4-7).

Avulsions are fractures caused by tensile loading in which a tendon or ligament pulls a small chip of bone away from the rest of the bone. Explosive throwing and jumping movements may result in avulsion fractures of the medial epicondyle of the humerus and the calcaneus.

Excessive bending and torsional loads can produce spiral fractures of the long bones (Figure 4-7). The simultaneous application of forces from opposite directions at different points along a structure such as a long bone generate a torque known as a bending moment, which can cause bending and ultimately fracture of the bone. A bending moment is created on a football player's leg when the foot is anchored to the ground and tacklers apply forces at different points on the leg in opposite directions. When bending is present the structure is loaded in tension on one side and in compression on the opposite side, as discussed in Chapter 3. Because bone is stronger in resisting compression than in resisting tension, the side of the bone loaded in tension will fracture first.

Torque applied about the long axis of a structure such as a long bone causes torsion, or twisting of the structure. Torsion creates shear stress

fracture
disruption in the continuity of a bone

Under excessive bending loads, bone tends to fracture on the side loaded in tension.

FIGURE 4-7

Types of fractures. From Shier, Butler, and Lewis. *Hole's Human Anatomy and Physiology,* © 1996. Reprinted by permission of The McGraw-Hill Companies, Inc.

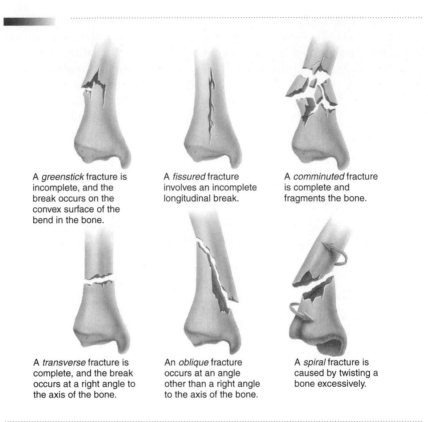

A *greenstick* fracture is incomplete, and the break occurs on the convex surface of the bend in the bone.

A *fissured* fracture involves an incomplete longitudinal break.

A *comminuted* fracture is complete and fragments the bone.

A *transverse* fracture is complete, and the break occurs at a right angle to the axis of the bone.

An *oblique* fracture occurs at an angle other than a right angle to the axis of the bone.

A *spiral* fracture is caused by twisting a bone excessively.

impacted
pressed together by a compressive load

throughout the structure, as explained in Chapter 3. When a skier's body rotates with respect to one boot and ski during a fall, torsional loads can cause a spiral fracture of the tibia. In such cases, a combined loading pattern of shear and tension produces failure at an oblique orientation to the longitudinal axis of the bone (41).

Since bone is stronger in resisting compression than in resisting tension and shear, acute compression fractures of bone (in the absence of osteoporosis) are rare. However, under combined loading a fracture resulting from a torsional load may be also be **impacted** by the presence of a compressive load. An impacted fracture is one in which the opposite sides of the fracture are compressed together. Fractures that result in depression of bone fragments into the underlying tissues are termed depressed.

Since the bones of children contain relatively larger amounts of collagen than adult bones, they are more flexible and more resistant to fracture under day-to-day loading than adult bones. Consequently, greenstick fractures, or incomplete fractures, are more common in children than in adults (Figure 4-7). A greenstick fracture is an incomplete fracture caused by bending or torsional loads.

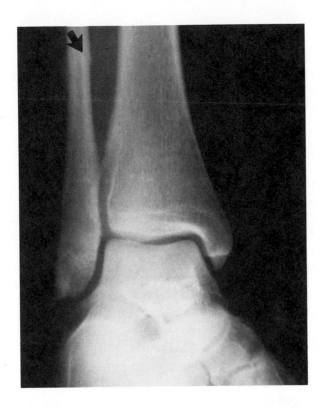

This X ray shows a stress fracture toward the distal end of the fibula. (Courtesy: Lester Cohn, MD.)

Stress fractures, also known as fatigue fractures, result from low-magnitude forces sustained on a repeated basis. Any increase in the magnitude or frequency of bone loading produces a **stress reaction,** which may involve microdamage (23). Bone responds to microdamage by remodeling: first osteoclasts resorb the damaged tissue; then osteoblasts deposit new bone at the site. When there is not time for the repair process to complete before additional microdamage occurs, the condition can progress to a stress fracture. Stress fractures begin as a small disruption in the continuity of the outer layers of cortical bone, but can worsen over time, eventually resulting in complete cortical fracture.

In runners, a group particularly prone to stress fractures, about 50% of fractures occur in the tibia and approximately 20% of fractures are in the metatarsals, with fractures of the femoral neck and pubis also reported (27, 36). Increases in training duration or intensity that do not allow enough time for bone remodeling to occur are the primary culprits. Other factors that predispose runners to stress fractures include muscular fatigue and abrupt changes in either the running surface or the running direction (36).

stress fracture
fracture resulting from repeated loading of relatively low magnitude

stress reaction
progressive bone pathology associated with repeated loading

Epiphyseal Injuries

About 10% of acute skeletal injuries in children and adolescents involve the epiphysis (33). Epiphyseal injuries include injuries to the cartilaginous epiphyseal plate, the articular cartilage, and the apophysis. The apophyses are the sites of tendon attachments to bone, where bone shape is influenced by the tensile loads to which these sites are subjected. The epiphyses of long bones are termed pressure epiphyses and the apophyses are called traction epiphyses, after the types of physiological loading present. Both acute and repetitive loading can injure the growth plate, potentially resulting in premature closure of the epiphyseal junction and termination of bone growth.

Injuries of an epiphyseal plate can terminate bone growth early.

Another form of epiphyseal injury, osteochondrosis, involves disruption of blood supply to an epiphysis, with associated tissue necrosis and potential deformation of the epiphysis. The cause of the condition is poorly understood. Osteochondrosis occurs most commonly between the ages of 3 and 10 and is more prevalent among boys than girls (52).

Osteochondrosis of an apophysis, known as apophysitis, is often associated with traumatic avulsions. Common sites for apophysitis are the calcaneus and the tibial tubercle at the site of the patellar tendon attachment, where the disorder is referred to respectively as Sever's disease and Osgood-Schlatter's disease.

SUMMARY

Bone is an important and dynamic living tissue. Its mechanical functions are to support and protect other body tissues and to act as a system of rigid levers that can be manipulated by the attached muscles.

Bone's strength and resistance to fracture depend on its material composition and organizational structure. Minerals contribute to a bone's hardness and compressive strength, and collagen provides its flexibility and tensile strength. Cortical bone is stiffer and stronger than trabecular bone, whereas trabecular bone has greater shock-absorbing capabilities.

Bone is an extremely dynamic tissue that is continually being remodeled in accordance with Wolff's law. Although bones grow in length only until the epiphyseal plates close at adolescence, bones continually change in density, and to some extent in size and shape, through the actions of osteoblasts and osteoclasts.

Osteoporosis, a disorder characterized by excessive loss of bone mineral mass and strength, is extremely prevalent among the elderly. It affects women at an earlier age and more severely than men. It is also present in an alarming frequency among young, eating-disordered, amenorrheic female athletes. Although the cause of osteoporosis remains unknown, the condition can often be improved through hormone therapy, avoidance of negative lifestyle factors, and a regular exercise program.

1. Explain why the bones of the human body are stronger in resisting compression than in resisting tension and shear.
2. In the human femur, bone tissue is strongest in resisting compressive force, approximately half as strong in resisting tensile force, and only about one-fifth as strong in resisting shear force. If a tensile force of 8000 N is sufficient to produce a fracture, how much compressive force will produce a fracture? How much shear force will produce a fracture? (Answer: compressive force = 16,000 N; shear force = 3200 N)
3. Explain why bone density is related to an individual's body weight.
4. Rank the following activities according to their effect on increasing bone density: running, backpacking, swimming, cycling, weight lifting, polo, tennis. Write a paragraph providing the rationale for your ranking.
5. Why is bone tissue organized differently (cortical versus trabecular bone)?
6. What kinds of fractures are produced by compression, tension, and shear, respectively?
7. What kinds of fractures are produced only by combined loading? (Identify fracture types with their associated causative loadings.)
8. Approximately 56% of body weight is supported by the fifth lumbar vertebra. How much stress is present on the 22 cm^2 surface area of that vertebra in an erect 756 N man? (Assume that the vertebral surface is horizontal.) (Answer: 19.2 N/cm^2)
9. In Problem 8, how much total stress is present on the fifth lumbar vertebra if the individual holds a 222 N weight bar balanced across his shoulders? (Answer: 29.3 N/cm^2)
10. Why are men less prone than women to vertebral compression fractures?

1. Hypothesize about the way or ways in which each of the following bones is loaded when a person stands in anatomical position. Be as specific as possible in identifying which parts of the bones are loaded.
   a. Femur
   b. Tibia
   c. Scapula
   d. Humerus
   e. Third lumbar vertebra
2. Outline a 6-week exercise program that might be used with a group of osteoporotic elderly persons who are ambulatory.
3. Speculate about what exercises or other strategies may be employed in outer space to prevent the loss of bone mineral density in humans.

4. Hypothesize as to the ability of bone to resist compression, tension, and shear, compared to the same properties in wood, steel, and plastic.
5. How are the bones of birds and fish adapted for their methods of transportation?
6. Why is it important to correct eating disorders?
7. Hypothesize as to why men are less prone to osteoporosis than women.
8. When an impact force is absorbed by the foot, the soft tissues at the joints act to lessen the amount of force transmitted upward through the skeletal system. If a ground reaction force of 1875 N is reduced 15% by the tissues of the ankle joint and 45% by the tissues of the knee joint, how much force is transmitted to the femur? (Answer: 750 N)
9. How much compression is exerted on the radius at the elbow joint when the biceps brachii, oriented at a 30° angle to the radius exerts a tensile force of 200 N? (Answer: 173 N)
10. If the anterior and posterior deltoids both insert at an angle of 60° on the humerus and each muscle produces a force of 100 N, how much force is acting perpendicular to the humerus? (Answer: 173.2 N)

LABORATORY EXPERIENCES

1. Using an anatomical model, review the bones of the human skeleton. Select a specific bone from each of the four shape categories and explain how each bone's size, shape, and internal structure are suited to its biomechanical function.
2. Select three bones on an anatomical model and study each bone's shape. What do the bone shapes indicate about the probable locations of muscle tendon attachments and the directions in which the muscles exert force?
3. Study bone specimens from an unknown animal source. Given each bone's size and shape, hypothesize as to the bone's biomechanical function.
4. Using a paper soda straw as a model of a long bone, progressively apply compression to the straw by loading it with weights until it buckles. Using a system of clamps and a pulley, repeat the experiment, progressively loading straws in tension and shear to failure. Record the weight at which each straw failed and write a paragraph discussing your results and relating them to long bones.
5. Visit some of the World Wide Web sites listed at the end of this chapter and locate a picture showing a surgical repair of a bone injury of interest. Write a paragraph summarizing how the surgical repair was performed.

REFERENCES

1. Abbott TA III, Lawrence BJ, and Wallach S: Osteoporosis: the need for comprehensive treatment guidelines, Clin Ther 18:127, 1996.
2. Adrian MJ and Cooper JM: *Biomechanics of human movement*, Indianapolis, 1989, Benchmark Press, Inc.
3. Ali NS and Bennett SJ: Postmenopausal women. Factors in osteoporosis preventive behaviors, J Gerontol Nurs 18:23, 1992.
4. Allen SH: Primary osteoporosis. Methods to combat bone loss that accompanies aging, Postgrad Med 93:43, 1993.
5. American Academy of Orthopedic Surgeons Position Statement: Osteoporosis as a National Public Health Priority, Document Number 1113, 1993.
6. Anderson JJ, Rondano P, and Holmes A: Roles of diet and physical activity in the prevention of osteoporosis, Scand J Rheumatol Suppl 103:65, 1996.
7. Anonymous: ACSM position stand on osteoporosis and exercise, Med Sci Sports Exerc 27:I, 1995.
8. Bachrach LK: Bone mineralization in childhood and adolescence, Curr Opin Pediatr 5:467, 1993.
9. Barlet JP, Coxam V, and Davicco MJ: Physical exercise and the skeleton, Arch Physiol Biochem 103:681, 1995.
10. Birkenhager-Frenkel DH et al: Age-related changes in cancellous bone structure, Bone Mineral 4:197, 1988.
11. Bloomfield SA: Changes in musculoskeletal structure and function with prolonged bed rest, Med Sci Sports Exerc 29:197, 1997.
12. Bloomfield SA et al: Non-weightbearing exercise may increase lumbar spine bone mineral density in healthy postmenopausal women, Am J Phys Med Rehabil 72:204, 1993.
13. Bockman RS: Osteoporosis and its management in older women, J Am Med Women's Assoc 52:121, 1997.
14. Bonjour JP, Burckhardt P, Dambacher M, Kraenzlin ME, and Wimpfheimer C: Epidemiology of osteoporosis, Schweiz Med Wochenschr 127:659, 1997.
15. Brewer V et al: Role of exercise in prevention of involutional bone loss, Med Sci Sports Exerc 15:445, 1983.
16. Carrié-Fässler AL and Bonjour JP: Osteoporosis as a pediatric problem, Pediatr Clin North Am 42:811, 1995.
17. Dalen N and Olsson KE: Bone mineral content and physical activity, Acta Orthop Scand 45:170, 1974.
18. Deal CL: Osteoporosis: prevention, diagnosis, and management, Am J Med 102:35S, 1997.
19. Drinkwater BL: Exercise and aging: the female masters athlete. In Puhl J, Brown CH, and Voy RL: *Sport science perspectives for women*, Champaign, Ill, 1987, Human Kinetics Publishers, Inc.
20. Eisman JA et al: Peak bone mass and osteoporosis prevention, Osteoporos Int 3 (Suppl 1):56, 1993.
21. Frost HM: Defining osteopenias and osteoporoses: another view (with insights from a new paradigm), Bone 20:385, 1997.
22. Gibson LJ: The mechanical behaviour of cancellous bone, Biomech 18:317, 1985.
23. Gromston SK and Zernicke RF: Exercise-related stress responses in bone, J Appl Biomech 9:2, 1993.
24. Haller E: Eating disorders. A review and update, West J Med 157:658, 1992.

25. Hetland ML, Haarbo J, and Christiansen C: Running induces menstrual disturbances but bone mass is unaffected, except in amenorrheic women, Am J Med 95:53, 1993.

26. Hopper JL and Seeman E: The bone density of female twins discordant for tobacco use, New Eng J Med 330:387, 1994.

27. Hulkko A and Orava S: Stress fractures in athletes. Int J Sports Med 8:221, 1987.

28. Ito M, Yamada M, Hayashi K, Ohki M, Uetani M, and Nakamura T: Relation of early menarche to high bone mineral density, Calcif Tissue Int 57:11, 1995.

29. Krolner B and Pors Nielsen S: Bone mineral content of the lumbar spine in normal and osteoporotic women: cross-sectional and longitudinal studies, Clin Sci 62:329, 1982.

30. LeBlanc AD and Schneider VS: Exp Gerontol 26:189, 1991.

31. LeBlanc AD et al: Bone mineral loss and recovery after 17 weeks of bed rest, J Bone Miner Res 8:843, 1990.

32. Lips P: Epidemiology and predictors of fractures associated with osteoporosis, Am J Med 103:3S, 1997.

33. Maffulli N: Intensive training in young athletes. The orthopaedic surgeon's viewpoint, Sports Med 9:229, 1990.

34. Marchigiano G: Osteoporosis: primary prevention and intervention strategies for women at risk, Home Care Provid 2:76, 1997.

35. Marie P: Growth factors and bone formation in osteoporosis: roles for IGF_I and TGF-beta, Rev Rhum Engl Ed 64:44, 1997.

36. Matheson GO et al: Stress fractures in athletes: a study of 320 cases, Am J Sports Med 15:46, 1987.

37. Maugars YM, Berthelot JM, Forestier R, Mammar N, Lalande S, Venisse JL, and Prost AM: Follow-up of bone mineral density in 27 cases of anorexia nervosa, Eur J Endocrinol 135:591, 1996.

38. Mosekilde Li. and Mosekilde Le.: Sex differences in age-related changes in vertebral body size, density and biomechanical competence in normal individuals, Bone 11:67, 1990.

39. Mosekilde Li. and Mosekilde Le.: Iliac crest bone volume as a predictor for vertebral compressive strength, ash density, and trabecular bone volume in normal individuals, Bone 9:195, 1988.

40. Nilsson B and Westline N: Bone density in athletes, Clin Orthop 77:179, 1971.

41. Nordin M and Frankel VH: Biomechanics of bone. In Nordin M and Frankel VH: *Basic biomechanics of the musculoskeletal system* (2nd ed.), Philadelphia, 1989, Lea & Febiger.

42. Notelovitz M: Osteoporosis: screening, prevention, and management, Fertil Steril 59:707, 1993.

43. Orwoll ES et al: The relationship of swimming exercise to bone mass in men and women, Arch Intern Med 149:2197, 1989.

44. Pirnay FM et al: Bone mineral content and physical activity, Int J Sport Med 8:331, 1987.

45. Prince R: The calcium controversy revisited: Implications of new data, Med J Aust 159:404, 1993.

46. Prodhonchukov AA et al: Comparative study of effects of weightlessness and artificial gravity on density, ash, calcium, and phosphorous content of calcified tissues, STAR Pub N80-34064, 1980.

47. Recker RR: Osteoporosis, Contemp Nutr 8:1, 1983.
48. Recker RR, Davies KM, Hinders SM, et al: Bone gain in young adult women, JAMA 268:2403, 1992.
49. Risser WL et al: Bone density in eumenorrheic female college athletes, Med Sci Sports Exerc 22:570, 1990.
50. Ross JA: Hypertrophy of the little finger, Brit Med J 2:987, 1950.
51. Ross PD: Clinical consequences of vertebral fractures, Am J Med 103:30S, 1997.
52. Salter RB: *Textbook of disorders and injuries of the musculoskeletal system* (2nd ed.), Baltimore, 1983, Williams & Wilkins.
53. Skolnick AA: Medical news and perspectives: 'Female athlete triad' risk for women, JAMA 270:921, 1993.
54. Snow-Harter C and Marcus R: Exercise, bone mineral density, and osteoporosis, Exerc Sport Sci Rev 19:351, 1991.
55. Sowers M: Epidemiology of calcium and vitamin D in bone loss, J Nutr 123 (2 Suppl):413, 1993.
56. Thomas WC Jr: Exercise, age, and bones, South Med J 87:S23, 1994.
57. Vico L and Alexandre C: Microgravity and bone adaptation at the tissue level, J Bone Miner Res 7(Suppl 2):S445, 1992.
58. Warren MP et al: Lack of bone accretion and amenorrhea: evidence for a relative osteopenia in weight-bearing bones, J Clin Endocrinol Metab 72:847, 1991.
59. Watson RC: Bone growth and physical activity. In Mazess RB, ed: International conference on bone measurements, DHEW Pub No NIH 75-683, Washington, DC, 1973.
60. Whalen RT: Preface to NASA symposium on the influence of gravity and activity on muscle and bone, J Biomech 24 (Suppl.1):v, 1991.
61. Whalen RT, Carter DR, and Steele CR: The relationship between physical activity and bone density, Trans Orthop Res Soc 33rd Mtg 12:464, 1987.
62. Wickman LA and Luna B: Locomotion while load-carrying in reduced gravities, Aviat Space Environ Med 67:940, 1996.
63. Wolff JD: *Das geretz der transformation der knochen*, Berlin, 1892, Hirschwald.

ANNOTATED READINGS

Martin RB: Determinants of the mechanical properties of bones, J Biomech 24 (Suppl 1):79, 1991.
    Provides a thorough technical description of current knowledge on bone material properties, structural organization, determinants of structural strength, and techniques for assessing bone strength and stiffness.
Nordin M and Frankel VH: Biomechanics of bone. In Nordin M and Frankel VH: *Basic biomechanics of the musculoskeletal system* (2nd ed.), Philadel-phia, 1989, Lea & Febiger.
    Concisely describes the mechanical properties of bones, the loading conditions to which bone is subjected, and some of the factors affecting bone strength and stiffness.
Smith AD: The female athlete triad: causes, diagnosis, and treatment, Phys Sportsmed 24:7, 1996.
    Presents current information related to causes, diagnosis, and treatment of the female athlete triad in a readable, easy-to-understand format.

Snow-Harter C and Marcus R: Exercise, bone mineral density, and osteoporosis, Exerc Sport Sci Rev 19:351, 1991.
Reviews current knowledge related to bone strength, hypothesized mechanisms of bone loss, the influence of physical activity on bone mass accretion, and the role of exercise in preventing and treating osteoporosis.

RELATED WEB SITES

Alfred I. DuPont Institute, Department of Pediatric Orthopaedics
http://gait.aidi.udel.edu/res695/homepage/pd_ortho/orthhome.htm
Contains numerous case presentations with images, divided into clinical, gait lab, and pediatric sections.

American Academy of Orthopaedic Surgeons
http://www.aaos.org
Includes information for the public, as well as for orthopods, with links to current news, a library, and research.

The Bone Home
http://www.bonehome.com
Orthopedic site that includes descriptions of cases, courses, protocols, news, a bookstore, an on-line chat, an image gallery, and excellent links to other bone-related sites.

Click the Bones and They Will Speak
http://www.cs.brown.edu/people/oa/Bin/skeleton.html
Provides audio identification of individual bones of the skeleton when they are mouse-clicked on.

Journal Archives
http://www.chiro.org/journals/home.html
Provides access to a number of medical, chiropractic, and other scientific journals, with the ability to download some full-length papers free of charge.

Orthopaedic Research Laboratories
http://orl-inc.com
Provides links to orthopaedic biomechanics courses and research.

Wheeless' Textbook of Orthopaedics
http://www.medmedia.com/Welcome.html
A comprehensive review of orthopaedics, with a search engine to locate topics related to fracture, joints, muscles, nerves, etc.

World Ortho
http://www.worldortho.com
Includes links to on-line textbooks, case studies, core orthopaedic topics, lecture notes, and quizzes.

THE BIOMECHANICS OF HUMAN SKELETAL ARTICULATIONS

After completing this chapter, the reader will be able to:

Categorize joints based on structure and movement capabilities.

Explain the functions of articular cartilage and fibrocartilage.

Describe the material properties of articular connective tissues.

Identify factors contributing to joint stability and flexibility.

Explain advantages and disadvantages of different approaches to increasing or maintaining joint flexibility.

The joints of the human body largely govern the directional motion capabilities of body segments. The anatomical structure of a given joint, such as the uninjured knee, varies little from person to person, as do the directions in which the attached body segments, such as the thigh and lower leg, are permitted to move at the joint. However, differences in the relative tightness or laxity of the surrounding soft tissues result in differences in joint ranges of movement. This chapter discusses the biomechanical aspects of joint function, including the concepts of joint stability and joint flexibility, and related implications for injury potential.

JOINT ARCHITECTURE

Classification of Joints

Anatomists have categorized joints in several ways: based on joint complexity, the number of axes present, joint geometry, or movement capabilities (24). Since this book focuses on human movement, a joint classification system based on motion capabilities is presented.

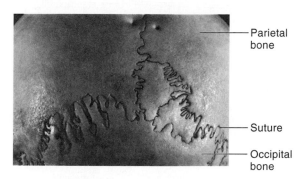

Parietal
bone

Suture

Occipital
bone

*Sutures between the
occipital and parietal
bones of the skull represent
synarthroses (immovable
joints).* From Shier, Butler,
and Lewis. *Hole's Human
Anatomy and Physiology,* © 1996.
Reprinted by permission of The
McGraw-Hill Companies, Inc.

1. *Synarthroses* (immovable) (*syn* = together; *arthron* = joint): These fibrous joints can attenuate force (absorb shock) but permit little or no movement of the articulating bones.
   a. *Sutures:* In these joints, the irregularly grooved articulating bone sheets mate closely and are tightly connected by fibers that are continuous with the periosteum. The fibers begin to ossify in early adulthood and are eventually replaced completely by bone. The only example in the human body is the sutures of the skull.
   b. *Syndesmoses* (*syndesmosis* = held by bands): In these joints, dense fibrous tissue binds the bones together, permitting extremely limited movement. Examples include the coracoacromial, mid-radioulnar, mid-tibiofibular, and inferior tibiofibular joints.
2. *Amphiarthroses* (slightly movable) (*amphi* = on both sides): These cartilaginous joints attenuate applied forces and permit more motion of the adjacent bones than synarthrodial joints.
   a. *Synchondroses* (*synchondrosis* = held by cartilage): In these joints the articulating bones are held together by a thin layer of hyaline cartilage. Examples include the sternocostal joints and the epiphyseal plates (before ossification).
   b. *Symphyses:* In these joints, thin plates of hyaline cartilage separate a disc of fibrocartilage from the bones. Examples include the vertebral joints and the pubic symphysis.
3. *Diarthroses* or *synovial* (freely movable) (*diarthrosis* = "through joint," indicating only slight limitations to movement capability): At these joints the articulating bone surfaces are covered with **articular cartilage,** an **articular capsule** surrounds the joint, and

articular cartilage
protective layer of dense white connective tissue covering the articulating bone surfaces at diarthrodial joints

articular capsule
double-layered membrane that surrounds every synovial joint

FIGURE 5-1

The knee is an example of a synovial joint, with a ligamentous capsule, an articular cavity, and articular cartilage. From Shier, Butler, and Lewis. *Hole's Human Anatomy and Physiology,* © 1996. Reprinted by permission of The McGraw-Hill Companies, Inc.

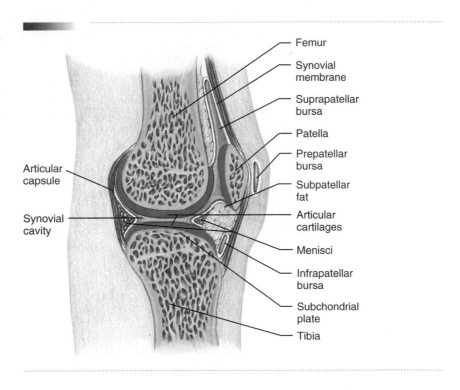

Femur
Synovial membrane
Suprapatellar bursa
Patella
Prepatellar bursa
Subpatellar fat
Articular cartilages
Menisci
Infrapatellar bursa
Subchondral plate
Tibia

Articular capsule
Synovial cavity

synovial fluid
clear, slightly yellow liquid that provides lubrication inside the articular capsule at synovial joints

a synovial membrane lining the interior of the joint capsule secretes a lubricant known as **synovial fluid** (Figure 5-1).

Two other synovial structures often associated with diarthrodial joints are *bursae* and *tendon sheaths.* Bursae are small capsules lined with synovial membranes and filled with synovial fluid that cushion the structures they separate. Most bursae separate tendons from bone, reducing the friction on the tendons during joint motion. Some bursae, such as the olecranon bursa of the elbow, separate bone from skin. Tendon sheaths are double-layered synovial structures that surround tendons positioned in close association with bones. Many of the long muscle tendons crossing the wrist and finger joints are protected by tendon sheaths.

Synovial joints vary widely in structure and movement capabilities, as shown in Figure 5-2. They are commonly categorized according to the number of axes of rotation present. Joints that allow motion around one, two, and three axes of rotation are referred to respectively as uniaxial, biaxial, and triaxial joints. A few joints where only limited motion is permitted in any direction are termed nonaxial joints. Joint motion capabilities are also sometimes described in terms of degrees of freedom (df) or the number of planes in which the joint allows motion. A uniaxial joint has one df, a biaxial joint has two df, and a triaxial joint has three df.

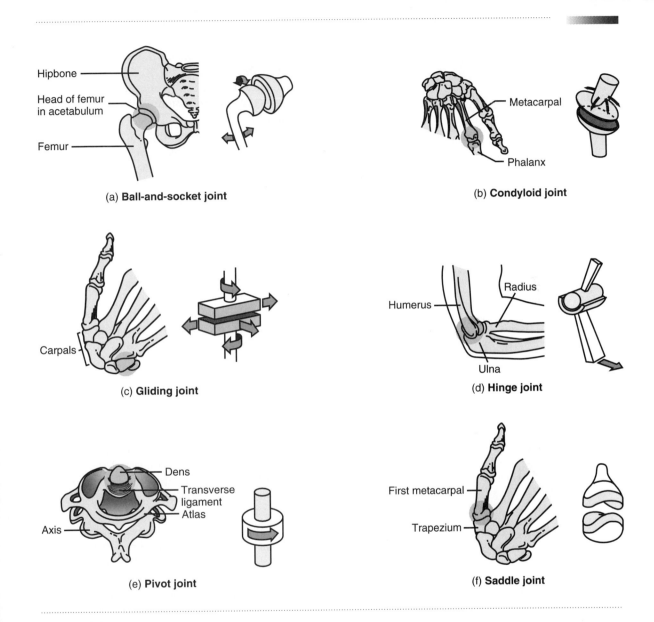

FIGURE 5-2

Examples of the synovial joints of the human body. From Shier, Butler, and Lewis. *Hole's Human Anatomy and Physiology*, © 1996. Reprinted by permission of The McGraw-Hill Companies, Inc.

a. *Gliding* (plane; arthrodial): In these joints the articulating bone surfaces are nearly flat, and the only movement permitted is non-axial gliding. Examples include the intermetatarsal, intercarpal, and intertarsal joints, and the facet joints of the vertebrae.

b. *Hinge* (ginglymus): One articulating bone surface is convex and the other is concave in these joints. Strong collateral ligaments restrict movement to a planar, hingelike motion. Examples include the ulnohumeral and interphalangeal joints.

c. *Pivot* (screw; trochoid): In these joints, rotation is permitted around one axis. Examples include the atlantoaxial joint and the proximal and distal radioulnar joints.

d. *Condyloid* (ovoid; ellipsoidal): One articulating bone surface is an ovular convex shape, and the other is a reciprocally shaped concave surface in these joints. Flexion, extension, abduction, adduction, and circumduction are permitted. Examples include the second through fifth metacarpophalangeal joints and the radiocarpal joints.

e. *Saddle* (sellar): The articulating bone surfaces are both shaped like the seat of a riding saddle in these joints. Movement capability is the same as that of the condyloid joint but with greater range of movement allowed. An example is the carpometacarpal joint of the thumb.

f. *Ball and socket* (spheroidal): In these joints the surfaces of the articulating bones are reciprocally convex and concave. Rotation in all three planes of movement is permitted. Examples include the hip and shoulder joints.

Articular Cartilage

The joints of a mechanical device must be properly lubricated if the movable parts of the machine are to move freely and not wear against each other. In the human body, a special type of dense, white connective tissue known as articular cartilage provides a protective lubrication. A 1- to 5-mm thick protective layer of this material coats the ends of bones articulating at diarthrodial joints (16). Articular cartilage serves two important purposes. First, it spreads loads at the joint over a wide area so that the amount of stress at any contact point between the bones is reduced. Second, it allows movement of the articulating bones at the joint with minimal friction and wear (16).

Cartilage can reduce the maximum contact stress acting at a joint by 50% or more (30). The lubrication supplied by the articular cartilage is so effective that the friction present at a joint is only approximately 17% to 33% of the friction of a skate on ice under the same load (2).

articular fibrocartilage
soft-tissue discs or menisci that intervene between articulating bones

Articular Fibrocartilage

At some joints, **articular fibrocartilage,** in the form of either a fibrocartilaginous disc or partial discs known as menisci, is also present be-

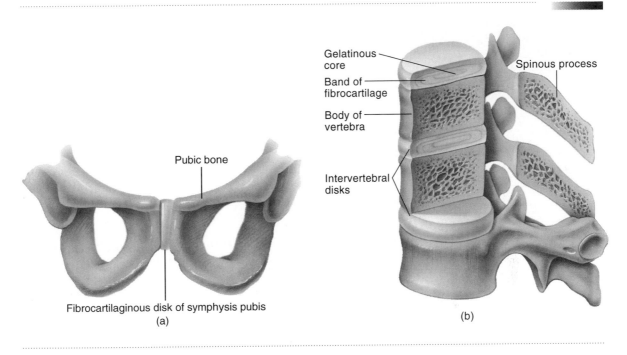

Pubic bone

Gelatinous core

Spinous process

Band of fibrocartilage

Body of vertebra

Intervertebral disks

Fibrocartilaginous disk of symphysis pubis
(a)

(b)

FIGURE 5-3

Fibrocartilage is present in (a) the symphysis pubis that separates the pubic bones and (b) the intervertebral discs between adjacent vertebrae. From Shier, Butler, and Lewis. *Hole's Human Anatomy and Physiology*, © 1996. Reprinted by permission of The McGraw-Hill Companies, Inc.

tween the articulating bones. The intervertebral discs (Figure 5-3) and the menisci of the knee (Figure 5-4) are examples. Although the function of discs and menisci is not clear, possible roles include the following (29):

1. Distribution of loads over the joint surfaces
2. Improvement of the fit of the articulating surfaces
3. Limitation of translation or slip of one bone with respect to another
4. Protection of the periphery of the articulation
5. Lubrication
6. Shock absorption

Intervertebral discs act as cushions between the vertebrae, reducing stress levels by spreading loads.

Articular Connective Tissue

Tendons, which connect muscles to bones, and ligaments, which connect bones to other bones, are passive tissues comprised primarily of collagen and elastic fibers. Tendons and ligaments do not have the ability to contract like muscle tissue, but they are slightly extensible. These

A material stretched beyond its elastic limit remains lengthened beyond its original length after tension is released.

FIGURE 5-4

The menisci at the knee joint help to distribute loads, lessening the stress transmitted across the joint.

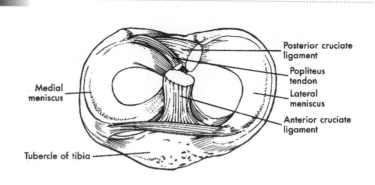

tissues are elastic and will return to their original length after being stretched, unless they are stretched beyond their elastic limits (see Chapter 4). A tendon or ligament stretched beyond its elastic limit during an injury remains stretched and can be restored to its original length only through surgery. The results of modeling studies suggest that tendons routinely undergo healing to repair internal microfailures over the course of the lifespan in order to remain intact (21).

Tendons and ligaments, like bone, respond to altered habitual mechanical stress by hypertrophying or atrophying. Research has shown that regular exercise over a period of time results in increased size and strength of both tendons (28) and ligaments (11), as well as in increased strength of the junctions between tendons or ligaments and bone (27).

JOINT STABILITY

The stability of an articulation is its ability to resist dislocation. Specifically, it is the ability to resist the displacement of one bone end with respect to another, while preventing injury to the ligaments, muscles, and muscle tendons surrounding the joint. Different factors influence **joint stability.**

joint stability
ability of a joint to resist abnormal displacement of the articulating bones

Shape of the Articulating Bone Surfaces

In many mechanical joints, the articulating parts are exact opposites in shape so that they fit tightly together (Figure 5-5). In the human body, the articulating ends of bones are usually shaped as mating convex and concave surfaces.

The articulating bone surfaces at all joints are of approximately matching (reciprocal) shapes.

Although most joints have reciprocally shaped articulating surfaces, these surfaces are not symmetrical, and there is typically one position of best fit in which the area of contact is maximum. This is known as the **close-packed position,** and it is in this position that joint stability

close-packed position
joint orientation for which the contact between the articulating bone surfaces is maximum

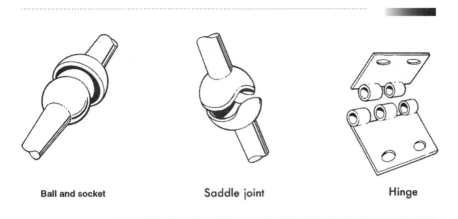

FIGURE 5-5
Mechanical joints are
often composed of
reciprocally shaped parts.

Ball and socket   Saddle joint   Hinge

is usually greatest (24). Any movement of the bones at the joint away from the close-packed position results in a **loose-packed position,** with reduction of the area of contact.

Some articulating surfaces are shaped so that in both close- and loose-packed positions, there is either a large or a small amount of contact area and consequently more or less stability. For example, the acetabulum provides a relatively deep socket for the head of the femur, and there is always a relatively large amount of contact area between the two bones, which is one reason the hip is a stable joint. At the shoulder, however, the small glenoid fossa has a vertical diameter that is approximately 75% of the vertical diameter of the humeral head and a horizontal diameter that is 60% of the size of the humeral head (16). Therefore, the area of contact between these two bones is relatively small, contributing to the relative instability of the shoulder complex. Slight anatomical variations in shapes and sizes of the articulating bone surfaces at any given joint among individuals are found; therefore, some people have joints that are more or less stable than average.

loose-packed position
any joint orientation other than the close-packed position

The close-packed position occurs for the knee, wrist, and interphalangeal joints at full extension and for the ankle at full dorsiflexion (24).

One factor enhancing the stability of the glenohumeral joint is a posteriorly tilted glenoid fossa and humeral head. Individuals with anteriorly tilted glenoids and humeral heads are predisposed to shoulder dislocation.

Arrangement of Ligaments and Muscles

Ligaments, muscles, and muscle tendons affect the relative stability of joints. At joints such as the knee and the shoulder, in which the bone configuration is not particularly stable, the tension in ligaments and muscles contributes significantly to joint stability by helping to hold the articulating bone ends together. If these tissues are weak from disuse or lax from being overstretched, the stability of the joint is reduced. Strong ligaments and muscles often increase joint stability. For example, strengthening of the quadriceps and hamstring groups enhances the stability of the knee (19). The complex array of ligaments and tendons crossing the knee is illustrated in Figure 5-6.

Stretching or rupturing of the ligaments at a joint can result in abnormal motion of the articulating bone ends, with subsequent damage to the articular cartilage.

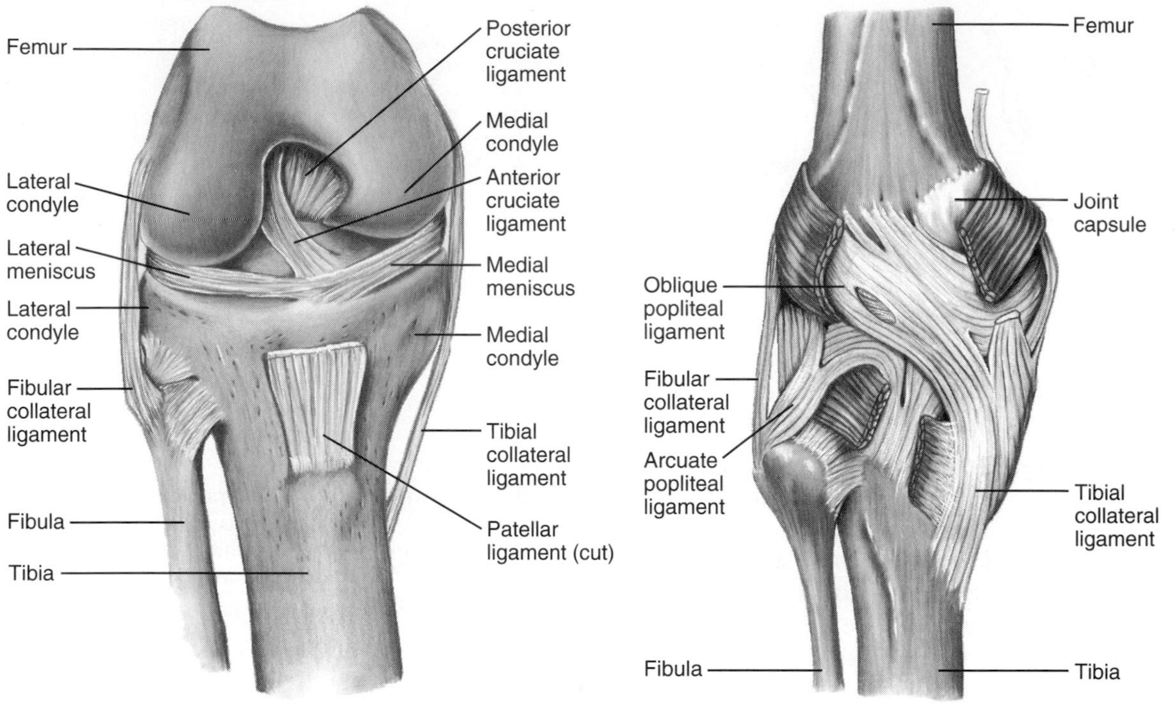

FIGURE 5-6

At the knee joint, stability is primarily derived from the tension in the ligaments and muscles that cross the joint. From Shier, Butler, and Lewis. *Hole's Human Anatomy and Physiology*, © 1996. Reprinted by permission of The McGraw-Hill Companies, Inc.

The angle of attachment of most tendons to bones is arranged so that when the muscle exerts tension, the articulating ends of the bones at the joint crossed are pulled closer together, enhancing joint stability. This situation is usually found when the muscles on opposite sides of a joint produce tension simultaneously. When muscles are fatigued, however, they are less able to contribute to joint stability. Rupture of the cruciate ligaments is most likely to occur when the tension in fatigued muscles surrounding the knee is inadequate to protect the cruciate ligaments from being stretched beyond their elastic limits (19). Gymnastics injuries also occur with greater frequency when the athletes are fatigued (18).

Engaging in athletic participation with fatigued muscles increases the likelihood of injury.

Other Connective Tissues

White fibrous connective tissue known as *fascia* surrounds muscles and the bundles of muscle fibers within muscles, providing protection and

FIGURE 5-7

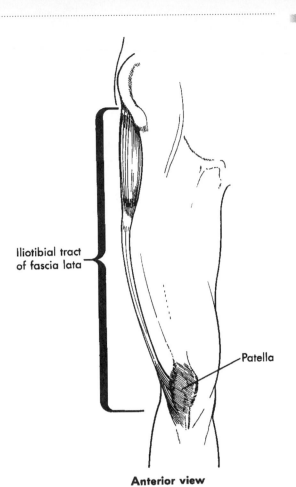

The iliotibial band is a
strong, thickened region
of the fascia lata that
crosses the knee,
contributing to the knee's
stability.

Iliotibial tract
of fascia lata

Patella

Anterior view

support. A particularly strong, prominent tract of fascia known as the
iliotibial band crosses the lateral aspect of the knee, contributing to its
stability (Figure 5-7). The fascia and the skin on the exterior of the body
are other tissues that contribute to joint integrity.

JOINT FLEXIBILITY

Joint flexibility is a term used to describe the **range of motion** (ROM)
allowed in each of the planes of motion at a joint. *Static flexibility* refers
to the ROM present when a body segment is passively moved (by an exercise partner or clinician), whereas *dynamic flexibility* refers to the ROM
that can be achieved by actively moving a body segment by virtue of muscle contraction (22). Static flexibility is considered to be the better
indicator of the relative tightness or laxity of a joint in terms of implications for injury potential. Dynamic flexibility, however, must be sufficient
not to restrict the ROM needed for daily living, work, or sport activities.

joint flexibility
a term representing the
relative ranges of motion
allowed at a joint

range of motion
angle through which a joint
moves from anatomical
position to the extreme
limit of segment motion in a
particular direction

The range of motion for flexion at the hip is typically measured with the individual supine. (This is also a good position for stretching the hamstrings using PNF techniques.)

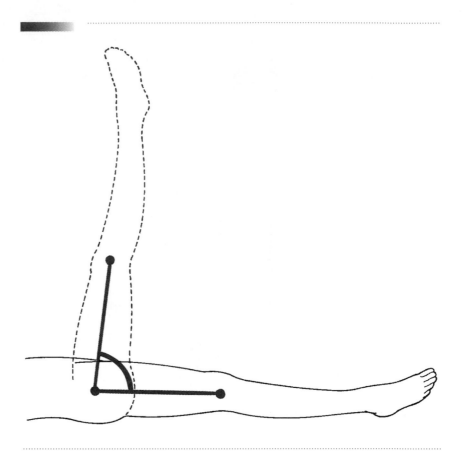

Although people's general flexibility is often compared, flexibility is actually joint-specific. That is, an extreme amount of flexibility at one joint does not guarantee the same degree of flexibility at all joints.

Measuring Joint Range of Motion

Joint ROM is measured directionally in units of degrees. In anatomical position, all joints are considered to be at zero degrees. The ROM for flexion at the hip is therefore considered to be the size of the angle through which the extended leg moves from zero degrees to the point of maximum flexion (Figure 5-8). The ROM for extension (return to anatomical position) is the same as that for flexion, with movement past anatomical position in the other direction quantified as the ROM for hyperextension. Devices used for measuring joint range of motion are shown in Figures 5-9 and 5-10.

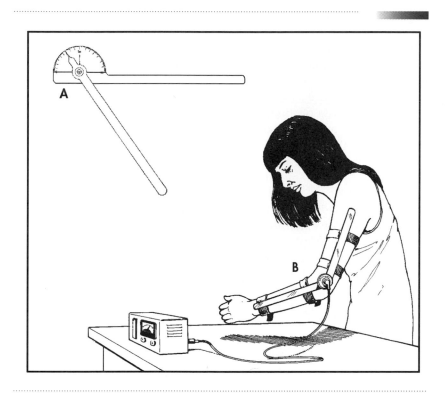

FIGURE 5-9
A, A goniometer is basically a protractor with two arms. The point where the arms intersect is aligned over the joint center while the arms are aligned with the longitudinal axes of the body segments to measure the angle present at a joint. **B,** An electrogoniometer is a goniometer with a potentiometer located at its axis of rotation so that as the angle between the goniometer arms changes, the amount of electrical current output to a recorder changes. Electrogoniometers are used for monitoring changes in joint angle during the performance of different activities.

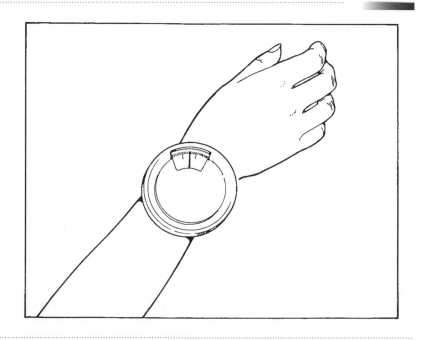

FIGURE 5-10
A device for measuring changes in the orientation of a body segment is the Leighton flexometer, which consists of a weighted needle attached to a 360-degree scale.

Factors Influencing Joint Flexibility

Different factors influence joint flexibility. The shapes of the articulating bone surfaces and intervening muscle or fatty tissue may terminate movement at the extreme of a ROM. When the elbow is in extreme hyperextension, for example, contact of the olecranon of the ulna with the olecranon fossa of the humerus restricts further motion in that direction. Muscle and/or fat on the anterior aspect of the arm may terminate elbow flexion. Regular participants in bilaterally asymmetrical sports such as tennis are likely to have less range of motion for the dominant arm than for the nondominant arm at the glenohumeral joint of the shoulder (3).

For most individuals, joint flexibility is primarily a function of the relative laxity and/or extensibility of the collagenous tissues and muscles crossing the joint. Tight ligaments and muscles with limited extensibility are the most common inhibitors of a joint's ROM. When these tissues are not stretched, their extensibility usually diminishes. The fluid content of the cartilaginous discs present at some joints also influences the mobility at these joints.

Laboratory studies have shown that the extensibility of collagenous tissues increases slightly with temperature elevation (20). Although this finding suggests that "warm-up" exercises should increase joint ROM, this has not been well documented in humans. Fifteen minutes of stationary cycling has been shown to decrease resting tension in the hamstrings, with an accompanying increase in hip range of motion (31). More research, however, is needed to identify the specific mechanism responsible for the effects of warm-up on joint ROM.

Flexibility and Injury

Research shows that the risk of injury is heightened when joint flexibility is extremely low, extremely high, or significantly imbalanced between dominant and nondominant sides of the body (8). Severely limited joint flexibility is undesirable because if the collagenous tissues and muscles crossing the joint are tight, the likelihood of their tearing or rupturing if the joint is forced beyond its normal ROM increases. Tight ligaments and muscles were found to be related to lower extremity injury incidence among a group of male, but not female, college athletes, possibly because the female athletes studied were more flexible and less tight at the lower extremity joints (9). In a study of competitive female gymnasts, those in a highly injury-prone category had less flexibility of the shoulder, elbow, wrist, hip, and knee joints than those in a low injury incidence category (25). Alternatively, an extremely loose, lax joint is lacking in stability and, therefore, prone to displacement-related injuries. Among U.S. Army infantry recruits assessed for hip/low back flex-

A joint with an unusually large range of motion is termed hypermobile.

Gymnastics is a sport requiring a large amount of flexibility at the major joints of the body.

ibility with the sit-and-reach test, both the least flexible and the most flexible were over two times as likely to get injured as soldiers in the middle of the flexibility range. Female college athletes with a hip extension flexibility *imbalance* of 15% or more were also 2.6 times more likely to suffer lower extremity injuries (7).

The desirable amount of joint flexibility is largely dependent on the activities in which an individual wishes to engage. Gymnasts and dancers obviously require greater joint flexibility than nonathletes. However, athletes such as gymnasts and dancers also require strong muscles, tendons, and ligaments to perform well and avoid injury. Although large, bulky muscles may inhibit joint ROM, an extremely strong, stable joint can also enable large ROMs.

Although people usually become less flexible as they age, this phenomenon appears to be primarily related to decreased levels of physical activity rather than to changes inherent in the aging process. Regardless of the age of the individual, if the collagenous tissues and muscles crossing a joint are not stretched, they will shorten. Conversely, when these tissues are regularly stretched, they lengthen and flexibility

FIGURE 5-11

Schematic representation of a Golgi tendon organ. From Shier, Butler, and Lewis. *Hole's Human Anatomy and Physiology*, © 1996. Reprinted by permission of The McGraw-Hill Companies, Inc.

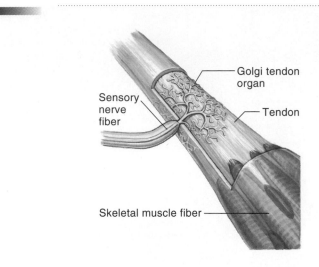

Golgi tendon organ

Sensory nerve fiber

Tendon

Skeletal muscle fiber

is increased. The results of several studies indicate that flexibility can be significantly increased among elderly individuals who participate in a program of regular stretching and exercise (5, 10, 14).

TECHNIQUES FOR INCREASING JOINT FLEXIBILITY

Increasing joint flexibility is often an important component of therapeutic and rehabilitative programs and programs designed to train athletes for a particular sport. Increasing or maintaining flexibility involves stretching the ligaments and muscles that limit the ROM at a joint. Several approaches for stretching these tissues can be used, with some being more effective than others because of differential neuromuscular responses elicited.

Neuromuscular Response to Stretch

Golgi tendon organ
sensory receptor that inhibits tension development in a muscle and initiates tension development in antagonist muscles

Sensory receptors known as **Golgi tendon organs** (GTOs) are located in the muscle-tendon junctions and in the tendons at both ends of muscles (Figure 5-11). Approximately 10–15 muscle fibers are connected in direct line, or in series, with each GTO. These receptors are stimulated by tension in the muscle-tendon unit, both tension produced by muscle contraction and tension produced by passive muscle stretch. The GTOs respond through their neural connections by inhibiting tension development in the muscle (promoting muscle relaxation), and by initiating tension development in the antagonist muscles.

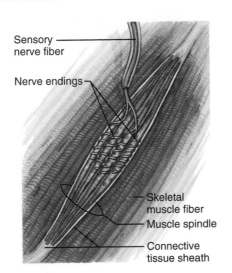

Sensory nerve fiber

Nerve endings

Skeletal muscle fiber

Muscle spindle

Connective tissue sheath

FIGURE 5-12

Schematic representation of a muscle spindle. From Shier, Butler, and Lewis. *Hole's Human Anatomy and Physiology,* © 1996. Reprinted by permission of The McGraw-Hill Companies, Inc.

Other sensory receptors are interspersed throughout the fibers of muscles. These receptors, which are oriented parallel to the fibers, are known as **muscle spindles** because of their shape (Figure 5-12). Each muscle spindle is composed of approximately 3–10 small muscle fibers, termed intrafusal fibers, that are encased in a sheath of connective tissue.

Functionally, there are two types of muscle spindles: primary and secondary. The primary spindles respond to both the amount of muscle lengthening and the rate of muscle lengthening (dynamic response). The secondary spindles respond only to the amount of lengthening (static response). Because the dynamic response is much stronger than the static response, a slow rate of stretching does not activate the muscle spindle response.

The spindle response includes activation of the **stretch reflex** and inhibition of tension development in the antagonist muscle group, a process known as **reciprocal inhibition.** The stretch reflex, also known as the *myotatic reflex,* is provoked by the activation of the spindles in a stretched muscle. This rapid response involves neural transmission across a single synapse, with afferent nerves carrying stimuli from the spindles to the spinal cord and efferent nerves, returning an excitatory signal directly from the spinal cord to the muscle, resulting in tension development in the muscle. The knee-jerk test, a common neurological test of motor function, is an example of muscle spindle function producing a quick, brief contraction in a stretched muscle. A tap on the patellar tendon initiates

muscle spindle
sensory receptor that provokes reflex contaction in a stretched muscle and inhibits tension development in antagonist muscles

stretch reflex
monosynaptic reflex initiated by stretching of muscle spindles and resulting in immediate development of muscle tension

reciprocal inhibition
inhibition of tension development in the antagonist muscles resulting from activation of muscle spindles

FIGURE 5-13

The myotatic (stretch reflex) is initiated by stretching of the muscle spindles. From Shier, Butler, and Lewis. *Hole's Human Anatomy and Physiology*, © 1996. Reprinted by permission of The McGraw-Hill Companies, Inc.

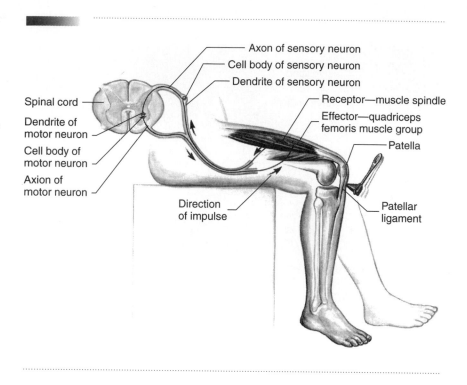

the stretch reflex, resulting in the jerk caused by the immediate development of tension in the quadriceps group (Figure 5-13).

Because muscle spindle activation produces tension development in stretching muscle, whereas GTO activation promotes relaxation of stretching muscle, the general goals of any procedure for stretching muscle are minimizing the spindle effect and maximizing the GTO effect. A summary comparison of GTOs and muscle spindles is presented in Figure 5-14.

Active and Passive Stretching

Stretching can be done either actively or passively. **Active stretching** is produced by contraction of the antagonist muscles (those on the side of the joint opposite the muscles, tendons, and ligaments to be stretched). Thus, to actively stretch the hamstrings (the primary knee flexors), the quadriceps (primary knee extensors) should be contracted. **Passive stretching** involves the use of gravitational force, force applied by another body segment, or force applied by another person, to move a body segment to the end of the ROM. Active stretching provides the

active stretching
stretching of muscles, tendons, and ligaments produced by active development of tension in the antagonist muscles

passive stretching
stretching of muscles, tendons, and ligaments produced by a stretching force other than tension in the antagonist muscles

FIGURE 5-14

GOLGI TENDON ORGANS (GTOS) AND MUSCLE SPINDLES: HOW DO THEY COMPARE?		
	GOLGI TENDON ORGANS	MUSCLE SPINDLES
Location	Within tendons near the muscle-tendon junction in series with muscle fibers	Interspersed among muscle fibers in parallel with the fibers
Stimulus	Increase in muscle tension	Increase in muscle length
Response	1) Inhibits tension development in stretched muscle 2) Initiates tension development in antagonist muscles	1) Initiates rapid contraction of stretched muscle 2) Inhibits tension development in antagonist muscles
Overall effect	Promotes stretch in muscle being stretched	Inhibits stretch in muscle being stretched

advantage of exercising the muscle groups used to develop force. With passive stretching, movement can be carried farther beyond the existing ROM than with active stretching, but with the concomitant disadvantage of increased injury potential.

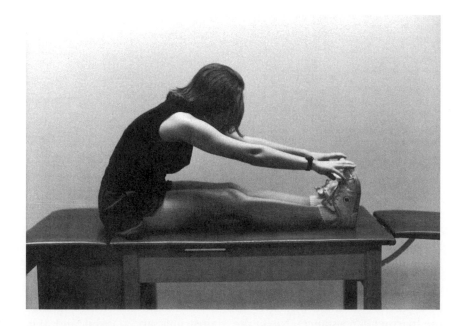

Active, static stretching involves holding a position at the extreme of the range of motion.

Ballistic and Static Stretching

ballistic stretching
a series of quick, bouncing-type stretches

Ballistic, bouncing types of stretches can be dangerous because they tend to promote contraction of the muscles being stretched and the momentum generated may carry the body segments far enough beyond the normal ROM to tear or rupture collagenous tissues.

static stretching
maintaining a slow, controlled, sustained stretch over time, usually about 30 seconds

Ballistic stretching, or performance of bouncing stretches, makes use of the momentum of body segments to repeatedly extend joint position to or beyond the extremes of the ROM. Because a ballistic stretch activates the stretch reflex and results in the immediate development of tension in the muscle being stretched, microtearing of the stretched muscle tissue may occur. Because the extent of the stretch is not controlled, the potential for injury to all of the stretched tissues is heightened.

With **static stretching** the movement is slow, and when the desired joint position is reached, it is maintained statically, usually from about 10 to 30 seconds. This period of time is sufficient to stimulate the GTOs, which override the muscle spindle response and promote relaxation of the muscles being stretched (22). There is some controversy over the optimal length of time to hold a static stretch. Whereas one group of researchers found that static hamstring stretches of 30 seconds are equally as effective as stretches of one minute (1), the results of another investigation showed that resistance to stretch in the hamstrings decreases progressively during the first 45 seconds of a 90-second static stretch (12). There seems to be general agreement that for optimal effect, the static stretch of each muscle group should be sequentially repeated three to five times (12, 16).

Although ballistic and static stretching are equally as effective in increasing joint ROM, static stretching is preferred because it tends to promote the GTO response that facilitates stretching, whereas ballistic stretching can activate the muscle spindle response that inhibits stretching. Both forms of stretching can induce soreness in muscles that are not habitually stretched, however (23).

Proprioceptive Neuromuscular Facilitation

proprioceptive neuromuscular facilitation
a group of stretching procedures involving alternating contraction and relaxation of the muscles being stretched

The most effective stretching procedures are known collectively as **proprioceptive neuromuscular facilitation** (PNF). PNF techniques were originally used by physical therapists for treating patients with neuromuscular paralysis (22). All PNF procedures involve some pattern of alternating contraction and relaxation of agonist and antagonist muscles designed to take advantage of the GTO response. All PNF techniques require a partner or clinician. Stretching the hamstrings from a supine position provides a good illustration for several of the popular PNF approaches (see Figure 5-12).

The contract-relax-antagonist-contract technique (also referred to as slow-reversal-hold-relax), involves passive static stretch of the hamstrings by a partner, followed by active contraction of the hamstrings against the partner's resistance. Next, the hamstrings are relaxed and the

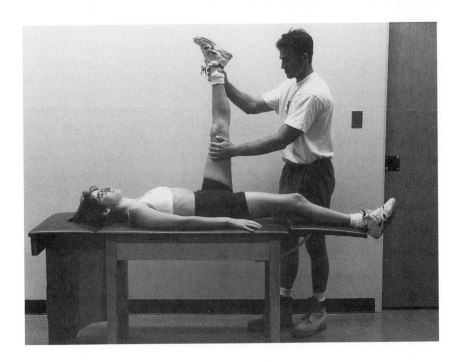

Passive stretching can be accomplished with the assistance of a partner.

quadriceps are contracted as the partner pushes the leg into increasing flexion at the hip. There is then a phase of complete relaxation, with the leg held in the new position of increased hip flexion. Each phase of this process is typically maintained for a duration of 5–10 seconds, and the entire sequence is carried out at least four times.

The contract-relax and hold-relax procedures begin as in the slow-reversal-hold method, with a partner applying passive stretch to the hamstrings, followed by active contraction of the hamstrings against the partner's resistance. With the contract-relax approach, the contraction of the hamstrings is isotonic, resulting in slow movement of the leg in the direction of hip extension. In the hold-relax method, the contraction of the hamstrings is isometric against the partner's unmoving resistance. Following contraction, both methods involve relaxation of the hamstrings and quadriceps while the hamstrings are passively stretched. Again, the duration of each phase is usually 5–10 seconds and the entire sequence is repeated several times.

The agonist-contract-relax method is another PNF variation, with 5- to 20-second sequential phases. This procedure begins with active, maximal contraction of the quadriceps to extend the knee, followed by relaxation as the partner manually supports the leg in the position actively attained.

PNF techniques can significantly increase joint range of motion over a single stretching session and studies show PNF to be superior to static and ballistic stretching techniques over an extended training period (22). However, the results of one investigation show that when stretching the hamstrings, maintaining anterior pelvic tilt as opposed to posterior pelvic tilt is more important than the stretching technique used (26).

Among PNF procedures, research indicates that contract-relax-agonist-contract is superior to contract-relax (4). However, one study also reveals that contract-relax and agonist-contract-relax procedures with contraction periods of 5 seconds duration do not elicit GTO responses sufficient to overcome the myotatic reflex (17). This, coupled with the finding that 10-second contraction periods during PNF produce no larger ROM changes than 5-second contraction periods (15), suggests that more research on the appropriate duration for muscle contraction and relaxation periods during PNF is needed.

COMMON JOINT INJURIES AND PATHOLOGIES

The joints of the human body support weight, are loaded by muscle forces, and at the same time provide range of movement for the body segments. They are consequently subject to both acute and overuse injuries, as well as to infection and degenerative conditions.

Sprains

Sprains are injuries caused by abnormal displacement or twisting of the articulating bones that results in stretching or tearing of ligaments, tendons, and connective tissues crossing a joint. Lateral ankle sprains are particularly common, because the ankle is a major weightbearing joint and because there is less ligamentous support on the lateral than on the medial side of the ankle. Pain and swelling are the symptoms of joint sprains.

Dislocations

Displacement of the articulating bones at a joint is termed a dislocation. These injuries usually result from falls or other mishaps involving a large magnitude of force. Common sites for dislocations include the shoulders, fingers, knees, elbows, and jaw. Symptoms include visible joint deformity, pain, swelling, and usually some loss of joint movement capability.

Bursitis

Bursitis is an overuse injury caused by excessive use of a joint that produces frictional irritation and inflammation of one or more bursae. For example, runners who increase training mileage too abruptly may experience inflammation of the bursa between the Achilles tendon and the calcaneous. Pain and possibly some swelling are symptoms of bursitis.

Arthritis

Arthritis is a pathology involving joint inflammation accompanied by pain and swelling. It is extremely common with aging, with over 100 different types of arthritis identified.

Rheumatoid Arthritis

The most debilitating and painful form of arthritis is rheumatoid arthritis, an autoimmune disorder that involves the body's immune system attacking healthy tissues. It is more common in adults, but there is also a juvenile rheumatoid arthritis. Characteristics include inflammation and thickening of the synovial membranes and breakdown of the articular cartilage, resulting in limitation of motion and eventually ossification or fusing of the articulating bones. Other symptoms include anemia, fatigue, musclar atrophy, osteoporosis, and other systemic changes.

Osteoarthritis

Osteoarthritis is the most common form of noninflammatory degenerative arthritis. In the early stages of the disorder, the joint cartilage loses its smooth glistening appearance and becomes rough and irregular. Eventually the cartilage completely wears away, leaving the articulating bone surfaces bare. Pain, swelling, range of motion restriction, and stiffness are all symptoms, with the pain typically relieved by rest, and joint stiffness improved by activity. The cause of osteoarthritis is unknown, but mechanical abuse is often a factor, along with heredity and aging. Recent research suggests that degenerative joint disease may actually stem from remodeling and related vascular insufficiency in the underlying subchondral bone (6, 13).

SUMMARY

The anatomical configurations of the joints of the human body govern the directional movement capabilities of the articulating body segments. From the perspective of movements permitted, there are three major categories of joints: synarthroses (immovable joints), amphiarthroses (slightly movable joints), and diarthroses (freely movable joints). Each major category is further subdivided into classes of joints with common anatomical characteristics.

The ends of bones articulating at diarthrodial joints are covered with articular cartilage, which reduces contact stress and regulates joint lubrication. Fibrocartilaginous discs or menisci present at some joints also may contribute to these functions.

Tendons and ligaments are strong collagenous tissues that are slightly extensible and elastic. These tissues are similar to muscle and bone in that they adapt to levels of increased or decreased mechanical stress by hypertrophying or atrophying.

Joint stability is the ability of the joint to resist displacement of the articulating bones. The major factors influencing joint stability are the size and shape of the articulating bone surfaces, and the arrangement and strength of the surrounding muscles, tendons, and ligaments.

Joint flexibility is primarily a function of the relative tightness of the muscles and ligaments that span the joint. If these tissues are not stretched, they tend to shorten. Approaches for increasing flexibility include active versus passive stretching, and static versus dynamic stretching. PNF is a particularly effective procedure for stretching muscles and ligaments.

INTRODUCTORY PROBLEMS

(Reference may be made to Chapters 7 to 9 for additional information on specific joints.)

1. Construct a table that identifies joint type and the plane or planes of allowed movement for the shoulder (glenohumeral joint), elbow, wrist, hip, knee, and ankle.
2. Describe the directions and approximate ranges of movement that occur at the joints of the human body during each of the following movements:
   a. Walking
   b. Running
   c. Performing a jumping jack
   d. Rising from a seated position
3. What factors contribute to joint stability?
4. Explain why athletes' joints are often taped before they participate in an activity. What are some possible advantages and disadvantages of taping?
5. What factors contribute to flexibility?
6. What degree of joint flexibility is desirable?
7. How is flexibility related to the likelihood of injury?
8. Discuss the relationship between joint stability and joint flexibility.
9. Explain why grip strength diminishes as the wrist is hyperextended.
10. Why is ballistic stretching contraindicated?

1. Construct a table that identifies joint type and the plane or planes of movement for atlanto-occipital joint, the L5-S1 vertebral joint, the metacarpophalangeal joints, the interphalangeal joints, the carpometacarpal joint of the thumb, the radioulnar joint, and the talocrural joint.
2. Identify the position (for example, full extension, 90° of flexion) for which each of the following joints is close packed:
    a. Shoulder
    b. Elbow
    c. Knee
    d. Ankle
3. How is articular cartilage similar to and different from ordinary sponge? (You may wish to consult the annotated readings.)
4. Comparatively discuss the properties of muscle, tendon, and ligament. (You may wish to consult the annotated readings.)
5. Discuss the relative importance of joint stability and joint mobility for athletes participating in each of the following sports:
    a. Gymnastics
    b. Football
    c. Swimming
6. What specific exercises would you recommend for increasing the stability of each of the following joints:
    a. Shoulder
    b. Knee
    c. Ankle
7. What specific exercises would you recommend for increasing the flexibility of each of the following joints:
    a. Hip
    b. Shoulder
    c. Ankle
8. In which sports are athletes more likely to incur injuries that are related to insufficient joint stability? Explain your answer.
9. In which sports are athletes likely to incur injuries related to insufficient joint flexibility? Explain your answer.
10. What exercises would you recommend for senior citizens interested in maintaining an appropriate level of joint flexibility?

1. Using a skeleton or anatomical model, locate an example of each type of joint.
2. Using a skeleton or anatomical model, observe the amount of bone contact for close-packed and loose-packed positions of the elbow and knee.

3. With a partner, use a goniometer to measure the range of motion for hip flexion with the leg fully extended before and after a 30 s active static hamstring stretch. Explain your results.

4. With a partner, use a goniometer to measure the range of motion for hip flexion with the leg fully extended before and after a 30 s passive static hamstring stretch. Explain your results.

5. With a partner, use a goniometer to measure the range of motion for hip flexion, with the leg fully extended before and after stretching the hamstrings with one of the PNF techniques described in the chapter. Explain your results.

REFERENCES

1. Brandy WD and Irion JM: The effect of time on static stretch on the flexibility of the hamstring muscles, Phys Ther 74:845, 1994.

2. Brand RA: Joint lubrication. In Albright JA and Brand RA, eds: *The scientific basis of orthopedics,* New York, 1979, Appleton-Century-Crofts.

3. Ellenbecker TS, Roetert EP, Piorkowski PA, and Schulz DA: Glenohumeral joint internal and external range of motion in elite junior tennis players, J Orthop Sports Phys Ther 24:336, 1996.

4. Etnyre BR and Abraham LD: Gains in range of ankle dorsiflexion using three popular stretching techniques, Am J Phys Med 65:189, 1986.

5. Girouard CK and Hurley BF: Does strength training inhibit gains in range of motion from flexibility training in older adults?, Med Sci Sports Exerc 27:1444, 1995.

6. Imhof H, Breitenseher M, Kainberger F, and Trattnig S: Degenerative joint disease: cartilage or vascular disease?, Skeletal Radiol 26:398, 1997.

7. Knapik JJ, et al: Preseason strength and flexibility imbalances associated with athletic injuries in female college athletes, Am J Sports Med 19: 76, 1991.

8. Knapik JJ, Jones BH, Bauman CL and Harris JM: Strength, flexibility and athletic injuries, Sports Med 14:277, 1992.

9. Krivickas LS and Feinberg JH: Lower extremity injuries in college athletes: relation between ligamentous laxity and lower extremity muscle tightness, Arch Phys Med Rehabil 77:1139, 1996.

10. Lesser M: The effects of rhythmic exercise on the range of motion in older adults, Am Corrective Ther J 32:4, 1978.

11. Loitz BJ and Frank CB: Biology and mechanics of ligament and ligament healing, Exerc Sport Sci Rev 21:33, 1993.

12. Magnusson SP, Simonsen EB, Aagaard P, Gleim GW, McHugh MP, and Kjaer M: Viscoelastic response to repeated static stretching in the human hamstring muscle, Scand J Med Sci Sports 5:342, 1995.

13. Matsui H, Shimizu M, and Tsuji H: Cartilage and subchondral bone interaction in osteoarthrosis of human knee joint: a histological and histomorphometric study, Microsc Res Tech 37:333, 1997.

14. Munns K: Effects of exercise on the range of joint motion in elderly subjects. In Smith EL and Serfass RC, eds: *Exercise and aging: The scientific basis,* Hillside, NJ, 1981, Enslow Publishers.

15. Nelson KC and Cornelius WL: The relationship between isometric contraction durations and improvement in shoulder joint range of motion, J Sports Med Phys Fitness 31:385, 1991.

16. Nordin M and Frankel VH: *Basic biomechanics of the skeletal system* (2nd ed.), Philadelphia, 1989, Lea & Febiger.

17. Osternig LR, Robertson R, Troxel R, and Hansen P: Muscle activation during proprioceptive neuromuscular facilitation (PNF) stretching techniques, Am J Phys Med 66:298, 1987.

18. Pettrone FA and Ricciardelli E: Gymnastic injuries: The Virginia experience, Am J Sports Med 15:59, 1987.

19. Radin EL: Role of muscles in protecting athletes from injury, Acta Med Scand Suppl 711:143, 1986.

20. Safran MR, Garrett Jr WE, Seaber AV, Glisson PR, and Ribbeck BM: The role of warm-up in muscular injury prevention. Am J Sports Med 16:123, 1988.

21. Schechtman H and Bader DL: In vitro fatigue of human tendons, J Biomech 30:829, 1997.

22. Shellock FG and Prentice WE: Warming-up and stretching for improved physical performance and prevention of sports-related injuries, Sports Med 2:267, 1985.

23. Smith LL et al.: The effects of static and ballistic stretching on delayed onset muscle soreness and creatine kinase, Res Q Exerc Sport 64:103, 1993.

24. Soderberg GL: *Kinesiology: Application to pathological motion*, Baltimore, 1986, Williams & Wilkins.

25. Steele VA and White JA: Injury prediction in female gymnasts, Br J Sports Med 20:31, 1986.

26. Sullivan MK, Dejulia JJ, and Worrell TW: Effect of pelvic position and stretching method on hamstring muscle flexibility, Med Sci Sports Exerc 24:1383, 1992.

27. Tipton CM et al: The influence of physical activity on ligaments and tendons, Med Sci Sports Exerc 7:165, 1975.

28. Tipton CM et al: Influence of exercise on strength of medial collateral ligaments of dogs, Am J Physiol 218:894, 1970.

29. Warwick R and Williams PL: *Gray's anatomy* ed 36, Philadelphia, 1980, WB Saunders Co.

30. Weightman BO and Kempson GE: Load carriage. In Freeman MAR, ed: *Adult articular cartilage*, London, 1979, Pitman.

31. Wiemann K and Hahn K: Influences of strength, stretching and circulatory exercises on flexibility parameters of the human hamstrings, Int J Sports Med 18:340, 1997.

ANNOTATED READINGS

Carlstedt CA and Nordin M: Biomechanics of tendons and ligaments. In Frankel VH and Nordin M, eds: *Basic biomechanics of the skeletal system* (2nd ed.), Philadelphia, 1989, Lea & Febiger.
Describes the material properties of tendons and ligaments, including excellent graphs and illustrations.

Loitz BJ and Frank CB: Biology and mechanics of ligament and ligament healing, Exerc Sport Sci Rev 21:33, 1993.
Reviews the scientific literature on properties of healthy and damaged ligament.

Mow VC, Proctor CS, and Kelly MA: Biomechanics of articular cartilage. In Frankel VH and Nordin M, eds: *Basic biomechanics of the skeletal system* (2nd ed.), Philadelphia, 1989, Lea & Febiger.
Provides in-depth information from the research literature on the structure and function of joint cartilage. Extensive reference list included.

Soderberg GL: Articular mechanics and function. In Soderberg GL, ed: *Kinesiology: application to pathological motion*, Baltimore, 1986, Williams & Wilkins.
Provides an overview of joint function from a mechanical perspective.

RELATED WEB SITES

Calvert Orthopaedic and Sports Medicine Center
http://www.calvertorthoandsports.com/index.html
Provides basic information and color graphics of the anatomy and common injuries of the ankle, spine, knee, and shoulder.

The Center for Othopaedics and Sports Medicine
http://www.arthroscopy.com/sports.htm
Includes information and color graphics on the anatomy and function of the upper extremity, foot and ankle, and knee, as well as description of knee surgery techniques and articular surface grafting.

Rothman Institute
http://www.rothmaninstitute.com/index.html
Includes information on common sport injuries to the knee, shoulder and elbow, arthritis of the hip and knee, total joint replacements, spinal anatomy and spine abnormalities and pathologies, the foot and ankle, plus a QuickTime animation of a herniated disc.

University of Washington Bone and Joint Sources
http://www.orthop.washington.edu
Provides radiographs and information on common injuries and pathological conditions for the neck, back/spine, hand/wrist, hip, knee, and ankle/foot. Clicking on the radiographs brings up structure labels. Also includes numerous movies showing therapeutic exercises, examination procedures, and surgical techniques.

The Virtual Hospital: Joint Fluoroscopy
http://www.vh.org/Providers/Textbooks/JointFluoro/JointFluoroHP.html
Provides detailed descriptions of anatomical features and includes multiple-view electric joint fluoroscopies of the ankle, elbow, lateral elbow, hip, knee, shoulder, and wrist.

The "Virtual" Medical Center: Anatomy & Histology Center
http://www-sci.lib.uci.edu/HSG/MedicalAnatomy.html
Contains numerous images, movies, and course links for human anatomy.

Wheeless' Textbook of Orthopaedics
http://www.medmedia.com/med.htm
Provides comprehensive, detailed information, graphics, and related literature for all the joints.

THE BIOMECHANICS OF HUMAN SKELETAL MUSCLE

After completing this chapter, the reader will be able to:

Identify the basic behavioral properties of the musculotendinous unit.

Explain the relationships of fiber types and fiber architecture to muscle function.

Explain how skeletal muscles function to produce coordinated movement of the human body.

Discuss the effects of the force-velocity and length-tension relationships and electromechanical delay on muscle function.

Discuss the concepts of strength, power, and endurance from a biomechanical perspective.

What enables some athletes to excel at endurance events such as the marathon and others to dominate in power events such as the shot put or sprinting? What characteristics of the neuromuscular system contribute to quickness of movement? What exercises tend to cause muscular soreness? From a biomechanical perspective, what is muscular strength?

Muscle is the only tissue capable of actively developing tension. This characteristic enables skeletal, or striated, muscle to perform the important functions of maintaining upright body posture, moving the body limbs, and absorbing shock. Because muscle can only perform these functions when appropriately stimulated, the human nervous system and muscular system are often referred to collectively as the neuromuscular system. This chapter discusses the behavioral properties of muscle tissue, the functional organization of skeletal muscle, and the biomechanical aspects of muscle function.

BEHAVIORAL PROPERTIES OF THE MUSCULOTENDINOUS UNIT

The four behavioral properties of muscle tissue are extensibility, elasticity, irritability, and the ability to develop tension. These properties are common to all muscle, including the cardiac, smooth, and skeletal muscle of human beings, as well as the muscles of other mammals, reptiles, amphibians, birds, and insects.

The characteristic behavioral properties of muscle are extensibility, elasticity, irritability, and the ability to develop tension.

Extensibility and Elasticity

The properties of extensibility and elasticity are common to many biological tissues. As shown in Figure 6-1, extensibility is the ability to be stretched or to increase in length, and elasticity is the ability to return to normal length after a stretch. Muscle's elasticity returns it to normal resting length following a stretch and provides for the smooth transmission of tension from muscle to bone.

The elastic behavior of muscle has been described as consisting of two major components (32, 45) (Figure 6-2). The **parallel elastic component** (PEC), provided by the muscle membranes, supplies resistance when a muscle is passively stretched. The **series elastic component** (SEC), residing in the tendons, acts as a spring to store elastic energy when a tensed muscle is stretched. These components of muscle elasticity are so named because the membranes and tendons are respectively parallel to and in series (or in line) with the muscle fibers, which provide the **contractile component.**

The elasticity of human skeletal muscle is believed to be due primarily to the SEC. When a tensed muscle is stretched, the SEC causes an elastic recoil effect, and the stretch reflex simultaneously initiates the development of tension in the muscle. Thus, a stretch promotes subsequent forceful shortening of the muscle. This pattern of eccentric contraction followed immediately by concentric contraction is known as the **stretch-shortening cycle.** This phenomenon contributes to effective development of concentric muscular force in many sport activities. Football quarterbacks and baseball pitchers typically initiate a forceful stretch of the shoulder flexors and horizontal adductors immediately before throwing the ball. The same action occurs in muscle groups of the trunk and shoulders at the peak of the backswings of a golf club and a baseball bat. The stretch-shortening cycle also promotes storage and use of elastic energy during running, particularly with the

parallel elastic component
passive elastic property of muscle derived from the muscle membranes

series elastic component
passive elastic property of muscle derived from the tendons

contractile component
muscle property enabling tension development by stimulated muscle fibers

stretch-shortening cycle
eccentric contraction followed immediately by concentric contraction

FIGURE 6-1

The characteristic properties of muscle tissue are extensibility, elasticity, contractility, and irritability (*not shown*).

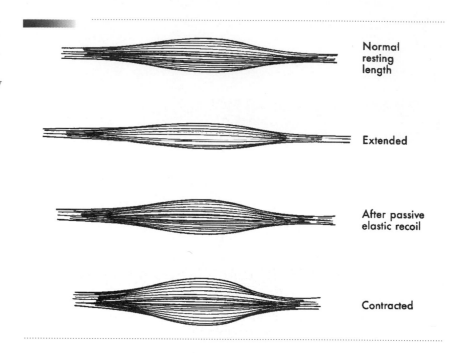

Normal resting length

Extended

After passive elastic recoil

Contracted

FIGURE 6-2

From a mechanical perspective, the musculotendinous unit behaves as a contractile component (the muscle fibers) in parallel with one elastic component (the muscle membranes) and in series with another elastic component (the tendons).

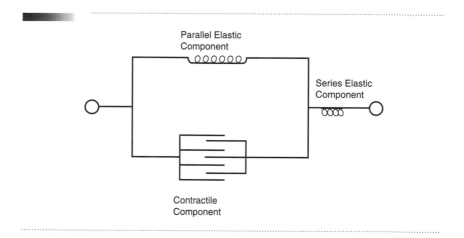

Parallel Elastic Component

Series Elastic Component

Contractile Component

alternating eccentric and concentric tension present in the gastrocnemius (36). Researchers have found that the muscles, tendons, and ligaments in the lower extremity behave very much like a spring during running, with higher stride frequencies associated with increased spring stiffness (19). Research also shows that positioning the starting blocks at a 30° or a 50° angle as compared to a 70° angle to the track also increases the velocity of the start during sprinting because of the added stretch placed on the Achilles tendon (29).

Both the SEC and the PEC have a viscous property that enables muscle to stretch and recoil in a time-dependent fashion. When a static stretch of a muscle group such as the hamstrings is maintained over a period of time, the muscle progressively lengthens, increasing joint range of motion. Likewise, after a muscle group has been stretched, it does not recoil to resting length immediately, but shortens gradually over a period of time. This **viscoelastic** response of muscle is independent of gender (50).

Muscle's viscoelastic property enables it to progressively lengthen over time when stretched.

viscoelastic
having the ability to stretch or shorten over time

Irritability and the Ability to Develop Tension

Another of muscle's characteristic properties, irritability, is the ability to respond to a stimulus. Stimuli affecting muscles are either electrochemical, such as an action potential from the attaching nerve, or mechanical, such as an external blow to a portion of a muscle. When activated by a stimulus, muscle responds by developing tension.

The ability to develop tension is the one behavioral characteristic unique to muscle tissue. Historically, the development of tension by muscle has been referred to as contraction, or the contractile component of muscle function. Contractility is the ability to shorten in length. However, as discussed in a later section, tension in a muscle may not result in the muscle's shortening.

STRUCTURAL ORGANIZATION OF SKELETAL MUSCLE

There are approximately 434 muscles in the human body, making up 40% to 45% of the body weight of most adults. Muscles are distributed in pairs on the right and left sides of the body. About 75 muscle pairs are responsible for body movements and posture, with the remainder involved in activities such as eye control and swallowing. When tension is developed in a muscle, biomechanical considerations such as the magnitude of the force generated, the speed with which the force is developed, and the length of time that the force may be maintained are affected by the particular anatomical and physiological characteristics of the muscle.

Baseball pitchers initiate a forceful stretch of the shoulder flexors and horizontal adductors immediately before throwing the ball. The stretch reflex then contributes to forceful tension development in these muscles.

Muscle Fibers

A single muscle cell is termed a *muscle fiber* because of its threadlike shape. The membrane surrounding the muscle fiber is sometimes called the sarcolemma and the specialized cytoplasm is termed sarcoplasm. The sarcoplasm of each fiber contains a number of nuclei and mitochondria, as well as numerous threadlike myofibrils that are aligned parallel to one another. The myofibrils contain two types of protein

FIGURE 6-3

The sarcoplasm of a muscle fiber contains parallel, threadlike myofibrils, each composed of myosin and actin filaments. From Shier, Butler, and Lewis. *Hole's Human Anatomy and Physiology.* © 1996. Reprinted by permission of The McGraw-Hill Companies, Inc.

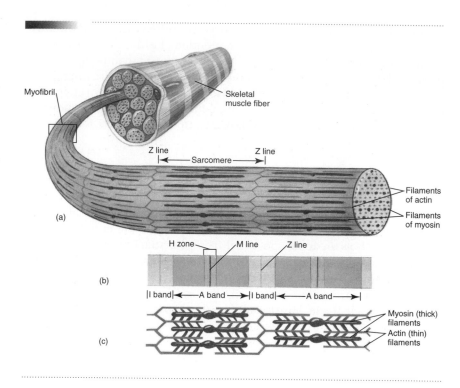

filaments whose arrangement produces the characteristic striated pattern after which skeletal, or striated, muscle is named.

Observations through the microscope of the changes in the visible bands and lines in skeletal muscle during muscle contraction has prompted the naming of these structures for purposes of reference (Figure 6-3). The sarcomere, compartmentalized between two Z lines, is the basic structural unit of the muscle fiber (Figure 6-4). Each sarcomere is bisected by an M line. The A bands contain thick, rough myosin filaments, each of which is surrounded by six thin, smooth actin filaments. The I bands contain only thin actin filaments. In both bands the protein filaments are held in place by attachment to Z lines, which adhere to the sarcolemma. In the center of the A bands are the H zones, which contain only the thick myosin filaments. (See Table 6-1 for the origins of the names of these bands.)

During muscle contraction, the thin actin filaments from either end of the sarcomere slide toward each other. As viewed through a microscope, the Z lines move toward the A bands, which maintain their original size, while the I bands narrow and the H zone disappears. Projections from the myosin filaments called cross-bridges form physical linkages with the actin filaments during muscle contraction,

Sarcomere

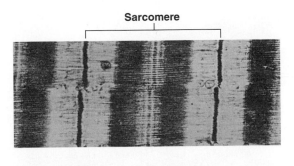

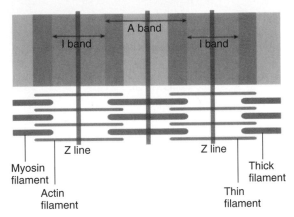

I band

A band

I band

Z line

Z line

Myosin
filament

Actin
filament

Thick
filament

Thin
filament

FIGURE 6-4
The sarcomere is
composed of alternating
dark and light bands that
give muscle its striated
appearance. From Shier,
Butler, and Lewis. *Hole's Human
Anatomy and Physiology.* © 1996.
Reprinted by permission of The
McGraw-Hill Companies, Inc.

TABLE 6-1

HOW DID THE STRUCTURES WITHIN THE SARCOMERES GET THEIR NAMES?

STRUCTURE	HISTORICAL DERIVATION OF NAME
A bands	Polarized light is *anisotropic* as it passes through this region.
I bands	Polarized light is *isotropic* as it passes through this region.
Z lines	The German word *Zwischenscheibe* means "between disc."
H zones	Those were discovered by Hensen.
M band	The German word *Mittelscheibe* means middle or intermediate.

FIGURE 6-5
The sarcoplasmic
reticulum and transverse
tubules provide channels
for movement of
electrolytes. From Shier,
Butler, and Lewis. *Hole's Human
Anatomy and Physiology.* © 1996.
Reprinted by permission of The
McGraw-Hill Companies, Inc.

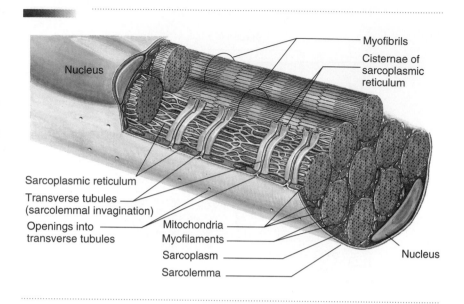

Nucleus

Myofibrils

Cisternae of
sarcoplasmic
reticulum

Sarcoplasmic reticulum

Transverse tubules
(sarcolemmal invagination)

Openings into
transverse tubules

Mitochondria

Myofilaments

Sarcoplasm

Sarcolemma

Nucleus

with the number of linkages proportional to both force production
and energy expenditure.

A network of membranous channels known as the sarcoplasmic reticulum is associated with each fiber externally (Figure 6-5). Internally, the
fibers are transected by tiny tunnels called transverse tubules that pass
completely through the fiber and open only externally. The sarcoplasmic reticulum and transverse tubules provide the channels for transport
of the electrochemical mediators of muscle activation.

Several layers of connective tissue provide the superstructure for muscle fiber organization (Figure 6-6). Each fiber membrane, or sarcolemma, is surrounded by a thin connective tissue called the endomysium. Fibers are bundled into fascicles by connective tissue sheaths
referred to as the perimysium. Groups of fascicles forming the whole
muscles are then surrounded by the epimysium, which is continuous
with the muscle tendons.

Considerable variation in the length and diameter of muscle fibers
within muscles is seen in adults. Some fibers may run the entire length
of a muscle, whereas others are much shorter. Fibers over 30 cm in
length have been identified in the sartorius (4). Skeletal muscle fibers
grow in length and diameter from birth to adulthood, with a fivefold
increase in fiber diameter during this period (47). Fiber diameter can
also be increased by resistance training with few repetitions of large loads
in adults of all ages (20, 62).

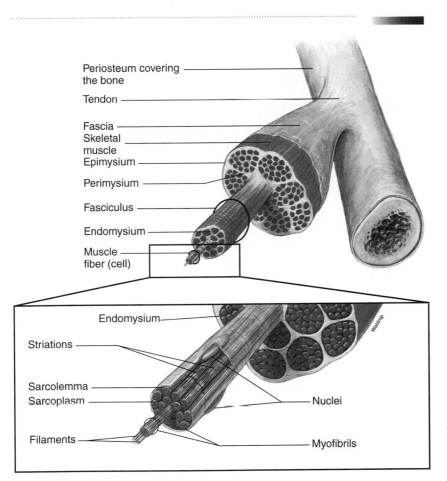

Periosteum covering the bone

Tendon

Fascia

Skeletal muscle

Epimysium

Perimysium

Fasciculus

Endomysium

Muscle fiber (cell)

Endomysium

Striations

Sarcolemma

Sarcoplasm

Filaments

Nuclei

Myofibrils

FIGURE 6-6
Muscle is compartmentalized by a series of connective tissue membranes. From Fox, *Human Physiology.* © 1999. Reprinted by permission of The McGraw-Hill Companies, Inc.

In animals such as amphibians the number of muscle fibers present also increases with the age and size of the organism. However, this does not appear to occur in human beings. The number of muscle fibers present in humans is genetically determined and varies from person to person. The same number of fibers present at birth is apparently maintained throughout life, except for the occasional loss from injury (24). The increase in muscle size after resistance training is generally believed to represent an increase in fiber diameters rather than in the number of fibers (40). There is, however, a possibility that hyperplasia, or increase in fiber number, may occur among some individuals in response to training (49).

FIGURE 6-7

A motor unit consists of a single neuron and all muscle fibers innervated by that neuron. From Fox, *Human Physiology.* © 1999. Reprinted by permission of The McGraw-Hill Companies, Inc.

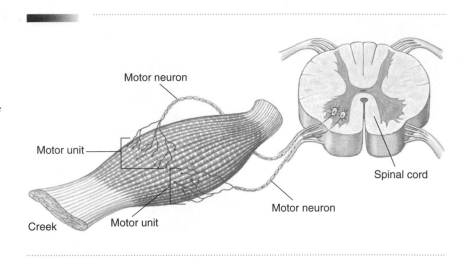

Motor Units

Muscle fibers are organized into functional groups of different sizes. Composed of a single motor neuron and all fibers innervated by it, these groups are known as **motor units** (Figure 6-7). The axon of each motor neuron subdivides many times so that each individual fiber is supplied with a motor end plate (Figure 6-8). Typically, there is only

motor unit

a single motor neuron and all fibers it innervates

FIGURE 6-8

Each muscle fiber in a motor unit receives a motor end plate from the motor neuron. From Fox, *Human Physiology.* © 1999. Reprinted by permission of The McGraw-Hill Companies, Inc.

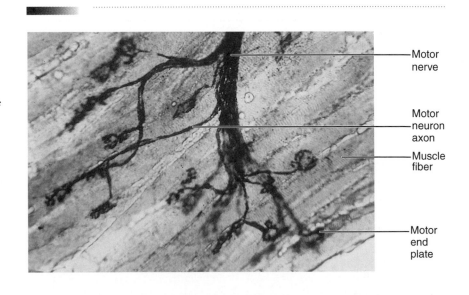

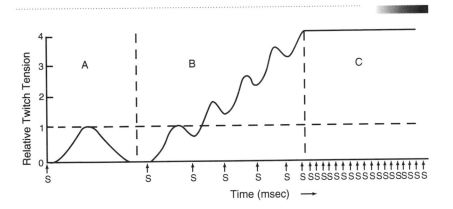

FIGURE 6-9
Tension developed in a
muscle fiber *(A)* in
response to a single
stimulus, *(B)* in response
to repetitive stimulation,
and *(C)* in response to
high frequency stimulation
(tetanus).

one end plate per fiber, although multiple innervation of fibers has been reported in vertebrates other than humans (30). The fibers of a motor unit may be spread over a several-centimeter area and be interspersed with the fibers of other motor units. With rare exceptions (18), motor units are confined to a single muscle and are localized within that muscle (11). A single mammalian motor unit may contain from less than 100 to nearly 2000 fibers, depending on the type of movements the muscle executes (9). Movements that are precisely controlled, such as those of the eyes or fingers, are produced by motor units with small numbers of fibers. Gross, forceful movements, such as those produced by the gastrocnemius, are usually the result of the activity of large motor units.

Most skeletal motor units in mammals are composed of *twitch type* cells that respond to a single stimulus by developing tension in a twitch-like fashion. The tension in a twitch fiber following the stimulus of a single nerve impulse rises to a peak value in less than 100 msec and then immediately declines.

In the human body, however, motor units are generally activated by a volley of nerve impulses. When rapid, successive impulses activate a fiber already in tension, **summation** occurs and tension is progressively elevated until a maximum value for that fiber is reached (Figure 6-9). A fiber repetitively activated so that its maximum tension level is maintained for a time is in **tetanus.** The tension present during tetanus may be as much as four times peak tension during a single twitch (58). As tetanus is prolonged, fatigue causes a gradual decline in the level of tension produced.

Not all human skeletal motor units are of the twitch type. Motor units of the *tonic type* are found in the oculomotor apparatus. These motor units require more than a single stimulus before the initial development of tension.

summation
building in an additive fashion

tetanus
state of muscle producing sustained maximal tension resulting from repetitive stimulation

FIGURE 6-10
Fast twitch fibers both
reach peak tension and
relax more quickly than
slow twitch fibers.

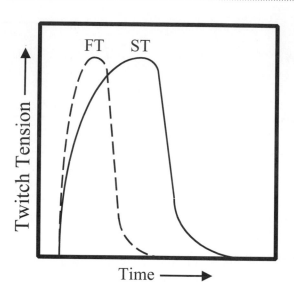

Fiber Types

Skeletal muscle fibers exhibit many different structural, histochemical,
and behavioral characteristics. Because these differences have direct implications for muscle function, they are of particular interest to many
scientists. The fibers of some motor units contract to reach maximum
tension more quickly than others after being stimulated. Based on this
distinguishing characteristic, fibers may be divided into the umbrella
categories of **fast twitch** (FT) and **slow twitch** (ST). It takes FT fibers
only about one-seventh the time required by ST fibers to reach peak
tension (Figure 6-10) (22). However, a wide range of twitch times to
achieve maximum tension exists within both categories (10). This difference in time to peak tension is attributed to higher concentrations
of myosin-ATPase in FT fibers. The FT fibers are also larger in diameter than ST fibers. Because of these and other differences, FT fibers usually fatigue more quickly than ST fibers. Although intact FT and ST muscles generate approximately the same amount of peak isometric force
per cross-sectional area of muscle, individuals with a high percentage of
FT fibers are able to generate higher magnitudes of torque and power
during movement than those with more ST fibers (22).

FT fibers are divided into two categories based on histochemical properties. The first type of FT fiber shares the resistance to fatigue that characterizes ST fibers. The second type of FT fiber is larger in diameter, contains fewer mitochondria, and fatigues more rapidly than the first type (16).

fast twitch fiber
a fiber that reaches peak
tension relatively quickly

slow twitch fiber
a fiber that reaches peak
tension relatively slowly

TABLE 6-2

SKELETAL MUSCLE FIBER CHARACTERISTICS

CHARACTERISTIC	TYPE I SLOW-TWITCH OXIDATIVE (SO)	TYPE IIA FAST-TWITCH OXIDATIVE GLYCOLYTIC (FOG)	TYPE IIB FAST-TWITCH GLYCOLYTIC (FG)
Contraction speed	slow	fast	fast
Fatigue rate	slow	intermediate	fast
Diameter	small	intermediate	large
ATPase concentration	low	high	high
Mitochondrial concentration	high	high	low
Glycolytic enzyme concentration	low	intermediate	high

Researchers have proposed several categorization schemes based on metabolic and contractile properties of the three different general types of fiber (Table 6-2). In one scheme, ST fibers are referred to as *Type I,* and the FT fibers are called *Type IIa* and *Type IIb* (8). Another system terms the ST fibers as *slow-twitch oxidative* (SO), with FT fibers divided into *fast-twitch oxidative glycolytic* (FOG) and *fast-twitch glycolytic* (FG) fibers (54). An additional categorization includes ST fibers, and fast-twitch fatigue resistant (FFR) and fast-twitch fast fatigue (FF) fibers (11). These systems of classification are based on different fiber properties, however, and are *not* interchangeable (25).

Although all fibers in a motor unit are the same type, most skeletal muscles contain both FT and ST fibers, with the relative amounts varying from muscle to muscle and individual to individual. For example, the soleus, which is generally used only for postural adjustments, primarily contains ST fibers. In contrast, the overlying gastrocnemius may contain more FT than ST fibers.

The FT fibers are important contributors to a performer's success in events requiring fast, powerful muscular contraction, such as sprinting and jumping. Endurance events such as distance running, cycling, and swimming require effective functioning of the more fatigue-resistant ST fibers. Using muscle biopsies, researchers have shown that highly successful athletes in events requiring strength and power tend to have unusually high proportions of FT fibers (6), and that elite endurance athletes usually have abnormally high proportions of ST fibers (42).

A high percentage of fast twitch fibers is advantageous for generating fast movements and a high percentage of slow twitch fibers is beneficial for activities requiring endurance.

Elite sprint cyclists are likely to have muscles composed of a high percentage of FT fibers.

Although these findings suggest that athletic training may cause fibers to convert from ST to FT or the reverse, this has not been found to be the case. Endurance exercise training has been shown to increase the contraction velocity of the predominantly ST soleus by 20%. However, this increase was associated with elevated concentration of fiber ATPase, rather than with an increase in the percentage of fast twitch fibers present in the muscle (22). Within the FT fibers, though, conversions from Type IIb to Type IIa fibers has been found to occur with heavy resistance (strength) training, endurance training, and concentric and eccentric isokinetic training (1, 14, 34).

Individuals genetically endowed with a high percentage of FT fibers may gravitate to sports requiring strength, and those with a high percentage of ST fibers may choose endurance sports. However, the fiber type distributions of both elite strength-trained and elite endurance-trained athletes fall within the range of fiber type compositions found in untrained individuals (24). Within the general population, a bell-shaped distribution of FT versus ST muscle composition exists, with most people having an approximate balance of FT and ST fibers, and relatively small percentages having a much greater number of FT or ST fibers.

Two factors known to affect muscle fiber type composition are age and obesity. There is a progressive, age-related reduction in number of motor units and muscle fibers and in the size of Type II fibers that is not related to gender or to training (46, 48, 56, 57). A longitudinal study of 28 male distance runners showed a significantly increased proportion of Type I fibers over a 20-year period, presumably due to selective loss of Type II fibers (64). Infants and young children, on the other hand, also have significantly smaller proportions of Type IIb fibers than adults, and significantly lower proportions of Type IIb fibers are found in obese than in non-obese adults (44).

Fiber Architecture

parallel fiber arrangement
pattern of fibers within a muscle in which the fibers are roughly parallel to the longitudinal axis of the muscle

pennate fiber arrangement
pattern of fibers within a muscle with short fibers attaching to one or more tendons

Another variable influencing muscle function is the arrangement of fibers within a muscle. The orientations of fibers within a muscle and the arrangements by which fibers attach to muscle tendons vary considerably among the muscles of the human body. These structural considerations affect the strength of muscular contraction and the range of motion through which a muscle group can move a body segment.

The two umbrella categories of muscle fiber arrangement are termed **parallel** and **pennate.** Although numerous subcategories of parallel and pennate fiber arrangements have been proposed (23), the distinction between these two umbrella categories is sufficient for discussing biomechanical features.

FIGURE 6-11

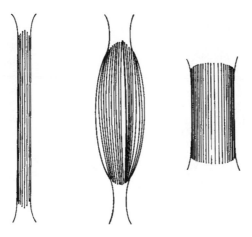

Parallel fiber arrangements

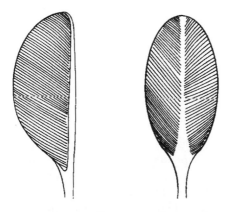

Pennate fiber arrangements

In a parallel fiber arrangement, the fibers are oriented largely in parallel with the longitudinal axis of the muscle (Figure 6-11). The sartorius, rectus abdominis, and biceps brachii have parallel fiber orientations.

A pennate fiber arrangement is one in which the fibers lie at an angle to the muscle's longitudinal axis. Each fiber in a pennate muscle attaches to one or more tendons, some of which extend the entire length of the muscle. The fibers of a muscle may exhibit more than one angle of pennation (angle of attachment) to a tendon. The tibialis posterior, rectus femoris, and deltoid muscles have pennate fiber arrangements.

FIGURE 6-12
The angle of pennation
increases as tension
progressively increases in
the muscle fibers.

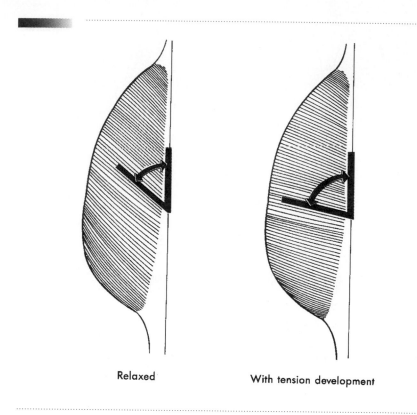

Relaxed   With tension development

*Pennate fiber arrangement
promotes muscle force
production and parallel fiber
arrangement facilitates muscle
shortening.*

When tension is developed in a parallel-fibered muscle, any shortening of the muscle is primarily the result of the shortening of its fibers. When the fibers of a pennate muscle shorten, they rotate about their tendon attachment or attachments, progressively increasing the angle of pennation (61) (Figure 6-12). As demonstrated in the problem shown in Figure 6-13, the greater the angle of pennation, the smaller the amount of effective force actually transmitted to the tendon or tendons to move the attached bones. Once the angle of pennation exceeds 60°, the amount of effective force transferred to the tendon is less than one-half of the force actually produced by the muscle fibers.

Although pennation reduces the effective force generated at a given level of fiber tension, this arrangement allows the packing of more fibers than the amount that can be packed into a longitudinal muscle occupying equal space. Because pennate muscles contain more fibers per unit of muscle volume, they can generate more force than parallel-fibered muscles of the same size. However, the parallel fiber arrangement enables greater shortening of the entire muscle than is possible

FIGURE 6-13

S A M P L E P R O B L E M I

How much force is exerted by the tendon of a pennate muscle when
the tension in the fibers is 100 N, given that the angle of pennation is

a. 40°?
b. 60°?
c. 80°?

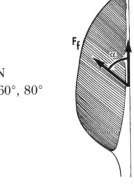

Known:

$$F_{fibers} = 100 \text{ N}$$
Angle of pennation = 40°, 60°, 80°

Solution
Wanted: F_{tendon}

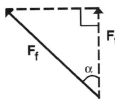

The relationship between the tension in the fibers and the tension in
the tendon is

$$F_{tendon} = F_{fibers} \cos \alpha$$

a. For $\alpha = 40°$, $F_{tendon} = (100 \text{ N}) (\cos 40)$

$$F_{tendon} = 76.6 \text{ N}$$

b. For $\alpha = 60°$, $F_{tendon} = (100 \text{ N}) (\cos 60)$

$$F_{tendon} = 50 \text{ N}$$

c. For $\alpha = 80°$, $F_{tendon} = (100 \text{ N}) (\cos 80)$

$$F_{tendon} = 17.4 \text{ N}$$

with a pennate arrangement. Parallel-fibered muscles can move body
segments through larger ranges of motion than comparably sized
pennate-fibered muscles. Increasing research findings point to differences in regional structural organization and regional functional differences within a given muscle (27).

SKELETAL MUSCLE FUNCTION

When an activated muscle develops tension, the amount of tension present is constant throughout the length of the muscle, in both tendons, and at the sites of the musculotendinous attachments to bone. The tensile force developed by the muscle pulls on the attached bones and creates torque at the joints crossed by the muscle. As discussed in Chapter 3, the magnitude of the torque generated is the product of the muscle force and the force's moment arm (Figure 6-14). In keeping with the laws of vector addition, the net torque present at a joint determines the direction of any resulting movement. The weight of the attached body segment, external forces acting on the body, and tension in any muscle crossing a joint can all generate torques at that joint (Figure 6-15).

The net torque at a joint is the vector sum of the muscle torque and the resistive torque.

Recruitment of Motor Units

The central nervous system exerts an elaborate system of control that enables matching of the speed and magnitude of muscle contraction to the requirements of the movement so that smooth, delicate, and precise movements can be executed. The neurons that innervate ST motor units generally have low thresholds and are relatively easy to activate, whereas FT motor units are supplied by nerves more difficult to activate. Consequently, the ST fibers are the first to be activated, even when the resulting limb movement is rapid (15).

Slow twitch motor units always produce tension first, whether the resulting movement is slow or fast.

As the force requirement, speed requirement, and/or duration of the activity increases, motor units with higher thresholds are progressively activated, with Type IIa, or FOG, fibers added before the Type IIb, or FG, fibers (23). Within each fiber type, a continuum of ease of activation exists, and the central nervous system may selectively activate more or fewer motor units.

During low-intensity exercise, the central nervous system may recruit ST fibers almost exclusively. As activity continues and fatigue sets in, Type IIa and then Type IIb motor units are activated until all motor units are involved (26).

Change in Muscle Length with Tension Development

When muscular tension produces a torque larger than the resistive torque at a joint, the muscle shortens, causing a change in the angle at the joint. When a muscle shortens, the contraction is **concentric** and the resulting joint movement is in the same direction as the net torque generated by the muscles. A single muscle fiber is capable of shortening to approximately one-half of its normal resting length.

concentric
contraction involving shortening of a muscle

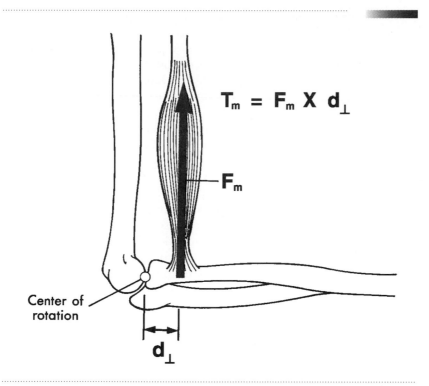

$$T_m = F_m \times d_\perp$$

F_m

Center of
rotation

$d_\perp$

FIGURE 6-14
Torque (T_m) produced by
a muscle at the joint
center of rotation is the
product of muscle force
(F_m) and muscle moment
arm ($d_\perp$).

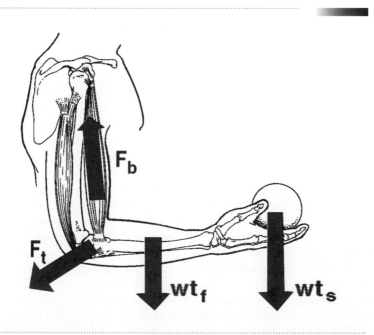

F_b

F_t

wt_f

wt_s

FIGURE 6-15
The torque exerted by the
biceps brachii (F_b) must
counteract the torques
created by the force
developed in the triceps
brachii (F_t), the weight of
the forearm and hand
(wt_f), and the weight of
the shot held in the hand
(wt_s).

isometric
contraction involving no change in muscle length

eccentric
contraction involving lengthening of a muscle

Body builders commonly develop isometric tension in their muscles to display muscle size and definition.

agonist
role played by a muscle acting to cause a movement

antagonist
role played by a muscle acting to slow or stop a movement

Muscles can also develop tension without shortening. If the opposing torque at the joint crossed by the muscle is equal to the torque produced by the muscle (with zero net torque present), muscle length remains unchanged and no movement occurs at the joint. When muscular tension develops but no change in muscle length occurs, the contraction is **isometric.** Because the development of tension increases the diameter of the muscle, body builders develop isometric tension to display their muscles when competing. Developing isometric tension simultaneously in muscles on opposite sides of a limb, such as in the triceps brachii and the biceps brachii, enlarges the cross-sectional area of the tensed muscles, although no movement occurs at either the shoulder or the elbow joints.

When opposing joint torque exceeds that produced by tension in a muscle, the muscle lengthens. When a muscle lengthens as it is being stimulated to develop tension, the contraction is **eccentric** and the direction of joint motion is opposite that of the net muscle torque. Eccentric tension occurs in the elbow flexors during the elbow extension or weight-lowering phase of a curl exercise. The eccentric tension acts as a braking mechanism to control movement speed. Without the presence of eccentric tension in the muscles, the forearm, hand, and weight would drop uncontrolled because of the force of gravity. Research indicates that enhanced ability to develop tension under concentric, isometric, and eccentric conditions is best achieved by training in the same respective exercise mode (68).

Roles Assumed by Muscles

An activated muscle can do only one thing: develop tension. Because one muscle rarely acts in isolation, however, we sometimes speak in terms of the function or role that a given muscle is carrying out when it acts in concert with other muscles crossing the same joint.

When a muscle contracts and causes movement of a body segment at a joint, it is acting as an **agonist** or mover. Because several different muscles often contribute to a movement, the distinction between primary and assistant agonists is sometimes also made. For example, during the elbow flexion phase of a forearm curl, the brachialis and the biceps brachii act as the primary agonists, with the brachioradialis, extensor carpi radialis longus, and pronator teres serving as assistant agonists.

Muscles with actions opposite those of the agonists can act as **antagonists** or opposers by developing eccentric tension at the same time that the agonists are causing movement. Agonists and antagonists are typically positioned on opposite sides of a joint. During elbow flexion when the brachialis and the biceps brachii are primary agonists, the triceps could act as antagonists by developing resistive tension. Conversely, during elbow extension when the triceps are the agonists, the brachialis and

During the elbow flexion phase of a forearm curl, the brachialis and the biceps brachii act as the primary agonists, with the brachioradialis, extensor carpi radialis longus, and pronator teres serving as assistant agonists.

biceps brachii could perform as antagonists. Although skillful movement is not characterized by continuous tension in antagonist muscles, antagonists often provide controlling or braking actions, particularly at the end of fast, forceful movements. Whereas agonists are particularly active during acceleration of a body segment, antagonists are primarily active during deceleration, or negative acceleration (38). When a person runs down a hill, for example, the quadriceps functions eccentrically as an antagonist to control the amount of knee flexion occurring.

Another role assumed by muscles involves stabilizing a portion of the body against a particular force. The force may be internal, from tension in other muscles, or external, such as the weight of an object being lifted. The rhomboids act as **stabilizers** by developing tension to stabilize the scapulae against the pull of the tow rope during water skiing.

A fourth role assumed by muscles is that of **neutralizer.** Neutralizers prevent unwanted accessory actions that normally occur when agonists develop concentric tension. For example, if a muscle causes both flexion and abduction at a joint but only flexion is desired, the action of a neutralizer causing adduction can eliminate the unwanted abduction. When the biceps brachii develops concentric tension, it produces both flexion at the elbow and supination of the forearm. If only elbow flexion is desired, the pronator teres act as a neutralizer to counteract the supination of the forearm.

Performance of human movements typically involves the cooperative actions of many muscle groups acting sequentially and in concert. For

stabilizer
role played by a muscle acting to stabilize a body part against some other force

neutralizer
role played by a muscle acting to eliminate an unwanted action produced by an agonist

example, even the simple task of lifting a glass of water from a table requires several different muscle groups to function in different ways. Stabilizing roles are performed by the scapular muscles and both flexor and extensor muscles of the wrist. The agonist function is performed by the flexor muscles of the fingers, elbow, and shoulder. Because the major shoulder flexors, the anterior deltoid and pectoralis major, also produce horizontal adduction, horizontal abductors such as the middle deltoid and supraspinatus act as neutralizers. Movement speed during the motion may also be partially controlled by antagonist activity in the elbow extensors. When the glass of water is returned to the table, gravity serves as the prime mover, with antagonist activity in the elbow and shoulder flexors controlling movement speed.

Two-Joint and Multijoint Muscles

Many muscles in the human body cross two or more joints. Examples are the biceps brachii, the long head of the triceps brachii, the hamstrings, the rectus femoris, and a number of muscles crossing the wrist and all finger joints. Since the amount of tension present in any muscle is essentially constant throughout its length as well as at the sites of its tendinous attachments to bone, these muscles affect motion at both or all of the joints crossed simultaneously. The effectiveness of a two-joint or multijoint muscle in causing movement at any joint crossed depends on the location and orientation of the muscle's attachment relative to the joint, the tightness or laxity present in the musculotendinous unit, and the actions of other muscles that cross the joint. During power-based activities such as jumping and sprinting, the biarticular muscles crossing the hip and knee have been shown to be particularly effective in converting body segment rotations into the desired translational motion of the total body center of gravity (37).

Two-joint muscles can fail to produce force when slack (active insufficiency) and can restrict range of motion when fully stretched (passive insufficiency).

There are also, however, two disadvantages associated with the function of two-joint and multijoint muscles. They are incapable of shortening to the extent required to produce a full range of motion at all joints crossed simultaneously, a limitation that is termed *active insufficiency*. For example, the finger flexors cannot produce as tight a fist when the wrist is in flexion as when it is in a neutral position (Figure 6-16). Some two-joint muscles are not able to produce force at all when the positions of both joints crossed place the muscles in a severely slackened state (31). A second problem is that for most people, two-joint and multijoint muscles cannot stretch to the extent required for full range of motion in the opposite direction at all joints crossed. This problem is referred to as *passive insufficiency* (58). For example, a larger range of hyperextension is possible at the wrist when the fingers are not fully extended (Figure 6-17). Likewise, a larger range of ankle dorsiflexion can be accomplished when the knee is in flexion due to the change in the tightness of the gastrocnemius.

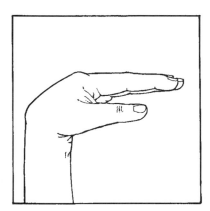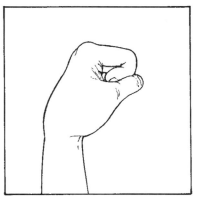

FIGURE 6-16
When the wrist is fully flexed, the finger flexors (that cross the wrist) are placed on slack and cannot develop sufficient tension to form a fist until the wrist is extended to a more neutral position. The inability to develop tension in a two- or multi-joint muscle is referred to as *active insufficiency*.

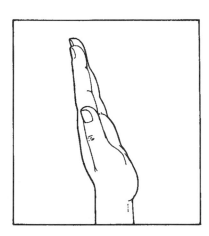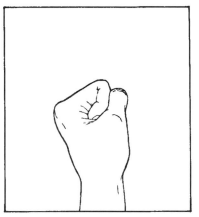

FIGURE 6-17
When the finger flexors are maximally stretched with both the wrist and fingers in full extension, the range of motion for wrist extension is restricted. Flexion of the fingers enables further extension at the wrist. Restriction of range of motion at a joint because of tightness in a two- or multijoint muscle is referred to as *passive insufficiency*.

FACTORS AFFECTING MUSCULAR FORCE GENERATION

The magnitude of the force generated by muscle is also related to the velocity of muscle shortening, the length of the muscle when it is stimulated, and the period of time since the muscle received a stimulus. Because these factors are significant determiners of muscle force, they have been extensively studied by scientists.

Force-Velocity Relationship

The classical force-velocity relationship for concentric tension development in muscle tissue was first documented by Hill in 1938 (33). The

The force-velocity relationship for muscle tissue. When the resistance (force) is negligible, muscle contracts with maximal velocity. As the load progressively increases, concentric contraction velocity slows to zero at isometric maximum. As the load increases further, the muscle lengthens eccentrically.

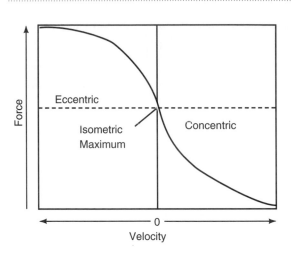

relationship between the concentric force exerted by a muscle and the velocity at which the muscle is capable of shortening is inverse, as shown in the portion of the curve below the point of isometric contraction in Figure 6-18. When a muscle develops concentric tension against a high load, the velocity (speed) of muscle shortening must be relatively slow. When the resistance is low, the velocity of shortening can be relatively fast.

The force-velocity relationship does *not* imply that it is impossible to move a heavy resistance at a fast speed. The stronger the muscle is, the greater the magnitude of maximum isometric tension (shown in the center of Figure 6-18). This is the maximum amount of force that a muscle can generate before actually lengthening as the resistance is increased. However, the general shape of the force-velocity curve remains the same, regardless of the magnitude of maximum isometric tension.

The force-velocity relationship also does *not* imply that it is impossible to move a light load at a slow speed. Most activities of daily living require slow, controlled movements of submaximal loads. The relationship does indicate that for a given load or muscular force requirement, the maximum velocity of muscle shortening is fixed in accordance with the pattern shown in the graph of Figure 6-18. At submaximal movement speeds the velocity of muscle shortening is subject to volitional control. Only the number of motor units required are activated. For example, a pencil can be picked up from a desktop quickly or slowly, depending on the controlled pattern of motor unit recruitment in the muscle groups involved.

The force-velocity relationship has been tested for skeletal, smooth, and cardiac muscle in humans, as well as for muscle tissues from other species (32). The general pattern holds true for all types of muscle, even the tiny muscles responsible for the rapid fluttering of insect wings. Maximum values of force at zero velocity and maximum values of velocity at a minimal load vary with the size and type of muscle. Although the physiological basis for the force-velocity relationship is not completely understood, the shape of the concentric portion of the curve corresponds to the rate of energy production in a muscle.

The force-velocity relationship for muscle loaded beyond the isometric maximum is shown in the top half of Figure 6-18 (39). At loads greater than the isometric maximum, the muscle is forced to lengthen in accordance with the pattern shown in the graph. At loads less than the isometric maximum, the velocity of muscle lengthening is subject to volitional control.

Eccentric strength training involves the use of resistances that are greater than the athlete's maximum isometric force generation capability. As soon as the load is assumed, the muscle begins to lengthen. Research shows this type of training to be more effective than concentric training in increasing muscle size and strength (34). As compared with concentric and isometric training, however, eccentric training is also associated with increased muscle soreness and structural damage (13).

The stronger a muscle, the greater the magnitude of its isometric maximum on the force-velocity curve.

Length-Tension Relationship

The amount of maximum isometric tension a muscle is capable of producing is partly dependent on the muscle's length. In single muscle fibers and isolated muscle preparations, force generation is at its peak when the muscle is at normal resting length (neither stretched nor contracted). When the length of the muscle increases or decreases beyond resting length, the maximum force the muscle can produce decreases, following the form of a bell-shaped curve (12).

Within the human body, however, force generation capability increases when the muscle is slightly stretched. Parallel-fibered muscles produce maximum tensions at just over resting length, and pennate-fibered muscles generate maximum tensions at between 120% and 130% of resting length (28). This phenomenon is due to the contribution of the elastic components of muscle (primarily the SEC), which add to the tension present in the muscle when the muscle is stretched. Figure 6-19 shows the pattern of maximum tension development as a function of muscle length, with the active contribution of the contractile component and the passive contribution of the SEC and PEC indicated. Research indicates that following eccentric exercise there may be a slight, transient increase in muscle length that impairs force development when joint angle does not place the muscle in sufficient stretch (60).

The total tension present in a stretched muscle is the sum of the active tension provided by the muscle fibers and the passive tension provided by the tendons and muscle membranes.

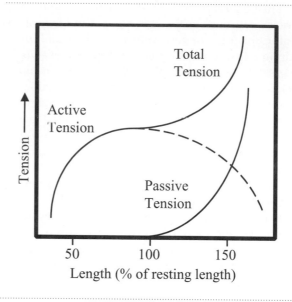

Electromechanical Delay

electromechanical delay
time between arrival of neural stimulus and tension development by the muscle

When a muscle is stimulated, a brief period of time elapses before the muscle begins to develop tension (Figure 6-20). Referred to as **electromechanical delay** (EMD), this period of time is believed to be needed for the contractile component of the muscle to stretch the SEC. During this time, muscle laxity is eliminated. Once the SEC is sufficiently stretched, tension development proceeds.

The length of EMD varies considerably among human muscles, with values of from 20 to 100 msec reported (40). Researchers have found shorter EMDs produced by muscles with high percentages of FT fibers as compared to muscles with high percentages of ST fibers (52). Development of higher contraction forces is also associated with shorter EMDs (66). Factors such as muscle length, contraction type, contraction velocity, and fatigue, however, do not appear to affect EMD (66).

The time required for a muscle to develop maximum isometric tension may be a full second following EMD (40). Shorter maximum force development times are associated with a high percentage of FT fibers in the muscle and with a trained state (65). EMD in children is significantly longer than that in adults (3).

MUSCULAR STRENGTH, POWER, AND ENDURANCE

In practical evaluations of muscular function, the force-generating characteristics of muscle are discussed within the concepts of muscular strength, power, and endurance. These characteristics of muscle func-

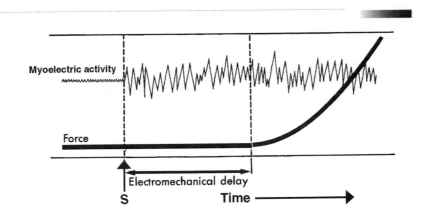

FIGURE 6-20
The brief time (20–100 msec) that elapses between the stimulation (S) of a muscle and the initiation of tension development is known as electromechanical delay.

tion have significant implications for success in different forms of strenuous physical activity, such as splitting wood, throwing a javelin, or hiking up a mountain trail. Among senior citizens and individuals with neuromuscular disorders or injuries, maintaining adequate muscular strength and endurance is essential for carrying out daily activities and avoiding injury.

Muscular Strength

When scientists excise a muscle from an experimental animal and electrically stimulate it in a laboratory, they can directly measure the force generated by the muscle. It is largely from controlled experimental work of this kind that our understandings of the force-velocity and length-tension relationships for muscle tissue are derived.

In the human body, however, it is not convenient to directly assess the force produced by a given muscle. The most direct assessment of "muscular strength" commonly practiced is a measurement of the maximum torque generated by an entire muscle group at a joint. Muscular strength, then, is measured as a function of the collective force-generating capability of a given functional muscle group. More specifically, muscular strength is the ability of a given muscle group to generate torque at a particular joint.

As discussed in Chapter 3, torque is the product of force and the force's moment arm, or the perpendicular distance at which the force acts from an axis of rotation. Resolving a muscle force into two orthogonal components, perpendicular and parallel to the attached bone, provides a clear picture of the muscle's torque-producing effect (Figure 6-21). Because the component of muscle force directed perpendicular to the attached bone produces torque, or a rotary effect, this component is termed the rotary component of muscle force. The size of the

Muscular strength is most commonly measured as the amount of torque a muscle group can generate at a joint.

FIGURE 6-21
The component of muscular force that produces torque at the joint crossed (F_t) is directed perpendicular to the attached bone.

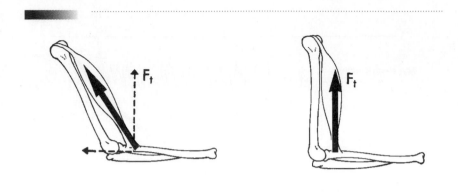

rotary component is maximum when the muscle is oriented at 90° to the bone, with change in angle of orientation in either direction progressively diminishing it. Isokinetic resistance machines are designed to match the size of the rotary component of muscle force throughout the joint range of motion. The sample problem shown in Figure 6-22 demonstrates how the torque generated by a given muscle force changes as the angle of the muscle's attachment to the bone changes.

The component of muscle force acting parallel to the attached bone does not produce torque, since it is directed through the joint center and therefore has a moment arm of zero (Figure 6-23). This component, however, can provide either a stabilizing influence or a dislocating influence, depending on whether it is directed toward or away from the joint center. Actual dislocation of a joint rarely occurs from the tension developed by a muscle, but if a dislocating component of muscle force is present, a tendency for dislocation occurs. If the elbow is at an acute angle in greater than 90° of flexion, for example, tension produced by the biceps tends to pull the radius away from its articulation with the humerus, thereby lessening the stability of the elbow in that particular position.

Therefore, muscular strength is derived both from the amount of tension the muscles can generate and from the moment arms of the contributing muscles with respect to the joint center. Both sources are affected by several factors.

The tension-generating capability of a muscle is related to its cross-sectional area and its training state. The force generation capability per cross-sectional area of muscle is approximately 90 N/cm^2 (53), as illustrated in the sample problem shown in Figure 6-24. With both concentric and eccentric strength training, gains in strength over approximately the first 12 weeks appear to be related more to improved innervation of the trained muscle than to the increase in its cross-sectional area (43). This notion is further strengthened by the finding that unila-

S A M P L E   P R O B L E M   2

FIGURE 6-22

How much torque is produced at the elbow by the biceps brachii inserting at an angle of 60° on the radius when the tension in the muscle
is 400 N? (Assume that the muscle attachment to the radius is 3 cm from
the center of rotation at the elbow joint.)

Known:

$F_m = 400$ N
$\alpha = 60°$
$d_\perp = 0.03$ m

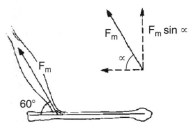

Solution

Wanted: T_m

Only the component of muscle force perpendicular to the bone generates torque at the joint. From the diagram, the perpendicular component of muscle force is

$$F_p = F_m \sin\alpha$$
$$F_p = (400 \text{ N}) (\sin 60)$$
$$F_p = 346.4 \text{ N}$$

$$T_m = F_p d_\perp$$
$$T_m = (346.4 \text{ N}) (0.03 \text{ m})$$
$$T_m = 10.4 \text{ N-m}$$

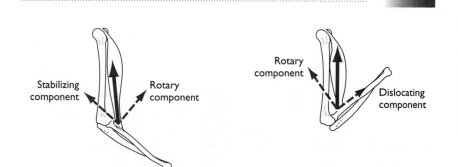

FIGURE 6-23
Contraction of the biceps
brachii produces a
component of force at the
elbow that may tend to be
stabilizing or dislocating
depending on the angle
present at the elbow when
contraction occurs.

FIGURE 6-24

S A M P L E   P R O B L E M   3

How much tension may be developed in muscles with the following cross-sectional areas:

a. 4 cm^2?
b. 10 cm^2?
c. 12 cm^2?

Known:
Muscle cross-sectional areas = 4 cm^2, 10 cm^2, and 12 cm^2

Cross-sectional area

Solution
Wanted: Tension development capability
The tension-generating capability of muscle tissue is 90 N/cm^2. The force produced by a muscle is the product of 90 N/cm^2 and the muscle's cross-sectional area. So,

a. F = (90 N/cm^2) (4 cm^2)
   F = 360 N

b. F = (90 N/cm^2) (10 cm^2)
   F = 900 N

c. F = (90 N/cm^2) (12 cm^2)
   F = 1080 N

teral strength training also produces strength gains in the untrained contralateral limb (35).

A muscle's moment arm is affected by two equally important factors. The first is the distance between the muscle's anatomical attachment to bone and the axis of rotation at the joint center, and the second is the angle of the muscle's attachment to bone, which is typically a function of relative joint angle. The greatest amount of torque is produced by maximum tension in a muscle that is oriented at a 90° angle to the bone, and anatomically attached as far from the joint center as possible.

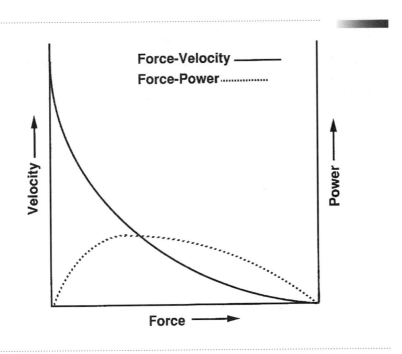

FIGURE 6-25
The relationships among concentric tension, shortening velocity, and power for muscle.

Muscular Power

Mechanical power (discussed in Chapter 12) is the product of force and velocity. Muscular power is therefore the product of muscular force and the velocity of muscle shortening. Maximum power occurs at approximately one-third of maximum velocity (33) and at approximately one-third of maximum concentric force (40) (Figure 6-25). Research indicates that training designed to increase muscular power over a range of resistance occurs most effectively with loads of one-third of one maximum repetition (51).

Because neither muscular force nor the speed of muscle shortening can be directly measured in an intact human being, muscular power is more generally defined as the rate of torque production at a joint, or the product of the net torque and the angular velocity at the joint. Accordingly, muscular power is affected by both muscular strength and movement speed.

Muscular power is an important contributor to activities requiring both strength and speed. The strongest shot putter on a team is not necessarily the best shot putter, because the ability to accelerate the shot is a critical component of success in the event. Sports that require explosive movements, such as power lifting, throwing, jumping, and sprinting, are based on the ability to generate muscular power.

Since FT fibers develop tension more rapidly than ST fibers, a large percentage of FT fibers in a muscle are an asset for an individual training

Explosive movements require muscular power.

for a muscular, power–based event. Individuals with a predominance of FT fibers generate more power at a given load than individuals with a high percentage of ST compositions. Those with primarily FT compositions also develop their maximum power at faster velocities of muscle shortening (63). The ratio for mean peak power production by Type IIb, Type IIa, and Type I fibers in human skeletal muscle is 10:5:1 (22).

Muscular Endurance

Muscular endurance is the ability of the muscle to exert tension over a period of time. The tension may be constant, as when a gymnast performs an iron cross, or varying cyclically, as during rowing, running, and cycling. The longer the time tension is exerted, the greater the endurance. Although maximum muscular strength and maximum muscular power are relatively specific concepts, muscular endurance is less well understood because the force and speed requirements of the activity dramatically affect the length of time it can be maintained.

Training for muscular endurance typically involves large numbers of repetitions against relatively light resistance. This type of training does not increase muscle fiber diameter.

Muscle Fatigue

The research literature typically addresses muscular endurance from the standpoint of muscle fatigue. Fatiguability is the opposite of endurance. The more rapidly a muscle fatigues, the less endurance it has. A complex array of physiological and neurological factors affect the rate at which a muscle fatigues. There is some evidence, in fact, that mechanisms of fatigue may be muscle-specific and/or exercise duration–specific (7). Moreover, within a given muscle, fiber type composition and the pattern of motor unit activation play a role in determining the rate at which a muscle fatigues.

Characteristics of muscle fatigue include reduction in muscle force production capability and shortening velocity, as well as prolonged relaxation of motor units between recruitment (2). High-intensity muscle activity over time also results in prolonged twitch duration and a prolonged sarcolemma action potential of reduced amplitude (21).

A muscle fiber reaches absolute fatigue once it is unable to develop tension when stimulated by its motor axon. Fatigue may also occur in the motor neuron itself, rendering it unable to generate an action potential. FG fibers fatigue more rapidly than FOG fibers, and SO fibers are the most resistant to fatigue. Researchers have shown that the proportion of ST fibers in the vastus lateralis is directly related to the length of time that a level of 50% of maximum isometric tension can be maintained (41).

Sprinting requires muscular power, particularly in the hamstrings and the gastrocnemius.

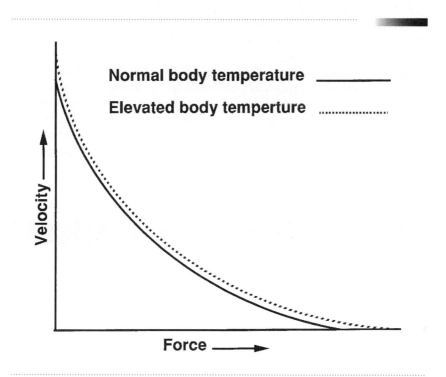

Normal body temperature ————
Elevated body temperture ⋯⋯⋯⋯⋯

Velocity ⟶

Force ⟶

FIGURE 6-26
When muscle temperature is slightly elevated, the force-velocity curve is shifted. This is one benefit of warming up before an athletic endeavor.

The specific causes of muscle fatigue are not well understood. However, a growing body of evidence indicates that reduction in the rate of intracellular calcium release and uptake by the sarcoplasmic reticulum is involved (67). As many as three different mechanisms of reduced calcium release have been identified, but are incompletely understood (2). Some experimental evidence suggests that sliding of the actin and myosin filaments during repeated muscle contraction reduces the affinity for calcium at uptake sites on the thin actin filaments (17). A variety of other factors have also been implicated in the development of fatigue, including increases in muscle acidity and intracellular potassium (5) and decreases in muscle energy supplies and intracellular oxygen (55).

Effect of Muscle Temperature

As body temperature elevates, the speeds of nerve and muscle functions increase. This causes a shift in the force-velocity curve, with a higher value of maximum isometric tension and a higher maximum velocity of shortening possible at any given load (Figure 6-26). At an elevated temperature, the activation of fewer motor units is needed to sustain a given load (59). The metabolic processes supplying oxygen and removing waste products for the working muscle also quicken with higher body temperatures. These benefits result in increased muscular strength, power, and endurance, and provide the rationale for warming up before an athletic endeavor.

Muscle function is most efficient at 38.5°C (101°F)(4). Elevation of body temperature beyond this point may occur during strenuous exercise under conditions of high ambient temperature and/or humidity and can be extremely dangerous, possibly resulting in heat exhaustion or heat stroke. Organizers of long-distance events involving running or cycling should be particularly cognizant of the potential hazards associated with competition in such environments.

SUMMARY

Muscle is elastic and extensible and responds to stimulation. Most important, however, it is the only biological tissue capable of developing tension.

The functional unit of the neuromuscular system is the motor unit, consisting of a single motor neuron and all the fibers it innervates. The fibers of a given motor unit are either slow twitch, fast-twitch fatigue resistant, or fast-twitch fast fatigue. Both ST and FT fibers are typically found in all human muscles, although the proportional fiber composition varies. The number and distribution of fibers within muscles appear to be genetically determined and related to age. Within human skeletal muscles, fiber arrangements are parallel or pennate. Pennate fiber arrangements promote force production, whereas parallel fiber arrangement enables greater shortening of the muscle.

Muscle responds to stimulation by developing tension. Depending on what other forces act, however, the resulting action can be concentric, eccentric, or isometric, for muscle shortening, lengthening, or remaining unchanged in length. The central nervous system directs the recruitment of motor units such that the speed and magnitude of muscle tension development are well matched to the requirements of the activity.

There are well-defined relationships between muscle force output and the velocity of muscle shortening, the length of the muscle at the time of stimulation, and the time since the onset of the stimulus. Because of the added contribution of the elastic components of muscle, force production is enhanced when a muscle is stretched.

Muscle performance is typically described in terms of muscular strength, power, and endurance. From a biomechanical perspective, strength is the ability of a muscle group to generate torque at a joint, power is the rate of torque production at a joint, and endurance is resistance to fatigue.

1. List three examples of activities requiring concentric muscle action and three examples of activities requiring eccentric muscle action, and identify the specific muscles or muscle groups involved.
2. List five movement skills for which a high percentage of fast twitch muscle fibers are an asset and list five movement skills for which a high percentage of slow twitch fibers are an asset. Provide brief statements of rationale for each of your lists.
3. Hypothesize about the pattern of recruitment of motor units in the major muscle group or groups involved during each of the following activities:
   a. Walking up a flight of stairs
   b. Sprinting up a flight of stairs
   c. Throwing a ball
   d. Cycling in a 100 km race
   e. Threading a needle
4. Identify three muscles that have parallel fiber arrangements and explain the ways in which the muscles' functions are enhanced by this arrangement.
5. Answer Problem 4 for pennate fiber arrangement.
6. How is the force-velocity curve affected by muscular strength training?
7. Write a paragraph describing the biomechanical factors determining muscular strength.
8. List five activities in which the production of muscular force is enhanced by the series elastic component and the stretch reflex.
9. Muscle can generate approximately 90 N of force per square centimeter of cross-sectional area. If a biceps brachii has a cross-sectional area of 10 cm^2, how much force can it exert? (Answer: 900 N)
10. Using the same force/cross-sectional area estimate as in Problem 9, and estimating the cross-sectional area of your own biceps brachii, how much force should the muscle be able to produce?

1. Identify the direction of motion (flexion, extension, etc.) at the hip, knee, and ankle, and the source of the force(s) causing motion at each joint for each of the following activities:
   a. Sitting down in a chair
   b. Taking a step up on a flight of stairs
   c. Kicking a ball
2. Considering both the force-length relationship and the rotary component of muscle force, sketch what you would hypothesize to be

the shape of a force versus joint angle curve for the elbow flexors. Write a brief rationale in support of the shape of your graph.

3. Certain animals, such as the kangaroo and the cat, are well known for their jumping abilities. What would you hypothesize about the biomechanical characteristics of their muscles?

4. Identify the functional roles played by the muscle groups that contribute to each of the following activities:
   a. Carrying a suitcase
   b. Throwing a ball
   c. Rising from a seated position

5. If the fibers of a pennate muscle are oriented at a 45° angle to a central tendon, how much tension is produced in the tendon when the muscle fibers contract with 150 N of force? (Answer: 106 N)

6. How much force must be produced by the fibers of a pennate muscle aligned at a 60° angle to a central tendon to create a tensile force of 200 N in the tendon? (Answer: 400 N)

7. What must be the effective minimal cross-sectional areas of the muscles in Problems 5 and 6 above, given an estimated 90 N of force-producing capacity per square centimeter of muscle cross-sectional area? (Answer: 1.2 cm^2; 4.4 cm^2)

8. If the biceps brachii, attaching to the radius 2.5 cm from the elbow joint, produces 250 N of tension perpendicular to the bone and the triceps brachii, attaching 3 cm away from the elbow joint, exerts 200 N of tension perpendicular to the bone, how much net torque is present at the joint? Will there be flexion, extension, or no movement at the joint? (Answer: 0.25 N-m; flexion)

9. Calculate the amount of torque generated at a joint when a muscle attaching to a bone 3 cm from the joint center exerts 100 N of tension at the following angles of attachment:
   a. 30°
   b. 60°
   c. 90°
   d. 120°
   e. 150°
   (Answer: a. 1.5 N-m; b. 2.6 N-m; c. 3 N-m; d. 2.6 N-m; e. 1.5 N-m)

10. Write a quantitative problem of your own involving the following variables: muscle tension, angle of muscle attachment to bone, distance of the attachment from the joint center, and torque at the joint. Provide a solution for your problem.

LABORATORY EXPERIENCES

1. With a partner, use a goniometer to measure the ankle range of motion for dorsiflexion and plantar flexion both when the knee is fully extended and comfortably flexed. Explain your results.

2. Using a series of dumbbells, determine your maximum weight for the forearm curl exercise when the elbow is at angles of 5°, 90°, and 140°. Explain your findings.
3. Using electromyography apparatus with surface electrodes positioned over the biceps brachii, perform a forearm curl exercise with light and heavy weights. Explain the changes evident in the electromyogram.
4. Using electromyography apparatus with surface electrodes positioned over the pectoralis major and triceps brachii, perform bench presses with wide, medium, and narrow grip widths on the bar. Explain the differences in muscle contributions evident.
5. Using electromyography apparatus with surface electrodes positioned over the biceps brachii, perform a forearm curl exercise to fatigue. What changes are evident in the electromyogram with fatigue? Explain your results.

REFERENCES

1. Allemeier CA, Fry AC, Johnson P, Hikida RS, Hagerman FC, and Staron RS: Effects of spring cycle training on human skeletal muscle, J Appl Physiol 77:2385, 1994.
2. Allen DG, Lännergren J, and Westerblad H: Muscle cell function during prolonged activity: Cellular mechanisms of fatigue, Exp Physiol 80:497, 1995.
3. Asai H and Aoki J: Force development of dynamic and static contractions in children and adults, Int J Sports Med 17:170, 1996.
4. Astrand PO and Rodahl K: *Textbook of work physiology,* 2nd ed, New York, 1977, McGraw-Hill, Inc.
5. Bangsbo J, Madsen K, Kiens B, and Richter EA: Effect of muscle acidity on muscle metabolism and fatigue during intense exercise in man, J Physiol 495:587, 1996.
6. Bar-Or O et al: Anaerobic capacity and muscle fiber type distribution in man, Int J Sports Med 10:82, 1980.
7. Behm DG and St Pierre DM: Effects of fatigue duration and muscle type on voluntary and evoked contractile properties, J Appl Physiol 82:1654, 1997.
8. Brooke MH and Kaiser KK: The use and abuse of muscle histochemistry, Ann NY Acad Sci 228:121, 1974.
9. Buchthal F and Schalbruch H: Motor unit of mammalian muscle, Physiol Rev 60:90, 1980.
10. Burke RE: Motor units: Anatomy, physiology, and functional organization. In Brookhart JM et al, eds: *Handbook of physiology.* The nervous system, motor control, part 2, Baltimore, 1981, Williams & Wilkins.
11. Burke RE and Tsairis P: The correlation of physiological properties with histochemical characteristics in single motor units, Ann NY Acad Sci 228:145, 1974.
12. Chapman AE: The mechanical properties of human muscle, Exerc Sport Sci Rev 13:443, 1985.

13. Clarkson PM and Newham DJ: Associations between muscle soreness, damage, and fatigue, Adv Exp Med Biol 384:457, 1995.
14. Delecluse C: Influence of strength training on sprint running performance. Current findings and implications for training, Sports Med 24:147, 1997.
15. Desmedt JE and Godaux E: Fast motor units are not preferentially activated in rapid voluntary contractions in man, Nature 267:717, 1977.
16. Edgerton VR and Roy RR: Regulation of skeletal muscle fiber size, shape, and function, J Biomech, 24 (Suppl. 1):123, 1991.
17. Edman KA: Fatigue vs. shortening-induced deactivation in striated muscle, Acta Physiol Scand 156:183, 1996.
18. Emonet-Denand F, Laporte Y, and Proske V: Contraction of muscle fibers in two adjacent muscles innervated by branches of the same motor axon, J Neurophysiol 34:132, 1971.
19. Farley CT and González O: Leg stiffness and stride frequency in human running, J Biomech 29:181, 1996.
20. Fielding RA: The role of progressive resistance training and nutrition in the preservation of lean body mass in the elderly, J Am Col Nutr 14:587, 1995.
21. Fitts RH: Muscle fatigue: The cellular aspects, Am J Sports Med 24:S, 1996.
22. Fitts RH, and Widrick JJ: Muscle mechanics; adaptations with exercise-training, Exerc Sport Sci Rev 24:427, 1996.
23. Gans C: Fiber architecture and muscle function, Exerc Sport Sci Rev 10:160, 1982.
24. Gollnick PD: Muscle characteristics as a foundation of biomechanics. In Matsui H and Kobayashi K, eds: *Biomechanics VIII-A,* Champaign, Ill, 1983, Human Kinetics Publishers, Inc.
25. Gollnick PD and Hodgson DR: The identification of fiber types in skeletal muscle: A continual dilemma, Exerc Sport Sci Rev 14:81, 1986.
26. Gollnick PD, Piehl K, and Saltin B: Selective glycogen depletion pattern in human muscle fibres after exercise of varying intensity and at varying pedalling rates, J Physiol 241:45, 1974.
27. Gordon T and Pattullo MC: Plasticity of muscle fiber and motor unit types, Exerc Sports Sci Rev 21:331, 1993.
28. Gowitzke BA and Milner M: *Understanding the scientific bases of human movement,* 2nd ed, Baltimore, 1980, Williams & Wilkins.
29. Guissard N, Duchateau J, and Hainaut K: EMG and mechanical changes during sprint starts at different front block obliquities, Med Sci Sports Exerc 24:1257, 1992.
30. Guthe K: Reptilian muscle: Fine structure and physiological parameters. In Gans C and Parsons TS, eds: *Biology of the reptilia, vol 11,* London, 1981, Academic Press, Inc.
31. Herzog W, Abrahamse SK, and ter Keurs HE, Pflugers Arch 416:113, 1990.
32. Hill AV: *First and last experiments in muscle mechanics,* Cambridge, Mass, 1970, Cambridge University Press.
33. Hill AV: The heat of shortening and the dynamic constants of muscle, Proc R Soc Lond B126:136, 1938.
34. Hortobágyi T, Hill JP, Houmard JA, Fraser DD, Labert NJ, and Israel RG: Adaptive responses to muscle lengthening and shortening in humans: J Appl Physiol 80:765, 1996.
35. Housh DJ and Housh TJ: The effects of unilateral velocity-specific concentric strength training, J Orthop Sports Phys Ther 17:252, 1993.

36. Jacobs R, et al: Function of mono- and biarticular muscles in running, Med Sci Sports Exerc, 25:1163, 1993.

37. Jacobs R, Bobbert MF, and vanIngen Schenau GJ: Mechanical output from individual muscles during explosive leg extensions: the role of biarticular muscles, J Biomech 29:513, 1996.

38. Jaric S, Radovanovik S, Milanovic S, Ljubisavljevic M, and Anastasijevic R: A comparison of the effects of agonist and antagonist muscle fatigue on performance of rapid movements, Eur J Appl Physiol 76:41, 1997.

39. Katz B: The relation between force and speed in muscular contraction, J Physiol (Lond) 96:45, 1939.

40. Komi PV: Physiological and biomechanical correlates of muscle function: Effects of muscle structure and stretch-shortening cycle on force and speed, Exerc Sport Sci Rev 12:81, 1984.

41. Komi PV et al: Effects of heavy resistance and explosive-type strength training methods on mechanical, functional, and metabolic aspects of performance. In Komi PV, ed: *Exercise and sport biology,* Champaign, Ill, 1982, Human Kinetics Publishers, Inc.

42. Komi PV et al: Anaerobic performance capacity in athletes, Acta Physiol Scand 100:107, 1977.

43. Kraemer WJ, Fleck SJ, and Evans WJ: Strength and power training: Physiological mechanisms of adaptation, Exerc Sport Sci Rev 24:363, 1996.

44. Kriketos AD, Baur LA, O'Connor J, Carey D, King S, Caterson ID, and Storlien LH: Muscle fibre type composition in infant and adult populations and relationships with obesity, Int J Obes Relat Metab Disord 21:796, 1997.

45. Levin A and Wyman J: The viscous elastic properties of muscle, Proc R Soc Lond B101:218, 1927.

46. Lexell J: Human aging, muscle mass, and fiber type composition, J Gerontol A Biol Sci Med 50 Spec No: 11, 1995.

47. Lockhart RD: Anatomy of muscles and their relation to movement and posture. In Bourne GH, ed: *The structure and function of muscle,* vol 1, New York, 1973, Academic Press, Inc.

48. Manta P, Kalfakis N, Kararizou E, Vassilopoulos D, Papageorgiou C: Size and proportion of fiber types in human muscle fascicles, Clin Neuropathol 15:116, 1996.

49. McCall GE, Byrnes WC, Dickinson A, Pattany PM, and Fleck SJ: Muscle fiber hypertrophy, hyperplasia, and capillary density in college men after resistance training, J Appl Physiol 81:2004, 1996.

50. McHugh MP, et al: Viscoelastic stress relaxation in human skeletal muscle, Med Sci Sports Exerc 24:1375, 1992.

51. Moss BM, Refsnes PE, Abildgaard A, Nicolaysen K, and Jensen J: Effects of maximal effort strength training with different loads on dynamic strength, cross-sectional area, load-power and load-velocity relationships, Eur J Appl Physiol 75:193, 1997.

52. Nilsson J, Tesch P, and Thorstensson A: Fatigue and EMG of repeated fast and voluntary contractions in man, Acta Physiol Scand 101:194, 1977.

53. Norman RW: The use of electromyography in the calculation of dynamic joint torque, doctoral dissertation, University Park, Pa, 1977, Pennsylvania State University.

54. Peter JB et al: Metabolic profiles of three fiber types of skeletal muscle in guinea pigs and rabbits, Biochemistry 11:2627, 1972.

55. Pitcher JB and Miles TS: Influence of muscle blood flow on fatigue during intermittent human hand-grip exercise and recovery, Clin Exp Pharmacol Physiol 24:471, 1997.

56. Porter NM, Vandervoort AA, and Lexell J: Aging of human muscle: Structure, function and adaptability, Scand J Med Sci Sports 5:129, 1995.

57. Proctor DN, Sinning WE, Walro JM, Sieck GC, and Lemon PW: Oxidative capacity of human muscle fiber types: Effects of age and training status, J Appl Physiol 78:2033, 1995.

58. Rasch PJ: *Kinesiology and applied anatomy*, 7th ed., Philadelphia, 1989, Lea & Febiger.

59. Rosenbaum D and Henning EM: The influence of stretching and warm-up exercises on Achilles tendon reflex activity, J Sports Sci 13:481, 1995.

60. Saxton JM and Donnelly AE: Length-specific impairment of skeletal muscle contractile function after eccentric muscle actions in man, Clin Sci (Colch) 90:119, 1996.

61. Scott SH and Winter DA: A comparison of three muscle pennation assumptions and their effect on isometric and isotonic force, J Biomech 24:163, 1991.

62. Taaffe DR, Pruitt L, Pyka G, Guido D, and Marcus R: Comparative effects of high- and low-intensity resistance training on thigh muscle strength, fiber area, and tissue composition in elderly women, Clin Physiol 16:381, 1996.

63. Tihanyi J, Apor P, and Fekete GY: Force-velocity—power characteristics and fiber composition in human knee extensor muscles, Eur J Appl Physiol 48:331, 1982.

64. Trappe SW, Costill DL, Fink WJ, and Pearson DR: Skeletal muscle characteristics among distance runners: A 20-yr follow-up study, J Appl Physiol 78:823, 1995.

65. Viitasalo JT and Komi PV: Interrelationships between electromyographic, mechanical, muscle structure and reflex time measurements in man, Acta Physiol Scand 111:97, 1981.

66. Vos EJ, Harlaar J, and Van Ingen Schenau GJ: Electromechanical delay during knee extensor contractions, Med Sci Sports Exerc 23:1187, 1991.

67. Williams JH and Klug GA: Calcium exchange hypothesis of skeletal muscle fatigue: A brief review, Muscle Nerve 18:421, 1995.

68. Wilson GJ, Murphy AJ, and Giorgi A: Weight and plyometric training: Effects on eccentric and concentric force production, Can J Apl Physiol 21:301, 1996.

ANNOTATED READINGS

Chapman AE: The mechanical properties of human muscle, Exerc Sport Sci Rev 13:443, 1985.
   Reviews the mechanical models associated with human muscle functioning in a format written specifically for students of human movement.

Fitts RH and Widrick JJ: Muscle mechanics: Adaptations with exercise-training, Exerc Sport Sci Rev 24:427, 1996.
   Provides detailed review of the literature related to adaptations in muscle resulting from exercise training.

Kraemer WJ, Fleck SJ, and Evans WJ: Strength and power training: Physiological mechanisms of adaptation, Exerc Sport Sci Rev 24:363, 1996.

Comprehensively reviews current knowledge about adaptations occurring in the neuromuscular, cardiovascular, and endocrine systems with resistance training.

Rogers MA and Evans WJ: Changes in skeletal muscle with aging: Effects of exercise training, Exerc Sport Sci Rev 21:65, 1993.

Discusses changes in body composition, muscle function, muscle morphology, metabolic capacity with aging, as well as responses to exercise training in the population.

RELATED WEB SITES

E-Tech: An Orthopaedics & Biomechanics Resource
http://dspace.dial.pipex.com/town/square/fk14/index.htm

Includes links to additional pages on a number of orthopaedic and biomechanic topics, including linear viscoelasticity.

The Nicholas Institute of Sports Medicine and Athletic Trauma
http://www.nismat.org

Provides links to pages on skeletal muscle anatomy and physiology.

The "Virtual" Medical Center: Anatomy & Histology Center
http://www-sci.lib.uci.edu/HSG/MedicalAnatomy.html

Contains numerous images, movies, and course links for human anatomy.

Wheeless' Textbook of Orthopaedics
http://www.medmedia.com/Welcome.html

A comprehensive review of orthopaedics, with a search engine to locate topics related to fracture, joints, muscles, nerves, etc.

THE BIOMECHANICS OF THE HUMAN UPPER EXTREMITY

After completing this chapter, the reader will be able to:

Explain how anatomical structure affects movement capabilities of upper extremity articulations.

Identify factors influencing the relative mobility and stability of upper extremity articulations.

Identify muscles that are active during specific upper extremity movements.

Describe the biomechanical contributions to common injuries of the upper extremity.

Pitching a ball requires the coordination of the muscles of the entire upper extremity.

The capabilities of the upper extremity are varied and impressive. With the same basic anatomical structure of the arm, forearm, hand, and fingers, major league baseball pitchers hurl fast balls at 40 m/s, swimmers cross the English Channel, gymnasts perform the iron cross, travelers carry suitcases, seamstresses thread needles, and students type on computer keyboards. This chapter reviews the anatomical structures enabling these different types of movement and examines the ways in which the muscles cooperate to achieve the diversity of movement of which the upper extremity is capable.

STRUCTURE OF THE SHOULDER

The shoulder is the most complex joint in the human body, largely because it includes four separate articulations: the glenohumeral joint, the sternoclavicular joint, the acromioclavicular joint, the coracoclavicular joint, and the scapulothoracic joint. The glenohumeral joint is the

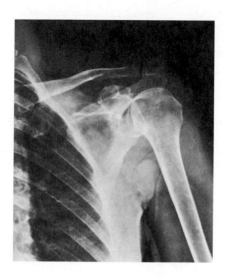

Front view of left shoulder.
From Shier, Butler, and Lewis.
*Hole's Human Anatomy and
Physiology.* © 1996. Reprinted by
permission of The McGraw-Hill
Companies, Inc.

articulation between the head of the humerus and the glenoid fossa of the scapula, which is the ball and socket joint typically considered as *the* major shoulder joint. The sternoclavicular and acromioclavicular joints provide mobility for the clavicle and the scapula—the bones of the shoulder girdle.

The glenohumeral joint is considered to be the shoulder joint.

Sternoclavicular Joint

The proximal end of the clavicle articulates with the clavicular notch of the manubrium of the sternum and with the cartilage of the first rib to form the **sternoclavicular joint.** This joint provides the major axis of rotation for movements of the clavicle and scapula (Figure 7-1). The sternoclavicular joint is a modified ball and socket, with frontal and transverse plane motion freely permitted and some forward and backward sagittal plane rotation allowed. A fibrocartilaginous articular disk improves the fit of the articulating bone surfaces and serves as a shock absorber. Rotation occurs at the sternoclavicular joint during motions such as shrugging the shoulders, elevating the arms above the head, and swimming. The close-packed position for the SC joint occurs with maximal shoulder elevation.

sternoclavicular joint modified ball and socket joint between the proximal clavicle and the manubrium of the sternum

The clavicles and the scapulae make up the shoulder girdle.

Most of the motion of the shoulder girdle takes place at the sternoclavicular joints.

FIGURE 7-1

The sternoclavicular joint.

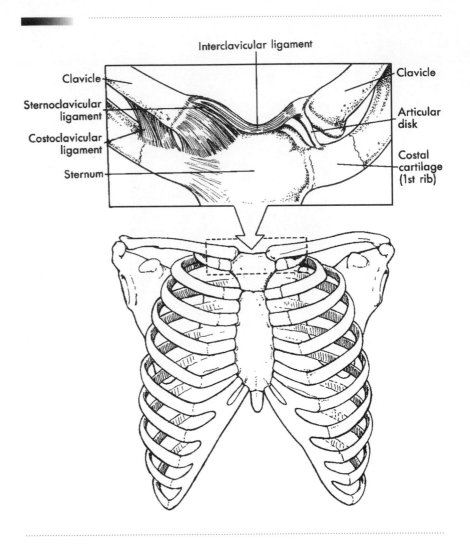

Interclavicular ligament

Clavicle

Sternoclavicular ligament

Costoclavicular ligament

Sternum

Clavicle

Articular disk

Costal cartilage (1st rib)

Acromioclavicular Joint

acromioclavicular joint
irregular joint between the acromion process of the scapula and the distal clavicle

The articulation of the acromion process of the scapula with the distal end of the clavicle is known as the **acromioclavicular joint.** It is classified as an irregular diarthrodial joint, although the joint's structure allows limited motion in all three planes. Rotation occurs at the acromioclavicular joint during arm elevation. The close-packed position of the AC joint occurs when the humerus is abducted to 90°.

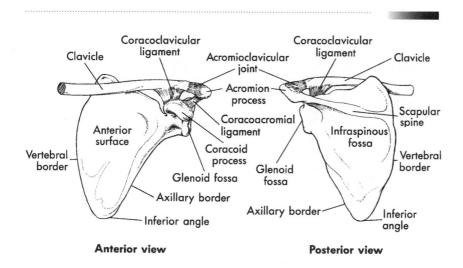

FIGURE 7-2
The acromioclavicular and coracoclavicular joints.

Coracoclavicular Joint

The **coracoclavicular joint** is a syndesmosis, formed where the coracoid process of the scapula and the inferior surface of the clavicle are bound together by the coracoclavicular ligament. This joint permits little movement. The coracoclavicular and acromioclavicular joints are shown in Figure 7-2.

coracoclavicular joint
syndesmosis with the coracoid process of the scapula bound to the inferior clavicle by the coracoclavicular ligament

Glenohumeral Joint

The **glenohumeral joint** is the most freely moving joint in the human body, enabling flexion, extension, hyperextension, abduction, adduction, horizontal abduction and adduction, and medial and lateral rotation of the humerus (Figure 7-3). The almost hemispherical head of the humerus has three to four times the amount of surface area as the shallow glenoid fossa of the scapula with which it articulates. The glenoid fossa is also less curved than the surface of the humeral head, enabling the humerus to move linearly across the surface of the glenoid fossa in addition to its extensive rotational capability (31). With passive rotation of the arm, large translations of the humeral head on the glenoid fossa are present at the extremes of the range of motion (21). The muscle forces during active rotation tend to limit ranges of motion at the shoulder, thereby limiting the humeral translation that occurs (21).

At the perimeter of the glenoid fossa is a lip or **labrum** composed of part of the joint capsule, the tendon of the long head of the biceps brachii, and the glenohumeral ligaments. The labrum deepens the fossa and adds stability to the joint. The capsule surrounding the glenohumeral

glenohumeral joint
ball and socket joint in which the head of the humerus articulates with the glenoid fossa of the scapula

The extreme mobility of the glenohumeral joint is achieved at the expense of joint stability.

glenoid labrum
rim of soft tissue located on the periphery of the glenoid fossa that adds stability to the glenohumeral joint

FIGURE 7-3

The glenohumeral joint.

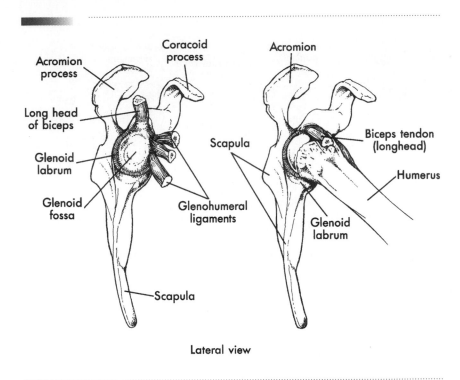

Lateral view

joint is shown in Figure 7-4. Several ligaments merge with the gleno-humeral joint capsule, including the superior, middle, and inferior gleno-humeral ligaments on the anterior side of the joint and the coraco-humeral ligament on the superior side.

The tendons of four muscles—subscapularis, supraspinatus, infra-spinatus, and teres minor—also join the joint capsule. These are known as the **rotator cuff** muscles because they contribute to rotation of the humerus and their tendons form a collagenous cuff around the gleno-humeral joint. Tension in the rotator cuff muscles pulls the head of the humerus toward the glenoid fossa, contributing significantly to the joint's minimal stability. The joint is most stable in its close-packed po-sition, when the humerus is abducted and laterally rotated.

rotator cuff
band of tendons of subscapularis, supraspinatus, infraspinatus, and the teres minor, which attach to the humeral head

Scapulothoracic Joint

Because the scapula can move in both sagittal and frontal planes with respect to the trunk, the region between the anterior scapula and the thoracic wall is sometimes referred to as the scapulothoracic joint. The muscles attaching to the scapula perform two functions. First, they can contract to stabilize the shoulder region. For example, when a suitcase is lifted from the floor, the levator scapula, trapezius, and rhomboids

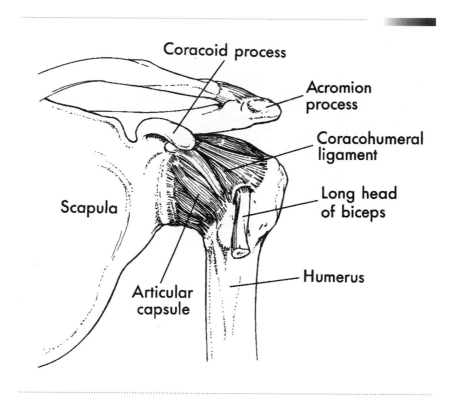

FIGURE 7-4
The capsule surrounding
the glenohumeral joint
contributes to joint
stability.

develop tension to support the scapula, and in turn the entire shoulder, through the acromioclavicular joint. Second, the scapular muscles can facilitate movements of the upper extremity through appropriate positioning of the glenohumeral joint. During an overhand throw, for example, the rhomboids contract to move the entire shoulder posteriorly as the humerus is horizontally abducted and externally rotated during the preparatory phase. As the arm and hand then move forward to execute the throw, tension in the rhomboids is released to permit forward movement of the glenohumeral joint.

Bursae

Several small fibrous sacs that secrete synovial fluid internally similar to a joint capsule are located in the shoulder region. These sacs, known as **bursae,** cushion and reduce friction between layers of collagenous tissues. The shoulder is surrounded by several bursae, including the subcoracoid, subscapularis, and subacromial.

The subacromial bursa lies in the subacromial space, between the acromion process of the scapula and the coracoacromial ligament

bursae
sacs secreting synovial fluid internally that lessen friction between soft tissues around joints

FIGURE 7-5
The four muscles of the
rotator cuff.

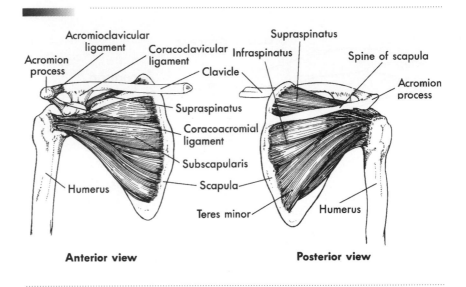

Anterior view   Posterior view

(above) and the glenohumeral joint (below). This bursa cushions the
rotator cuff muscles, particularly the supraspinatus, from the overlying
bony acromion (Figure 7-5). The subacromial bursa may become irri-
tated when repeatedly compressed during overhead arm action.

MOVEMENTS OF THE SHOULDER COMPLEX

Although some amount of glenohumeral motion may occur while the
other shoulder articulations remain stabilized, movement of the
humerus more commonly involves some movement at all three shoulder joints (Figure 7-6). As the arm is elevated in both abduction and
flexion, rotation of the scapula accounts for part of the total humeral
range of motion. Although the absolute positions of the humerus and
scapula vary due to anatomical variations among individuals, a general
pattern persists (14). During about the first 30 degrees of humeral elevation, the contribution of the scapula is only about one-fifth that of the
glenohumeral joint (31). As elevation proceeds beyond 30°, the scapula
rotates approximately 1 degree for every 2 degrees of movement of the
humerus (7, 17, 35). This important coordination of scapular and
humeral movements, known as **scapulohumeral rhythm,** enables a
much greater range of motion at the shoulder than if the scapula were
fixed. During the first 90° of arm elevation (in sagittal, frontal,
or diagonal planes), the clavicle is also elevated through approximately
35° to 45° of motion at the sternoclavicular joint (31). Rotation at the
acromioclavicular joint occurs during the first 30° of humeral elevation

scapulohumeral rhythm
a regular pattern of scapular
rotation that accompanies
and facilitates humeral
abduction

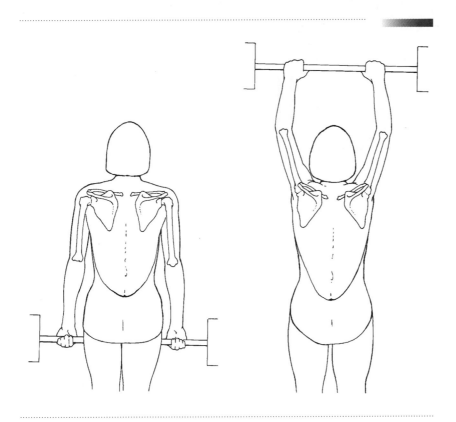

FIGURE 7-6
Elevation of the arm is
accompanied by rotation
of the clavicle and scapula.

and again as the arm is moved from 135° to maximum elevation (22).
Positioning of the humerus is further facilitated by motions of the spine.

Muscles of the Scapula

The muscles that attach to the scapula are the levator scapula, rhomboids, serratus anterior, pectoralis minor, subclavius, and the four parts
of the trapezius. Figures 7-7 and 7-8 show the directions in which each
of these muscles exert force on the scapula when contracting. Scapular
muscles have two general functions. First, they stabilize the scapula so
that it forms a rigid base for muscles of the shoulder during the development of tension. For example, when a person carries a suitcase, the
levator scapula, trapezius, and rhomboids stabilize the shoulder against
the added weight. Second, scapular muscles facilitate movements of the
upper extremity by positioning the glenohumeral joint appropriately.
For example, during an overhand throw the rhomboids contract to move
the entire shoulder posteriorly as the arm and hand move posteriorly

*The scapular muscles perform
two functions: 1. stabilizing the
scapula when the shoulder
complex is loaded, and 2.
moving and positioning the
scapula to facilitate movement
at the glenohumeral joint.*

FIGURE 7-7
Actions of the scapular
muscles.

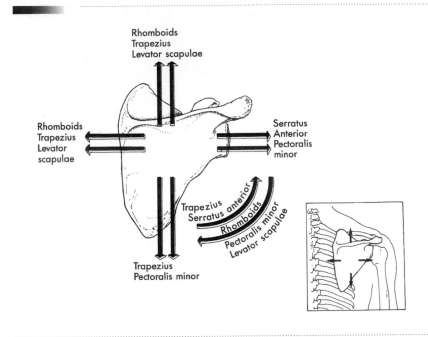

The development of tension in one shoulder muscle must frequently be accompanied by the development of tension in an antagonist to prevent dislocation of the humeral head.

during the preparatory phase. As the arm and hand move anteriorly to deliver the throw, tension in the rhomboids subsides to permit forward movement of the shoulder, facilitating outward rotation of the humerus.

Muscles of the Glenohumeral Joint

Many muscles cross the glenohumeral joint. Because of their attachment sites and lines of pull, some muscles contribute to more than one action of the humerus. A further complication is that the action produced by the development of tension in a muscle may change with the orientation of the humerus because of the shoulder's large range of motion. With the basic instability of the structure of the glenohumeral joint, a significant portion of the joint's stability is derived from tension in the muscles and tendons crossing the joint. However, when one of these muscles develops tension, tension development in an antagonist may be required to prevent dislocation of the joint. A review of the muscles of the shoulder is presented in Table 7-1.

Flexion at the Glenohumeral Joint

The muscles crossing the glenohumeral joint anteriorly participate in flexion at the shoulder (Figure 7-9). The prime flexors are the anterior

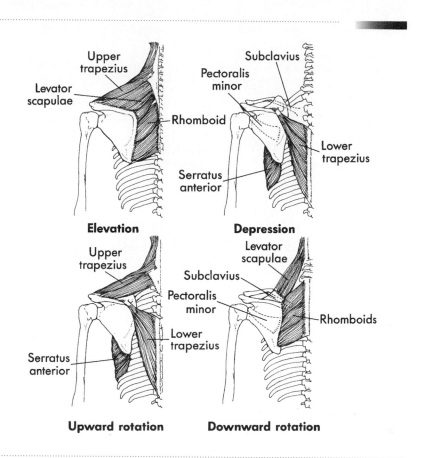

FIGURE 7-8
The muscles of the scapula.

Elevation

Depression

Upward rotation

Downward rotation

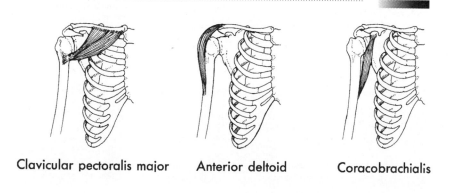

FIGURE 7-9
The major flexor muscles of the shoulder.

Clavicular pectoralis major   Anterior deltoid   Coracobrachialis

TABLE 7-1

MUSCLES OF THE SHOULDER

MUSCLE	PROXIMAL ATTACHMENT	DISTAL ATTACHMENT	PRIMARY ACTIONS ABOUT THE SHOULDER
Deltoid (Anterior)	Outer third of the clavicle	Deltoid tuberosity of the humerus	Flexion, horizontal of adduction
(Middle)	Top of the acromion	Deltoid tuberosity of the humerus	Abduction, horizontal abduction
(Posterior)	Scapular spine	Deltoid tuberosity of the humerus	Extension, horizontal abduction
Perctoralis major (Clavicular)	Medial half of the clavicle	Lateral aspect of the humerus just below the head	Flexion, horizontal adduction
(Sternal)	Anterior sternum and cartilage of first six ribs	Lateral aspect of the humerus just below the head	Extension, adduction, horizontal adduction
Supraspinatus	Suprasipinous fossa	Greater tuberosity of the humerus	Abduction, assists with lateral rotation
Coracobrachialis	Coracoid process of the scapula	Medial anterior humerus	Flexion, adduction, horizontal adduction
Latissimus dorsi	Lower six thoracic and all lumbar vertebrae, posterior sacrum, iliac crest, lower three ribs	Anterior humerus	Extension, adduction, medial rotation
Teres major	Lower, lateral, dorsal scapula	Anterior humerus	Extension, adduction, medial rotation
Infraspinatus	Infraspinous fossa	Greater tubercle of the humerus	Lateral rotation, horizontal abduction
Teres minor	Posterior, lateral border of scapula	Greater tubercle and adjacent shaft of humerus	Lateral rotation, horizontal abduction
Subscapularis	Entire anterior surface of scapula	Lesser tubercle of the humerus	Medial rotation
Biceps brachii (long head)	Upper rim of the glenoid fossa	Tuberosity of the radius	Assists with abduction
(short head)	Coracoid process of the scapula	Tuberosity of the radius	Assists with flexion, adduction, medial rotation, and horizontal adduction
Triceps brachii (long head)	Just inferior to the glenoid fossa	Olecranon process of the ulna	Assists with extension and adduction

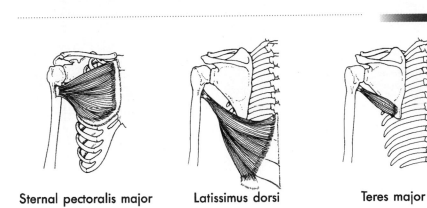

FIGURE 7-10
The major extensor
muscles of the shoulder.

Sternal pectoralis major   Latissimus dorsi   Teres major

deltoid and the clavicular portion of the pectoralis major. The small coracobrachialis assists with flexion, as does the short head of the biceps
brachii. Because the biceps also crosses the elbow joint, it is more effective in its actions at the shoulder when the elbow is in full extension.

Extension at the Glenohumeral Joint

When shoulder extension is unresisted, gravitational force is the primary mover, with eccentric contraction of the flexor muscles controlling or braking the movement. When resistance is present, contraction
of the muscles posterior to the glenohumeral joint, particularly the sternocostal pectoralis, latissimus dorsi, and teres major, extend the
humerus. The posterior deltoid assists in extension, especially when the
humerus is externally rotated. The long head of the triceps brachii also
assists, and because the muscle crosses the elbow, its contribution is
slightly more effective when the elbow is in flexion. The shoulder extensors are illustrated in Figure 7-10.

Abduction at the Glenohumeral Joint

The middle deltoid and supraspinatus are the major abductors of the
humerus. Both muscles cross the shoulder superior to the glenohumeral
joint (Figure 7-11). The supraspinatus, which is active through approximately the first 110° of motion, initiates abduction. During the contribution of the middle deltoid (occurring from approximately 90° to 180°
of abduction) the infraspinatus, subscapularis, and teres minor neu-
tralize the superiorly dislocating component of force produced by the
middle deltoid.

FIGURE 7-11
The major abductor
muscles of the shoulder.

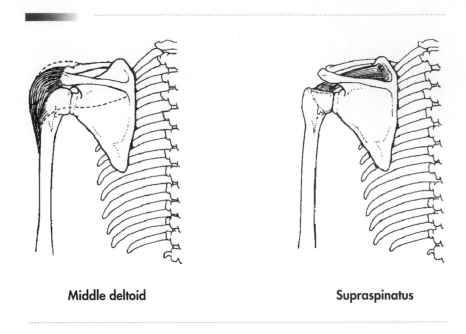

Middle deltoid   **Supraspinatus**

FIGURE 7-11
The major abductor
muscles of the shoulder.

FIGURE 7-12
The major adductor
muscles of the shoulder.

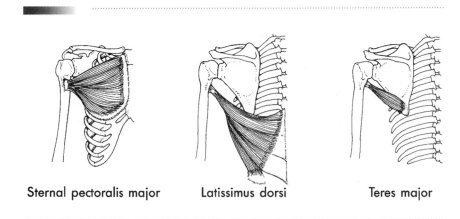

Sternal pectoralis major   **Latissimus dorsi**   **Teres major**

Adduction at the Glenohumeral Joint

As with extension at the shoulder, adduction in the absence of resistance
results from gravitational force, with the abductors controlling the speed
of motion. With resistance added, the primary adductors are the latissimus dorsi, teres major, and sternocostal pectoralis, which are located
on the inferior side of the joint (Figure 7-12). The short head of the biceps and the long head of the triceps contribute minor assistance, and
when the arm is elevated above 90°, the coracobrachialis and subscapularis also assist.

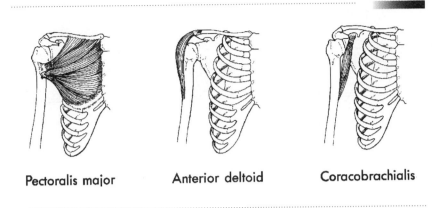

Pectoralis major   Anterior deltoid   Coracobrachialis

FIGURE 7-13
The major horizontal
adductors of the shoulder.

Medial and Lateral Rotation of the Humerus

Medial or inward rotation of the humerus results primarily from the action of the subscapularis and teres major, both attaching to the anterior side of the humerus. Both portions of the pectoralis major, the anterior deltoid, the latissimus dorsi, and the short head of the biceps brachii assist. Muscles attaching to the posterior aspect of the humerus, particularly infraspinatus and teres minor, produce lateral or outward rotation, with some assistance from the posterior deltoid.

Horizontal Adduction and Abduction at the Glenohumeral Joint

The muscles anterior to the joint, including both heads of pectoralis major, the anterior deltoid, and coracobrachialis, produce horizontal adduction, with the short head of the biceps brachii assisting. Muscles posterior to the joint axis affect horizontal abduction. The major horizontal abductors are the middle and posterior portions of the deltoid, infraspinatus, and teres minor, with assistance provided by teres major and the latissimus dorsi. The major horizontal adductors and abductors are shown in Figures 7-13 and 7-14.

LOADS ON THE SHOULDER

Because the articulations of the shoulder girdle are interconnected, they function to some extent as a unit in bearing loads and absorbing shock. However, because the glenohumeral joint provides direct mechanical support for the arm, it sustains much greater loads than the other shoulder joints.

As indicated in Chapter 3, when analyzing the effect of body position, body weight may be assumed to act at the body's center of gravity.

FIGURE 7-14
The major horizontal abductors of the shoulder.

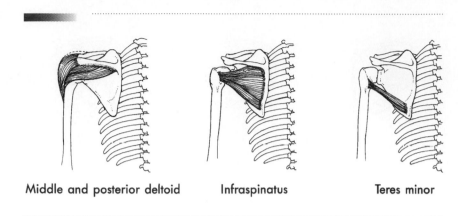

Middle and posterior deltoid   Infraspinatus   Teres minor

Likewise, when analyzing the effect of the positions of body segments on a joint such as the shoulder, the weight of each body segment is assumed to act at the segmental center of mass. The moment arm for the entire arm segment with respect to the shoulder is therefore the perpendicular distance between the weight vector (acting at the arm's center of gravity) and the shoulder (Figure 7-15). When the elbow is in flexion, the effects of the upper arm and the forearm/hand segments must be analyzed separately (Figure 7-16). The sample problem in Figure 7-17 illustrates the effect of arm position on shoulder loading.

Although the weight of the arm is only approximately 5% of body weight, the length of the horizontally extended arm creates large segment moment arms and therefore large torques that must be countered by the shoulder muscles. When these muscles contract to support the extended arm, the glenohumeral joint sustains compressive forces estimated to reach 50% of body weight (30). Although this load is reduced by about half when the elbow is maximally flexed due to the shortened moment arms of the forearm and hand, this can place a rotational torque on the humerus that requires the activation of additional shoulder muscles (Figure 7-18).

Because of the effect of arm position on shoulder loading, ergonomists recommend that workers seated at a desk or a table attempt to position the arms with 20° or less of abduction and 25° or less of flexion (4). Workers who are required to hold the arms in a sustained position overhead are particularly susceptible to degenerative tendinitis in the biceps and supraspinatus (11).

Muscles that attach to the humerus at small angles with respect to the glenoid fossa contribute primarily to shear as opposed to compression at the joint. These muscles serve the important role of stabilizing the humerus in the glenoid fossa against the contractions of powerful

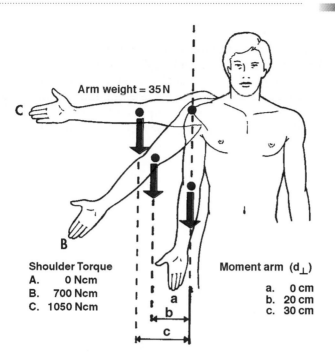

FIGURE 7-15
The torque created at the shoulder by the weight of the arm is the product of arm weight and the perpendicular distance between the arm's center of gravity and the shoulder (the arm's moment arm). Adapted from Chaffin DB and Andersson GBJ, *Occupational Biomechanics* (2nd ed.), New York, John Wiley and Sons, 1991.

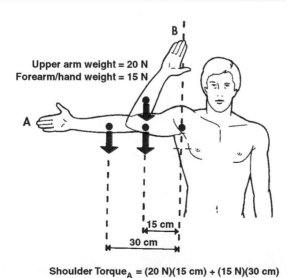

FIGURE 7-16
The torque created at the shoulder by each arm segment is the product of the segment's weight and the segment's moment arm.

FIGURE 7-17

S A M P L E   P R O B L E M   1

Using the simplifying assumptions of Poppen and Walker (31), a free body diagram of the arm and shoulder can be constructed as shown below. If the weight of the arm is 33 N, the moment arm for the total arm segment is 30 cm, and the moment arm for the deltoid muscle (F_m) is 3 cm, how much force must be supplied by the deltoid to maintain the arm in this position. What is the magnitude of the joint reaction force (R)?

Known
wt = 33 N
d_{wt} = 30 cm
d_m = 3 cm

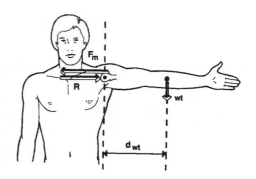

Solution
The torque at the shoulder created by the muscle force must equal the torque at the shoulder created by arm weight, yielding a net shoulder torque of 0.

$$\sum T_s = 0$$

$$\sum T_s = (F_m)(d_m) - (wt)(d_{wt})$$

$$0 = (F_m)(3 \text{ cm}) - (33 \text{ N})(30 \text{ cm})$$

$$F_m = \frac{(33 \text{ N})(30 \text{ cm})}{3 \text{ cm}}$$

$$F_m = 330 \text{ N}$$

Since the joint reaction force (R) and F_m are the only two horizontal forces present, and since the arm is stationary, these forces must be equal and opposite. The magnitude of R is therefore the same as the magnitude of F_m.

$$R = 330 \text{ N}$$

muscles that might otherwise dislocate the joint. For example, when the arm is elevated, the deltoid and the rotator cuff muscles act in tandem, with the deltoid producing an upward shear force that is counteracted by the downward shear produced by the rotator cuff (28) (Figure 7-19). Maximum shear force has been found to be present at the glenohumeral joint when the arm is elevated approximately 60° (41).

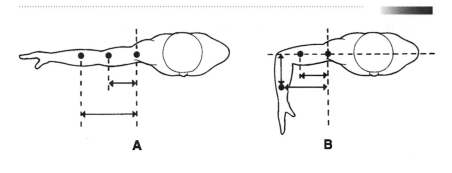

A   **B**

FIGURE 7-18
A, The weight of the arm segments creates a frontal plane torque at the shoulder, with moment arms as shown. **B**, The weight of the upper arm segment creates a frontal plane torque at the shoulder. The weight of the forearm/hand creates both frontal plane and sagittal plane torques at the shoulder, with moment arms as shown. Adapted from Chaffin DB and Andersson GBJ, *Occupational Biomechanics* (2nd ed.), New York, John Wiley and Sons, 1991.

COMMON INJURIES OF THE SHOULDER

The shoulder is susceptible to both traumatic and overuse types of injuries, including 8% to 13% of all sport-related injuries (19).

Dislocations

The loose structure of the glenohumeral joint enables extreme mobility but provides little stability, and dislocations may occur in anterior, posterior, and inferior directions. The strong coracohumeral ligament usually prevents displacement in the superior direction. Glenohumeral dislocations typically occur when the humerus is abducted and externally

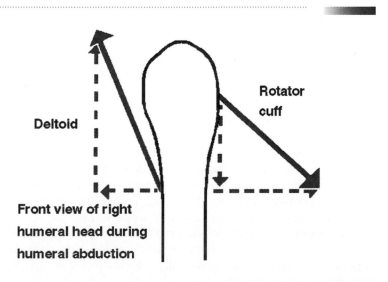

Deltoid

Rotator cuff

Front view of right humeral head during humeral abduction

FIGURE 7-19
Abduction of the humerus requires the cooperative action of the deltoid and the rotator cuff muscles. Because the vertical components of muscle force largely cancel each other, the oppositely directed horizontal components produce rotation of the humerus.

rotated, with anterior-inferior dislocations more common than those in other directions. Factors that predispose the joint to dislocations include inadequate size of the glenoid fossa, anterior tilt of the glenoid fossa, inadequate retroversion of the humeral head, and deficits in the rotator cuff muscles (35).

Glenohumeral dislocation may result from sustaining a large external force during an accident, such as in cycling, or during participation in a contact sport such as wrestling or football. Unfortunately, once the joint has been dislocated, the stretching of the surrounding collagenous tissues beyond their elastic limits commonly predisposes it to subsequent dislocations. Glenohumeral capsular laxity may also be present due to genetic factors. Individuals with this condition should strengthen their shoulder muscles before athletic participation (10).

Dislocations or separations of the acromioclavicular joint are also common among wrestlers and football players. When a rigidly outstretched arm sustains the force of a full-body fall, either acromioclavicular separation or fracture of the clavicle is likely to result.

When a glenohumeral joint dislocation occurs, the supporting soft tissues are often stretched beyond their elastic limits, thus predisposing the joint to subsequent dislocations.

Rotator Cuff Damage

A common injury among workers and athletes who engage in forceful overhead movements typically involving abduction or flexion along with medial rotation is *rotator cuff impingement syndrome* or shoulder impingement syndrome. This syndrome is caused by progressive pressure on the rotator cuff tendons by the surrounding bone and soft tissue structures. Symptoms include hypermobility of the anterior shoulder capsule, hypomobility of the posterior capsule, excessive external rotation coupled with limited internal rotation of the humerus, and general ligamentous laxity of the glenohumeral joint (42). This can result in inflammation of the underlying tendons and bursae, or in severe cases, rupture of one of the rotator cuff tendons. The muscle most commonly affected is the supraspinatus, possibly because its blood supply is the most susceptible to pressure (36). This condition is accompanied by pain and tenderness in the superior and anterior shoulder regions, and sometimes by associated shoulder weakness. The symptoms are exacerbated by rotary movements of the humerus, especially those involving elevation and internal rotation.

Activities that may promote the development of shoulder impingement syndrome include throwing (particularly an implement like the javelin), serving in tennis, and swimming (especially the freestyle, backstroke, and butterfly) (1, 29, 37). Among competitive swimmers, the syndrome is known as *swimmer's shoulder.* Reports indicate shoulder pain complaints in up to 50% of competitive swimmers (34). Older golfers also frequently develop impaired rotator cuff function secondary to degenerative changes such as osteophyte formation that impinges upon the subacromial space (2).

Physicians have proposed two theories regarding the biomechanical cause of most rotator cuff problems (26). The impingement theory suggests that a genetic factor results in the formation of too narrow a space between the acromion process of the scapula and the head of the humerus. In this situation the rotator cuff and associated bursae are pinched between the acromion, the acromioclavicular ligament, and the humeral head each time the arm is elevated, with the resulting friction causing irritation and wear. An alternative theory proposes that the major factor is inflammation of the supraspinatus tendon caused by repeated overstretching of the muscle-tendon unit. When the rotator cuff tendons become stretched and weakened, they cannot perform their normal function of holding the humeral head in the glenoid fossa. Consequently, the deltoid muscles pull the humeral head up too high during abduction, resulting in impingement and subsequent wear and tear on the rotator cuff.

Another theory has been suggested regarding rotator cuff damage among swimmers. Research has shown that during the recovery phase of swimming, the serratus anterior rotates the scapula so that the supraspinatus, infraspinatus, and middle deltoid may freely abduct the humerus (27). The serratus develops nearly maximum tension to accomplish this task. It has been hypothesized that if the serratus becomes fatigued, the scapula may not be rotated sufficiently to abduct the humerus freely, and impingement may develop (27).

Rotational Injuries

Tears of the labrum, the rotator cuff muscles, and the biceps brachii tendon are among the injuries that may result from repeated, forceful rotation at the shoulder. Throwing, serving in tennis, and spiking in volleyball are examples of forceful rotational movements. If the attaching muscles do not sufficiently stabilize the humerus, it can articulate with the glenoid labrum rather than with the glenoid fossa, contributing to wear on the labrum. Most tears are located in the anterior-superior region of the labrum. Tears of the rotator cuff, primarily of the supraspinatus, have been attributed to the extreme tension requirements placed on the muscle group during the deceleration phase of a vigorous rotational activity. Tears of the biceps brachii tendon at the site of its attachment to the glenoid fossa may result from the forceful development of tension in the biceps when it negatively accelerates the rate of elbow extension during throwing (24).

Other pathologies of the shoulder attributed to throwing movements are calcifications of the soft tissues of the joint and degenerative changes in the articular surfaces (32). Bursitis, the inflammation of one or more bursae, is another overuse syndrome, generally caused by friction within the bursa (33).

Serving and hitting in volleyball are activities that often result in overuse injuries of the shoulder.

humeroulnar joint
hinge joint in which the humeral trochlea articulates with the trochlear fossa of the ulna

The humeroulnar hinge joint is considered to be the elbow joint.

humeroradial joint
gliding joint in which the capitellum of the humerus articulates with the proximal end of the radius

radioulnar joints
the proximal and distal radioulnar joints are pivot joints; the middle radioulnar joint is a syndesmosis

When pronation and supination of the forearm occurs, the radius pivots around the ulna.

Subscapular Neuropathy

A shoulder injury that sometimes occurs among competitive volleyball players is subscapular neuropathy. This condition involves denervation of the infraspinatus, with accompanying loss of strength during external rotation of the humerus. It has been attributed to the repeated stretching of the nerve during the serving motion (8).

STRUCTURE OF THE ELBOW

The elbow encompasses three articulations: the humeroulnar, humeroradial, and proximal radioulnar joints. All are enclosed in the same joint capsule, which is reinforced by the anterior and posterior radial collateral and ulnar collateral ligaments.

Humeroulnar Joint

The hinge joint at the elbow is the **humeroulnar joint,** where the ovular trochlea of the humerus articulates with the reciprocally shaped trochlear fossa of the ulna (Figure 7-20). Flexion and extension are the primary movements, although in some individuals, a small amount of hyperextension is allowed. The joint is most stable in the close-packed position of extension.

Humeroradial Joint

The **humeroradial joint** is immediately lateral to the humeroulnar joint and is formed between the spherical capitellum of the humerus and the proximal end of the radius (Figure 7-20). Although the humeroradial articulation is classified as a gliding joint, the immediately adjacent humeroulnar joint restricts motion to the sagittal plane. The close-packed position is with the elbow flexed at 90° and the forearm supinated about 5°.

Proximal Radioulnar Joint

The annular ligament binds the head of the radius to the radial notch of the ulna, forming the proximal **radioulnar joint.** This is a pivot joint, with forearm pronation and supination occurring as the radius rolls medially and laterally over the ulna (Figure 7-21). The close-packed position is at 5° of forearm supination.

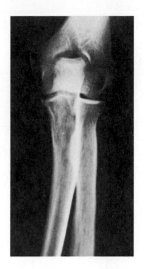

Posterior view of the left elbow. From Shier, Butler, and Lewis. *Hole's Human Anatomy and Physiology.* © 1996. Reprinted by permission of The McGraw-Hill Companies, Inc.

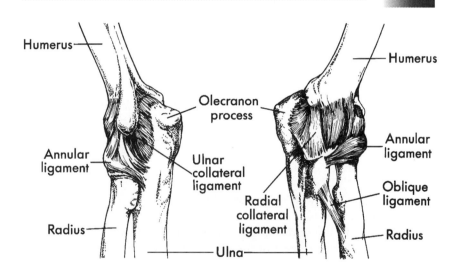

FIGURE 7-20
The major ligaments of the elbow.

MOVEMENTS AT THE ELBOW

Muscles Crossing the Elbow

Numerous muscles cross the elbow, including those that also cross the shoulder or extend into the hands and fingers. The muscles classified as primary movers of the elbow are summarized in Table 7-2.

FIGURE 7-21

From Shier, Butler, and Lewis. *Hole's Human Anatomy and Physiology.* © 1996. Reprinted by permission of The McGraw-Hill Companies, Inc.

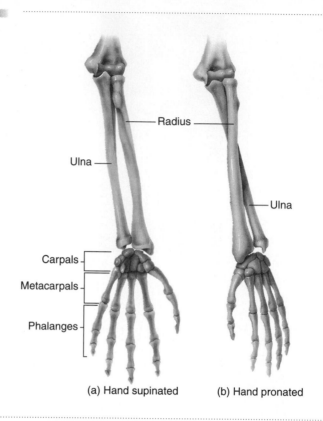

(a) Hand supinated   (b) Hand pronated

Flexion and Extension

Muscles crossing the anterior side of the elbow are the elbow flexors (Figure 7-22). The strongest of the elbow flexors is the brachialis. Since the distal attachment of the brachialis is the coronoid process of the ulna, the muscle is equally effective when the forearm is in supination or pronation. Because it is the major forearm flexor, the brachialis has been called the workhorse of the elbow (18).

Another elbow flexor is the biceps brachii, with both long and short heads attached to the radial tuberosity by a single common tendon. The muscle contributes effectively to flexion when the forearm is supinated, because it is slightly stretched. When the forearm is pronated, the muscle is less taut and consequently less effective.

The brachioradialis is a third contributor to flexion at the elbow. This muscle is most effective when the forearm is in a neutral position (midway between full pronation and full supination) because of its distal attachment to the base of the styloid process on the lateral radius. In this

TABLE 7-2

MAJOR MUSCLES OF THE ELBOW

MUSCLE	PROXIMAL ATTACHMENT	DISTAL ATTACHMENT	PRIMARY ACTIONS ABOUT THE ELBOW
Biceps brachii			
(Long head)	Superior rim of the glenoid fossa	Tuberosity of the radius	Flexion, supination
(Short head)	Coracoid process of the scapula	Tuberosity of the radius	Flexion, supination
Brachioradialis	Upper two-thirds lateral supra-condylar ridge of humerus	Styloid process of the radius	Flexion
Brachialis	Anterior lower half of the humerus	Anterior coronoid process of ulna	Flexion
Pronator teres			
(Humeral head)	Medial epicondyle of the humerus	Lateral midpoint of the radius	Assists with flex-ion and prona-tion
(Ulnar head)	Coronoid process of the ulna	Lateral midpoint of the radius	Assists with flex-ion and pronation
Pronator quadratus	Lower fourth of the anterior ulna	Lower fourth of the anterior radius	Pronation
Triceps brachii			
(Long head)	Just inferior to the glenoid fossa	Olecranon process of the ulna	Extension
(Lateral head)	Upper half of the posterior humerus	Olecranon process of the ulna	Extension
(Medial head)	Lower two-thirds of posterior humerus	Olecranon process of the ulna	Extension
Anconeus	Posterior, lateral epicondyle of the humerus	Lateral olecranon and posterior ulna	Assists with extension
Supinator	Lateral epicondyle of humerus and adjacent ulna	Lateral upper third of radius	Supination

FIGURE 7-22
The major flexor muscles
of the elbow.

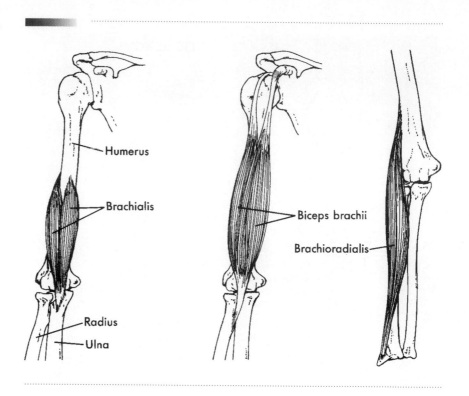

position the muscle is in slight stretch, and the radial attachment is centered in front of the elbow joint.

The major extensor of the elbow is the triceps, which crosses the posterior aspect of the joint (Figure 7-23). Although the three heads have separate proximal attachments, they attach to the olecranon process of the ulna through a common distal tendon. Even though the distal attachment is relatively close to the axis of rotation at the elbow, the size and strength of the muscle make it effective as an elbow extensor. The relatively small anconeus muscle, which courses from the posterior surface of the lateral epicondyle of the humerus to the lateral olecranon and posterior proximal ulna, also assists with extension.

Pronation and Supination

Pronation and supination of the forearm involve rotation of the radius around the ulna. There are three radioulnar articulations: the proximal, middle, and distal radioulnar joints. Both the proximal and the distal joints are pivot joints, and the middle radioulnar joint is a syndesmosis at which an elastic, interconnecting membrane permits supination and pronation but prevents longitudinal displacement of the bones.

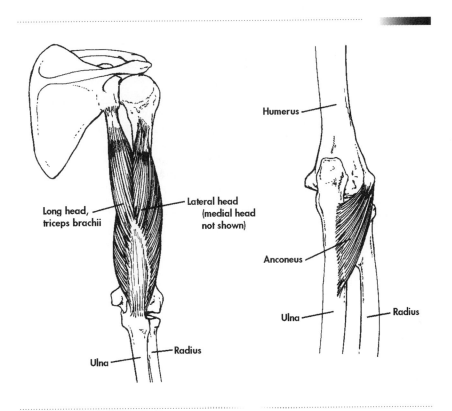

FIGURE 7-23
The major extensor muscles of the elbow.

The major pronator is the pronator quadratus, which attaches to the distal ulna and radius (Figure 7-24). When pronation is resisted or rapid, the pronator teres crossing the proximal radioulnar joint assists.

As the name suggests, the supinator is the muscle primarily responsible for supination (Figure 7-25). It is attached to the lateral epicondyle of the humerus and to the lateral proximal third of the radius. When the elbow is in flexion, tension in the supinator lessens, and the biceps assists with supination. When the elbow is flexed to 90° or less, the biceps is positioned to serve as a strong supinator (38).

LOADS ON THE ELBOW

Although the elbow is not considered to be a weightbearing joint, it regularly sustains large loads during daily activities. For example, research shows that the compressive load at the elbow reaches an estimated 300 N (67 lb) during activities such as dressing and eating, 1700 N (382 lb) when the body is supported by the arms when rising from a chair,

FIGURE 7-24
The major pronator
muscle is the pronator
quadratus.

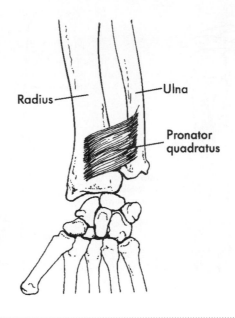

FIGURE 7-25
The major supinator
muscle is the supinator.

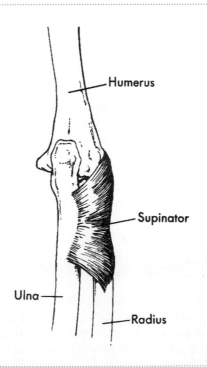

and 1900 N (427 lb) when pulling a table across the floor (43). During push-up exercises peak forces on each elbow are calculated to reach 45% of body weight (6).

It is likely that even greater loads are present during the execution of selected sport skills. Large forces are generated by the muscles crossing the elbow during forceful pitching and throwing motions, as well as during many resistance training exercises and weight lifting. During the execution of gymnastic skills such as the handspring and the vault, the elbow functions as a weightbearing joint. Research indicates that maximal isometric flexion when the elbow is fully extended can produce joint compression forces of as much as two times body weight (16).

Since the attachment of the triceps tendon to the ulna is closer to the elbow joint center than the attachments of the brachialis on the ulna and the biceps on the radius, the extensor moment arm is shorter than the flexor moment arm. This means that the elbow extensors must generate more force than the elbow flexors to produce the same amount of joint torque. This translates to larger compression forces at the elbow during extension than during flexion when movements of comparable speed and force requirements are executed. The sample problem in Figure 7-26 illustrates the relationship between moment arm and torque at the elbow.

Because of the shape of the olecranon process, the triceps moment arm also varies with the position of the elbow. As shown in Figure 7-27, the triceps moment arm is larger when the arm is fully extended than when it is flexed past 90°.

COMMON INJURIES OF THE ELBOW

Although the elbow is a stable joint reinforced by large strong ligaments, the large loads placed upon the joint during daily activities and sport participation render it susceptible to dislocations and overuse injuries.

Sprains and Dislocations

Forced hyperextension of the elbow can cause posterior displacement of the coronoid process of the ulna with respect to the trochlea of the humerus. Such displacement stretches the ulnar collateral ligament, which may rupture (sprain) anteriorly.

Continued hyperextension of the elbow can cause the distal humerus to slide over the coronoid process of the ulna, resulting in dislocation. Although dislocations of the elbow do not occur as frequently as dislocations of the glenohumeral joint, elbow dislocations occurring in individuals under the age of 30 are most likely to arise during participation in sports (20). The mechanism involved is typically falling on an outstretched hand or a forceful, twisting blow. The subsequent stability

FIGURE 7-26

S A M P L E   P R O B L E M   2

How much force must be produced by the brachioradialis and biceps (F_m) to maintain the 15 N forearm and hand in the position shown below given moment arms of 5 cm for the muscles and 15 cm for the forearm/hand weight? What is the magnitude of the joint reaction force?

Known
wt = 15 N
d_{wt} = 15 cm
d_m = 5 cm

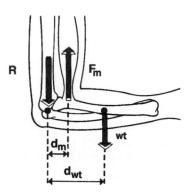

Solution
The torque at the elbow created by the muscle force must equal the torque at the elbow created by forearm/hand weight, yielding a net elbow torque of 0.

$$\sum T_e = 0$$

$$\sum T_e = (F_m)(d_m) - (wt)(d_{wt})$$
$$0 = (F_m)(5\ cm) - (15\ N)(15\ cm)$$
$$F_m = \frac{(15\ N)(15\ cm)}{5\ cm}$$

$$F_m = 45N$$

Since the arm is stationary, the sum of all of the acting vertical forces must be equal to 0. In writing the force equation, it is convenient to regard upward as the positive direction.

$$\sum F_v = 0$$

$$\sum F_v = F_m - wt - R$$

$$\sum F_v = 45\ N - 15\ N - R$$

$$R = 30\ N$$

of a once dislocated elbow is impaired, particularly if the dislocation was accompanied by humeral fracture or rupture of the ulnar collateral ligament (20). Because of the large number of nerves and blood vessels passing through the elbow, elbow dislocations are particularly dangerous.

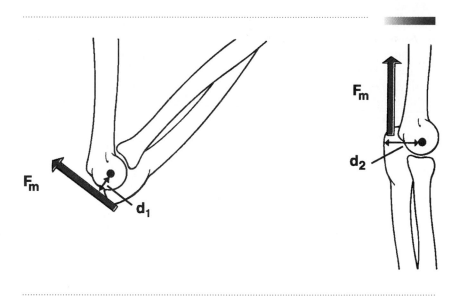

FIGURE 7-27
Because of the shape of
the olecranon process of
the ulna, the moment arm
for the triceps tendon is
shorter when the elbow is
in flexion.

Elbow dislocations in children aged 1–3 years are sometimes referred
to as "nursemaid's elbow" or "pulled elbow." Adults should avoid lifting
or swinging young children by the hands, wrists, or forearms, as this type
of injury can result.

Overuse Injuries

With the exception of the knee, the elbow is the joint most commonly
affected by overuse injuries (18). Stress injuries to the collagenous tissues at the elbow are progressive. The first symptoms are inflammation
and swelling, followed by scarring of the soft tissues. If the condition
progresses further, calcium deposits accumulate and ossification of the
ligaments ensues.

Lateral **epicondylitis** involves inflammation or microdamage to the
tissues on the lateral side of the distal humerus, including the tendinous
attachment of the extensor carpi radialis brevis and possibly that of the
extensor digitorum. Although a host of factors may contribute to the
development of the condition, overuse of the wrist extensors is cited as
a major culprit (12).

Because of the relatively high incidence of lateral epicondylitis among
tennis players, the injury is commonly referred to as tennis elbow. A reported 30% to 40% of tennis players develop lateral epicondylitis, with
onset typically in players between the ages of 35 and 50 (18). The amount
of force to which the lateral aspect of the elbow is subjected during tennis play increases with poor technique and improper equipment. For

epicondylitis
inflammation and sometimes
microrupturing of the
collagenous tissues on either
the lateral or the medial
side of the distal humerus;
believed to be an overuse
injury

example, hitting off-center shots and using an overstrung racquet increase the amount of force transmitted to the elbow (12). Other activities, such as swimming, fencing, and the act of hammering, can contribute to lateral epicondylitis as well (13).

Medial epicondylitis, which has been called "Little Leaguer's elbow," is the same type of injury to the tissues on the medial aspect of the distal humerus. During pitching, the strain imparted to the medial aspect of the elbow during the initial stage when the trunk and shoulder are brought forward ahead of the forearm and hand contributes to development of the condition. Medial epicondyle avulsion fractures have also been attributed to forceful terminal wrist flexion during the follow-through phase of the pitch (18). Throwing curve balls and sliders requires added forceful rotation of the arm and hand during the delivery and increases the stresses placed on the elbow and shoulder. For this reason these pitches are contraindicated for Little League pitchers. It is recommended that young athletes not progress more than 10% per week in the number of balls thrown (9).

Medial and lateral epicondylitis occur with about equal frequency in golfers, particularly amateurs (2). Among right-handed golfers, lateral epicondylitis occurs more often on the left side and medial epicondylitis is found more often on the right side (2). Lateral epicondylitis may be related to gripping the club with excessive pronation of the right hand, while medial epicondylitis appears to be associated with repeatedly striking the ground with the club (2).

STRUCTURE OF THE WRIST

radiocarpal joints
condyloid articulations between the radius and the three carpal bones

The radiocarpal joints make up the wrist.

The wrist is composed of radiocarpal and intercarpal articulations (Figure 7-28). Most wrist motion occurs at the **radiocarpal joint,** a condyloid joint where the radius articulates with the scaphoid, the lunate, and the triquetrum. The joint allows sagittal plane motions (flexion, extension, and hyperextension) and frontal plane motions (radial deviation and ulnar deviation), as well as circumduction. Its close-packed position is in extension with radial deviation. A cartilaginous disc separates the distal head of the ulna from the lunate and triquetral bones and the radius. Although this articular disk is common to both the radiocarpal joint and the distal radioulnar joint, the two articulations have separate joint capsules. The radiocarpal joint capsule is reinforced by the volar radiocarpal, dorsal radiocarpal, radial collateral, and ulnar collateral ligaments. The intercarpal joints are gliding joints that contribute little to wrist motion.

retinacula
fibrous bands of fascia

The fascia around the wrist is thickened into strong fibrous bands called **retinacula** that form protective passageways through which tendons, nerves, and blood vessels pass. The flexor retinaculum protects the extrinsic flexor tendons and the median nerve where they cross the palmar side of the wrist. On the dorsal side of the wrist the extensor retinaculum provides a passageway for the extrinsic extensor tendons.

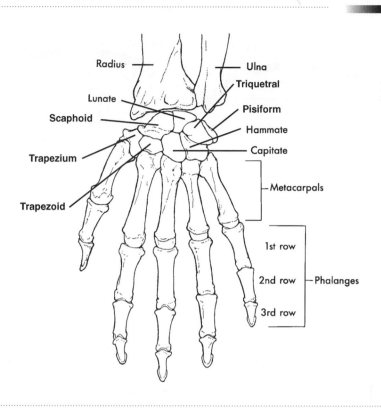

FIGURE 7-28
The bones of the wrist.

MOVEMENTS OF THE WRIST

The wrist is capable of sagittal and frontal plane movements, as well as rotary motion (Figure 7-29). Flexion is motion of the palmar surface of the hand toward the anterior forearm. Extension is the return of the hand to anatomical position, and in hyperextension, the dorsal surface of the hand approaches the posterior forearm. Movement of the hand toward the thumb side of the arm is radial deviation, with movement in the opposite direction designated as ulnar deviation. Rotational movement of the hand through all four directions produces circumduction.

Flexion

The muscles responsible for flexion at the wrist are the flexor carpi radialis and the powerful flexor carpi ulnaris (Figure 7-30). The palmaris longus, which is often absent in one or both forearms, contributes to flexion when present. All three muscles have proximal

FIGURE 7-29
Movements occurring at
the wrist.

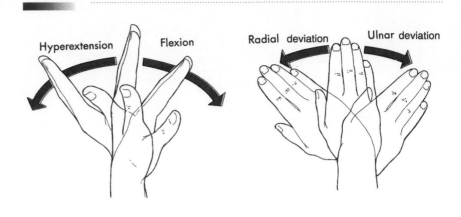

FIGURE 7-30
The major flexor muscles
of the wrist.

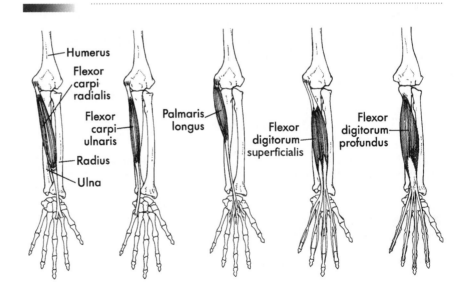

attachments on the medial epicondyle of the humerus. The flexor digitorum superficialis and flexor digitorum profundus can assist with
flexion at the wrist when the fingers are completely extended, but when
the fingers are in flexion these muscles cannot develop sufficient tension due to active insufficiency.

Extension and Hyperextension

Extension and hyperextension at the wrist result from contraction of
the extensor carpi radialis longus, extensor carpi radialis brevis, and ex-

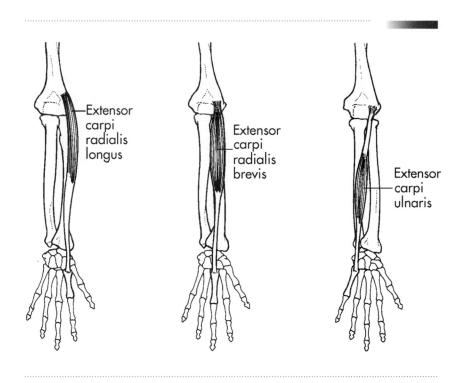

FIGURE 7-31
The major extensor muscles of the wrist.

Extensor carpi radialis longus

Extensor carpi radialis brevis

Extensor carpi ulnaris

tensor carpi ulnaris (Figure 7-31). These muscles originate on the lateral epicondyle of the humerus. The other posterior wrist muscles may also assist with extension, particularly when the fingers are in flexion. Included in this group are the extensor pollicis longus, extensor indicis, extensor digiti minimi, and extensor digitorum (Figure 7-32).

Radial and Ulnar Deviation

Cooperative action of both flexor and extensor muscles produces lateral deviation of the hand at the wrist. The flexor carpi radialis and extensor carpi radialis longus and brevis contract to produce radial deviation, and the flexor carpi ulnaris and extensor carpi ulnaris cause ulnar deviation.

STRUCTURE OF THE JOINTS OF THE HAND

A large number of joints are required to provide the extensive motion capabilities of the hand. Included are the carpometacarpal (CM), intermetacarpal, metacarpophalangeal (MP), and interphalangeal (IP) joints (Figure 7-33). The fingers are referred to as digits one through five, with the first digit being the thumb.

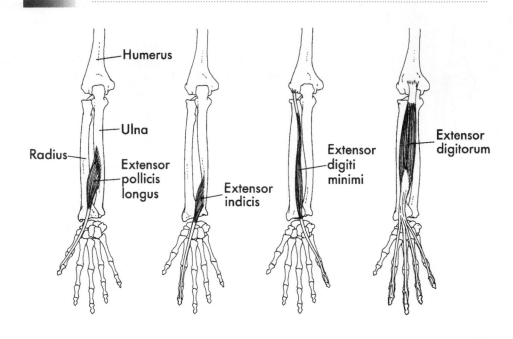

FIGURE 7-32

Muscles that assist with extension of the wrist.

Carpometacarpal and Intermetacarpal Joints

The carpometacarpal joint of the thumb, the articulation between the trapezium and the first metacarpal, is a classic saddle joint. The other carpometacarpal joints are generally regarded as gliding joints, although some anatomists have described them as modified saddle joints (23). All carpometacarpal joints are surrounded by joint capsules, which are reinforced by the dorsal, volar, and interosseous carpometacarpal ligaments. The irregular intermetacarpal joints share these joint capsules.

Metacarpophalangeal Joints

The metacarpophalangeal joints are the condyloid joints between the rounded distal heads of the metacarpals and the concave proximal ends of the phalanges. These joints form the knuckles of the hand. Each joint is enclosed in a capsule that is reinforced by strong collateral ligaments. A dorsal ligament also merges with the metacarpophalangeal joint of the thumb. Close-packed positions of the MP joints in the fingers and thumb are full flexion and opposition, respectively.

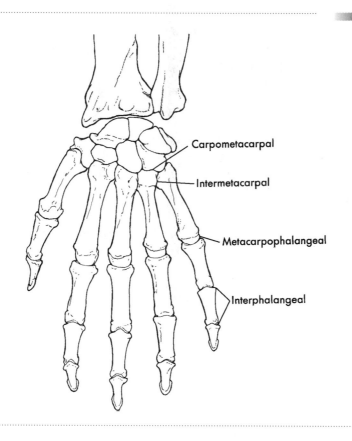

FIGURE 7-33
The bones of the hand.

Carpometacarpal

Intermetacarpal

Metacarpophalangeal

Interphalangeal

Interphalangeal Joints

The proximal and distal interphalangeal joints of the fingers and the single interphalangeal joint of the thumb are all hinge joints. An articular capsule joined by volar and collateral ligaments surrounds each interphalangeal joint. These joints are most stable in the close-packed position of full extension.

MOVEMENTS OF THE HAND

The carpometacarpal joint of the thumb allows a large range of movement similar to that of a ball and socket joint (Figure 7-34). Motion at carpometacarpal joints two through four is slight due to constraining ligaments, with somewhat more motion permitted at the fifth carpometacarpal joint.

The large range of movement of the thumb compared to that of the fingers is derived from the structure of the thumb's carpometacarpal joint.

FIGURE 7-34
Movements of the thumb.

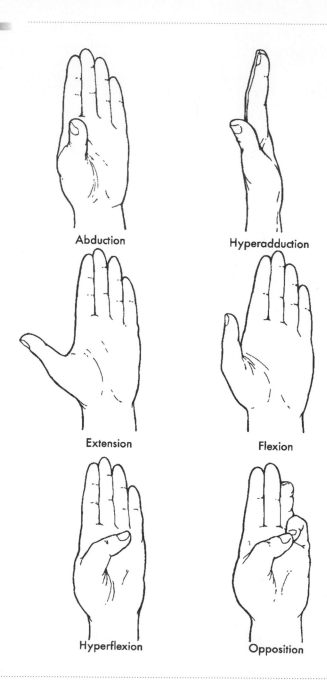

Abduction

Hyperadduction

Extension

Flexion

Hyperflexion

Opposition

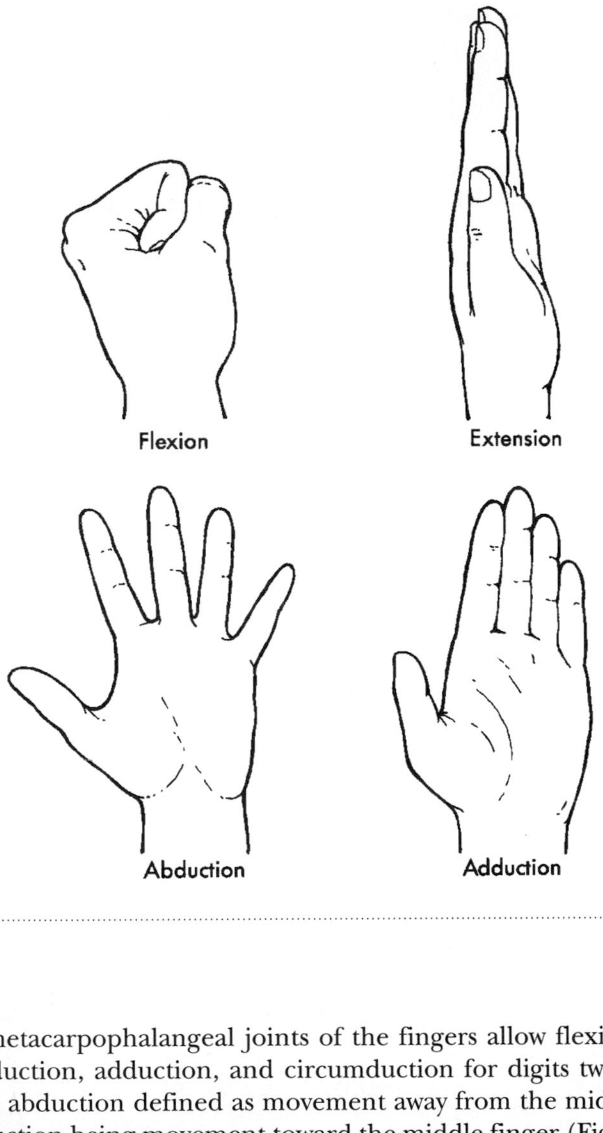

FIGURE 7-35
Movements of the fingers.

Flexion   Extension

Abduction   Adduction

The metacarpophalangeal joints of the fingers allow flexion, extension, abduction, adduction, and circumduction for digits two through five, with abduction defined as movement away from the middle finger and adduction being movement toward the middle finger (Figure 7-35). Because the articulating bone surfaces at the metacarpophalangeal joint of the thumb are relatively flat, the joint functions more as a hinge joint, allowing only flexion and extension.

The interphalangeal joints permit flexion and extension, and in some individuals, slight hyperextension. These are classic hinge joints.

A relatively large number of muscles are responsible for the many precise movements performed by the hand and fingers (Table 7-3). There are nine **extrinsic muscles** that cross the wrist and ten **intrinsic muscles** with both of their attachments distal to the wrist.

extrinsic muscles
muscles with proximal attachments located proximal to the wrist and distal attachments located distal to the wrist

intrinsic muscles
muscles with both attachments distal to the wrist

TABLE 7-3

MAJOR MUSCLES OF THE HAND AND FINGERS

The Extrinsic Muscles

MUSCLE	PROXIMAL ATTACHMENT	DISTAL ATTACHMENT	PRIMARY ACTIONS
Extensor pollicis longus	Middle dorsal ulna	Dorsal distal phalanx of thumb	Extension at MP & IP joints of thumb, abduction of hand
Extensor pollicis brevis	Middle dorsal radius	Dorsal proximal phalanx of thumb	Extension at MP & CM joints of thumb, abduction of hand
Flexor pollicis longus	Middle palmar radius	Palmar distal phalanx of thumb	Flexion at IP & MP joints of thumb
Abductor pollicis longus	Middle dorsal ulna and radius	Radial base of first metacarpal	Extension at CM joint of thumb, abduction of hand
Extensor indicis	Distal dorsal ulna	Ulnar side of the extensor digitorum tendon	Extension at MP joint of 2nd digit
Extensor digitorum	Lateral epicondyle and humerus	Base of 2nd & 3rd phalanges, digits 2-5	Extension at MP, proximal & distal IP joints, digits 2-5
Extensor digiti minimi	Proximal tendon of extensor digitorum	Tendon of extensor digitorum distal to 5th MP joint	Extension at 5th MP joint
Flexor digitorum profundus	Proximal 3/4 ulna	Base of distal phalanx, digits 2-5	Flexion at distal & proximal IP joints and MP joints, digits 2-5, and hand
Flexor digitorum superficialis	Medial epicondyle of humerus	Base of middle phalanx, digits 2-5	Flexion at proximal IP and MP joints, digits 2-5, and hand

(Continued)

MAJOR MUSCLES OF THE HAND AND FINGERS (CONTINUED)

The Intrinsic Muscles

MUSCLE	PROXIMAL ATTACHMENT	DISTAL ATTACHMENT	PRIMARY ACTIONS
Flexor pollicis brevis	Ulnar side, first metacarpal	Ulnar, palmar base of proximal phalanx	Flexion and abduction at MP joint of the thumb
Abductor pollicis brevis	Trapezium and scaphoid bones	Radial base of first phalanx of thumb	Abduction at first CM joint
Opponens pollicis	Trapezium	Radial side first metacarpal	Opposition at CM joint of the thumb
Adductor pollicis	Capitate, distal 2nd & 3rd metacarpals	Ulnar proximal phalanx of thumb	Adduction at CM joint of thumb
Abductor digiti minimi	Pisiform bone	Ulnar base of proximal phalanx, 5th digit	Abduction & flexion at 5th MP joint
Flexor digiti minimi brevis	Hamate bone	Ulnar base of proximal phalanx, 5th digit	Flexion at 5th MP joint
Opponens digiti minimi	Hamate bone	Ulnar metacarpal of 5th metacarpal	Opposition at 5th CM joint
Dorsal interossei (four muscles)	Sides of metacarpals, all digits	Base of proximal phalanx, all digits	Abduction at 2nd & 4th MP joints, radial & ulnar deviation of 3rd MP joint, flexion of 2nd, 3rd, & 4th MP joints
Palmar interossei (three muscles)	2nd, 4th, & 5th metacarpals	Base of proximal phalanx, digits 2, 4, & 5	Adduction & flexion at MP joints, digits 2, 4, & 5
Lumbricales (four muscles)	Tendons of flexor digitorum profundus, digits 2-5	Tendons of extensor digitorum, digits 2-5	Flexion at MP joints of digits 2-5

The extrinsic flexor muscles of the hand are more than twice as strong as the strongest of the extrinsic extensor muscles (40). This should come as little surprise, given that the flexor muscles of the hand are used extensively in everyday activities involving gripping, grasping, or pinching movements, while the extensor muscles rarely exert much force. The most powerful of all the wrist muscles is the flexor carpi ulnaris (39). Contraction of this muscle moves the wrist in the directions of flexion and ulnar deviation.

COMMON INJURIES OF THE WRIST AND HAND

The hand is used almost continuously in daily activities and in many sports. Wrist sprains or strains are fairly common and are occasionally accompanied by dislocation of a carpal bone or the distal radius. These types of injuries often result from the natural tendency to sustain the force of a fall on the hyperextended wrist. Fractures of the scaphoid and lunate bones are relatively common for the same reason.

Certain hand/wrist injuries are characteristic of participation in a given sport. Examples are metacarpal (boxer's) fractures and mallet or drop finger deformity resulting from injury at the distal interphalangeal joints among football receivers and baseball catchers. Forced abduction of the thumb leading to ulnar collateral ligament injury often results from wrestling, football, hockey, and skiing (19). The most common injuries encountered in skateboarding and snowboarding are fractures of or close to the wrist (5). In sport rock climbing, 62% of all injuries are to the elbow, forearm, wrist, and hand, with many injuries specific to the handholds employed (15).

The wrist is the most frequently injured joint among golfers, with right-handed golfers tending to injure the left wrist (2). Both overuse injuries such as De Quervains disease (tendinitis of the extensor pollicis brevis and the abductor pollicis longus) and impact-related injuries are common (25). According to one study, golfers with overuse injuries of the wrist use a larger-than-average range of motion of the wrists during the swing (3).

Carpal tunnel syndrome is a fairly common disorder that occurs most often in women of middle age or older (36). The carpal tunnel is a passageway between the carpal bones and the flexor retinaculum on the palmar side of the wrist. Any swelling caused by acute or chronic trauma in the region can compress the median nerve that passes through the carpal tunnel, thus bringing on the syndrome. Symptoms include pain and numbness along the median nerve, clumsiness of finger function, and eventually weakness and atrophy of the muscles supplied by the median nerve. Workers at jobs requiring repeated forceful wrist flexion and office workers who habitually rest the arms on the palmar sides of the wrist are especially vulnerable to carpal tunnel syndrome.

SUMMARY

The shoulder is the most complex joint in the human body, with four different articulations contributing to movement. The glenohumeral joint is a loosely structured ball and socket joint in which range of movement is substantial and stability is minimal. The sternoclavicular joint enables some movement of the bones of the shoulder girdle, clavicle, and scapula. Movements of the shoulder girdle contribute to optimal positioning of the glenohumeral joint for different humeral movements. Small movements are also provided by the acromioclavicular and coracoclavicular joints.

The humeroulnar articulation controls flexion and extension at the elbow. Pronation and supination of the forearm occur at the proximal and distal radioulnar joints.

The structure of the condyloid joint between the radius and the three carpal bones controls motion at the wrist. Flexion, extension, radial flexion, and ulnar flexion are permitted. The joints of the hand at which most movements occur are the carpometacarpal joint of the thumb, the metacarpophalangeal joints, and the hinges at the interphalangeal articulations.

INTRODUCTORY PROBLEMS

1. Construct a chart listing all muscles crossing the glenohumeral joint according to whether they are superior, inferior, anterior, or posterior to the joint center. Note that some muscles may fall into more than one category. Identify the action or actions performed by muscles in each of the four categories.

2. Construct a chart listing all muscles crossing the elbow joint according to whether they are medial, lateral, anterior, or posterior to the joint center with the arm in anatomical position. Note that some muscles may fall into more than one category. Identify the action or actions performed by muscles in each of the four categories.

3. Construct a chart listing all muscles crossing the wrist joint according to whether they are medial, lateral, anterior, or posterior to the joint center with the arm in anatomical position. Note that some muscles may fall into more than one category. Identify the action or actions performed by muscles in each of the four categories.

4. List the muscles that develop tension to stabilize the scapula during each of the following activities:
   a. Carrying a suitcase
   b. Waterskiing
   c. Performing a push-up
   d. Performing a pull-up

5. List the muscles used as agonists, antagonists, stabilizers, and neutralizers during the performance of a push-up.

6. Explain how the use of an overhand as compared to an underhand grip affects an individual's ability to perform a pull-up.

7. Select a familiar activity and identify the muscles of the upper extremity that are used as agonists during the activity.

8. Using the diagram in Figure 7-26 as a model, calculate the tension required in the deltoid with a moment arm of 3 cm from the shoulder, given the following weights and moment arms for the upper arm (u), forearm (f), and hand (h) segments: $wt_u = 19$ N, $wt_f = 11$ N, $wt_h = 4$ N, $d_u = 12$ cm, $d_f = 40$ cm, $d_h = 64$ cm. (308 N)

9. Which of the three segments in problem 8 creates the largest torque about the shoulder when the arm is horizontally extended? Explain your answer and discuss the implications for positioning the arm for shoulder level tasks.

10. Solve sample Problem 2 with the addition of a 10 kg bowling ball held in the hand at a distance of 35 cm from the elbow. (Remember that kg is a unit of mass, not weight!) ($F_m = 732$ N, R = 619 N)

ADDITIONAL PROBLEMS

1. Identify the sequence of movements that occur at the scapulothoracic, shoulder, elbow, and wrist joints during the performance of an overhand throw.

2. Which muscles are most likely to serve as agonists to produce the movements identified in your answer to Problem 1?

3. Select a familiar racket sport and identify the sequence of movements that occur at the shoulder, elbow, and wrist joints during the execution of forehand and backhand strokes.

4. Which muscles are most likely to serve as agonists to produce the movements identified in your answer to Problem 5?

5. Select five resistance-based exercises for the upper extremity and identify which muscles are the primary movers and which muscles assist during the performance of each exercise.

6. Discuss the importance of the rotator cuff muscles as stabilizers of the glenohumeral joint and movers of the humerus.

7. Discuss possible mechanisms of rotator cuff injury. Include in your discussion the implications of the force-length relationship for muscle (described in Chapter 6).

8. How much tension (F_m) must be supplied by the triceps to stabilize the arm against an external force (F_e) of 200 N, given $d_m = 2$ cm and $d_e = 25$ cm as shown in the diagram below? What is the magnitude of the joint reaction force (R)? (Since the forearm is vertical, its weight does not produce torque at the elbow.) (Fm = 2500 N, R = 2700 N)

9. What is the length of the moment arm between the dumbbell and the shoulder when the extended 50 cm arm is positioned as shown below? (43.3 cm)

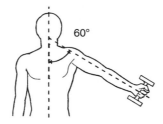

10. The medial deltoid attaches to the humerus at an angle of 15° as shown below. What are the sizes of the rotary and stabilizing components of muscle force when the total muscle force is 500 N? (rotary component = 129 N, stabilizing component = 483 N)

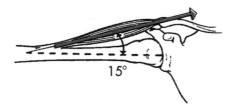

LABORATORY EXPERIENCES

1. Study anatomical models of the shoulder, elbow, and wrist. Be able to locate the major bones, muscle attachments, and ligaments.
2. With a partner, use a goniometer to measure wrist range of motion in flexion and hyperextension, with the fingers both completely flexed and completely extended. Explain your results.
3. Perform pull-ups using an overhand grip and using an underhand grip. Explain which is easier in terms of muscle function.
4. Perform push-ups using wide, medium, and narrow hand placements. Explain which is easiest and which is hardest in terms of muscle function.

5. With a partner, use a goniometer to measure both active and passive shoulder range of motion in horizontal abduction. Explain your results.

REFERENCES

1. Allegrucci M, Whitney SL, and Irrgang JJ: Clinical implications of secondary impingement of the shoulder in freestyle swimmers, J Orthop Sports Phys Ther 20: 307, 1994.
2. Batt ME: Golfing injuries: An overview, Sports Med 16:64, 1993.
3. Calahan TD, et al: Biomechanics of the golf swing in players with pathological conditions of the forearm, wrist, and hand, Am J Sports Med 19:288, 1991.
4. Chaffin DB and Andersson GBJ: *Occupational biomechanics,* 2nd ed, New York, 1991, John Wiley & Sons, Inc.
5. Dingerkus ML, Imhoff A, and Hipp E: Snowboard sports technique, injury pattern, prevention, Fortschr Med 115:26, 1997.
6. Donkers MJ, et al: Hand position affects elbow joint load during push-up exercise, J Biomech 26:625, 1993.
7. Doody SG, Freedman L, and Waterland JC: Shoulder movement during abduction in the scapular plane, Arch Phys Med Rehabil 51:595, 1970.
8. Ferretti A, Cerullo G, and Russo G: Subscapular neuropathy in volleyball players, J Bone Joint Surg 69-A:260, 1987.
9. Gill TJ IV and Micheli LJ: The immature athlete. Common injuries and overuse syndromes of the elbow and wrist, Clin Sports Med 15:401, 1996.
10. Goldberg B and Boiardo R: Profiling children for sports participation, Clin Sports Med 3:153, 1984.
11. Hagberg M: Shoulder pain pathogenesis. In Hadler N, ed: *Clinical concepts in regional musculoskeletal illness,* Orlando, 1988, Grune and Stratton, Inc.
12. Hennig EM, Rosenbaum D, and Milani TL: Transfer of tennis racket vibrations onto the human forearm, Med Sci Sports Exerc, 24:1134, 1992.
13. Hinson MM: *Kinesiology,* 2nd ed, Dubuque, Iowa, 1981, Wm C Brown Group.
14. Hogfors C, et al.: Biomechanical model of the human shoulder joint—II. The shoulder rhythm, J Biomech 24:699, 1991.
15. Holtzhausen LM and Noakes TD: Elbow, forearm, wrist, and hand injuries among sport rock climbers, Clin J Sport Med 6:196, 1996.
16. Hui FC, Chao EY, and An KN: Muscle and joint forces at the elbow during isometric lifting, Abstracts of the 24th annual orthopedic research society, Dallas, 1978.
17. Inman VT, Saunders JBdeCM, and Abbott LC: Observations on the function of the shoulder joint, J Bone Joint Surg 26A:1, 1944.
18. Jobe FW and Nuber G: Throwing injuries of the elbow, Clin Sports Med 5:621, 1986.
19. Johnson RE and Rust RJ: Sports related injury: An anatomic approach, part 2, Minn Med 68:829, 1985.
20. Josefsson PO and Nilsson BE: Incidence of elbow dislocation, Acta Orthop Scand 57:537, 1986.
21. Karduna AR, Williams GR, Illiams JL, and Iannotti JP: Kinematics of the glenohumeral joint: Influences of muscle forces, ligamentous constraints, and articular geometry, J Orthop Res 14:986, 1996.

22. Kent BE: Functional anatomy of the shoulder complex: A review, J Am Phys Ther Assoc 51:867, 1971.

23. Luttgens K and Wells KF: Kinesiology: *Scientific basis of human motion,* 7th ed, Philadelphia, 1982, WB Saunders Co.

24. McLeod WD and Andrews JR: Mechanisms of shoulder injuries, Phys Ther 66:1901, 1986.

25. Murray PM and Cooney WP: Golf-induced injuries of the wrist, Clin Sports Med 15:85, 1996.

26. Nash HL: Rotator cuff damage: Reexamining the causes and treatments, Physician Sportsmed 16:129, 1988.

27. Nuber GW et al: Fine wire electromyography analysis of muscles of the shoulder during swimming, Am J Sports Med 14:7, 1986.

28. Perry J and Glousman R: Biomechanics of throwing. In Nicholas JA and Hershman EB, eds: *The upper extremity in sports medicine,* St Louis, 1990, CV Mosby.

29. Plancher KD, Litchfield R, and Hawkins RJ: Rehabilitation of the shoulder in tennis players, Clin Sports Med 14:111, 1995.

30. Poppen KN and Walker PS: Forces at the glenohumeral joint in abduction, Clin Orthop, 135:165, 1978.

31. Poppen KN and Walker PS: Normal and abnormal motion of the shoulder, J Bone Joint Surg [Am] 58:195, 1976.

32. Rafii M et al: Athlete shoulder injuries: CT arthrographic findings, Radiology 162:559, 1987.

33. Renstrom P and Johnson RJ: Overuse injuries in sports, Sports Med 2:316, 1985.

34. Richardson AB, Jobe FW, and Collins HR: The shoulder in competitive swimming, Am J Sports Med 8:159, 1980.

35. Saha AK: Dynamic stability of the glenohumeral joint, Acta Orthop Scand 42:491, 1971.

36. Salter RB: *Textbook of disorders and injuries of the musculoskeletal system,* 2nd ed, Baltimore, 1983, Williams & Wilkins.

37. Stocker D, Pink M, and Jobe FW: Comparison of shoulder injury in collegiate- and master's-level swimmers, Clin J Sport Med 5:4, 1995.

38. Tichauer ER: *The biomechanical basis of ergonomics,* New York, 1978, Wiley-Interscience.

39. Volz RG, Lieb M, and Benjamin J: Biomechanics of the wrist, Clin Orthop 149:112, 1980.

40. Von Lanz T and Wachsmuth W: Functional anatomy. In Boyes JH, ed, *Bunnell's surgery of the hand,* 5th ed, Philadelphia, 1970, JB Lippincott.

41. Walker PS and Poppen NK: Biomechanics of the shoulder joint during abduction in the plane of the scapula, Bull Hosp Joint Dis 38:107, 1977.

42. Wilk KE and Arrigo C: Current concepts in the rehabilitation of the athletic shoulder, J Orthop Sports Phys Ther 18:365, 1993.

43. Zuckerman JD and Matsen FA: Biomechanics of the elbow. In Nordin M and Frankel VH, eds: *Basic biomechanics of the musculoskeletal system,* 2nd ed, Philadelphia, 1989, Lea & Febiger.

ANNOTATED READINGS

Brunet ME, Hoddad RJ, and Porche EB: Rotator cuff impingement syndrome in sports, Physician Sportsmed 11:79, 1983.

Discusses common causes, symptoms, and treatment modalities for the rotator cuff impingement syndrome.

Soderberg GL: Shoulder. In Soderberg GL: *Kinesiology: Application to pathological motion*, Baltimore, 1986, Williams & Wilkins.

Includes an interesting discussion of evolutionary changes in the anatomical structures of the shoulder and a section on pathokinesiology of the shoulder.

Zuckerman JD and Matsen FA: Biomechanics of the elbow. In Nordin M and Frankel VH, eds: *Basic biomechanics of the musculoskeletal system*, Philadelphia, 1989, Lea & Febiger.

Describes kinematics and kinetics of the elbow, with sample calculations of joint reaction forces.

Zuckerman JD and Matsen FA: Biomechanics of the shoulder. In Nordin M and Frankel VH, eds: *Basic biomechanics of the musculoskeletal system*, Philadelphia, 1989, Lea & Febiger.

Describes kinematics and kinetics of the joints of the shoulder complex.

RELATED WEB SITES

Calvert Orthopaedic and Sports Medicine Center

http://www.calvertorthoandsports.com/index.html

Provides basic information and color graphics of the anatomy and common injuries of the shoulder.

The Center for Orthopaedics and Sports Medicine

http://www.arthroscopy.com/sports.htm

Includes information and color graphics on the anatomy and function of the upper extremity.

Eaton's Hand Surgery Links

http://www.eatonhand.com/hom/hom000.htm

Presents a comprehensive list of links related to anatomy, injury, treatment, and research on the hand, including a clip art image gallery and hand surgery slide lecture archives.

MedFacts Play Doctor! MedLib

http://www.medfacts.com/medlib.htm

Provides links to pages on anterior shoulder instability, lateral epicondylitis, rotator cuff tear/tendinitis, shoulder impingement syndrome, shoulder arthroscopy, and shoulder replacement surgery. Includes a 3-D video of the shoulder.

Northern Rockies Orthopaedics Specialists

http://www.orthopaedic.com

Provides links to pages describing descriptions of injuries, diagnostic tests, and surgical procedures for the shoulder, elbow, wrist, and hand.

Orthopedic Medical Information

http://www.opendoor.com/albert/one/orthopedic.html

Includes links to pages on shoulder disorders and carpal tunnel syndrome.

Rothman Institute

http://rothmaninstitute.com/index.html

Includes information on common sport injuries to the shoulder and elbow.

Southern California Orthopaedic Institute

http://www.scoi.com

Includes links to anatomical descriptions and labeled photographs of the shoulder, elbow, wrist, and hand.

University of Washington Orthopaedic Physicians

http://www.orthop.washington.edu

Provides extensive descriptions of scientific and clinical aspects of the shoulder, hand, and wrist. Clicking on the radiographs brings up structure labels. Also includes numerous movies showing therapeutic exercises, examination procedure, and surgical techniques.

The Virtual Hospital: Joint Fluoroscopy

http://www.vh.org/Providers/Textbooks/JointFluoro/JointFluoroHP.html

Provides detailed descriptions of anatomical features and includes multiple-view electric joint fluoroscopies of the elbow, lateral elbow, shoulder, and wrist.

The "Virtual" Medical Center: Anatomy & Histology Center

http://www-sci.lib.uci.edu/HSG/Medical Anatomy.html

Contains numerous images, movies, and course links for human anatomy.

West Michigan Hand Center

http://www.westmihand.com

Includes links to pages on sports injuries of the shoulder, elbow, and wrist, as well as carpal tunnel syndrome.

Wheeless' Textbook of Orthopaedics

http://www.medmedia.com/med.htm

Provides comprehensive, detailed information, graphics, and related literature for all the joints.

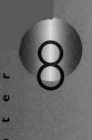
THE BIOMECHANICS OF THE HUMAN LOWER EXTREMITY

After completing this chapter, the reader will be able to:

Explain how anatomical structure affects movement capabilities of lower extremity articulations.

Identify factors influencing the relative mobility and stability of lower extremity articulations.

Explain the ways in which the lower extremity is adapted to its weightbearing function.

Identify muscles that are active during specific lower extremity movements.

Describe the biomechanical contributions to common injuries of the lower extremity.

The lower extremity is well structured for its functions of weight bearing and locomotion.

Although there are some similarities between the joints of the upper and the lower extremities, the upper extremity is more specialized for activities requiring large ranges of motion. In contrast, the lower extremity is well equipped for its functions of weight bearing and locomotion. Beyond these basic functions, activities such as kicking a field goal in football, performing a long jump or a high jump, and maintaining balance *en pointe* by ballet dancers reveal some of the more specialized capabilities of the lower extremity. This chapter examines the joint and muscle functions that enable lower extremity movements.

STRUCTURE OF THE HIP

The hip is a ball and socket joint (Figure 8-1). The ball is the head of the femur, which forms approximately two-thirds of a sphere. The socket is the concave acetabulum, which is angled obliquely in an anterior, lateral, and inferior direction. Joint cartilage covers both articulating surfaces. The cartilage on the acetabulum is thicker around its periphery, where it merges with a rim, or labrum, of fibrocartilage that contributes to the stability of the joint. The acetabulum also provides a much deeper socket than the glenoid fossa of the shoulder joint, and the bony structure of the hip is therefore much more stable or less likely to dislocate than that of the shoulder.

The hip is inherently more stable than the shoulder because of bone structure and the number and strength of the muscles and ligaments crossing the joint.

Several large, strong ligaments also contribute to the stability of the hip (Figure 8-2). The extremely strong iliofemoral or Y ligament and the pubofemoral ligament strengthen the joint capsule anteriorly, with posterior reinforcement from the ischiofemoral ligament. Tension in these major ligaments acts to twist the head of the femur into the acetabulum during hip extension, as when a person rises from sitting to an upright standing position. Inside the joint capsule, the ligamentum teres supplies a direct attachment from the rim of the acetabulum to the head of the femur.

As with the shoulder joint, several bursae are present in the surrounding tissues to assist with lubrication. The most prominent among the bursae are the iliopsoas bursa and the deep trochanteric bursa. The iliopsoas bursa is positioned between the iliopsoas and the articular capsule, serving to reduce the friction between these structures. The deep trochanteric bursa provides a cushion between the greater trochanter of the femur and the gluteus maximus at the site of its attachment to the iliotibial tract.

The femur is a major weightbearing bone and is the longest, largest, and strongest bone in the body. Its weakest component is the femoral neck, which is smaller in diameter than the rest of the bone, and weak internally because it is primarily composed of trabecular bone. The femur angles medially downward from the hip during the support phase of walking and running, enabling single-leg support beneath the body's center of gravity.

MOVEMENTS AT THE HIP

Although movements of the femur are primarily due to rotation occurring at the hip joint, the **pelvic girdle** has a function similar to that of the shoulder girdle in positioning the hip joint for effective limb movement. Unlike the shoulder girdle, the pelvis is a single nonjointed structure, but it can rotate in all three planes of movement. The pelvis

pelvic girdle
the two hip bones plus the sacrum, which can be rotated forward, backward, and laterally to optimize positioning of the hip joint

FIGURE 8-1
The bony structure of the hip.

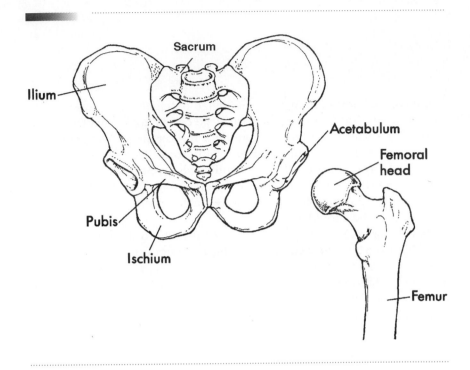

FIGURE 8-2
The ligaments of the hip.

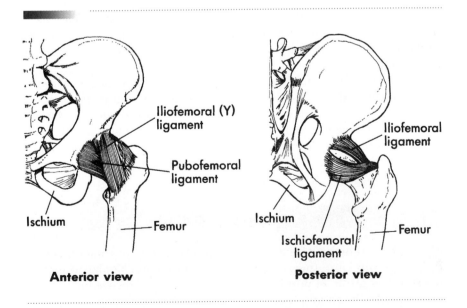

facilitates movement of the femur by rotating so that the acetabulum is positioned toward the direction of impending femoral movement. For example, posterior pelvic tilt, with the anterior superior iliac spine tilted backward with respect to the acetabulum, positions the head of the femur in front of the hipbone to enable ease of flexion. Likewise, anterior pelvic tilt promotes femoral extension, and lateral pelvic tilt toward the opposite side facilitates lateral movements of the femur. Movement of the pelvic girdle also coordinates with certain movements of the spine (see Chapter 9).

Muscles of the Hip

A number of large muscles cross the hip, further contributing to its stability. The locations and functions of the muscles of the hip are summarized in Table 8-1.

Flexion

The six muscles primarily responsible for flexion at the hip are those crossing the joint anteriorly: the iliacus, psoas major, pectineus, rectus femoris, sartorius, and tensor fascia latae. Of these, the large iliacus and psoas major (often referred to jointly as the **iliopsoas** because of their common attachment to the femur) are the major hip flexors (Figure 8-3). Other major hip flexors are shown in Figure 8-4. Because the rectus femoris is a two-joint muscle active during both hip flexion and knee extension, it functions more effectively as a hip flexor when the knee is in flexion, as when a person kicks a ball. The thin, strap-like sartorius, or *tailor's muscle*, is also a two-joint muscle. Crossing from the superior anterior iliac spine to the medial tibia just below the tuberosity, the sartorius is the longest muscle in the body.

iliopsoas
the psoas major and iliacus muscles with a common insertion on the lesser trochanter of the femur

Extension

The hip extensors are the gluteus maximus and the three **hamstrings—** the biceps femoris, semitendinosus, and semimembranosus (Figure 8-5). The gluteus maximus is a massive, powerful muscle that is usually active only when the hip is in flexion, as during stair climbing or cycling, or when extension at the hip is resisted (Figure 8-6). The hamstrings derive their name from their prominent tendons, which can readily be palpated on the posterior aspect of the knee. These two-joint muscles contribute to both extension at the hip and flexion at the knee and are active during standing, walking, and running.

hamstrings
the biceps femoris, semimembranosus, and semitendinosus

Two-joint muscles function more effectively at one joint when the position of the other joint stretches the muscle slightly.

TABLE 8-1 ■

MUSCLES OF THE HIP

MUSCLE	PROXIMAL ATTACHMENT	DISTAL ATTACHMENT	PRIMARY ACTION(S) ABOUT THE HIP
Rectus femoris	Anterior inferior iliac spine	Patella	Flexion
Iliopsoas			
(Iliacus)	Iliac fossa and adjacent sacrum	Lesser trochanter of the femur	Flexion
(Psoas)	12th thoracic and lumbar vertebrae and lumbar discs	Lesser trochanter of the femur	Flexion
Sartorius	Anterior superior iliac spine	Upper medial tibia	Assists with flexion, abduction, and lateral rotation
Pectineus	Pectineal crest of pubic ramus	Medial femur	Flexion, adduction and lateral rotation
Tensor fascia latae	Crest of the ilium	Iliotibial band	Assists with flexion, abduction, and medial rotation
Gluteus maximus	Posterior ilium, iliac crest, sacrum, and coccyx	Gluteal tuberosity of the femur and iliotibial band	Extension, lateral rotation
Gluteus medius	Between posterior and anterior gluteal lines on the posterior ilium	Superior, lateral greater trochanter	Abduction, medial rotation

Abduction

The gluteus medius is the major abductor acting at the hip, with the gluteus minimus assisting. These muscles stabilize the pelvis during the support phase of walking and running and when an individual stands on one leg. For example, when body weight is supported by the right foot during walking, the right hip abductors contract isometrically and eccentrically to prevent the left side of the pelvis from being pulled downward by the weight of the swinging left leg. This allows the left leg to move freely through the swing phase. If the hip abductors are too weak to perform this function, then lateral pelvic tilt and dragging of

MUSCLES OF THE HIP (CONTINUED)

MUSCLE	PROXIMAL ATTACHMENT	DISTAL ATTACHMENT	PRIMARY ACTION(S) ABOUT THE HIP
Gluteus minimus	Between anterior and inferior gluteal lines on the posterior ilium	Anterior surface of the greater trochanter	Abduction, medial rotation
Gracilis	Anterior, inferior symphysis pubis	Medial proximal tibia	Adduction
Adductor magnus	Inferior ramus of pubis and ischium	Entire linea aspera	Adduction, lateral rotation
Adductor longus	Anterior pubis	Middle linea aspera	Adduction, lateral rotation, assists with flexion
Adductor brevis	Inferior ramus of the pubis	Upper linea aspera	Adduction, lateral rotation
Semitendinosus	Medial ischial tuberosity	Proximal, medial tibia	Extension
Semimembranosus	Lateral ischial tuberosity	Proximal, medial tibia	Extension
Biceps femoris (long head)	Lateral ischial tuberosity	Posterior lateral condyle of tibia, head of fibula	Extension of thigh
The six outward rotators	Sacrum, ilium, and ischium	Posterior greater trochanter	Outward rotation

the swing foot occurs with every step during gait. The hip abductors are also active during the performance of dance movements such as the *grande ronde jambe*.

Contraction of the hip abductors is required during the swing phase of gait to prevent the swinging foot from dragging.

Adduction

The hip adductors are those muscles that cross the joint medially and include the adductor longus, adductor brevis, adductor magnus, and gracilis (Figure 8-7). The hip adductors are regularly active during the swing phase of the gait cycle to bring the foot beneath the body's center of gravity for placement during the support phase. The gracilis is a long, relatively weak strap muscle that also contributes to flexion of the

FIGURE 8-3
The iliopsoas complex is the major flexor of the hip.

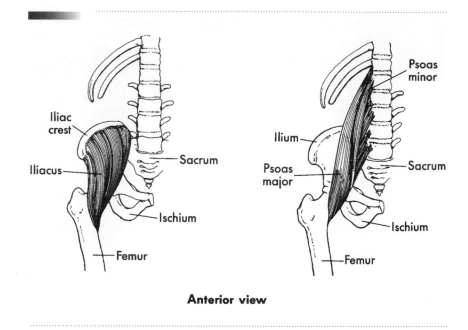

Anterior view

FIGURE 8-4
Assistant flexor muscles of the hip.

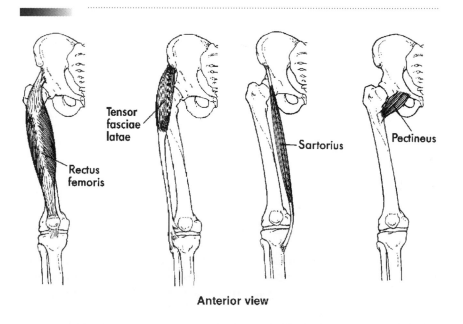

Anterior view

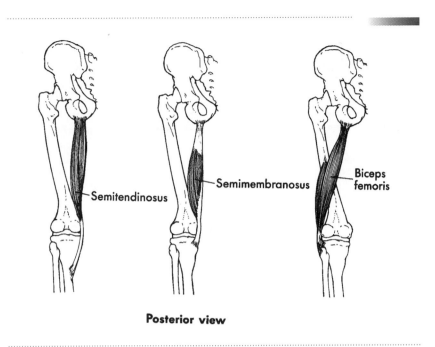

Posterior view

FIGURE 8-5
The hamstrings are major hip extensors and knee flexors.

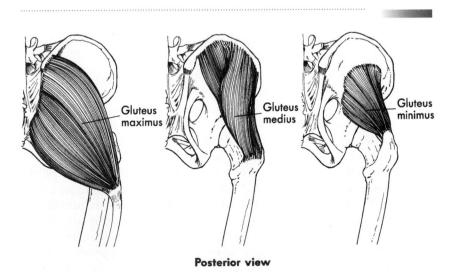

Posterior view

FIGURE 8-6
The three gluteal muscles.

FIGURE 8-7
Adductor muscles of the hip.

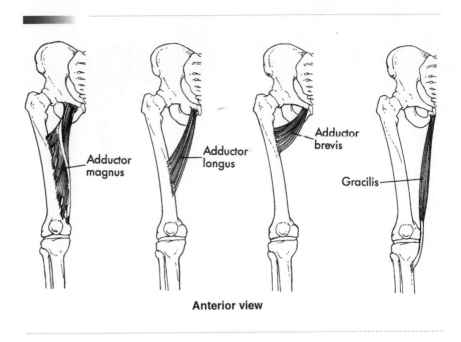

Anterior view

leg at the knee. The other three adductor muscles also contribute to flexion and lateral rotation at the hip, particularly when the femur is medially rotated.

Medial and Lateral Rotation of the Femur

During the gait cycle, lateral and medial rotation of the femur occur in coordination with pelvic rotation.

Although a number of muscles contribute to lateral rotation of the femur, six muscles function solely as lateral rotators. These are the piriformis, gemellus superior, gemellus inferior, obturator internus, obturator externus, and quadratus femoris (Figure 8-8). Although we tend to think of walking and running as involving strictly sagittal plane movement at the joints of the lower extremity, outward rotation of the femur also occurs with every step to accommodate the rotation of the pelvis.

The major medial rotator of the femur is the gluteus minimus, with assistance from the tensor fascia latae, semitendinosus, semimembranosus, and gluteus medius. Medial rotation of the femur is usually not a resisted motion requiring a substantial amount of muscular force. The medial rotators are weak in comparison to the lateral rotators, with the estimated strength of the medial rotators only approximately one-third that of the lateral rotators (23).

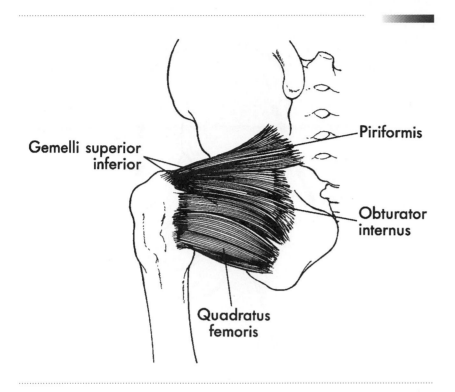

FIGURE 8-8
The lateral rotator muscles
of the femur.

Horizontal Abduction and Adduction

Horizontal abduction and adduction of the femur occur when the hip
is in 90° of flexion while the femur is either abducted or adducted. These
actions require the simultaneous, coordinated actions of several muscles. Tension is required in the hip flexors for elevation of the femur.
The hip abductors can then produce horizontal abduction and, from a
horizontally abducted position, the hip adductors can produce horizontal adduction. The muscles located on the posterior aspect of the
hip are more effective as horizontal abductors and adductors than the
muscles on the anterior aspect because the former are stretched when
the femur is in 90° of flexion, whereas tension in the anterior muscles
is usually reduced with the femur in this position.

LOADS ON THE HIP

The hip is a major weightbearing joint. When body weight is evenly distributed across both legs during upright standing, the weight supported
at each hip is one-half the weight of the body segments above the hip,

FIGURE 8-9

The major forces on the hip during static stance are the weight of the body segments above the hip (with $1/2$ of the weight on each hip), tension in the hip abductor muscles (F_m), and the joint reaction force (R).

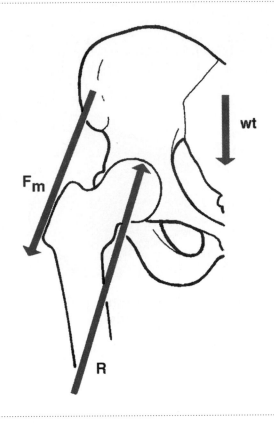

During the support phase of walking, compression force at the hip reaches a magnitude of three to four times body weight.

or about one-third of total body weight. However, the total load on each hip in this situation is greater than the weight supported, because tension in the large, strong hip muscles further adds to compression at the joint (Figure 8-9).

Because of muscle tension, compression on the hip is approximately the same as body weight during the swing phase of walking (39). During the support phase, peak joint forces can range from 300%–400% of body weight at normal walking speed, to 550% of body weight during fast walking and jogging, and as high as 870% of body weight during stumbling (7). Stair climbing and descending increase the loads on the hip by approximately 10% and 20%, respectively (6). Hip loading is also increased with wearing of hard-soled shoes (8). Carrying a load such as a suitcase of 25% body weight on one side produces a 167% increase in loading on the contralateral hip as compared to the hip on the loaded side (5).

As gait speed increases, the load on the hip increases during both swing and support phases. Hip loads during jogging can be reduced with a smooth gait pattern and soft heel strikes (8). In summary, body weight, impact forces translated upward through the skeleton from the foot, and muscle tension all contribute to this large compressive load on the hip, as demonstrated in the sample problem in Figure 8-10.

Use of a crutch or cane on the side opposite an injured or painful hip is beneficial in that it serves to more evenly distribute the load between the legs throughout the gait cycle. During stance, a support opposite the painful hip reduces the amount of tension required of the powerful abductor muscles, thereby reducing the load on the painful hip. This reduction in load on the painful hip, however, increases the stress on the opposite hip (31).

COMMON INJURIES OF THE HIP

Fractures

Although the pelvis and the femur are large, strong bones, the hip is subjected to high, repetitive loads ranging from 4 to 7 times body weight during locomotion (23). Fractures of the femoral neck frequently occur during the support phase of walking among elderly individuals with osteoporosis, a condition of reduced bone mineralization and strength (see Chapter 4). These femoral neck fractures often result in loss of balance and a fall. A common misconception is that the fall always causes the fracture rather than the reverse, which may also be true. When the hipbones have good health and mineralization, they can sustain tremendous loads, as illustrated during many weight-lifting events. Research demonstrates that regular physical activity exerts a protective effect against the risk of hip fracture (20).

Fracture of the femoral neck (broken hip) is a seriously debilitating injury that occurs frequently among elderly individuals with osteoporosis.

Contusions

The muscles on the anterior aspect of the thigh are in a prime location for sustaining blows during participation in contact sports. The resulting internal hemorrhaging and appearance of bruises vary from mild to severe.

Strains

Since most daily activities do not require simultaneous hip flexion and knee extension, the hamstrings are rarely stretched unless exercises are performed for that specific purpose. The resulting loss of extensibility makes the hamstrings particularly susceptible to strain. Strains to these muscles most commonly occur during sprinting, particularly if the individual is fatigued and neuromuscular coordination is impaired. Researchers have proposed different explanations for mechanisms of hamstring

FIGURE 8-10

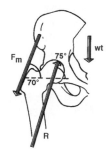

SAMPLE   PROBLEM   1

How much compression acts on the hip during two-legged standing, given that the joint supports 250 N of body weight and the abductor muscles are producing 600 N of tension?

Known
wt = 250
F_m = 600 N

Graphic Solution
Since the body is motionless, all vertical force components must sum to 0 and all horizontal force components must sum to 0. Graphically, this means that all acting forces can be transposed to form a closed force polygon (in this case, a triangle). The forces from the diagram of the hip above can be reconfigured to form the triangle shown on the right.

If the triangle is drawn to scale (perhaps 1 cm = 100 N), the amount of joint compression can be approximated by measuring the length of the joint reaction force (R).

R is about 750 N

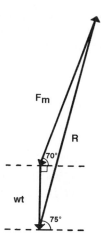

Mathematical Solution
The Law of Cosines can be used with the same triangle to calculate the length of R.

$$R^2 = F_m{}^2 + wt^2 - 2(F_m)\,(wt)\cos 160°$$
$$R^2 = 600\ N^2 + 250 N^2 - 2\,(600\ N)(250\ N)\cos 160°$$

R = 751 N

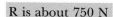

strains (46). According to one hypothesis, hamstring strains result from the overstretching of the muscle group, such as during overstriding. An alternative notion is that they result when the fully elongated muscle group must develop maximum tension. Strains to the groin area are also relatively common among athletes in sports in which forceful thigh abduction movements may overstretch the adductor muscles.

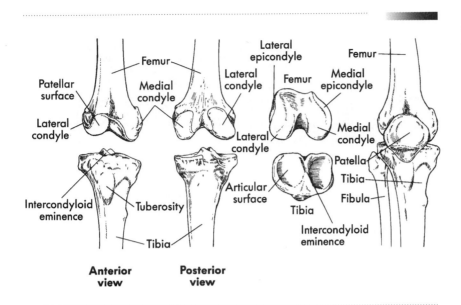

Anterior view

Posterior view

FIGURE 8-11
The bony structure of the tibiofemoral joint.

STRUCTURE OF THE KNEE

The structure of the knee permits the bearing of tremendous loads as well as the mobility required for locomotor activities. The knee is a large synovial joint, including three articulations within the joint capsule. The weightbearing joints are the two condylar articulations of the **tibiofemoral joint,** with the third articulation being the **patellofemoral joint.** Although not a part of the knee, the proximal tibiofibular joint has soft-tissue connections that also slightly influence knee motion.

Tibiofemoral Joint

The medial and lateral condyles of the tibia and the femur articulate to form two side-by-side condyloid joints (Figure 8-11). These joints function together primarily as a modified hinge joint because of the restricting ligaments, with some lateral and rotational motions allowed. The condyles of the tibia, known as the *tibial plateaus,* form slight depressions separated by a region known as the *intercondylar eminence.* Because the medial and lateral condyles of the femur differ somewhat in size, shape, and orientation, the tibia rotates laterally on the femur during the last few degrees of extension to produce "locking" of the knee. This phenomenon, known as the "screwing-home" mechanism, brings the knee into the close-packed position of full extension.

tibiofemoral joint
dual condyloid articulations between the medial and lateral condyles of the tibia and the femur, composing the main hinge joint of the knee

patellofemoral joints
articulation between the patella and the femur

The bony anatomy of the knee requires a small amount of lateral rotation of the tibia to accompany full extension.

FIGURE 8-12
The menisci of the knee.

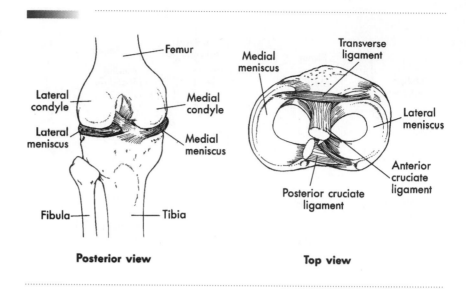

Posterior view   Top view

Menisci

menisci
cartilaginous discs located between the tibial and femoral condyles

The **menisci,** also known as semilunar cartilages after their half-moon shapes, are discs of fibrocartilage firmly attached to the superior plateaus of the tibia by the coronary ligaments and joint capsule (Figure 8-12). They are also joined to each other by the transverse ligament. The menisci are thickest at their peripheral borders, where fibers from the joint capsule solidly anchor them to the tibia (3). The medial semilunar disc is also directly attached to the medial collateral ligament. Medially, both menisci taper down to paper thinness, with the inner edges unattached to the bone.

The menisci distribute the load at the knee over a larger surface area and also help absorb shock.

The menisci deepen the articulating depressions of the tibial plateaus and assist with the absorption of force at the knee. The internal structure of the medial two-thirds of each meniscus is particularly well suited to resisting compression (3). The stress on the tibiofemoral joint can be an estimated three times higher during load bearing if the menisci have been removed (43). Injured knees, in which part or all of the menisci have been removed, may still function adequately but undergo increased wear on the articulating surfaces, significantly increasing the likelihood of the development of degenerative conditions at the joint.

Ligaments

Many ligaments cross the knee, significantly enhancing its stability (Figure 8-13). The location of each ligament determines the direction in which it is capable of resisting the dislocation of the knee.

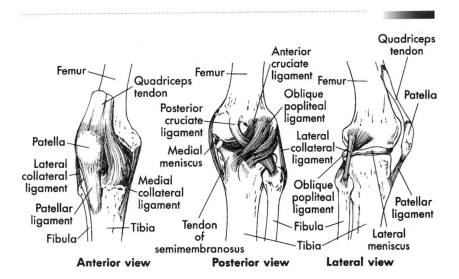

FIGURE 8-13
The ligaments of the knee.

The medial and lateral **collateral ligaments** prevent lateral motion at the knee, as do the collateral ligaments at the elbow. They are also respectively referred to as the *tibial* and *fibular collateral ligaments,* after their distal attachments. Fibers of the medial collateral ligament merge with the joint capsule and the medial meniscus to connect the medial epicondyle of the femur to the medial tibia. The attachment is just below the pes anserinus, the common attachment of semitendinosus, semimembranosus, and gracilis to the tibia, thereby positioning the ligament to resist medially directed shear (valgus) and rotational forces acting on the knee. The lateral collateral ligament connects the lateral epicondyle of the femur to the head of the fibula, contributing to lateral stability of the knee.

The anterior and posterior **cruciate ligaments** limit the forward and backward sliding of the femur on the tibial plateaus during knee flexion and extension, and also limit knee hyperextension. The name *cruciate* is derived from the fact that these ligaments cross each other; *anterior* and *posterior* refer to their respective tibial attachments. The anterior cruciate ligament stretches from the anterior aspect of the intercondyloid fossa of the tibia just medial and posterior to the anterior tibial spine in a superior, posterior direction to the posterior medial surface of the lateral condyle of the femur. The shorter and stronger posterior cruciate ligament runs from the posterior aspect of the tibial intercondyloid fossa in a superior, anterior direction to the lateral anterior medial condyle of the femur. These ligaments restrict the anterior and posterior sliding of the femur on the tibial plateaus during knee flexion and extension and limit knee hyperextension.

collateral ligaments
major ligaments that cross the medial and lateral aspects of the knee

cruciate ligaments
major ligaments that cross each other in connecting the anterior and posterior aspects of the knee

Several other ligaments contribute to the integrity of the knee. The oblique and arcuate popliteal ligaments cross the knee posteriorly, and the transverse ligament connects the two semilunar discs internally. Another restricting tissue is the **iliotibial band** or tract, a broad, thickened band of the fascia lata with attachments to the lateral condyle of the femur and the lateral tubercle of the tibia, which has been hypothesized to function as an anterolateral ligament of the knee (48).

Patellofemoral Joint

The patellofemoral joint consists of the articulation of the triangularly shaped patella, encased in the patellar tendon, with the trochlear groove between the femoral condyles. The posterior surface of the patella is covered with articular cartilage, which reduces the friction between the patella and the femur.

The patella improves the mechanical advantage of the knee extensors by as much as 50%.

The patella serves several biomechanical functions. Most notably, it increases the angle of pull of the quadriceps tendon on the tibia, thereby improving the mechanical advantage of the quadriceps muscles for producing knee extension by as much as 50% (15). It also centralizes the divergent tension from the quadriceps muscles that is transmitted to the patellar tendon. The patella also increases the area of contact between the patellar tendon and the femur, thereby decreasing patellofemoral joint contact stress. Finally, it also provides some protection for the anterior aspect of the knee and helps protect the quadriceps tendon from friction against the adjacent bones.

Joint Capsule and Bursae

The thin articular capsule at the knee is large and lax, encompassing both the tibiofemoral and patellofemoral joints. A number of bursae are located in and around the capsule to reduce friction during knee movements. The suprapatellar bursa, positioned between the femur and the quadriceps femoris tendon, is the largest bursa in the body. Other important bursae are the subpopliteal bursa, located between the lateral condyle of the femur and the popliteal muscle, and the semimembranosus bursa, situated between the medial head of the gastrocnemius and semimembranosus tendons.

Three other key bursae associated with the knee, but not contained in the joint capsule, are the prepatellar, superficial infrapatellar, and deep infrapatellar bursae. The prepatellar bursa is located between the skin and the anterior surface of the patella, allowing free movement of the skin over the patella during flexion and extension. The superficial infrapatellar bursa provides cushioning between the skin and the patellar tendon, and the deep infrapatellar bursa reduces friction between the tibial tuberosity and the patellar tendon.

MOVEMENTS AT THE KNEE

Muscles Crossing the Knee

Like the elbow, the knee is crossed by a number of two-joint muscles. The primary actions of the muscles crossing the knee are summarized in Table 8-2.

Flexion and Extension

Flexion and extension are the primary movements permitted at the tibiofemoral joint. For flexion to be initiated from a position of full extension, however, the knee must first be "unlocked." In full extension the articulating surface of the medial condyle of the femur is longer than that of the lateral condyle, rendering motion almost impossible. The service of locksmith is provided by the **popliteus,** which acts to medially rotate the tibia with respect to the femur, enabling flexion to occur (Figure 8-14). As flexion proceeds, the femur must slide forward on the tibia to prevent rolling off the tibial plateaus. Likewise, the femur must slide backwards on the tibia during extension.

The three hamstring muscles are the primary flexors acting at the knee. Muscles that assist with knee flexion are the gracilis, sartorius, popliteus, and gastrocnemius.

The **quadriceps** muscles, consisting of the rectus femoris, vastus lateralis, vastus medialis, and vastus intermedius, are the extensors of the knee (Figure 8-15). The rectus femoris is the only one of these muscles that also crosses the hip joint. All four muscles attach distally to the patellar tendon, which inserts on the tibia.

popliteus
muscle known as the unlocker of the knee because its action is lateral rotation of the femur with respect to the tibia

quadriceps
the rectus femoris, vastus lateralis, vastus medialis, and vastus intermedius

Rotation and Passive Abduction and Adduction

Rotation of the tibia relative to the femur is possible when the knee is in flexion and not bearing weight, with rotational capability greatest at approximately 90° of flexion. Tension development in the semimembranosus, semitendinosus, and popliteus produces medial rotation of the tibia, with the gracilis and sartorius assisting. The biceps femoris is solely responsible for lateral rotation of the tibia. A few degrees of passive abduction and adduction are also permitted.

Patellofemoral Joint Motion

During flexion and extension at the tibiofemoral joint, the patella glides inferiorly and superiorly against the distal end of the femur in a primarily vertical direction with an excursion of approximately 7 cm. During

TABLE 8-2 ▬▬

MUSCLES OF THE KNEE

MUSCLE	PROXIMAL ATTACHMENT	DISTAL ATTACHMENT	PRIMARY ACTION(S) ABOUT THE KNEE
Rectus femoris	Anterior inferior iliac spine	Patella	Extension
Vastus lateralis	Greater trochanter and lateral linea aspera	Patella	Extension
Vastus inter-medius	Anterior femur	Patella	Extension
Vastus medialis	Medial linea aspera	Patella	Extension
Semitendinosus	Medial ischial tuberosity	Proximal, medial tibia	Flexion and medial rotation
Semimembranosus	Lateral ischial tuberosity	Proximal, medial tibia	Flexion and medial rotation
Biceps femoris (short head)	Lateral linea aspera	Posterior lateral condyle of tibia, head of fibula	Flexion and lateral rotation
Sartorius	Anterior superior iliac spine	Upper medial tibia	Assists with flexion and lateral rotation of thigh
Gracilis	Anterior, inferior symphysis pubis	Medial, proximal tibia	Adduction of thigh, flexion of lower leg
Popliteus	Lateral condyle of the femur	Medial, posterior tibia	Medial rotation and flexion of lower leg
Gastrocnemius	Posterior medial & lateral condyles of the femur	Tuberosity of the calcaneus by the Achilles tendon	Flexion
Plantaris	Distal, posterior femur	Tuberosity of the calcaneus by the Achilles tendon	Flexion

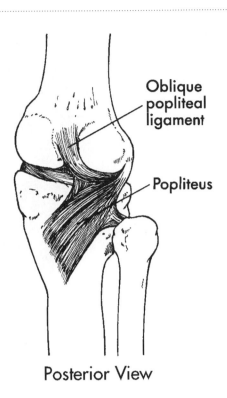

Oblique
popliteal
ligament

Popliteus

Posterior View

FIGURE 8-14
The popliteus muscle is
the locksmith of the knee.

normal tracking, the patella may also slightly shift and rotate in either medial or lateral directions (15).

Tracking of the patella against the femur is dependent on the direction of the net force produced by the attached quadriceps. The vastus lateralis tends to pull the patella laterally, while the vastus medialis oblique opposes the lateral pull of the vastus lateralis, keeping the patella centered in the patellofemoral groove.

LOADS ON THE KNEE

Because the knee is positioned between the body's two longest bony levers (the femur and the tibia), the potential for torque development at the joint is large. The knee is also a major weightbearing joint.

Forces at the Tibiofemoral Joint

The tibiofemoral joint is loaded in both compression and shear during daily activities. Weight bearing and tension development in the muscles

FIGURE 8-15
The quadriceps muscles extend the knee.

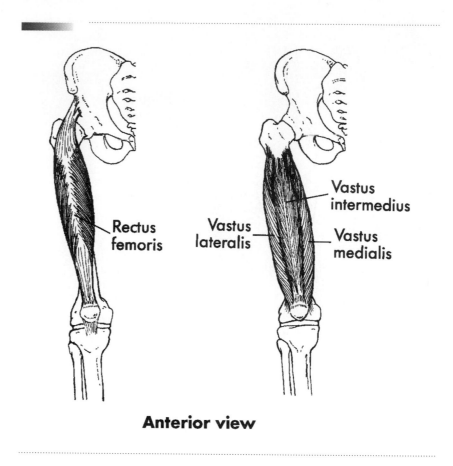

Rectus femoris

Vastus lateralis

Vastus intermedius

Vastus medialis

Anterior view

The medial tibial plateau is well adapted for its weightbearing function during stance, with greater surface area and thicker articular cartilage than the lateral plateau.

crossing the knee contribute to these forces, with compression dominating when the knee is fully extended.

Compressive force at the tibiofemoral joint has been reported to be slightly greater than three times body weight during the stance phase of gait, increasing up to approximately four times body weight during stair climbing (34). The medial tibial plateau bears most of this load during stance when the knee is extended, with the lateral tibial plateau bearing more of the much smaller loads imposed during the swing phase (34). Since the medial tibial plateau has a surface area roughly 60% larger than that of the lateral tibial plateau, the stress acting on the joint is less than if peak loads were distributed medially (24). The fact that the articular cartilage on the medial plateau is three times thicker than that on the lateral plateau also helps protect the joint from wear.

The menisci act to distribute loads at the tibiofemoral joint over a broader area, thus reducing the magnitude of joint stress. The menisci also directly assist with force absorption at the knee, bear-

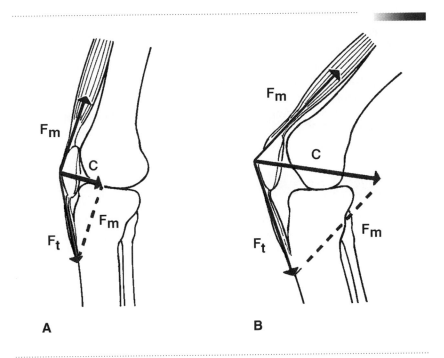

A   **B**

FIGURE 8-16
Compression at the patellofemoral joint is the vector sum of tension in the quadriceps and the patellar tendon. **A,** In extension, the compressive force is small because tension in the muscle group and tendon act nearly perpendicular to the joint. **B,** As flexion increases, compression increases because of 1. changed orientation of the force vectors and 2. increased tension requirement in the quadriceps to maintain body position.

ing as much as an estimated 45% of the total load (44). Since the menisci help protect the articulating bone surfaces from wear, knees that have undergone meniscectomies are more likely to develop degenerative conditions.

As knee flexion occurs and the angle at the joint increases to 90°, the shear component of joint force produced by weight bearing increases. Shear at the knee, which causes a tendency for the femur to displace anteriorly on the tibial plateaus, must be resisted by the ligaments and other supportive structures crossing the knee. Since these structures can be stretched or even ruptured under such stress, activities like deep knee bends and full squats that require load bearing during extreme knee flexion are not recommended.

Forces at the Patellofemoral Joint

Compressive force at the patellofemoral joint has been found to be one-half of body weight during normal walking gait, increasing up to over three times body weight during stair climbing (40). As shown in Figure 8-16, patellofemoral compression increases with knee flexion during weight bearing. There are two reasons for this. First, the increase in knee flexion increases the compressive component of force acting at the joint. Second, as flexion increases, a larger amount of quadriceps tension is required to prevent the knee from buckling against gravity. The squat exercise, known for being particularly stressful to the knee complex, produces a patellofemoral joint reaction force on the order of 7.6 times body weight (40). Given the small contact area between the articulating bone surfaces, the transmitted stress at the patellofemoral joint during such maneuvers is high. The sample problem in Figure 8-17 illustrates the relationship between quadriceps force and patellofemoral joint compression.

COMMON INJURIES OF THE KNEE AND LOWER LEG

The location of the knee between the long bones of the lower extremity, combined with its weightbearing and locomotion functions, make it susceptible to injury, particularly during participation in contact sports. A common injury mechanism involves the stretching or tearing of soft tissues on one side of the joint when a blow is sustained from the opposite side during weight bearing.

Ligament Injuries

Generally speaking, forces sustained by the knee from the anterior direction may strain or rupture the posterior cruciate ligament (PCL), and forces directed from the posterior of the knee may damage the anterior cruciate ligament (ACL), with injuries to the ACL occurring more frequently. Injuries to both cruciate ligaments, however, most often occur as a result of combined forms of loading. For the ACL, the most dangerous loading situation occurs when the knee is fully extended and weight bearing, and anteriorly directed tibial force is combined with internal tibial torque (28). The most devastating loading situation for the PCL is posteriorly directed tibial force combined with a valgus moment, resulting from a medially directed force at the knee (29).

Blows to the lateral side of the knee are much more common than blows to the medial side because the opposite leg commonly protects the medial side of the joint. When the foot is planted and a lateral blow of sufficient force is sustained, the result is sprain or rupture of the medial collateral ligament (MCL). Modeling studies suggest that the muscles crossing the knee are able to resist approximately 17% of external medial and lateral loads on the knee, with the remaining 83% sustained by the ligaments and other soft tissues (27). However, direct measurements of strain to the knee ligaments indicate that external tibial torque is more dangerous than medially directed tibial force for the MCL (19). In contact sports such as football the MCL is more frequently injured, while both medial and lateral collateral ligament sprains occur among wrestlers (52).

In contact sports, blows to the knee are most commonly sustained on the lateral side, with injury occurring to the stretched tissues on the medial side.

FIGURE 8-17

S A M P L E   P R O B L E M   2

How much compression acts on the patellofemoral joint when the quadriceps exerts 300 N of tension and the angle between the quadriceps and the patellar tendon is a) 160°, and b) 90°?

Known

$F_m = 300$ N

angle between F_m and F_t:

a) 160°

b) 90°

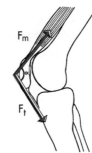

Graphic Solution

Vectors for F_m and F_t are drawn to scale (perhaps 1 cm: 100 N), with the angle between them first at 160° and then at 90°. The tip-to-tail method of vector composition is then used (see Chapter 3) to translate one of the vectors so that its tail is positioned on the tip of the other vector. The compression force is the resultant of F_m and F_t and is constructed with its tail on the tail of the original vector and its tip on the tip of the transposed vector.

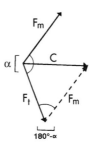

The amount of joint compression can be approximated by measuring the length of vector C.

a) C is about 100 N

b) C is about 420 N

Mathematical Solution

The angle between F_t and transposed vector F_m is 180° minus the size of the angle between the two original vectors, or a) 20° and b) 90°. The Law of Cosines can be used to calculate the length of C.

a) $C^2 = F_m{}^2 + F_t{}^2 - 2(F_m)(F_t)\cos 20$
$C^2 = 300$ N$^2 + 300$ N$^2 - 2(300$ N$)(300$ N$)\cos 20$
   C = 104 N

b) $C^2 = F_m{}^2 + F_t{}^2 - 2(F_m)(F_t)\cos 90$
$C^2 = 300$ N$^2 + 300$ N$^2 - 2(300$ N$)(300$ N$)\cos 90$
   C = 424 N

Note: This problem illustrates the extent to which patellofemoral compression can increase due solely to change in knee flexion.

Normally, there is also increased quadriceps force with increased knee flexion.

Ligament injuries constitute 25% of all injuries sustained during Alpine skiing, with the MCL and the ACL injured ten times more often than the lateral collateral ligament and the PCL (13). A common injury mechanism involves fixation of a ski when a tip is caught in the snow or when the ski tips cross, with the skier simultaneously twisting and falling.

To prevent knee ligament injuries, especially during contact sports, some athletes wear prophylactic knee braces. The wearing of such braces by healthy individuals has been a contentious issue since the American Academy of Orthopaedics issued a position statement against their use in 1987. Recent work on this topic shows that knee braces can protect the ACL against anterior and torsional loads on the tibia by significantly reducing the strain present in the ligament under these conditions (9). Knee braces can also contribute 20%–30% added resistance against lateral blows to the knee, with custom-fitted braces providing the best protection (2). A possible concern, however, is that knee braces act to change the pattern of lower extremity muscle activity during gait, with less work performed at the knee and more at the hip (12). Other documented problems that appear to affect some athletes more than others and may be brace-specific include reduced sprinting speed and earlier onset of fatigue (2).

Meniscus Injuries

Because the medial collateral ligament attaches to the medial meniscus, stretching or tearing of the ligament can also result in damage to the meniscus. A torn meniscus is the most common knee injury, with damage to the medial meniscus approximately 10 times as frequent as damage to the lateral meniscus. This is the case partly because the medial meniscus is more securely attached to the tibia, and therefore less mobile than the lateral meniscus. The mechanism of injury frequently involves the foot's being planted during weight bearing while the body undergoes rotation. The condition is problematic in that the unattached cartilage often slips from its normal position, interfering with normal joint mechanics. Symptoms include pain, which is sometimes accompanied by intermittent bouts of locking or buckling of the joint.

Iliotibial Band Friction Syndrome

The tensor fascia lata develops tension to assist with stabilization of the pelvis when the knee is in flexion during weight bearing. This can produce friction of the posterior edge of the iliotibial band (ITB) against the lateral condyle of the femur around the time of footstrike, primarily during foot contact with the ground (38). The result is inflammation of the knee joint capsule under the ITB, with accompanying symptoms of pain and tenderness over the lateral aspect of the knee (36). This condition is an overuse syndrome relatively common among distance runners, and is sometimes referred to as *runner's knee*. Both training errors and anatomical malalignments within the lower extremity increase the risk of iliotibial band syndrome. Excessive weekly running mileage and large braking forces during foot strike have been found to be the training factors most related to the incidence of this overuse syndrome (32). Predisposing malalignments include excessive femoral anteversion, increased Q angle, lateral tibial torsion, tibial genu varum or valgum, subtalar varus, and excessive pronation (25).

Breaststroker's Knee

A condition of pain and tenderness localized on the medial aspect of the knee is often associated with performance of the whip kick, the kick used with the breaststroke. The forceful whipping together of the lower legs that provides the propulsive thrust of the kick often forces the lower leg into slight abduction at the knee, with subsequent irritation to the medial collateral ligament and the medial border of the patellar tract. A survey of 391 competitive swimmers revealed incidences of knee pain among 73% of the breaststroke specialists and 48% of nonbreaststrokers (51). In a study of breaststroke kinematics, it was found that angles of hip abduction of less than 37° or greater than 42° at the initiation of the kick resulted in a dramatically increased incidence of knee pain (51) (Figure 8-18).

Chondromalacia

An imbalance in tension on the medial and lateral sides of the patella can cause mistracking of the patella during knee flexion/extension movements. Tissue irritation and pain result. This condition, known as *chondromalacia*, often results from an imbalance between the strength of the vastus medialis oblique (VMO) and the vastus lateralis (VL). More commonly, the patella deviates laterally during tracking, with weakness of the VMO implicated. Researchers have hypothesized that proper tracking of the patella may require maintenance of a certain threshold strength in the VMO (15). If the patellar misalignment is not extremely severe, simple strengthening of the quadriceps muscles can alleviate, or even eliminate, the symptoms. Patellofemoral alignment problems are more frequent in females than in males (53).

Painful lateral deviation in patellar tracking is generally caused by weakness of the vastus medialis oblique.

Shin Splints

Generalized pain along the anterolateral or posteromedial aspect of the lower leg is commonly known as *shin splints*. This is a loosely defined

Shin splints is a generic term often ascribed to any pain emanating from the anterior aspect of the lower leg.

The likelihood of acquiring breaststroker's knee is much lower if the angle of hip abduction is between 37° and 42° at the beginning of the propulsive phase of the kick.

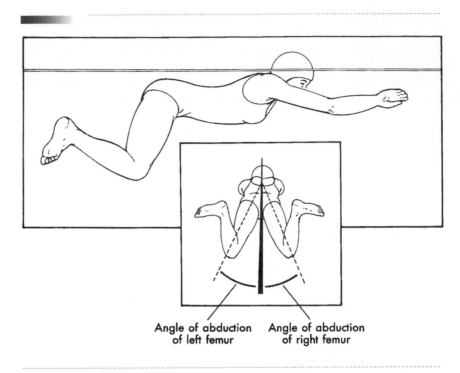

Angle of abduction of left femur   Angle of abduction of right femur

overuse injury often associated with running or dancing that may involve microdamage to muscle attachments on the tibia and/or inflammation of the periosteum (22). Common causes of the condition include running or dancing on a hard surface and running uphill. A change in workout conditions or rest usually alleviates shin splints.

STRUCTURE OF THE ANKLE

The ankle region includes the distal tibiofibular, tibiotalar, and fibulotalar joints (Figure 8-19). The distal tibiofibular joint is a syndesmosis where dense fibrous tissue binds the bones together. The joint is supported by the anterior and posterior tibiofibular ligaments, as well as by the crural interosseous tibiofibular ligament. Most motion at the ankle occurs at the tibiotalar hinge joint, where the convex surface of the superior talus articulates with the concave surface of the distal tibia. All three articulations are enclosed in a joint capsule that is thickened on the medial side and extremely thin on the posterior side. Three ligaments—the anterior and posterior talofibular and the calcaneofibular—reinforce the joint capsule laterally. The four bands of the deltoid ligament contribute to joint stability on the medial side. The ligamentous structure of the ankle is displayed in Figure 8-20.

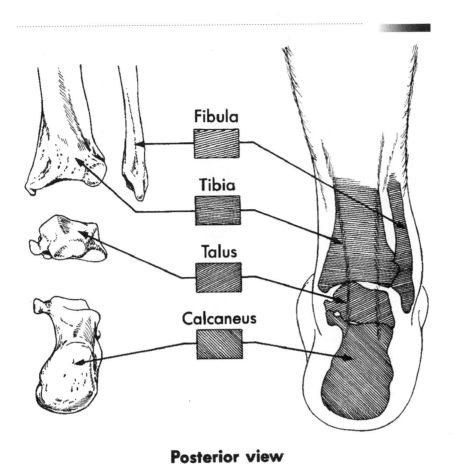

FIGURE 8-19
The bony structure of the ankle.

Fibula

Tibia

Talus

Calcaneus

Posterior view

MOVEMENTS AT THE ANKLE

The axis of rotation at the ankle is essentially frontal, although it is slightly oblique and its orientation changes somewhat as rotation occurs at the joint. Motion at the ankle occurs primarily in the sagittal plane, with the ankle functioning as a hinge joint during the stance phase of gait (1). Ankle flexion and extension are termed dorsiflexion and plantar flexion, respectively (see Chapter 2). Dorsiflexion of 25° also involves approximately 2.5° of external rotation of the foot, and plantar flexion to 35° is associated with slight internal rotation of the foot, on the order of less than 1° (33). In the loaded ankle, 10° of dorsiflexion is accompanied by 1.6° of eversion and 2.1° of internal tibial rotation, with 10° of plantar flexion resulting in 1.6° of inversion and 1.3° of external rotation of the tibia (18).

FIGURE 8-20
The ligaments of the
ankle.

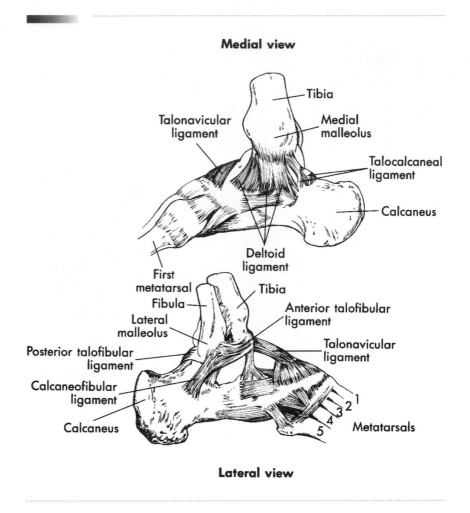

The malleoli serve as pulleys to channel muscle tendons anterior or posterior to the axis of rotation at the ankle; those anterior to the malleoli are dorsiflexors and those posterior to the malleoli serve as plantar flexors.

The medial and lateral malleoli serve as pulleys to channel the tendons of muscles crossing the ankle either posterior or anterior to the axis of rotation, thereby enabling their contributions to either dorsiflexion or plantar flexion.

The tibialis anterior, extensor digitorum longus, and peroneus tertius are the prime dorsiflexors of the foot. The extensor hallucis longus assists in dorsiflexion (Figure 8-21).

The major plantar flexors are the two heads of the powerful two-joint gastrocnemius and the soleus, which lies beneath the gastrocnemius (Figure 8-22). Assistant plantar flexors include the tibialis posterior, peroneus longus, peroneus brevis, plantaris, flexor hallucis longus, and

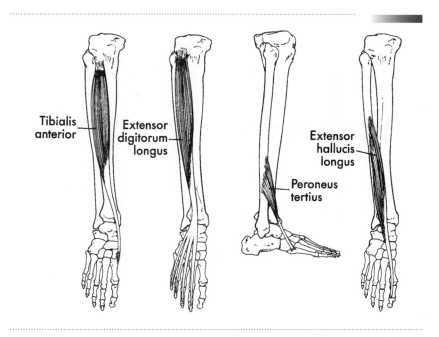

FIGURE 8-21
The dorsiflexors of the ankle.

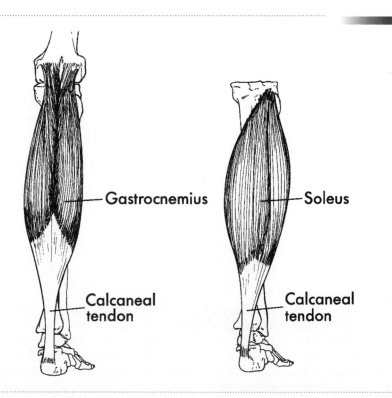

FIGURE 8-22
The major plantar flexors of the ankle.

FIGURE 8-23
Muscles with tendons passing posterior to the malleoli assist with plantar flexion of the ankle.

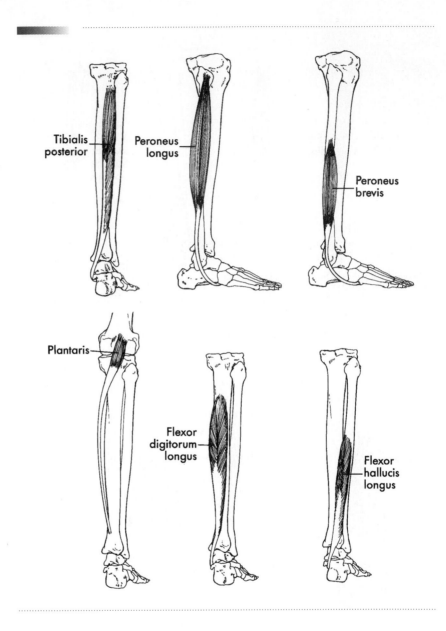

flexor digitorum longus (Figure 8-23). A new speed skate that allows the skater to forcefully plantar flex during the skating stride was recently developed by biomechanist Van Ingen Schenau (50). The advent of this skate during the 1998 Winter Olympic Games was responsible for the shattering of previous world records in every single one of the men's and women's speedskating events.

STRUCTURE OF THE FOOT

Like the hand, the foot is a multibone structure. It contains a total of 26 bones with numerous articulations (Figure 8-24). Included are the subtalar and midtarsal joints and several tarsometatarsal, intermetatarsal, metatarsophalangeal, and interphalangeal joints. Together, the bones and joints of the foot provide a foundation of support for the upright body, adapt to uneven terrain, and absorb shock.

Subtalar Joint

As the name suggests, the subtalar joint lies beneath the talus, where anterior and posterior facets of the talus articulate with the sustentaculum tali on the superior calcaneus. Four talocalcaneal ligaments join the talus and the calcaneus. The joint is essentially uniaxial, with an alignment slightly oblique to the conventional descriptive planes of motion.

Tarsometatarsal and Intermetatarsal Joints

Both the tarsometatarsal and intermetatarsal joints are nonaxial, with the bone shapes and the restricting ligaments permitting only gliding movements. These joints enable the foot to function as a semirigid unit or to adapt flexibly to uneven surfaces during weight bearing.

Metatarsophalangeal and Interphalangeal Joints

The metatarsophalangeal and interphalangeal joints are similar to their counterparts in the hand, with the former being condyloid joints and the latter being hinge joints. Numerous ligaments provide reinforcement for these joints. The toes function to smooth the weight shift to the opposite foot during walking and help maintain stability during weight bearing by pressing against the ground when necessary. The first digit is referred to as the hallux, or "great toe."

Plantar Arches

The tarsal and metatarsal bones of the foot form three arches. The medial and lateral longitudinal arches stretch from the calcaneus to the metatarsals and tarsals. The transverse arch is formed by the bases of the metatarsal bones.

Several ligaments and the plantar fascia support the plantar arches. The spring ligament is the primary supporter of the medial longitudinal arch, stretching from the sustentaculum tali on the calcaneus to the

FIGURE 8-24
The foot is composed of numerous articulating bones.

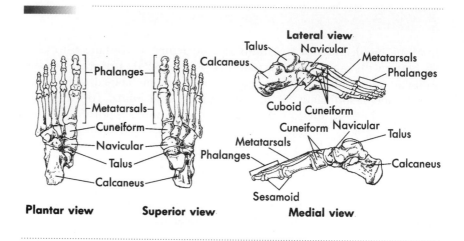

Plantar view   **Superior view**   **Medial view**

plantar fascia
thick bands of fascia that cover the plantar aspect of the foot

During weight bearing, mechanical energy is stored in the stretched ligaments, tendons, and plantar fascia of the arches of the foot, as well as in eccentrically contracting muscles. This stored energy is released to assist with push-off of the foot from the surface.

inferior navicular. The long plantar ligament provides the major support for the lateral longitudinal arch, with assistance from the short plantar ligament. Thick, fibrous, interconnected bands of connective tissue known as the **plantar fascia** extend over the plantar surface of the foot, assisting with support of the longitudinal arch (Figure 8-25). When muscle tension is present, the muscles of the foot, particularly the tibialis posterior, also contribute support to the arches and joints as they cross them.

As the arches deform during weight bearing, mechanical energy is stored in the stretched tendons, ligaments, and plantar fascia. Additional energy is stored in the gastrocnemius and soleus as they develop eccentric tension. During the push-off phase, the stored energy in all of these elastic structures is released, contributing to the force of push-off and actually reducing the metabolic energy cost of walking or running.

MOVEMENTS OF THE FOOT

Muscles of the Foot

The locations and primary actions of the major muscles of the ankle and foot are summarized in Table 8-3. As with the muscles of the hand, extrinsic muscles are those crossing the ankle, and intrinsic muscles have both attachments within the foot.

Toe Flexion and Extension

Flexion involves the curling under of the toes. Flexors of the toes include the flexor digitorum longus, flexor digitorum brevis, quadratus

FIGURE 8-25
The plantar fascia.

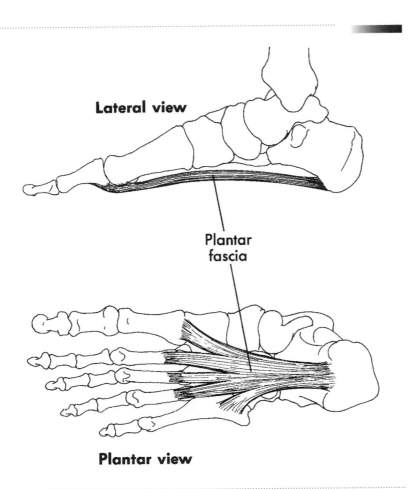

Lateral view

Plantar
fascia

Plantar view

plantae, lumbricals, and interossei. The flexor hallucis longus and brevis produce flexion of the hallux. Conversely, the extensor hallucis longus, extensor digitorum longus, and extensor digitorum brevis are responsible for extension of the toes.

Inversion and Eversion

Rotational movements of the foot in medial and lateral directions are termed inversion and eversion, respectively (see Chapter 2). These movements occur largely at the subtalar joint, although gliding actions among the intertarsal and tarsometatarsal joints also contribute. Inversion results in the sole of the foot turning inward toward the midline of the body. The tibialis posterior and tibialis anterior are the main muscles involved. Turning the sole of the foot outward is termed *eversion*.

TABLE 3-1

MUSCLES OF THE ANKLE AND FOOT

MUSCLE	PROXIMAL ATTACHMENT	DISTAL ATTACHMENT	PRIMARY ACTION(S)
Tibialis anterior	Upper two-thirds lateral tibia	Medial surface of first cuneiform and first metatarsal	Dorsiflexion, inversion
Extensor digitorum longus	Upper three-fourths anterior fibula	Second and third phalanges of four lesser toes	Dorsiflexion, eversion
Peroneus tertius	Lower third anterior fibula	Dorsal surface of fifth metatarsal	Dorsiflexion, eversion
Extensor hallucis longus	Middle anterior fibula	Dorsal surface of distal phalanx of great toe	Dorsiflexion, inversion, and hallux extension
Gastrocnemius	Posterior medial & lateral condyles of the femur	Tuberosity of the calcaneus by the Achilles tendon	Plantar flexion
Plantaris	Distal, posterior femur	Tuberosity of the calcaneus by the Achilles tendon	Assists with plantar flexion
Soleus	Posterior upper fibula and middle tibia	Tuberosity of the calcaneus by the Achilles tendon	Plantar flexion
Peroneus longus	Lateral upper two-thirds fibula	Lateral surface of first cuneiform and first metatarsal	Plantar flexion, eversion
Peroneus brevis	Distal two-thirds fibula	Lateral fifth metatarsal	Plantar flexion, eversion
Flexor digitorum longus	Posterior tibia	Distal phalanx of four lesser toes	Plantar flexion, inversion, toe flexion
Flexor hallucis longus	Lower two-thirds posterior fibula	Distal phalanx of the great toe	Plantar flexion, inversion, toe flexion
Tibialis posterior	Posterior upper two-thirds tibia and fibula	Cuboid, navicular, calcaneus, and all three cuneiforms	Plantar flexion, inversion

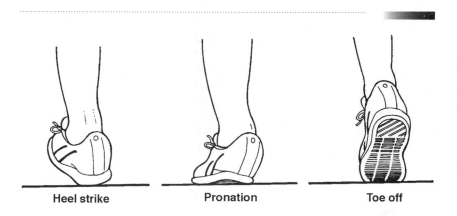

| Heel strike | Pronation | Toe off |

FIGURE 8-26
Rearfoot movement during running. Adapted from Nigg BM et al: Factors influencing kinetic and kinematic variables in running, in BM Nigg (ed.): *Biomechanics of running shoes,* Champaign, 1986, Human Kinetics Publishers.

The muscles primarily responsible for eversion are the peroneus longus and the peroneus brevis, both with long tendons coursing around the lateral malleolus. The peroneus tertius assists.

Pronation and Supination

During walking and running the foot and ankle undergo a cyclical sequence of movements (Figure 8-26). As the heel contacts the ground, the rear portion of the foot typically inverts to some extent. When the foot rolls forward and the forefoot contacts the ground, plantar flexion occurs. The combination of inversion, plantar flexion, and adduction of the foot is known as **supination** (see Chapter 2). While the foot supports the weight of the body during midstance, there is a tendency for eversion and abduction to occur as the foot moves into dorsiflexion. These movements are known collectively as **pronation.** Pronation serves to reduce the magnitude of the ground reaction force sustained during gait by increasing the time interval over which the force is sustained (11).

supination
combined conditions of plantar flexion, inversion, and adduction

pronation
combined conditions of dorsiflexion, eversion, and abduction

LOADS ON THE FOOT

Impact forces sustained during gait increase with body weight and with gait speed in accordance with Newton's third law of motion (see Chapter 3). The vertical ground reaction force applied to the foot during running is bimodal, with an initial impact peak followed almost immediately by a propulsive peak, as the foot pushes off against the ground (Figure 8-27). As running speed increases from 3.0 m/s (8:56 minutes per mile) to 5.0 m/s (5:22 minutes per mile), impact forces range from 1.6 to 2.3 times body weight and propulsive forces range from 2.5 to 2.8 times body weight (35).

FIGURE 8-27

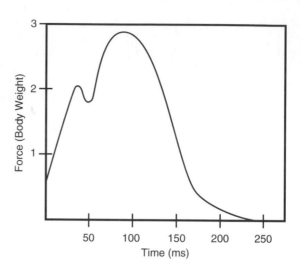

The structures of the foot are anatomically linked such that the load is evenly distributed over the foot during weight bearing. Approximately 50% of body weight is distributed through the subtalar joint to the calcaneus, with the remaining 50% transmitted across the metatarsal heads (42). The head of the first metatarsal sustains twice the load borne by each of the other metatarsal heads (42). A factor that influences this loading pattern, however, is the architecture of the foot. A pes planus (relatively flat arch) condition tends to reduce the load on the forefoot, with pes cavus (relatively high arch) significantly increasing the load on the forefoot (45).

COMMON INJURIES OF THE ANKLE AND FOOT

Because of the crucial roles played by the ankle and foot during locomotion, injuries to this region can greatly limit mobility. Injuries of the lower extremity, especially those of the foot and ankle, may result in weeks or even months of lost training time for athletes, particularly runners. Among dancers, the foot and the ankle are the most common sites of both chronic and acute injuries (16).

Ankle Injuries

Ankle sprains are among the most common of all injuries and constitute 20%–25% of all injuries associated with running and jumping sports that result in a loss of training time (22). Among classic ballet and modern dancers, ankle sprains are the most common traumatic injury, with the usual cause being forced inversion of the ankle during incorrect landing from a jump while the foot is plantar flexed (16).

Because the joint capsule and ligaments are stronger on the medial side of the ankle, inversion sprains involving stretching or rupturing of the lateral ligaments are much more common than eversion sprains of the medial ligaments. In fact, the bands of the deltoid ligament are so strong that excessive eversion is more likely to result in a fracture of the distal fibula than in rupturing of the deltoid ligament. Because of the protection by the opposite limb on the medial side, fractures in the ankle region also occur more often on the lateral than on the medial side.

Ankle sprains usually occur on the lateral side because of weaker ligamentous support than is present on the medial side.

Bracing and taping are two prophylactic measures often used to protect ankles from sprain during sport participation. Ankle braces are designed to preload the ankle and maintain it in a neutral position, thereby counteracting inversionally directed rotation (49). There are contradictory reports in the scientific literature regarding the efficacy of both bracing and taping of the ankle (10). One reported disadvantage of both bracing and taping of the ankle is a reduction in postural control, causing less stability and more touching down on the foot of the supported ankle (4).

Overuse Injuries

Achilles tendinitis involves inflammation and sometimes microrupturing of tissues in the Achilles tendon, typically accompanied by swelling. Two possible mechanisms for tendinitis have been proposed (41). The first is that repeated tension development results in fatigue and decreased flexibility in the muscle, increasing tensile load on the tendon even during relaxation of the muscle. The second theory is that repeated loading actually leads to failure or rupturing of the collagen threads in the tendon. Achilles tendinitis is usually associated with running and jumping activities and is extremely common among theatrical dancers (16). It has also been reported in skiers. Complete rupturing of the Achilles tendon occurs almost exclusively in male skiers, although incidence of the injury has decreased with the advent of high, rigid ski boots and effective release bindings (37).

Repetitive stretching of the plantar fascia can result in plantar fasciitis, a condition characterized by microtears and inflammation of the plantar fascia near its attachment to the calcaneus. The symptoms are pain in the heel and/or arch of the foot. The condition is the fourth most common cause of pain among runners, and also occurs with some frequency among basketball players, tennis players, gymnasts, and dancers (26). Anatomical factors believed to contribute to the development of plantar fasciitis include pes planus (flat foot), a rigid cavus (high-arch) foot, and a tight Achilles tendon, all of which reduce the foot's shock-absorbing capability (47).

Another factor linked to overuse injuries of the lower extremity is excessive pronation. Although walking normally involves approximately 6°–8° of subtalar pronation, individuals with pes planus undergo 10°–12° of pronation (41). Because pronation also causes a compensatory inward rotation of the tibia, excessive pronation can result in increased stress within the plantar fascia and Achilles tendon. Excessive pronation has been associated with running injuries, including shin splints, chondromalacia, plantar fasciitis, and Achilles tendinitis. Excessive pronation has been documented among 60% of one group of injured runners (21).

Stress fractures (see Chapter 4) occur relatively frequently in the bones of the lower extremity. Among a group of 320 athletes with bone-scan-positive stress fractures, the bone most frequently injured was the tibia (49.1%), followed by the tarsals (25.3%), metatarsals (8.8%), femur (7.2%), fibula (6.6%), and pelvis (1.6%) (30). The sites most frequently injured in the older athletes in the group were the femur and the tarsals, with the fibula and tibia most often injured among the younger athletes. Among runners, factors associated with stress fractures include training errors, forefoot striking (toe-heel gait), running on hard surfaces such as concrete, improper footwear, and alignment anomalies of the trunk and/or lower extremity (14). Stress fractures among dancers occur most frequently to the second and third metatarsals and appear to be related to dancing on overly hard surfaces (16). Dancing *en pointe* is particularly stressful to the second metatarsal, because tension in the peroneus longus and tibialis posterior needed for maintaining the *en pointe* position places the stressed second metatarsal in traction (17). Stress fractures among women runners, dancers, and gymnasts may be related to decreased bone mineral density secondary to oligomenorrhea (see Chapter 4).

Alignment Anomalies of the Foot

varus
condition of inward deviation in alignment from the proximal to the distal end of a body segment

valgus
condition of outward deviation in alignment from the proximal to the distal end of a body segment

Varus and **valgus** conditions (inward and outward lateral deviation, respectively, of a body segment) can occur in all of the major links of the lower extremity. These may be congenital or may arise from an imbalance in muscular strength.

In the foot, varus and valgus conditions can affect the forefoot, the rearfoot, and the toes. Forefoot varus and forefoot valgus refer to inversion and eversion misalignments of the metatarsals, and rearfoot varus and valgus involve inversion and eversion misalignments at the subtalar joint (Figure 8-28). Hallux valgus is a lateral deviation of the big toe often caused by wearing women's shoes with pointed toes (Figure 8-29).

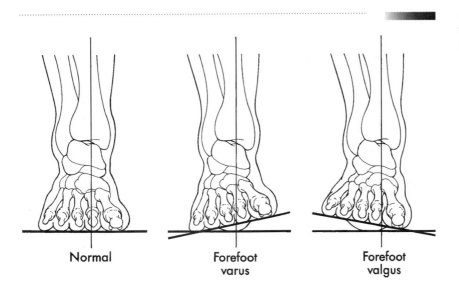

FIGURE 8-28
Varus and valgus
conditions of the forefoot.

Normal   Forefoot
          varus   Forefoot
                   valgus

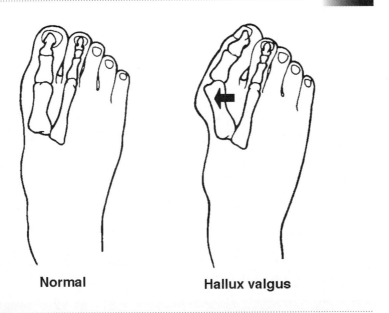

FIGURE 8-29
Wearing shoes with
pointed toes can cause
hallus valgus.

Normal   Hallux valgus

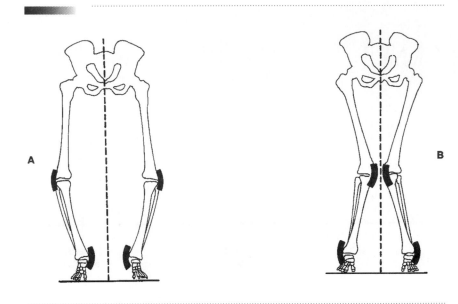

Varus and valgus conditions in the tibia and femur can alter the kinematics and kinetics of joint motion because they cause added tensile stress on the stretched side of the affected joint. For example, a combination of femoral varus and tibial valgus (a knock-knee condition) places added tension on the medial aspect of the knee (Figure 8-30). In contrast, the bow-legged condition of femoral valgus and tibial varus stresses the lateral aspect of the knee and is therefore a predisposing factor for iliotibial band friction syndrome. Unfortunately, lateral misalignments at one joint of the lower extremity are typically accompanied by compensatory misalignments at other lower extremity joints because of the nature of joint loading during weight bearing.

Misalignment at a lower extremity joint typically results in compensatory misalignments at one or more other joints because of the lower extremity's weightbearing function.

Depending on the cause of the misalignment problem, correctional procedures may involve exercises to strengthen or stretch specific muscles and ligaments of the lower extremity as well as the use of orthotics, specially designed inserts worn inside the shoe to provide added support for a portion of the foot.

SUMMARY

The lower extremity is well adapted to its functions of weight bearing and locomotion. This is particularly evident at the hip, where the bony structure and several large, strong ligaments provide considerable joint stability. The hip is a typical ball and socket joint, with flexion, extension, abduction, adduction, horizontal abduction, horizontal adduction, medial and lateral rotation, and circumduction of the femur permitted.

The knee is a large, complex joint composed of two side-by-side condyloid articulations. Medial and lateral menisci improve the fit between the articulating bone surfaces and assist in absorbing forces transmitted across the joint. Because of differences in the sizes, shapes, and orientations of the medial and lateral articulations, medial rotation of the tibia accompanies full knee extension. A number of ligaments cross the knee and restrain its mobility. The primary movements allowed at the knee are flexion and extension, although some rotation of the tibia is also possible when the knee is in flexion and not bearing weight.

The ankle includes the articulations of the tibia and the fibula with the talus. This is a hinge joint that is reinforced both laterally and medially by ligaments. Movements at the ankle joint are dorsiflexion and plantar flexion.

Like the hand, the foot is composed of numerous small bones and their articulations. Movements of the foot include inversion and eversion, abduction and adduction, and flexion and extension of the toes.

INTRODUCTORY PROBLEMS

1. Construct a chart listing all muscles crossing the hip joint according to whether they are anterior, posterior, medial, or lateral to the joint center. Note that some muscles may fall into more than one category. Identify the action or actions performed by the muscles listed in the four categories.

2. Construct a chart listing all muscles crossing the knee joint according to whether they are anterior, posterior, medial, or lateral to the joint center. Note that some muscles may fall into more than one category. Identify the action or actions performed by the muscles listed in the four categories.

3. Construct a chart listing all muscles crossing the ankle joint according to whether they are anterior or posterior to the joint center. Note that some muscles may fall into more than one category. Identify the action or actions performed by the muscles listed in the four categories.

4. Compare the structure of the hip (including bones, ligaments, and muscles) to the structure of the shoulder. What are the relative advantages and disadvantages of the two joint structures?

5. Compare the structure of the knee (including bones, ligaments, and muscles) to the structure of the elbow. What are the relative advantages and disadvantages of the two joint structures?

6. Describe sequentially the movements of the lower extremity that occur during kicking a ball. Identify the agonist muscle groups for each of these movements.

7. Describe sequentially the movements of the lower extremity that occur during performance of a vertical jump. Identify the agonist muscle groups for each of these movements.

8. Describe sequentially the movements of the lower extremity that occur during rising from a seated position. Identify the agonist muscle groups for each of these movements.

9. Using the diagram in Sample Problem 8-1 as a model, what is the magnitude of the reaction force at the hip when tension in the hip abductors is 750 N and 300 N of body weight is supported?

10. Using the diagram in Sample Problem 8-2 as a model, how much compression acts on the patellofemoral joint when the quadriceps exert 400 N of tension and the angle between the quadriceps and the patellar tendon is a) 140°, and b) 100°?

ADDITIONAL PROBLEMS

1. Explain the roles of two-joint muscles in the lower extremity, using the rectus femoris and the gastrocnemius as examples. How does the orientation of the limbs articulating at one joint influence the action of a two-joint muscle at the other joint?

2. Explain the sequencing of joint actions in the ankle and foot during the support phase of gait.

3. Describe the sequencing of contractions of the major muscle groups of the lower extremity during the gait cycle, indicating when contractions are concentric and eccentric.

4. Which muscles of the lower extremity are called on more for running uphill than for running on a level surface? For running downhill as compared to running on a level surface? Explain why.

5. The squat exercise with a barbell is sometimes performed with the heels elevated by a block of wood. Explain what effect this has on the function of the major muscle groups involved.

6. Explain why compression at the hip is higher than compression at the knee despite the fact that the knee supports more body weight during stance than the hip.

7. Construct a free body diagram that demonstrates how the use of a cane can alleviate compression on the hip.

8. Explain why lifting a heavy load with one or both knees in extreme flexion is dangerous. What structure(s) are placed at risk?

9. Explain how excessive pronation predisposes the individual to stress-related injuries of the Achilles tendon and plantar fascia.

10. What compensations during gait are likely to be made by individuals with genu valgum and genu varus?

1. Study anatomical models of the hip, knee, and ankle. Be able to locate the major bones, muscle attachments, and ligaments.
2. From the side view, videotape a volunteer walking at a slow pace, a normal pace, and a fast pace. Review the videotape several times and construct a chart that characterizes the kinematic differences among the three trials. Explain what differences in muscle activity are associated with the major kinematic differences.
3. Repeat Exercise 2 with a rear-view videotape of a volunteer on a treadmill.
4. Repeat Exercise 2 with a volunteer running at a slow pace and a fast pace.
5. From the side view, videotape a volunteer rising from a seated position. Review the videotape several times and construct a list indicating the sequencing and timing of the major joint actions and the associated activity of major muscle groups.

REFERENCES

1. Adachi K, Nishizawa S, and Endo B: The trajectory of the point of application of the resultant force of body mass at different walking speeds. Statistical analysis of human walking, Folia Primatol (Basel) 66:160, 1996.
2. Albright JP, Saterbak A, and Stokes J: Use of knee braces in sport. Current recommendations, Sports Med 20:281, 1995.
3. Beaupre A et al: Knee menisci: Correlation between microstructure and biomechanics, Clin Orthop 208:72, 1986.
4. Bennell KL and Goldie PA: The differential effects of external ankle support on postural control, J Orthop Sports Phys Ther 20:287, 1994.
5. Bergmann G, Graichen A, and Rohlmann A: Hip joint forces during load carrying, Clin Orthop 335: 190, 1997.
6. Bergmann G, Graichen F, and Rohlmann A: Is staircase walking a risk for the fixation of hip implants? J Biomech 28: 535, 1995.
7. Bergmann G, Graichen F, and Rohlmann A: Hip joint loading during walking and running, measured in two patients, J Biomech 26:969, 1993.
8. Bergmann G, Kniggendorf H, Graichen F, and Rohlmann A: Influence of shoes and heel strike on the loading of the hip joint, J Biomech 28:817, 1995.
9. Beynnon BD et al: The effect of functional knee bracing on the anterior cruciate ligament in the weightbearing and nonweightbearing knee, Am J Sports Med 25:353, 1997.
10. Callaghan MJ: Role of ankle taping and bracing in the athlete, Br J Sports Med 31:102, 1997.
11. Clarke TE, Frederick EC, and Hamill C: The effects of shoe design parameters on rearfoot control in running, Med Sci Sports Exerc 15:376, 1983.
12. DeVita P, Torry M, Glover KL, and Speroni DL: A functional knee alters joint to torque and power patterns during walking and running, J Biomech 29:583, 1996.
13. Figueras JM, et al: The evaluation of knee injuries in skiing accidents. In Johnson RJ and Mote CD: Skiing trauma and skiing safety: STP 860, Philadelphia, 1985, American Society for Testing and Materials.

14. Frey C: Footwear and stress fractures, Clin Sports Med 16:249, 1997.
15. Grabiner MD, Koh TJ, and Draganich LF: Neuromechanics of the patellofemoral joint, Med Sci Sports Exerc 26:10, 1994.
16. Hardaker WT, Margello S, and Goldner JL: Foot and ankle injuries in theatrical dancers, Foot Ankle 6:59, 1985.
17. Harrington T, et al: Overuse ballet injury of the base of the second metatarsal, Am J Sports Med 21:591, 1993.
18. Hintermann B and Nigg BM: In vitro kinematics of the axially loaded ankle complex in response to dorsiflexion and plantar flexion, Foot Ankle Int 16:514, 1995.
19. Hull ML, Berns GS, Varma H, and Patterson HA: Strain in the medial collateral ligament of the human knee under single and combined loads, J Biomech 29:199, 1996.
20. Jaglal SB, Kreiger N, and Darlington G: Past and recent physical activity and risk of hip fracture, Am J Epidemiol 138:107, 1993.
21. James SL, Bates BT, and Osternig LR: Injuries to runners, Am J Sports Med 6:40, 1978.
22. Johnson RE and Rust RJ: Sports related injury: An anatomic approach, part 2, Minn Med 68:829, 1985.
23. Johnston RC: Mechanical considerations of the hip joint, Arch Surg 107:411, 1973.
24. Kettlekamp DB and Jacobs AW: Tibiofemoral contact area determination and implications, J Bone Joint Surg 54A:349, 1972.
25. Krivickas LS: Anatomical factors associated with overuse sports injuries, Sports Med 24:132, 1997.
26. Leach RE, Seavey MS, and Salter DK: Results of surgery in athletes with plantar fascitis, Foot Ankle 7:156, 1986.
27. Lloyd DG and Buchanan TS: A model of load sharing between muscles and soft tissues at the human knee during static tasks, J Biomech Eng 118:367, 1996.
28. Markolf KL et al: Combined knee loading states that generate high anterior cruciate ligament forces, J Orthop Res 13:930, 1995.
29. Markolf KL, Slauterbeck JL, Armstrong KL, Shapiro MM, and Finnerman GA: Effects of combined knee loadings on posterior cruciate ligament force generation, J Orthop Res 14:633, 1996.
30. Matheson GO et al: Stress fractures in athletes, Am J Sports Med 15:46, 1987.
31. McGibbon CA, Krebs DE, and RW Mann: In vivo hip pressures during cane and load-carrying gait, Arthritis Care Res 10:300, 1997.
32. Messier SP et al: Etiology of iliotibial band friction syndrome in distance runners, Med Sci Sports Exerc 27:951, 1995.
33. Michelson JD and Helgemo SL Jr: Kinematics of the axially loaded ankle, Foot Ankle Int 16:577, 1995.
34. Morrison JB: The mechanics of the knee joint in relation to normal walking, J Biomech 3:51, 1970.
35. Munro CF, Miller DI, and Fuglevand AJ: Ground reaction forces in running: A reexamination, J Biomech 20:147, 1987.
36. Nemeth WC and Sanders BL: The lateral synovial recess of the knee: Anatomy and role in chronic iliotibial band friction syndrome, Arthroscopy 12:574, 1996.

37. Oden RR: Tendon injuries about the ankle resulting from skiing, Clin Orthop 216:63, 1987.

38. Orchard JW, Fricker PA, Abud AT and Mason BR: Biomechanics of iliotibial band friction syndrome in runners, Am J Sports Med 24:375, 1996.

39. Paul JP and McGrouther DA: Forces transmitted at the hip and knee joint of normal and disabled persons during a range of activities, Acta Orthop Belg, Suppl. 41:78, 1975.

40. Reilly DT and Martens M: Experimental analysis of the quadriceps muscle force and patello-femoral joint reaction force for various activities, Acta Orthop Scand 43:126, 1972.

41. Renstrom P and Johnson RJ: Overuse injuries in sports: A review, Sports Med 2:316, 1985.

42. Sammarco GJ: Biomechanics of the foot. In Nordin M and Frankel VH, eds: *Basic biomechanics of the musculoskeletal system*, 2nd ed, Philadelphia, 1989, Lea & Febiger.

43. Seedhom BB, Dowson D, and Wright V: The load-bearing function of the menisci: A preliminary study. In Ingwerson OS et al, eds: *The knee joint: Recent advances in basic research and clinical aspects*, Amsterdam, 1974, Excerpta Medica.

44. Shrive NG, O'Connor JJ, and Goodfellow JW: Load-bearing in the knee joint, Clin Orthop 131:279, 1978.

45. Sneyers CJ, Lysens R, Feys H, and Andries R: Influence of malalignment of feet on the plantar pressure pattern in running, Foot Ankle Int 16: 624, 1995.

46. Sutton G: Hamstrung by hamstring strains: A review of the literature, J Orthop Sports Phys Ther 5:184, 1984.

47. Tanner SM and Harvey JS: How we manage plantar fascitis, Physician Sportsmed 16:39, 1988.

48. Terry GC, Hughston JC, and Norwood LA: The anatomy of the iliopatellar band and iliotibial tract, Am J Sports Med 14:39, 1986.

49. Thonnard JL, Bragard D, Willems PA, and Plaghki L: Stability of the braced ankle. A biomechanical investigation, Am J Sports Med 24:356, 1996.

50. Van Ingen Schenau GJ et al: A new skate allowing powerful plantar flexions improves performance, Med Sci Sports Exerc 28:531, 1996.

51. Vizsolyi P et al: Breaststroker's knee, Am J Sports Med 15:63, 1987.

52. Wroble RR et al: Patterns of knee injuries in wrestling: A six year study, Am J Sports Med 14:55, 1986.

53. Yates C and Grana WA: Patellofemoral pain: A prospective study, Orthopedics 9:663, 1986.

ANNOTATED READINGS

Cavanagh PR, ed: Biomechanics of distance running, Champaign, 1990, Human Kinetics Publishers.
Includes chapters reviewing research on the biomechanics of distance running.

Grabiner MD, Koh TJ, and Draganich LF: Neuromechanics of the patellofemoral joint, Med Sci Sports Exerc 26:10, 1994.
Discusses current knowledge of patellar tracking, patellofemoral pressure, and neuromuscular control of patellofemoral agonists.

Reeder MT, Dick BH, Atkins JK, Pribis AB, and Martinez JM: Stress fractures. Current concepts of diagnosis and treatment, Sports Med 22:198, 1996.

Reviews current knowledge related to causes and treatment of stress fractures in athletes.

Wen DY, Puffer JC, and Schmalzried TP: Lower extremity alignment and risk of overuse injuries in runners, Med Sci Sports Exerc 29:1291, 1997.

Provides statistical relationships among numerous measures of lower extremity alignment, training factors, and overuse injury incidence in distance runners.

RELATED WEB SITES

Calvert Orthopaedic and Sports Medicine Center
http://www.calvertorthoandsports.com/index.html
Provides basic information and color graphics of the anatomy and common injuries of the knee and ankle.

The Center for Orthopaedics and Sports Medicine
http://www.arthroscopy.com/sports.htm
Includes information and color graphics on the foot and ankle and on knee surgery.

CMTnet: Basic Anatomy
http://www.ultranet.com/~smith/anatomy.html
Provides information on movements, bones, and muscles of the foot, with a link to 3-D images of the lower extremity that the user can rotate.

E.M. Hennig's Research Activities and Publications
http://www.uni-essen.de/~qpd800/research.html
Includes links to animations of a barefoot pressure isobarograph, barefoot walking, in-shoe pressures in two running shoes and others.

Link Orthopaedics
http://www.dundee.ac.uk/orthopaedics/link/teach.htm
Provides links to several pages on the knee, including an interactive page on knee evaluation, knee anatomy, and a virtual reality simulation of arthroscopic surgery.

The Lower Extremity Page
http://rivers.oscs.montana.edu/esg/kla/ta/thighleg.html
Provides photographs of cadaver sections, showing muscles and innervations, and providing quizes regarding the anatomical structures shown.

M & M Orthopaedics
http://mmortho.com
Includes information on ACL reconstruction, plantar fascitis, clubfoot, and hip and knee replacement, with QuickTime movies of the inside of the knee.

MedFacts Play Doctor! MedLib
http://www.medfacts.com/medlib.htm
Provides links to pages on ankle fracture and sprain, knee injuries, chondromalacia, meniscal tears and transplants, patellar subluxation, hip bursitis, shin splints, and osteoarthritis of the hip. Includes 3-D videos of the arthroscopic meniscus surgeries.

Northern Rockies Orthopaedics Specialists
http://www.orthopaedic.com
Provides links to pages describing descriptions of injuries, diagnostic tests, and surgical procedures for the shoulder, elbow, wrist, and hand.

Orthopedic Medical Information
http://www.opendoor.com/albert/one/orthopedic.html
Includes links to pages on knee disorders, joint replacements, and sprains and strains.

Patient's Guide to Foot and Ankle Problems
http://www.sechrest.com/mmg/foot/anatomy
Includes descriptive information and graphics on the anatomy of the foot and ankle,

diagnostic procedures for foot and ankle problems, and common foot and ankle syndromes.

Results of Hip Joint Force Measurements
http://www.medizin.fu-berlin.de/biomechanik/Resstair.htm
   Provides data on direct measurements of hip joint loading under different conditions.

Rothman Institute
http://rothmaninstitute.com/index.html
   Includes information on common sport injuries to the hip, knee, foot, and ankle.

Rush Gait Laboratory
http://www.ortho.rush.edu/gait/projects.htm
   Provides links to pages with detailed information on biomechanical aspects of total knee replacement, ACL function and reconstruction, knee osteoarthritis, and gait kinematics.

Southern California Orthopaedic Institute
http://www.scoi.com
   Includes links to anatomical descriptions and labeled photographs of the hip, knee, ankle, and toe.

University of Washington Orthopaedic Physicians
http://www.orthop.washington.edu
   Provides extensive descriptions of scientific and clinical aspects of the hip, knee, ankle, and foot. Clicking on the radiographs brings up structure labels. Also includes numerous movies showing therapeutic exercise, examination procedures, and surgical techniques.

The Virtual Hospital: Joint Fluoroscopy
http://www.vh.org/Providers/Textbooks/JointFluoro/JointFluoroHP.html
   Provides detailed descriptions of anatomical features and includes multiple-view electric joint fluoroscopies of the hip, knee, and ankle.

The "Virtual" Medical Center: Anatomy & Histology Center
http://www-sci.lib.uci.edu/HSG/MedicalAnatomy.html
   Contains numerous images, movies, and course links for human anatomy.

WebCAI Programs
http://www.kumc.edu/research/medicine/pharmacology/CAI/menu2.htm
   Provides interactive quizzes on all regions of the upper and lower extremities.

Wheeless' Textbook of Orthopaedics
http://www.medmedia.com/med.htm
   Provides comprehensive, detailed information, graphics, and related literature for all the joints.

Chapter

9

THE BIOMECHANICS
OF THE HUMAN SPINE

After completing this chapter, the reader will be able to:

Explain how anatomical structure affects movement capabilities of the spine.

Identify factors influencing the relative mobility and stability of different regions of the spine.

Explain the ways in which the spine is adapted to carry out its biomechanical functions.

Explain the relationship between muscle location and the nature and effectiveness of muscle action in the trunk.

Describe the biomechanical contributions to common injuries of the spine.

The spine is a complex and functionally significant segment of the human body. Providing the mechanical linkage between the upper and lower extremities, the spine enables motion in all three planes, yet still functions as a bony protector of the delicate spinal cord. To many researchers and clinicians, the lumbar region of the spine is of particular interest because low back pain is a major medical and socioeconomic problem of modern times.

STRUCTURE OF THE SPINE

Vertebral Column

The spine consists of a curved stack of 33 vertebrae divided structurally into five regions (Figure 9-1). Proceeding from superior to inferior,

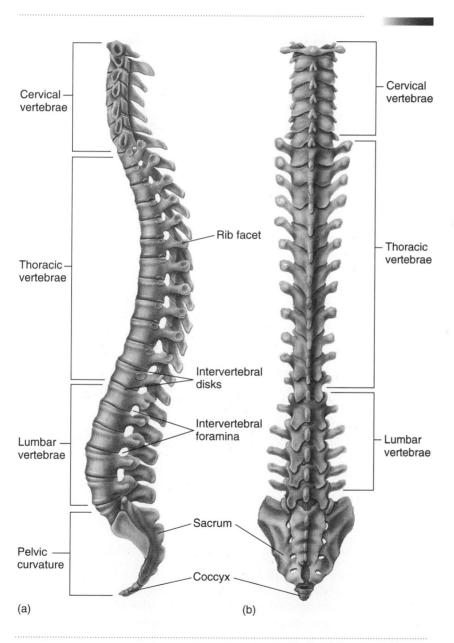

(a)

(b)

Cervical vertebrae

Thoracic vertebrae

Lumbar vertebrae

Pelvic curvature

Rib facet

Intervertebral disks

Intervertebral foramina

Sacrum

Coccyx

Cervical vertebrae

Thoracic vertebrae

Lumbar vertebrae

FIGURE 9-1

(a) Left lateral and
(b) posterior views of the
major regions of the spine.
From Shier, Butler, and Lewis.
*Hole's Human Anatomy and
Physiology.* © 1996. Reprinted by
permission of The McGraw-Hill
Companies, Inc.

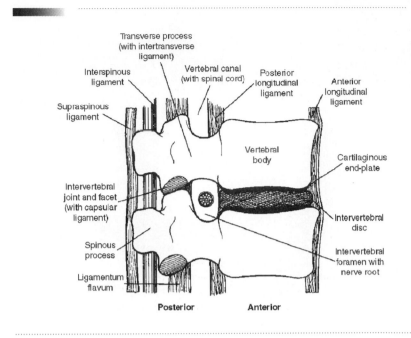

Transverse process
(with intertransverse ligament)

Interspinous ligament

Vertebral canal
(with spinal cord)

Posterior longitudinal ligament

Anterior longitudinal ligament

Supraspinous ligament

Vertebral body

Cartilaginous end-plate

Intervertebral joint and facet
(with capsular ligament)

Intervertebral disc

Spinous process

Intervertebral foramen with nerve root

Ligamentum flavum

Posterior   Anterior

there are 7 cervical vertebrae, 12 thoracic vertebrae, 5 lumbar vertebrae, 5 fused sacral vertebrae, and 4 small, fused coccygeal vertebrae. There may be one extra vertebra or one less, particularly in the lumbar region.

Because of structural differences and the ribs, varying amounts of movement are permitted between adjacent vertebrae in the cervical, thoracic, and lumbar portions of the spine. Within these regions, two adjacent vertebrae and the soft tissues between them are known as a **motion segment.** The motion segment is considered to be the functional unit of the spine (Figure 9-2).

Each motion segment contains three joints. The vertebral bodies separated by the intervertebral disc form a symphysis type of amphiarthrosis. The right and left facet joints between the superior and inferior articular processes are diarthroses of the gliding type that are lined with articular cartilage.

motion segment
two adjacent vertebrae and the associated soft tissues; the functional unit of the spine

The spine may be viewed as a triangular stack of articulations, with symphysis joints between vertebral bodies on the anterior side and two gliding diarthrodial facet joints on the posterior side.

Vertebrae

A typical vertebra consists of a body, a hollow ring known as the neural arch, and several bony processes (Figure 9-3). The vertebral bodies serve as the primary weightbearing components of the spine. The neural arches and posterior sides of the bodies and intervertebral discs form a protective passageway for the spinal cord and associated blood vessels known as the vertebral canal. From the exterior surface of each neural

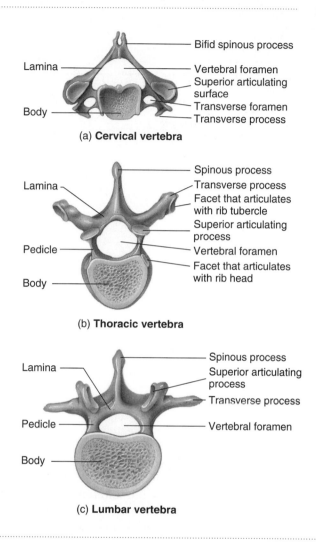

Lamina — Bifid spinous process

Lamina — Vertebral foramen
— Superior articulating surface

Body — Transverse foramen
— Transverse process

(a) **Cervical vertebra**

Lamina — Spinous process
— Transverse process
— Facet that articulates with rib tubercle
— Superior articulating process

Pedicle — Vertebral foramen

— Facet that articulates with rib head

Body

(b) **Thoracic vertebra**

Lamina — Spinous process
— Superior articulating process

— Transverse process

Pedicle — Vertebral foramen

Body

(c) **Lumbar vertebra**

FIGURE 9-3
Superior view of the typical vertebrae. From Shier, Butler, and Lewis. *Hole's Human Anatomy and Physiology.* © 1996. Reprinted by permission of The McGraw-Hill Companies, Inc.

arch, several bony processes protrude. The spinous and transverse processes serve as outriggers to improve the mechanical advantage of the attached muscles.

There is a progressive increase in vertebral size from the cervical region down through the lumbar region (Figure 9-3). The lumbar vertebrae, in particular, are larger and thicker than the vertebrae in the superior regions of the spine. This serves a functional purpose, since when the body is in an upright position each vertebra must support the weight of not only the arms and head, but all the trunk positioned above it. The increased surface area of the lumbar vertebrae reduces the amount of stress to which these vertebrae would otherwise be subjected. The

Although all vertebrae have the same basic shape, there is a progressive superior-inferior increase in the size of the vertebral bodies and a progression in the size and orientation of the articular processes.

The orientation of the facet joints determines the movement capabilities of the motion segment.

FIGURE 9-4

Approximate orientations of the facet joints. **A,** Lower cervical spine, with facets oriented 45° to the transverse plane and parallel to the frontal plane. **B,** Thoracic spine, with facets oriented 60° to the transverse plane and 20° to the frontal plane. **C,** Lumbar spine, with facets oriented 90° to the transverse plane and 45° to the frontal plane.

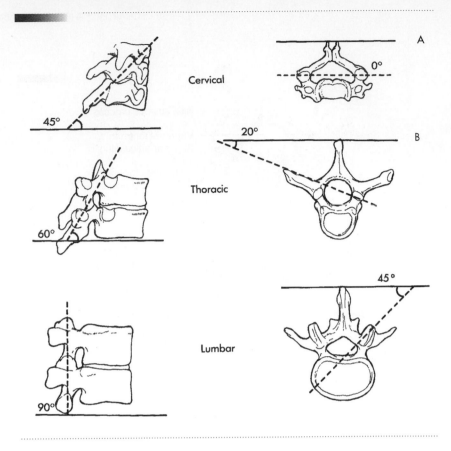

weightbearing surface area of the intervertebral disc also increases with the weight supported in all mammals (77).

The size and angulation of the vertebral processes vary throughout the spinal column (Figure 9-4). This changes the orientation of the facet joints, which limit range of motion in the different spinal regions. In addition to channeling the movement of the motion segment, the facet joints assist in load bearing. The facet joints and discs provide about 80% of the spine's ability to resist rotational torsion and shear, with half of this contribution from the facet joints (20, 34). The facet joints also sustain up to approximately 30% of the compressive loads on the spine, particularly when the spine is in hyperextension (Figure 9-5) (43). Contact forces are largest at the L5-S1 facet joints (66). Recent studies suggest that 15%–40% of chronic low back pain emanates from the facet joints (17).

Intervertebral Discs

The articulations between adjacent vertebral bodies are symphysis joints with intervening fibrocartilaginous discs that act as cushions. Healthy intervertebral discs in an adult account for approximately one-fourth of the height of the spine. When the trunk is erect, the differences in the

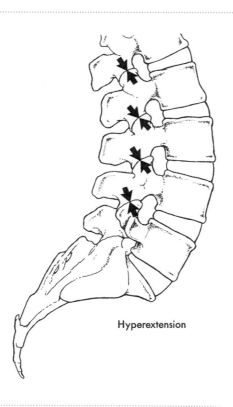

FIGURE 9-5
Hyperextension of the lumbar spine creates compression at the facet joints.

Hyperextension

anterior and posterior thicknesses of the discs produce the lumbar, thoracic, and cervical curves of the spine.

The intervertebral disc is composed of two functional structures: a thick outer ring composed of fibrous cartilage called the **annulus fibrosus** or annulus surrounds a central gelatinous material known as the **nucleus pulposus** or nucleus (Figure 9-6). The annulus consists of about 90 concentric bands of collagenous tissue that are bonded together. The collagen fibers of the annulus crisscross vertically at about 30° angles to each other, making the structure more sensitive to rotational strain than to compression, tension, and shear (20). The nucleus of a young, healthy disc is approximately 90% water, with the remainder being collagen and proteoglycans, specialized materials that chemically attract water (60). The extremely high fluid content of the nucleus makes it resistant to compression.

Mechanically, the annulus acts as a coiled spring whose tension holds the vertebral bodies together against the resistance of the nucleus pulposus, and the nucleus pulposus acts like a ball bearing composed of an incompressible gel (Figure 9-7) (51). During flexion and extension the vertebral bodies roll over the nucleus while the facet joints guide

annulus fibrosus
thick, fibrocartilaginous ring that forms the exterior of the intervertebral disc

nucleus pulposus
colloidal gel with a high fluid content, located inside the annulus fibrosus of the intervertebral disc

FIGURE 9-6

In the intervertebral disc, the annulus fibrosus, made up of laminar layers of criss-crossed collagen fibers, surrounds the nucleus pulposus.

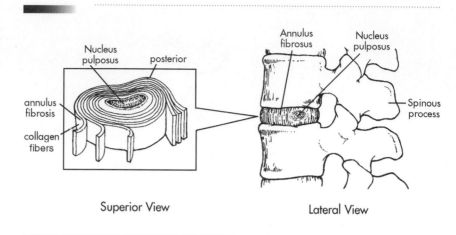

the movements. As shown in Figure 9-8, spinal flexion, extension, and lateral flexion produce compressive stress on one side of the discs and tensile stress on the other, whereas spinal rotation creates shear stress in the discs (Figure 9-9) (43). During daily activities, compression is the most common form of loading on the spine.

When a disc is loaded in compression, it tends to simultaneously lose water and absorb sodium and potassium until its internal electrolyte concentration is sufficient to prevent further water loss (41). When this

FIGURE 9-7

Mechanically, the annulus fibrosus behaves as a coiled spring, holding the vertebral bodies together, whereas the nucleus pulposus acts like a ball bearing that the vertebrae roll over during flexion/extension and lateral bending.

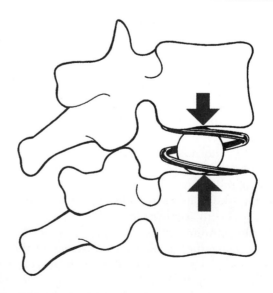

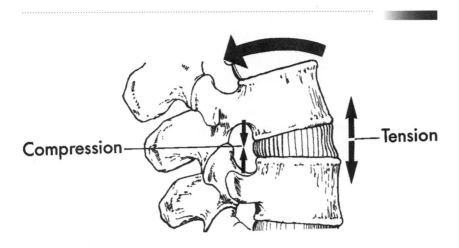

Compression—

—Tension

FIGURE 9-8
When the spine bends, a tensile load is created on one side of the discs and a compressive load is created on the other.

chemical equilibrium is achieved, internal disc pressure is equal to the external pressure (7). Continued loading over a period of several hours results in a further slight decrease in disc hydration (2). For this reason, the spine undergoes a height decrease of up to nearly 2 cm over the course of a day, with approximately 54% of this loss occurring during the first 30 minutes after getting up in the morning (62).

Once pressure on the discs is relieved, the discs quickly reabsorb water and disc volumes and heights are increased (41). Astronauts experience a temporary increase in spine height of approximately 5 cm while free from the influence of gravity (58). On earth, disc height and volume are typically greatest when a person first arises in the morning.

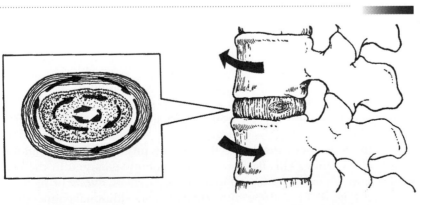

Superior view   **Lateral view**

FIGURE 9-9
Spinal rotation creates shear stress within the discs, with the greatest shear around the periphery of the annulus.

Because increased disc volume also translates to increased spinal stiffness, there appears to be a heightened risk of disc injury early in the morning (16).

The intervertebral discs have a blood supply up to about the age of 8 years, but after that the discs must rely on a mechanically based means for maintaining a healthy nutritional status. Intermittent changes in posture and body position alter internal disc pressure, causing a *pumping action* in the disc (53). The influx and outflux of water transports nutrients in and flushes out metabolic waste products, basically fulfilling the same function that the circulatory system provides for vascularized structures within the body. Maintaining even an extremely comfortable fixed body position over a period of time curtails this pumping action and can negatively affect disc health.

Injury and aging irreversibly reduce the water-absorption capacity of the discs, with a concomitant decrease in shock-absorbing capability. Magnetic resonance imaging (MRI) studies show degenerative changes to be the most common at L5-S1, the disc subjected to the most mechanical stress by virtue of its position (64). However, the fluid content of all discs begin to diminish around the second decade of life (7). A typical geriatric disc has a fluid content that is reduced by approximately 35% (76). As this normal degenerative change occurs, abnormal movements occur between adjacent vertebral bodies and more of the compressive, tensile, and shear loads on the spine must be assumed by other structures—particularly the facets and joint capsules. Results include reduced height of the spinal column, often accompanied by degenerative changes in the spinal structures that are forced to assume the discs' loads. Postural alterations may also occur. The normal lordotic curve of the lumbar region may be reduced as an individual attempts to relieve compression on the facet joints by maintaining a posture of spinal flexion (76). Factors such as habitual smoking and exposure to vibration can negatively affect disc nutrition, whereas regular exercise can improve it (60).

It is important not to maintain any one body position for too long, since the intervertebral discs rely on body movement to pump nutrients in and waste products out.

Although most of the load sustained by the spine is borne by the symphysis joints, the facet joints may play an assistive role, particularly when the spine is in hyperextension and when disc degeneration has occurred.

Ligaments

A number of ligaments support the spine, contributing to the stability of the motion segments (Figure 9-10). The powerful anterior longitudinal ligament and the weaker posterior longitudinal ligament connect the vertebral bodies in the cervical, thoracic, and lumbar regions. The supraspinous ligament attaches to the spinous processes throughout the length of the spine. This ligament is prominently enlarged in the cervical region, where it is referred to as the *ligamentum nuchae* or *ligament of the neck* (Figure 9-11). Adjacent vertebrae have additional connections between spinous processes, transverse processes, and laminae, supplied respectively by the interspinous ligaments, the intertransverse ligaments, and the ligamenta flava.

The enlarged cervical portion of the supraspinous ligament is the ligamentum nuchae or ligament of the neck.

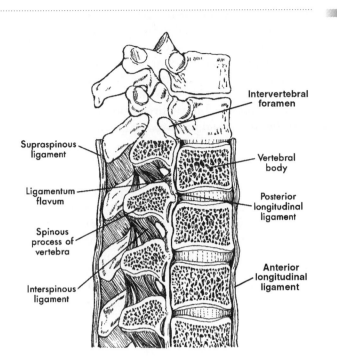

FIGURE 9-10
The major ligaments of the spine. (The intertransverse ligament is not visible in this medial section through the spine.)

Supraspinous ligament

Ligamentum flavum

Spinous process of vertebra

Interspinous ligament

Intervertebral foramen

Vertebral body

Posterior longitudinal ligament

Anterior longitudinal ligament

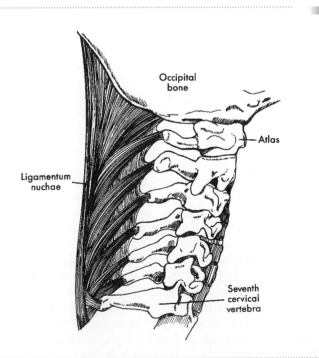

FIGURE 9-11
The supraspinous ligament is well developed in the cervical region, where it is referred to as the ligamentum nuchae.

Occipital bone

Ligamentum nuchae

Atlas

Seventh cervical vertebra

ligamentum flavum
yellow ligament that connects the laminae of adjacent vertebrae; distinguished by its elasticity

prestress
stress on the spine created by tension in the resting ligaments

primary spinal curves
curves that are present at birth

Note the relatively flat spine of this three year old. The lumbar curve does not reach full development until approximately age 17.

secondary spinal curves
curves that do not develop until the weight of the body begins to be supported in sitting and standing positions

lordosis
extreme lumbar curvature

kyphosis
extreme thoracic curvature

Another major ligament, the **ligamentum flavum,** connects the laminae of adjacent vertebrae. Although most spinal ligaments are composed primarily of collagen fibers that stretch only minimally, the ligamentum flavum contains a high proportion of elastic fibers, which lengthen when stretched during spinal flexion and shorten during spinal extension. The ligamentum flavum is in tension even when the spine is in anatomical position, enhancing spinal stability. This tension creates a slight, constant compression in the intervertebral discs, referred to as **prestress.**

Spinal Curves

As viewed in the sagittal plane, the spine contains four normal curves. The thoracic and sacral curves, which are concave anteriorly, are present at birth and are referred to as **primary curves.** The lumbar and cervical curves, which are concave posteriorly, develop from supporting the body in an upright position after young children begin to sit up and stand. Since these curves are not present at birth, they are known as the **secondary spinal curves.** Although the cervical and thoracic curves change little during the growth years, the curvature of the lumbar spine increases approximately 10% between the ages of 7 and 17 (74). Spinal curvature (posture) is influenced by heredity, pathological conditions, an individual's mental state, and the forces to which the spine is habitually subjected. Mechanically, the curves enable the spine to absorb more shock without injury than if the spine were straight (32).

As discussed in Chapter 4, bones are constantly modeled or shaped in response to the magnitudes and directions of the forces acting on them. Similarly, the four spinal curves can become distorted when the spine is habitually subjected to asymmetrical forces.

Exaggeration of the lumbar curve, or **lordosis,** is often associated with weakened abdominal muscles and anterior pelvic tilt (Figure 9-12). Causes of lordosis include congenital spinal deformity, weakness of the abdominal muscles, poor postural habits, and overtraining in sports requiring repeated lumbar hyperextension such as gymnastics, figure skating, javelin throwing, or swimming the butterfly stroke. Symptoms of lordosis vary with the severity of the condition. Because lordosis places added compressive stress on the posterior elements of the spine, low back pain is a common symptom. Lordosis accompanied by anterior pelvic tilt predisposes many adolescent athletes to the development of low back pain (26).

Another abnormality in spinal curvature is **kyphosis** (exaggerated thoracic curvature) (Figure 9-12). Kyphosis is the most frequent spinal disorder in adolescents, with about 25% of all adolescents experiencing some kyphosis-related difficulties (36). Kyphosis often results from Scheuermann's disease, in which one or more wedge-shaped vertebrae develop because of abnormal epiphyseal plate behavior. The condition has been called *swimmer's back* because it is frequently seen in adoles-

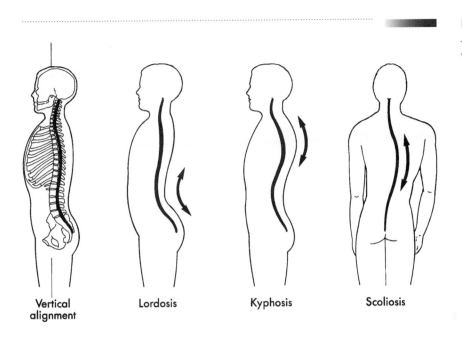

FIGURE 9-12
Abnormal spinal
curvatures.

| Vertical alignment | Lordosis | Kyphosis | Scoliosis |

cents who have trained heavily with the butterfly stroke (38). Kyphosis also often develops in elderly women with osteoporosis, as discussed in Chapter 4 (73). Occasionally the growth plate irregularities of Scheuermann's disease are found in the lumbar vertebrae as well. The condition is typically treated through bracing.

Lateral deviation or deviations in spinal curvature are referred to as **scoliosis** (Figure 9-12). The lateral deformity is coupled with rotational deformity of the involved vertebrae, with the condition ranging from mild to severe. Scoliosis may appear as either a "C" or an "S" curve involving the thoracic spine, the lumbar spine, or both.

A distinction is made between structural and nonstructural scoliosis. Structural scoliosis involves inflexible curvature that persists even with lateral bending of the spine. Nonstructural scoliotic curves are flexible and are corrected with lateral bending.

Scoliosis results from a variety of causes. Congenital abnormalities and selected cancers can contribute to the development of structural scoliosis. Nonstructural scoliosis may occur secondary to a leg length discrepancy or local inflammation. Small lateral deviations in spinal curvature are relatively common and may result from a habit such as carrying books or a heavy purse on one side of the body every day. Between approximately 70% and 90% of all scoliosis, however, is termed idiopathic, which means that the cause is unknown (31, 25). Idiopathic scoliosis is most commonly diagnosed between the ages of 10 and 13 years, but can be seen at any age. It is more common in females.

scoliosis
lateral spinal curvature

Symptoms associated with scoliosis vary with the severity of the condition. Mild cases may be nonsymptomatic and may self-correct with time. Although mild to moderate cases can be treated with appropriate strengthening exercises, severe scoliosis, characterized by extreme lateral deviation and localized rotation of the spine, can be painful and deforming and is treated with bracing and/or surgery.

MOVEMENTS OF THE SPINE

The movement capabilities of the spine as a unit are those of a ball and socket joint, with movement in all three planes, as well as circumduction, allowed.

As a unit the spine allows motion in all three planes of movement, as well as circumduction. Because the motion allowed between any two adjacent vertebrae is small, however, spinal movements always involve a number of motion segments. The range of motion allowed at each motion segment is governed by anatomical constraints that vary through the cervical, thoracic, and lumbar regions of the spine.

Flexion, Extension, and Hyperextension

The range of motion for flexion/extension of the motion segments is considerable in the cervical and lumbar regions, with representative values as high as 17° at the C5-C6 vertebral joint and 20° at L5-S1. In the thoracic spine, however, due to the orientation of the facets, the range of motion spans only approximately 4° at T1-T2 to approximately 10° at T11-T12 (75).

It is important not to confuse spinal flexion with hip flexion or anterior pelvic tilt, although all three motions occur during an activity such as touching the toes (Figure 9-13). Hip flexion consists of anteriorly directed sagittal plane rotation of the femur with respect to the pelvic girdle (or vice versa), and anterior pelvic tilt is anteriorly directed movement of the anterior superior iliac spine with respect to the pubic symphysis, as discussed in Chapter 8. Just as anterior pelvic tilt facilitates hip flexion, it also promotes spinal flexion.

Extension of the spine backward past anatomical position is termed hyperextension. The range of motion for spinal hyperextension is considerable in the cervical and lumbar regions. Lumbar hyperextension is required in the execution of many sport skills, including several swimming strokes, the high jump and pole vault, and numerous gymnastic skills. For example, during the execution of a back handspring the curvature normally present in the lower lumbar region may increase twentyfold (27).

Lateral Flexion and Rotation

Frontal plane movement of the spine away from anatomical position is termed lateral flexion. The largest range of motion for lateral flexion occurs in the cervical region, with approximately 9°–10° of motion

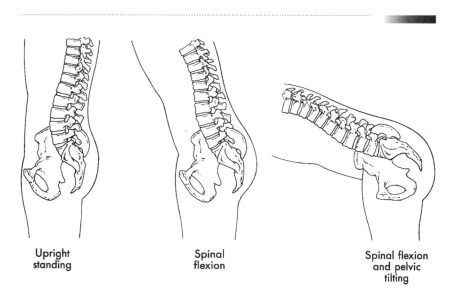

| Upright standing | Spinal flexion | Spinal flexion and pelvic tilting |

FIGURE 9-13
When the trunk is flexed, the first 50° to 60° of motion occurs in the lumbar spine, with additional motion resulting from anterior pelvic tilt.

allowed at C4-C5. Somewhat less lateral flexion is allowed in the thoracic region, where the range of motion between adjacent vertebrae is about 6°, except in the lower segments where lateral flexion capability may be as high as approximately 8°–9°. Lateral flexion in the lumbar spine is also on the order of 6°, except at L5-S1, where it is reduced to only about 3° (75).

Female gymnasts undergo extreme lumbar hyperextension during many commonly performed skills.

Spinal rotation in the transverse plane is again freest in the cervical region of the spine, with up to 12° of motion allowed at C1-C2. It is next freest in the thoracic region, where approximately 9° of rotation is permitted among the upper motion segments. From T7-T8 downward, the range of rotational capability progressively decreases, with only about 2° of motion allowed in the lumbar spine due to the interlocking of the articular processes there. At the lumbosacral joint, however, rotation on the order of 5° is allowed (75). Since the structure of the spine causes lateral flexion and rotation to be coupled, rotation is accompanied by slight lateral flexion to the same side, although this motion is not observable with the naked eye.

MUSCLES OF THE SPINE

The muscles of the neck and trunk are named in pairs, with one on the left and the other on the right side of the body. These muscles can cause lateral flexion and/or rotation of the trunk when they act unilaterally, and trunk flexion or extension when acting bilaterally. The primary functions of the major muscles of the spine are summarized in Table 9-1.

TABLE 9-1

MUSCLES OF THE SPINE

MUSCLE	PROXIMAL ATTACHMENT	DISTAL ATTACHMENT	PRIMARY ACTION(S) ABOUT THE SPINE
Prevertebral muscles (Rectus capitis anterior, rectus, capitis lateralis, longus capitis, longus coli)	Anterior aspect of occipital bone and cervical vertebrae	Anterior surfaces of cervical and first three thoracic vertebrae	Flexion, lateral flexion, rotation to opposite side
Rectus abdominis	Costal cartilage of ribs 5-7	Pubic crest	Flexion, lateral flexion
External oblique	External surface of lower eight ribs	Linea alba and anterior iliac crest	Flexion, lateral flexion, rotation to opposite side
Internal oblique	Linea alba and the lower four ribs	Inguinal ligament, iliac crest, and the lumbodorsal fascia	Flexion, lateral flexion, rotation to same side
The splenii (Splenius capitis and cervicis)	Mastoid process of the temporal bone, transverse processes of the first three cervical vertebrae	Lower half of the ligamentum nuchae, spinous processes of seventh cervical and upper six thoracic vertebrae	Extension, lateral flexion, rotation to same side

(Continued)

MUSCLES OF THE SPINE (CONTINUED)

MUSCLE	PROXIAL ATTACHMENT	DISTAL ATTACHMENT	PRIMARY ACTION(S) ABOUT THE SPINE
The suboccipals (Obliquus capitus superior and inferior, rectus capitis posterior major and minor)	Occipital bone, transverse process of the first cervical vertebra	Posterior surfaces of the first two cervical vertebrae	Extension, lateral flexion, rotation to same side
Erector spinae (Spinalis, longissimus, and iliocostalis)	Lower part of the ligamentum nuchae, posterior cervical, thoracic, and lumbar spine, lower nine ribs, iliac crest, posterior sacrum	Mastoid process of the temporal bone, posterior cervical, thoracic, and lumbar spine, twelve ribs	Extension, lateral flexion rotation to opposite side
Semispinalis (Capitis, cervicis, and thoracis)	Occipital bone, spinous processes of thoracic vertebrae 2-4	Transverse processes of the thoracic and seventh cervical vertebrae	Extension, lateral flexion, rotation to opposite side
The deep spinal muscles (Multifidi, rotatores, interspinales, intertransversarii, levatores costarum)	Posterior processes of all vertebrae, posterior sacrum	Spinous and transverse processes and laminae of vertebrae below those of the proximal attachment	Extension, lateral flexion, rotation to opposite side
Sternocleidomastoid	Mastoid process of the temporal bone	Superior sternum, inner third of the clavicle	Flexion of the neck, extension of the head, lateral flexion, rotation to opposite side
Levator scapulae	Transverse process of the first four cervical vertebrae	Vertebral border of the scapula	Lateral flexion
The scaleni (Scalenus anterior, medius, and posterior)	Transverse processes of the cervical vertebrae	Upper two ribs	Flexion, lateral flexion
Quadratus lumborum	Last rib, transverse processes of the first four lumbar vertebrae	Iliolumbar ligament, adjacent iliac crest	Lateral flexion
Psoas major	Sides of twelfth thoracic and all lumbar vertebrae	Lesser trochanter of the femur	Flexion

The rectus abdominis is a prominent abdominal muscle.

The prominent erector spinae muscle group—the major extensor and hyperextensor of the trunk—is the muscle group of the trunk most often strained.

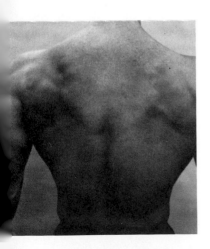

The superficial muscles of the posterior trunk.

Anterior Aspect

The major anterior muscle groups of the cervical region are the prevertebral muscles, including the rectus capitis anterior, rectus capitis lateralis, longus capitis, longus colli, and the eight pairs of hyoid muscles (Figures 9-14 and 9-15). Bilateral tension development by these muscles results in flexion of the head, although the main function of the hyoid muscles appears to be to move the hyoid bone during the act of swallowing. Unilateral tension development in the prevertebrals contributes to lateral flexion of the head toward the contracting muscles or to rotation of the head away from the contracting muscles, depending on which other muscles are functioning as neutralizers.

The main abdominal muscles are the rectus abdominis, the external obliques, and the internal obliques (Figures 9-16, 9-17, and 9-18). Functioning bilaterally, these muscles are the major spinal flexors and also reduce anterior pelvic tilt. Unilateral tension development by the muscles produces lateral flexion of the spine toward the tensed muscles. Tension development in the internal obliques causes rotation of the spine toward the same side. Tension development by the external obliques results in rotation toward the opposite side. If the spine is fixed, the internal obliques produce pelvic rotation toward the opposite side, with the external obliques producing rotation of the pelvis toward the same side. These muscles also form the major part of the abdominal wall, which protects the internal organs of the abdomen.

Posterior Aspect

The splenius capitis and splenius cervicis are the primary cervical extensors (Figure 9-19) (59). Bilateral tension development in the four suboccipitals—the rectus capitis posterior major and minor and the obliquus capitis superior and inferior—assist (Figure 9-20). When these posterior cervical muscles develop tension on one side only, they laterally flex or rotate the head toward the side of the contracting muscles.

The posterior thoracic and lumbar region muscle groups are the massive erector spinae (sacrospinalis), the semispinalis, and the deep spinal muscles. As shown in Figure 9-21, the erector spinae group includes the spinalis, longissimus, and iliocostalis muscles. The semispinalis, with its capitis, cervicis, and thoracis branches, is shown in Figure 9-22. The deep spinal muscles, including the multifidi, rotatores, interspinales, intertransversarii, and levatores costarum, are represented in Figure 9-23. The muscles of the erector spinae group are the major extensors and hyperextensors of the trunk. All posterior trunk muscles contribute to extension and hyperextension when contracting bilaterally and to lateral flexion when contracting unilaterally.

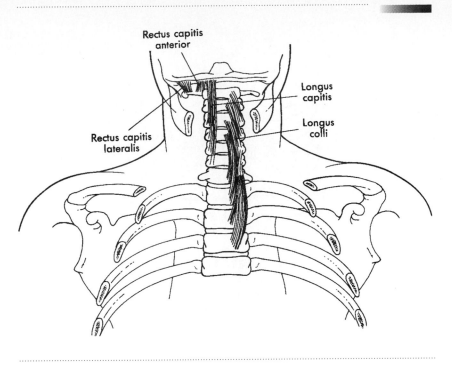

FIGURE 9-14
Anterior muscles of the
cervical region.

Rectus capitis
anterior

Longus
capitis

Longus
colli

Rectus capitis
lateralis

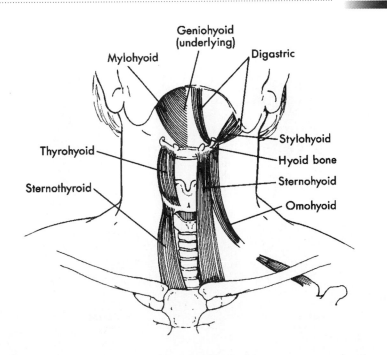

FIGURE 9-15
The hyoid muscles.

Geniohyoid
(underlying)

Mylohyoid

Digastric

Thyrohyoid

Stylohyoid

Hyoid bone

Sternohyoid

Sternothyroid

Omohyoid

FIGURE 9-16
The rectus abdominis.

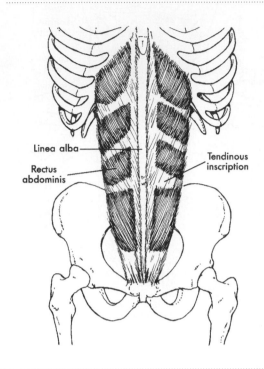

FIGURE 9-17
The external obliques.

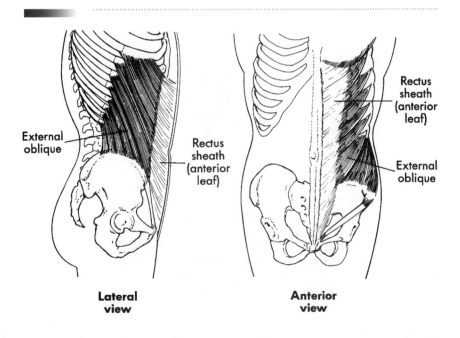

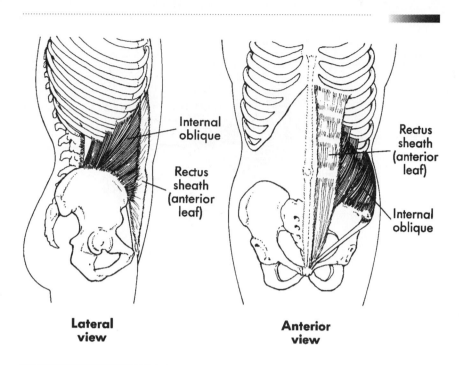

FIGURE 9-18
The internal obliques.

Lateral view

Anterior view

Internal oblique

Rectus sheath (anterior leaf)

Rectus sheath (anterior leaf)

Internal oblique

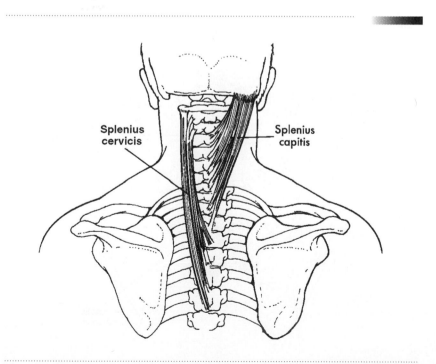

FIGURE 9-19
The major cervical extensors.

Splenius cervicis

Splenius capitis

FIGURE 9-20
The suboccipital muscles.

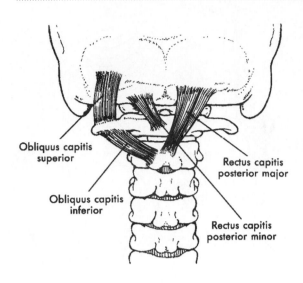

Obliquus capitis
superior

Obliquus capitis
inferior

Rectus capitis
posterior major

Rectus capitis
posterior minor

FIGURE 9-21
The erector spinae group.

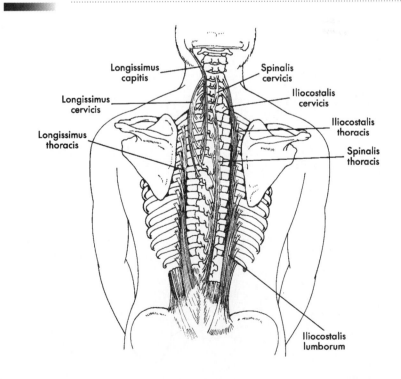

Longissimus
capitis

Longissimus
cervicis

Longissimus
thoracis

Spinalis
cervicis

Iliocostalis
cervicis

Iliocostalis
thoracis

Spinalis
thoracis

Iliocostalis
lumborum

FIGURE 9-22

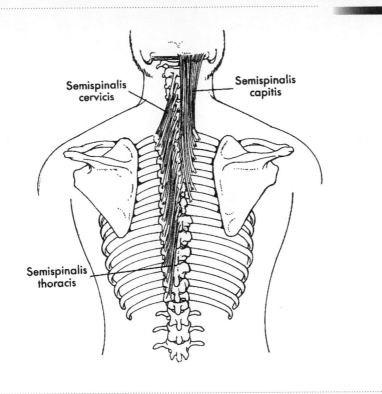

The semispinalis group.

Semispinalis
cervicis

Semispinalis
capitis

Semispinalis
thoracis

FIGURE 9-23

The deep spinal muscles.

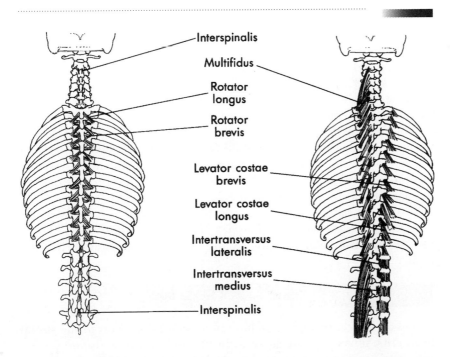

Interspinalis

Multifidus

Rotator
longus

Rotator
brevis

Levator costae
brevis

Levator costae
longus

Intertransversus
lateralis

Intertransversus
medius

Interspinalis

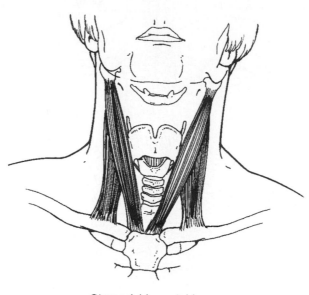

Sternocleidomastoid

Lateral Aspect

Many muscles of the neck and trunk cause lateral flexion when contracting unilaterally but either flexion or extension when contracting bilaterally.

Muscles on the lateral aspect of the neck include the prominent sterno-cleidomastoid, the levator scapulae, and the scalenus anterior, posterior, and medius (Figures 9-24, 9-25, and 9-26). Bilateral tension development in the sternocleidomastoid may result in either flexion of the neck or extension of the head, with unilateral contraction producing lateral flexion to the same side or rotation to the opposite side. The levator scapulae can also contribute to lateral flexion of the neck when contracting unilaterally with the scapula stabilized. The three scalenes assist with flexion and lateral flexion of the neck, depending on whether tension development is bilateral or unilateral.

In the lumbar region, the quadratus lumborum and psoas major are large, laterally oriented muscles (Figures 9-27 and 9-28). These muscles function bilaterally to flex and unilaterally to laterally flex the lumbar spine.

LOADS ON THE SPINE

Forces acting on the spine include body weight, tension in the spinal ligaments, tension in the surrounding muscles, intraabdominal pressure, and any applied external loads. When the body is in an upright

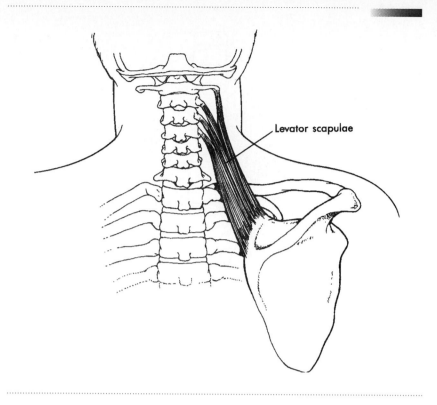

FIGURE 9-25
The levator scapulae.

Levator scapulae

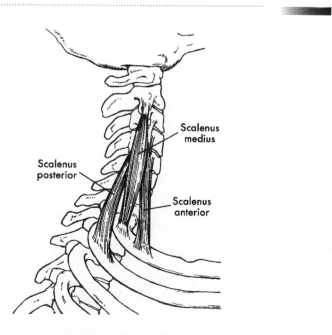

FIGURE 9-26
The scaleni muscles.

Scalenus
medius

Scalenus
posterior

Scalenus
anterior

Figure 9-27
The quadratus lumborum.

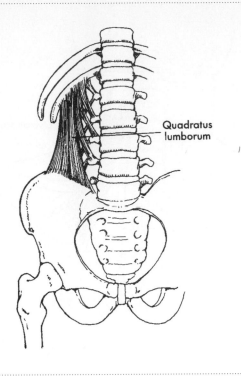

Figure 9-28
The psoas muscle.

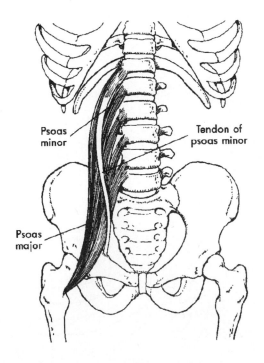

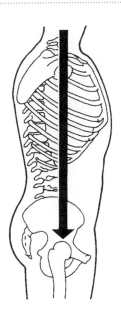

FIGURE 9-29
Because the line of gravity for the head, trunk, and upper extremity passes anterior to the vertebral column, a forward torque acts on the spine.

position, the major form of loading on the spine is axial. In this position body weight, the weight of any load held in the hands, and tension in the surrounding ligaments and muscles all contribute to spinal compression.

During erect standing, the total body center of gravity is anterior to the spinal column, placing the spine under a constant forward bending moment (Figure 9-29). To maintain body position, this torque must be counteracted by tension in the back extensor muscles. As the trunk or the arms are flexed, the increasing moment arms of these body segments contribute to increasing flexor torque and increasing compensatory tension in the back extensor muscles (Figure 9-30). Because the spinal muscles have extremely small moment arms with respect to the vertebral joints, they must generate large forces to counteract the torques produced about the spine by the weights of body segments and external loads (Figure 9-31). Consequently, the major force acting on the spine is usually that derived from muscle activity (65). In comparison to the load present during upright standing, compression on the lumbar spine increases with sitting, increases more with spinal flexion, and increases still further with a slouched sitting position (Figure 9-32).

During erect standing, body weight also loads the spine in shear. This is particularly true in the lumbar spine, where shear creates a tendency for vertebrae to displace anteriorly with respect to adjacent inferior vertebrae (Figure 9-33). Because very few of the fibers of the major spinal

Because the spinal muscles have small moment arms, they must generate large forces to counteract the flexion torques produced by the weight of body segments and external loads.

Body weight produces shear as well as compression on the lumbar spine.

FIGURE 9-30
The back muscles, with a moment arm of approximately 6 cm, must counter the torque produced by the weights of body segments plus any external load. This illustrates why it is advisable to lift and carry heavy objects close to the trunk.

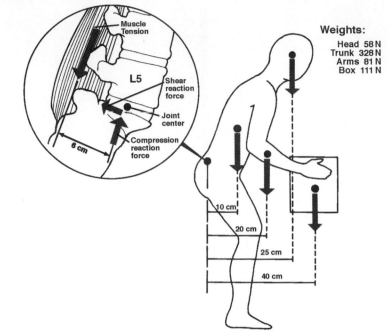

Weights:
Head 58 N
Trunk 328 N
Arms 81 N
Box 111 N

Torque at L5,S1 vertebral joint created by body segments and load:
T = (328 N)(10 cm) + (81 N)(20 cm) + (58 N)(25 cm) + (111 N)(40 cm)
   = 10,790 Ncm

extensors lie parallel to the spine, as tension in these muscles increases, both compression and shear on the vertebral joints and facet joints increase (50). Fortunately, however, the shear component produced by muscle tension in the lumbar region is directed posteriorly, so that it partially counteracts the anteriorly directed shear produced by body weight (61). Shear is a dominant force on the spine during flexion as well as during activities requiring backward lean of the trunk, such as rappelling and trapeezing during sailing (29). Although the relative significance of compression and shear on the spine is poorly understood, excessive shear stress is believed to contribute to disc herniation (32).

Tension in the trunk extensors increases with spinal flexion until the spine approaches full flexion, when it diminishes and finally disappears. At this point the posterior spinal ligaments completely support the flexion torque. The quiescence of the spinal extensors at full flexion is known as the **flexion relaxation phenomenon** (21). Unfortunately, when the spine is in full flexion, tension in the interspinous ligament contributes significantly to anterior shear force and increases facet joint loading (47).

flexion relaxation phenomenon
when the spine is in full flexion, the spinal extensor muscles relax and the flexion torque is supported by the spinal ligaments

FIGURE 9-31

S A M P L E   P R O B L E M   1

How much tension must be developed by the erector spinae with a moment arm of 6 cm from the L5–S1 joint center to maintain the body in the position shown below? (Segment weights are approximated for a 600 N (135 lb) person.)

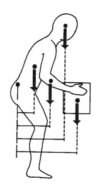

Known

Segment	Wt	Moment arm
head	50 N	22 m
trunk	280 N	12 cm
arms	65 N	25 cm
box	100 N	42 cm
F_m		6 cm

Solution

When the body is in a static position, the sum of the torques acting at any point is zero. At L5–S1:

$$\Sigma T_{L5,S1} = 0$$

$$0 = (F_m) \ (6 \ cm) - [(50 \ N) \ (22 \ cm) + (280 \ N) \ (12 \ cm)$$
$$+ \ (65 \ N) \ (25 \ cm) + (100 \ N) \ (42 \ cm)]$$
$$0 = (F_m) \ (6 \ cm) - 10,285 \ Ncm$$
$$\boxed{F_m = 1714 \ N}$$

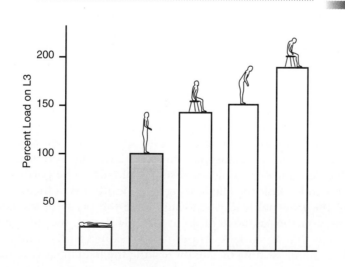

FIGURE 9-32

The load on the third lumbar disc during upright standing (100%) is markedly reduced in a supine position, but increases for each of the other positions shown. (Adapted from Nachemson A: Towards a better understanding of back pain; a review of the mechanics of the lumbar disc, Rheumatol Rehabil, 14:129, 1975.)

FIGURE 9-33
Body weight during upright standing produces shear (F_s) as well as compression (F_c) components on the lumbar spine. (Note that the vector sum of F_s and F_c is wt.)

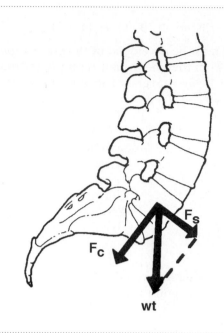

Lateral flexion and rotation create much larger spinal loads than are created by flexion.

When the spine undergoes lateral flexion and axial twisting, a more complex pattern of trunk muscle activation is required than for flexion and extension. Researchers estimate that whereas 50 Nm of extension torque places 800 N of compression on the L4-L5 vertebral joint, 50 Nm of lateral flexion and axial twisting torques respectively generate 1400 N and 2500 N of compression on the joint (48). Information derived from biomechanical modeling of the spine suggests that tension in antagonist trunk muscles produces a significant part of these increased loads (46). Asymmetrical frontal plane loading of the trunk also increases both compressive and shear loads on the spine because of the added lateral bending moment (18, 52).

Another factor affecting spinal loading is body movement speed. It has been shown that executing a lift in a very rapid, jerking fashion dramatically increases compression and shear forces on the spine, as well as tension in the paraspinal muscles (28). This is one of the reasons that resistance training exercises should always be performed in a slow, controlled fashion. Maximizing the smoothness of the motion pattern of the external load acts to minimize the peaks in the compressive force on the lumbosacral joint (33). However, when lifting moderate loads that are awkwardly positioned, skilled workers may be able to reduce spinal loading by initially jerking the load in close to the body

Lifting while twisting should be avoided because it places about three times more stress on the back than lifting in the sagittal plane.

and then transferring momentum from the rotating trunk to the load (see Chapter 14) (49).

The old adage to lift with the legs and not with the back refers to the advisability of minimizing trunk flexion and thereby minimizing the torque generated on the spine by body weight. However, either the physical constraints of the lifting task or the added physiological cost of leg-lifting as compared to back-lifting (24) often make this advice impractical. Recent research suggests that a more important focus of attention for people performing lifts may be maintaining the normal lumbar curve, rather than allowing the lumbar spine to flex, even when the spine is flexed from the hip (Figure 9-34) (48). This enables the active lumbar extensor muscles to partially offset the anterior shear produced by body weight (as discussed), and uniformly loads the lumbar discs rather than placing a tensile load on the posterior annulus of these discs. There is still controversy regarding the correct position for the spine during the performance of a lift (25).

A factor once believed to alleviate compression on the lumbar spine is **intraabdominal pressure.** Researchers hypothesized that intra-abdominal pressure works like a balloon inside the abdominal cavity to support the adjacent lumbar spine by creating a tensile force that partially offsets the compressive load (9) (Figure 9-35). This was corroborated

intraabdominal pressure pressure inside the abdominal cavity; believed to help stiffen the lumbar spine against buckling

FIGURE 9-34

It is advisable to maintain normal lumbar curvature rather than allowing the lumbar spine to flex when lifting, as discussed in the text.

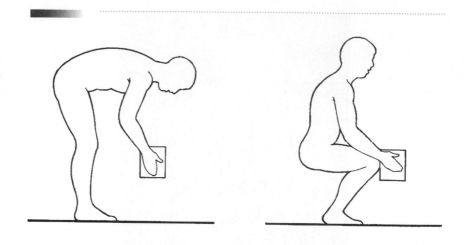

by the observation that intraabdominal pressure increases just prior to the lifting of a heavy load (28). More recently, however, scientists have discovered that pressure in the lumbar discs actually increases when intraabdominal pressure increases (55). It now appears that increased intraabdominal pressure may help to stiffen the trunk to prevent the spine from buckling under compressive loads (48).

Many activities of daily living are stressful to the low back. The constraints of the automobile trunk make it difficult to lift with the spine erect.

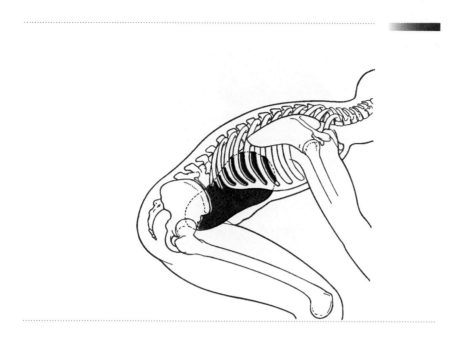

FIGURE 9-35
Intraabdominal pressure, which often increases during lifting, contributes to the stiffness of the lumbar spine to help prevent buckling.

COMMON INJURIES OF THE BACK AND NECK

Low Back Pain

Low back pain is an extremely prevalent problem, with 75%–80% of people experiencing low back pain at some time during life and more than half of the population having had back pain (22). Low back pain is second only to the common cold in causing absences from the workplace, and back injuries are the most frequent and the most expensive of all worker's compensation claims in the United States (14, 56). Most back injuries involve the lumbar or low back region (Figure 9-36).

Although psychological and social components are a factor in some low back pain cases, mechanical stress typically plays a significant causal role in the development of low back pain (23). Perhaps because of their predominance in occupations involving heavy materials handling, men experience low back pain about four times more frequently than women (42). However, some female-dominated groups, such as nurses' aides, register higher rates of low back injury than male workers in general (71).

The incidence of low back pain in children is nearly 30% (19). This incidence increases with age and approaches that found in adults by age 16, with back pain more common in boys than in girls (11). In contrast to the situation for adults, however, low back pain in children is associated with increased physical activity and stronger back flexor muscles

FIGURE 9-36

The majority of back injuries that result in lost work time involve the lumbar region.

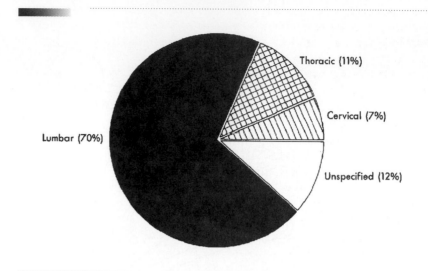

Thoracic (11%)

Cervical (7%)

Lumbar (70%)

Unspecified (12%)

(57). The main causes of low back pain in children are believed to be musculotendinous strains and ligamentous sprains (57).

High incidences of low back pain have been found in workers who sit for prolonged periods of time and in those unable to sit at all during the work day (44). High-risk occupations for the development of low back pain (in order of frequency) include miscellaneous laborers, truck drivers, garbage collectors, warehouse workers, miscellaneous mechanics, nursing aides, materials handlers, lumber workers, practical nurses, and construction laborers (70). Accidental low back pain is often associated with working in an unnatural posture, with sudden and unexpected motions, and with working single-handed (4, 78).

As discussed in Chapter 3, loading patterns that injure biological tissues may involve one or a few repetitions of a large load or numerous repetitions of a small load. Repeated loading, such as occurs in industrial work, in exercise performance, and in jobs such as truck driving that involve vibration, can all produce low back pain.

Although some known pathologies may cause low back pain, the majority of cases are nondiagnosable (53). The inability to specifically identify the anatomical structure or structures that are the source of the pain makes it more difficult to determine the biomechanical factors causing the development of pain. Unlike other major health problems that have declined in incidence in response to medical advances, the number of low back pain–disabled individuals in the United States continues to increase. Lumbar surgery rates increased more than 33% between 1979 and 1990 (15). However, most low back pain is self-limiting, and 90%– 95% of low back pain patients resolve their disability within three months of injury (5).

Despite this fact, clinicians often recommend abdominal exercises as both a prophylactic and a treatment for low back pain. The general rationale for such prescriptions is that an increase in the resting tension levels of the abdominal muscles may help prevent or reduce excessive anterior pelvic tilt and lumbar lordosis. As discussed previously, lumbar lordosis is believed to be a predisposing factor for low back pain. There is potential danger in performing sit-up exercises to prevent low back pain, however, because some versions of the sit-up also exercise the hip flexors. Among the hip flexors is the iliopsoas complex, which tends to increase lumbar lordosis when it is overdeveloped. According to one clinical report, the use of sit-up exercises appears to have actually contributed to low back pain development among a group of 29 exercisers (54). Research indicates that sit-up (or curl-up) exercises should be done with the knees in flexion, with the feet not supported, and with the trunk not elevated past approximately 30° to minimize iliopsoas involvement while maximizing the requirements placed on the abdominal muscles (30). Partial curl-up exercises have also been advocated as providing strong abdominal muscle challenge, with minimal spinal compression (8).

Soft Tissue Injuries

Contusions, muscle strains, and ligament sprains collectively compose the most common sports-related injury of the back (72). These types of injuries typically result from either sustaining a blow or overloading the muscles, particularly those of the lumbar region. Painful spasms and knotlike contractions of the back muscles may also develop as a sympathetic response to spinal injuries and may actually be only symptoms of the underlying problem. Researchers believe that a biochemical mechanism is responsible for these sympathetic muscle spasms (10), which act as a protective mechanism to immobilize the injured area (63).

Acute Fractures

Differences in the causative forces determine the type of vertebral fracture incurred. Transverse or spinous process fractures may result from extremely forceful contraction of the attached muscles or from the sustenance of a hard blow to the back of the spine, which may occur during participation in contact sports such as football, rugby, soccer, basketball, hockey, and lacrosse (63). The most common cause of cervical fractures is indirect trauma involving force applied to the head or trunk rather than to the cervical region itself (12). Fractures of the cervical vertebrae frequently occur from impacts to the head when people dive into shallow water or engage in gymnastics or trampolining activities without appropriate supervision (68, 69).

Large compressive loads (such as those encountered in the sport of weight lifting or in heavy materials handling) can cause fractures of the vertebral end plates. High levels of impact force may result in anterior compression fractures of the vertebral bodies. This type of injury is generally associated with vehicular accidents, although it can also result from hitting the boards during ice hockey, head-on blocks or tackles in football, or impacts during tobogganing, snowmobiling, and hot air ballooning (3). When a snowmobile drops 4 feet, the forces generated far exceed the level known to cause a compression fracture (37).

Because one function of the spine is to protect the spinal cord, acute spinal fractures are extremely serious, with possible outcomes including paralysis and death. Unfortunately, increased participation in leisure activities is associated with an increased prevalence of spinal injuries (67). Whenever a spinal fracture is a possibility, only trained personnel should move the victim.

Fractures of the ribs are generally caused by blows received during accidents or participation in contact sports. Rib fractures are extremely painful because pressure is exerted on the ribs with each inhalation. Damage to the underlying soft tissues is a potentially serious complication with this type of injury.

Stress Fractures

The most common type of vertebral fracture is a stress fracture of the pars interarticularis, the region between the superior and inferior articular facets, which is the weakest portion of the neural arch (Figure 9-37). A fracture of the pars is termed **spondylolysis,** with the severity ranging from hairline fracture to complete separation of bone. Although some pars defects may be congenital, they are also known to be caused by mechanical stress. One mechanism of injury appears to involve repeated axial loading of the lumbar spine when it is hyperextended.

A bilateral separation in the pars interarticularis, called **spondylolisthesis,** results in the anterior displacement of a vertebra with respect to the vertebra below it (Figure 9-37). The most common site of this injury is the lumbosacral joint, with 90% of the slips occurring at this level. Spondylolisthesis is often initially diagnosed in children between the ages of 10 and 15 years and is more commonly seen in boys. Risk factors for spondylolysis and spondylolisthesis in children include genetics, the growth spurt, repetitive stresses, and participating in sports for more than 15 hours per week (19). Whereas 2% of the general population have spondylolysis, 5% of those with spondylolysis also have spondylolisthesis (3).

Unlike most stress fractures, spondylolysis and spondylolisthesis do not typically heal with time, but tend to persist, particularly when there is no interruption in sport participation. Individuals whose sports or po-

spondylolysis
presence of a fracture in the pars interarticularis of the vertebral neural arch

spondylolisthesis
complete bilateral fracture of the pars interarticularis, resulting in the anterior slippage of the vertebra

Stress-related fractures of the pars interarticularis, the weakest section of the neural arch, are unusually common among participants in sports involving repeated hyperextension of the lumbar spine.

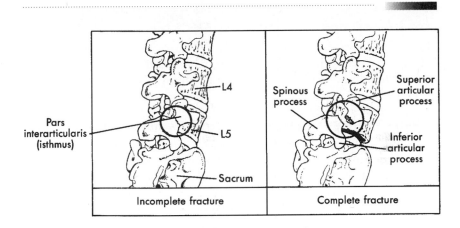

FIGURE 9-37
Stress fractures of the pars interarticularis may occur unilaterally or bilaterally and may or may not result in a complete separation.

sitions require repeated hyperextension of the lumbar spine are prime candidates for stress-related spondylolysis. Those particularly susceptible to this pathology include female gymnasts, interior football linemen, and weight lifters, with increased incidences also found among volleyball players, pole vaulters, wrestlers, and rowers (3, 45). A high incidence of spondylolysis has been found in prehistoric human remains from circa 1200–1500 A.D., a period characterized by the use of stone pillars for the construction of houses (6).

Disc Herniations

The source of approximately 1%–5% of back pain is a herniated disc, which consists of the protrusion of part of the nucleus pulposus from the annulus. Disc herniations may be either traumatic or stress related, and typically involve a disc that shows signs of previous degeneration (1). The most common sites of protrusions are between the fifth and sixth and the sixth and seventh cervical vertebrae, and between the fourth and fifth lumbar vertebrae and the fifth lumbar vertebra and the first sacral vertebra (39, 40). Most occur on the posterior or posterior-lateral aspect of the disc (7).

Although the disc itself is not innervated and therefore is incapable of generating a sensation of pain, sensory nerves do supply the anterior and posterior longitudinal ligaments, the vertebral bodies, and the articular cartilage of the facet joints (13). If the herniation presses on one or more of these structures, on the spinal cord, or on a spinal nerve, pain or numbness may result.

The term slipped disc *is often used to refer to a herniated or prolapsed disc. It is a misnomer, because the discs as intact units do not slip around.*

SUMMARY

The spine is composed of 33 vertebrae that are divided structurally into five regions: cervical, thoracic, lumbar, sacral, and coccygeal. Although most vertebrae adhere to a characteristic shape, there is a progression in vertebral size and in the orientation of the articular facets throughout the spinal column.

Within the cervical, thoracic, and lumbar regions, each pair of adjacent vertebrae with the intervening soft tissues is called a motion segment. Three joints interconnect the vertebrae of each motion segment. The intervertebral discs function as shock absorbers at the weightbearing vertebral joints. The right and left pairs of superior and inferior facet joints significantly influence the movement capabilities of the motion segments at different levels of the spine.

The muscles of the neck and trunk are named in pairs, with one on the left and the other on the right side of the body. These muscles can cause lateral flexion and/or rotation of the trunk when they act unilaterally, and trunk flexion or extension when acting bilaterally.

Forces acting on the spine include body weight, tension in the spinal ligaments, tension in the surrounding muscles, intraabdominal pressure, and any applied external loads. Because the spinal muscles have extremely small moment arms with respect to the vertebral joints, they must generate large forces to counteract the torques produced about the spine by the weights of body segments and external loads.

Because the spine serves as the protector of the spinal cord, spinal injuries are serious. Low back pain is a major modern-day health problem and a leading cause of lost working days.

INTRODUCTORY PROBLEMS

1. What regions of the spine contribute the most to flexion? Hyperextension? Lateral flexion? Rotation?
2. Construct a chart listing the muscles of the cervical region of the trunk according to whether they are anterior, posterior, medial, or lateral to the joint center. Note that some muscles may fall into more than one category. Identify the action or actions performed by the muscles in each category.
3. Construct a chart listing the muscles of the thoracic region of the trunk according to whether they are anterior, posterior, medial, or lateral to the joint center. Note that some muscles may fall into more than one category. Identify the action or actions performed by the muscles in each category.
4. Construct a chart listing the muscles of the lumbar region of the trunk according to whether they are anterior, posterior, medial, or

lateral to the joint center. Note that some muscles may fall into more than one category. Identify the action or actions performed by the muscles in each category.

5. How would trunk movement capability be affected if the lumbar region were immovable?
6. What are the postural consequences of having extremely weak abdominal muscles?
7. Weight training is used in conjunction with conditioning for numerous sports. What would you advise regarding spinal posture during weight training?
8. What exercises strengthen the muscles on the anterior, lateral, and posterior aspects of the trunk?
9. Why should twisting be avoided when performing a lift?
10. Solve Sample Problem 1 (Figure 9-31) using the following data:

SEGMENT	WT	MOMENT ARM
Head	50 N	22 cm
Trunk	280 N	12 cm
Arms	65 N	25 cm
Box	100 N	42 cm

ADDITIONAL PROBLEMS

1. Explain how pelvic movements facilitate spinal movements.
2. What exercises should be prescribed for individuals with scoliosis? Lordosis? Kyphosis?
3. Compare and contrast the major muscles that serve as agonists during performances of straight-leg and bent-knee sit-ups. Should sit-ups be prescribed as an exercise for a low back pain patient? Explain why or why not.
4. Why do individuals who work at a desk all day often develop low back pain?
5. Explain why maintaining a normal lumbar curve is advantageous during lifting.
6. Formulate a theory explaining why osteoporosis is often associated with increased thoracic kyphosis.
7. Formulate a theory explaining why a loss in spinal flexibility of approximately 50% is a result of the aging process.
8. What are the consequences of the loss in intervertebral disc hydration that accompanies the aging process?
9. What spinal exercises are appropriate for senior citizens? Provide a rationale for your choices.
10. Make up a problem similar to (but different from) Sample Problem 1 (Figure 9-31). Show a free body diagram and a solution for your problem.

LABORATORY EXPERIENCES

1. Using a skeleton or an anatomical model of the spine, carefully study the differences in vertebral size and shape among the cervical, thoracic, and lumbar regions. Construct a chart that characterizes the differences among regions.

2. Observe a male subject without a shirt at the extreme range of motion in flexion, hyperextension, lateral flexion, and rotation. Which anatomical region(s) of the spine contribute most to each of these motions?

3. From a side view, videotape a volunteer lifting objects of light, medium, and heavy weights. What differences do you observe in lifting kinematics? Write a short explanation of your findings.

4. Perform sit-up exercises, elevating the trunk to approximately 30°, under each of the following conditions: a) legs fully extended, b) legs flexed comfortably at the hip and knee, and c) legs flexed comfortably at the hip and knee, with a partner supporting the feet. Which exercise is easiest and which is hardest? Write a comparative description of these exercises from the standpoint of muscle involvement.

5. Observe the sitting posture of several individuals in a classroom or laboratory setting. Describe the similarities and differences you see and relate these to muscle activity and to the support characteristics of the chairs or desks in which your subjects sit.

REFERENCES

1. Adams MA and Hutton WC: Gradual disc prolapse, Spine 10:524, 1985.
2. Adams MA and Hutton WC: The effect of posture on the fluid content of lumbar intervertebral discs, Spine 8:665, 1983.
3. Alexander MJL: Biomechanical aspects of lumbar spine injuries in athletes: A review, Can J Applied Sport Sci 10:1, 1985.
4. Allread WG, Marras WS, and Parnianpour M: Trunk kinematics of one-handed lifting, and the effects of asymmetry and load weight, Ergonomics 39:322, 1996.
5. Anderson GBJ: Epidemiological aspects of low back pain in industry, Spine, 6:53, 1981.
6. Arriaza BT: Spondylolysis in prehistoric human remains from Guam and its possible etiology, Am J Phys Anthropol 104:393, 1997.
7. Ashton-Miller JA and Schultz AB: Biomechanics of the human spine and trunk, Exerc Sport Sci Rev 16:169, 1988.
8. Axler CT and McGill SM: Low back loads over a variety of abdominal exercises: Searching for the safest abdominal challenge, Med Sci Sports Exerc 29:804, 1997.
9. Bartelink DL: The role of abdominal pressure in relieving the pressure on the lumbar intervertebral discs, J Bone Joint Surg 39B:718, 1957.
10. Brennan GP et al: Physical characteristics of patients with herniated intervertebral lumbar discs, Spine 12:699, 1987.

11. Burton AK: Low back pain in children and adolescents: To treat or not? Bull Hosp Jt Dis 55:127, 1996.

12. Byun HS, Cantos EL, and Patel PP: Severe cervical injury due to break dancing: A case report, Orthopedics 9:550, 1986.

13. Cailliet R: *Low back pain syndrome,* 3rd ed, Philadelphia, 1981, FA Davis Co.

14. Cherkin DC, et al: An international comparison of back surgery rates, Spine 19:1201, 1994.

15. Davis H: Increasing rates of cervical and lumbar surgery in the United States, Spine 19:1117, 1994.

16. Dolan P, Benjamin E, and Adams M: Diurnal changes in bending and compressive stresses acting on the lumbar spine, J Bone Jt Surg (Suppl) 75B:22, 1993.

17. Dreyer SJ and Dreyfuss PH: Low back pain and the zygapophysial (facet) joints, Arch Phys Med Rehabil 77:290, 1996.

18. Drury CG, et al: Symmetric and asymmetric manual materials handling, part 2: Biomechanics, Ergonomics 32:565, 1989.

19. Duggleby T and Kumar S: Epidemiology of juvenile low back pain: A review, Disabil Rehabil 19:505, 1997.

20. Farfan H: *Mechanical disorders of the low back,* Philadelphia, 1973, Lea & Febiger.

21. Floyd WF and Silver PHS: The function of the erectores spinae muscles in certain movements and postures in man, J Physiol 129:184, 1955.

22. Frymoyer JW: Back pain and sciatica, N Eng J Med 318:291, 1988.

23. Frymoyer JW and Pope M: The role of trauma in low back pain: A review, J Trauma 18:628, 1978.

24. Garg A and Herrin G: Stoop or squat: A biomechanical and metabolic evaluation, Am Inst Ind Eng Trans 11:293, 1979.

25. Gassett RS, Hearne B, and Keelan B: Ergonomics and body mechanics in the work place, Orthop Clin North Am 27:861, 1996.

26. Goldberg B and Boiardo R: Profiling children for sports participation, Clin Sports Med 3:153, 1984.

27. Hall SJ: Mechanical contribution to lumbar stress injuries in female gymnasts, Med Sci Sports Exerc 18:599, 1986.

28. Hall SJ: Effect of attempted lifting speed on forces and torque exerted on the lumbar spine, Med Sci Sports Exerc 17:440, 1985.

29. Hall SJ, Kent JA, and Dickinson VR: Comparative assessment of novel sailing trapeze harness designs, Int J Sport Biomech 5:289, 1989.

30. Hall SJ, Lee J, and Wood TM: Evaluation of selected sit-up variations for the individual with low back pain, J Appl Sport Sci Res 4:42, 1990.

31. Heliovaara M: Body height, obesity, and risk of herniated lumbar intervertebral disc, Spine 12:469, 1987.

32. Hirsch C and Nachemson A: A new observation on the mechanical behavior of lumbar discs, Acta Orthop Scand 23:254, 1954.

33. Hsiang SM and McGorry RW: Three different lifting strategies for controlling the motion patterns of the external load, Ergonomics 40:928, 1997.

34. Hutton WC, Stott JRR, and Cyron BM: Is spondylolysis a fatigue fracture? Spine 2:202, 1977.

35. Jackson DW, Wiltse LL, and Cirincione RJ: Spondylolysis in the female gymnast, Clin Orthop 117:68, 1976.

36. Junghanns H: *Clinical implications of normal biomechanical stresses on spinal function*, Rockville, MD, 1990, Aspen Publishers, Inc.

37. Keene JS: Thoracolumbar fractures in winter sports, Clin Orthop 216:39, 1987.

38. Keene JS and Drummond DS: Mechanical back pain in the athlete, Compr Ther 11:7, 1985.

39. Kelsey J et al: Acute prolapsed lumbar intervertebral disc: An epidemiological study with special reference to driving automobiles and cigarette smoking, Spine 9:608, 1984.

40. Kelsey J et al: An epidemiological study of acute prolapsed cervical intervertebral disc, J Bone Joint Surg 66A:907, 1984.

41. Kraemer J, Kolditz D, and Gowin R: Water and electrolyte content of human intervertebral discs under variable load, Spine 10:69, 1985.

42. Kuwashima A, Aizawa Y, Nakamura K, Taniguchi S, and Watanabe M: National survey on accidental low back pain in workplace, Ind Health 35:187, 1997.

43. Lindh M: Biomechanics of the lumbar spine. In Nordin M and Frankel VH: *Basic biomechanics of the musculoskeletal system*, Philadelphia, 1989, Lea & Febiger.

44. Magora A: Investigation of the relationship between low back pain and occupation, Indust Med 41:5, 1972.

45. McCarroll JR, Miller JM, and Ritter MA: Lumbar spondylolysis and spondylolisthesis in college football players, Am J Sports Med 14:404, 1986.

46. McGill SM: A myoelectrically based dynamic three-dimensional model to predict loads on lumbar spine tissues during lateral bending, J Biomech 25:395, 1992.

47. McGill SM: Estimation of force and extensor moment contributions of the disc and ligaments at L4/L5, Spine 13:1395, 1988.

48. McGill SM and Norman RW: Low back biomechanics in industry: The prevention of injury through safer lifting. In Grabiner MD: *Current issues in biomechanics*, Champaign, 1993, Human Kinetics.

49. McGill SM and Norman RW: Dynamically and statically determined low back moments during lifting, J Biomech 18:877, 1985.

50. Macintosh JE and Bogduk N: The morphology of the lumbar erector spinae, Spine 12:658, 1987.

51. Macnab I and McCulloch J: *Backache*, 2nd ed, Baltimore, 1990, Williams & Wilkins.

52. Mital A and Kromodihardjo S: Kinetic analysis of manual lifting activities, part II: Biomechanical analysis of task variables, Int J Ind Ergonomics 1:91, 1986.

53. Mooney V: Where is the pain coming from? Spine 12:754, 1987.

54. Mutoh Y et al: The relation between sit-up exercises and the occurrence of low back pain. In Matsui H and Kobayashi K, eds: *Biomechanics VIII-A*, Champaign, 1983, Human Kinetics Publishing Co.

55. Nachemson A, Andersson GBJ, and Schultz AB: Valsalva manoeuvre biomechanics: Effects on lumbar trunk loads of elevated intra-abdominal pressure, Spine 11:476, 1986.

56. Neal C: The assessment of knowledge and application of proper body mechanics in the workplace, Orthop Nurs 16:66, 1997.

57. Newcomer K and Sinaki M: Low back pain and its relationship to back strength and physical activity in children, Acta Paediatr 85:1433, 1996.

58. Nixon J: Intervertebral disc mechanics: A review, J World Soc Med 79:100, 1986.

59. Nolan JP and Sherk HH: Biomechanical evaluation of the extensor musculature of the cervical spine, Spine 13:9, 1988.

60. Pope MH, Frymoyer JW, and Lehman TR: Structure and function of the lumbar spine. In Pope MH, et al, eds: *Occupational low back pain: Assessment, treatment and prevention,* St. Louis, 1991, Mosby Year Book.

61. Potvin JR, McGill SM, and Norman RW: Trunk muscle and lumbar ligament contributions to dynamic lifts with varying degrees of trunk flexion, Spine 16:1099, 1991.

62. Reilly T, Tynell A, and Troup JDG: Circadian variation in human stature, Chronobiology Int 1:121, 1984.

63. Rovere GD: Low back pain in athletes, Physician Sportsmed 15:105, 1987.

64. Savage RA, Whitehouse GH, and Roberts N: The relationship between magnetic resonance imaging appearance of the lumbar spine and low back pain, age and occupation in males, Eur Spine 6:106, 1997.

65. Schultz AB: Biomechanical analyses of loads on the lumbar spine. In Weinstein JN and Wiesel SW, eds: *The lumbar spine,* Philadelphia, 1990, WB Saunders Company.

66. Shirazi AA and Parnianpour M: Role of posture in mechanics of the lumbar spine in compression, J Spinal Disord 9:277, 1996.

67. Silver JR: Spinal injuries as a result of sporting accidents, Paraplegia 25:16, 1987.

68. Silver JR, Silver DD, and Godfrey JJ: Injuries of the spine sustained during gymnastic activities, Br Med J 293:861, 1986.

69. Silver JR, Silver DD, and Godfrey JJ: Trampolining injuries of the spine, Injury 17:117, 1986.

70. Snook S, Fine L, and Silverstein B: Musculoskeletal disorders. In Levy B and Wegman D, eds: *Occupational health: Recognizing and preventing work-related disease,* Boston, 1988, Little, Brown & Co, Inc.

71. Spengler D et al: Back injury in industry: A retrospective study, Spine 11:241, 1986.

72. Tall RL and DeVault W: Spinal injury in sport: Epidemiologic considerations, Clin Sports Med 12:441, 1993.

73. Thevenon A et al: Relationship between kyphosis, scoliosis, and osteoporosis in the elderly population, Spine 12:744, 1987.

74. Voutsinas SA and MacEwan GD: Sagittal profiles of the spine, Clin Orthop 210:235, 1986.

75. White AA and Panjabi MM: *Clinical biomechanics of the spine,* Philadelphia, 1978, JB Lippincott Co.

76. Wiesel SW, Bernini P, and Rothman RH: *The aging lumbar spine,* Philadelphia, 1982, WB Saunders Co.

77. Wilder DG, Krag MH, and Pope MH: *Atlas of mammalian lumbar vertebrae—pictorial and dimensional information,* Springfield, 1991, Charles C Thomas.

78. Yamamoto S: A new trend in the study of low back pain in workplaces, Ind Health 35:173, 1997.

ANNOTATED READINGS

McGill SM: The biomechanics of low back injury: Implications on current practice in industry and the clinic, J Biomech 30:465, 1997.
Reviews the spinal injury process, human variability, stoop versus squat lifting, motor control considerations, diurnal changes in the spine, the effects of sitting, and spinal memory.

McGill SM and Norman RW: Low back biomechanics in industry: The prevention of injury through safer lifting. In Grabiner MD: *Current issues in biomechanics*, Champaign, 1993, Human Kinetics.
Includes sections on biomechanical models for estimating low back loads, issues in low back mechanics specific to lifting, and recommendations for safer lifting.

Mooney V: Where is the pain coming from? Spine 12:754, 1987.
Discusses current lack of medical understanding of the causes and treatments of low back pain.

Pope MH, et al, eds: *Occupational low back pain: Assessment, treatment and prevention*, St. Louis, 1991, Mosby Year Book.
Includes chapters on structure and function of the lumbar spine, occupational biomechanics and the lumbar spine, and other related topics.

RELATED WEB SITES

Calvert Orthopaedic and Sports Medicine Center
http://www.calvertorthoandsports.com/index.html
Provides basic information and color graphics of the anatomy and common injuries of the spine.

CyberSpine—Anatomy
http://www.cyberspine.com/anatomy/anatomy.html
Describes spinal anatomy in lay terminology.

Exercises and Stretches
http://www.aomc.org/occmed.dir/exercise.html
Describes exercises for the neck, middle and lower back, for maintaining spinal health, for preventing back pain, for the desk-bound, for pregnancy, and for health and fitness.

Link Orthopaedics
http://www.dundee.ac.uk/orthopaedics/link/teach.htm
Provides links to pages on motor, sensory, and reflex testing of spinal levels.

Low Back Pain
http://www2.kumc.edu/people/gamundson/lowback.htm
Discusses incidence, impact, and causes of low back pain, as well as the use of diagnostic tools such as MRI, CT scans, bone scans, SPECT scans, and discography, and interbody cage fusion as a surgical procedure.

Low Back Pain—When is it serious?
http://www.primarycarecenter.com/backpain.htm
Discusses common causes and treatments for low back pain.

M & M Orthopaedics
http://mmortho.com/
Includes information on spinal anatomy, laminotomy, microdiscectomy, and scoliosis.

MedFacts Play Doctor! MedLib
http://www.medfacts.com/medlib.htm
Provides links to pages on arthritis of the spine, disc herniation, lumbar spine strain, and nerve root irritation.

Northern Rockies Orthopaedics Specialists
http://www.orthopaedic.com
> *Provides links to pages describing descriptions of injuries, diagnostic tests, and surgical procedures for the shoulder, elbow, wrist, and hand.*

Orthopedic Medical Information
http://www.opendoor.com/albert/one/orthopedic.html
> *Includes links to pages on low back pain and sprains and strains.*

Rothman Institute
http://rothmaninstitute.com/index.html
> *Includes information on spinal anatomy and problems such as scoliosis, spinal stenosis, and degenerative disc disease.*

Southern California Orthopaedic Institute
http://www.scoi.com
> *Includes links to anatomical descriptions and labeled photographs of the spine, as well as laminotomy and microdiscectomy and scoliosis.*

Trying to Explain the Unexplainable: A Conceptual Model for Low Back Disability
http://mir.med.ucalgary.ca/oemweb/gibson.htm
> *Provides a well-documented discussion of factors related to low back disability.*

University of Washington Orthopaedic Physicians
http://www.orthop.washington.edu
> *Provides extensive descriptions of scientific and clinical aspects of the spine. Clicking on the radiographs brings up structure labels. Also includes numerous movies showing therapeutic exercises, examination procedures, and surgical techniques.*

The "Virtual" Medical Center: Anatomy & Histology Center
http://www.sci.lib.uci.edu/HSG/MedicalAnatomy.html
> *Contains numerous images, movies, and course links for human anatomy.*

WebCAI Programs
http://www.kumc.edu/research/medicine/pharmacology/CAI/menu2.htm
> *Provides interactive quizzes on all regions of the spine.*

Wheeless' Textbook of Orthopaedics
http://www.medmedia.com/med.htm
> *Provides comprehensive, detailed information, graphics, and related literature for all joints.*

LINEAR KINEMATICS
OF HUMAN MOVEMENT

After completing this chapter, the reader will be able to:

Discuss the interrelationships among kinematic variables.

Correctly associate linear kinematic quantities with their units of measure.

Identify and describe the effects of factors governing projectile trajectory.

Explain why the horizontal and vertical components of projectile motion are analyzed separately.

Distinguish between average and instantaneous quantities and identify the circumstances under which each is a quantity of interest.

Select and use appropriate equations to solve problems related to linear kinematics. Why is a sprinter's acceleration close to zero in the middle of a race? How does the size of a dancer's foot affect the performance time that a choreographer must allocate for jumps? At what angle should a discus or a javelin be thrown to achieve maximum distance? Why does a ball, thrown horizontally, hit the ground at the same time as a ball dropped from the same height? These questions all relate to the kinematic characteristics of a pure form of movement: linear motion. This chapter introduces the study of human movement mechanics with a discussion of linear kinematic quantities and projectile motion.

LINEAR KINEMATIC QUANTITIES

kinematics
the form, pattern, or sequencing of movement with respect to time

Kinematics is the study of the geometry, pattern, or form of motion with respect to time. Kinematics, which describes the appearance of motion, is distinguished from kinetics, the study of the forces associated with motion. Linear kinematics involves the study of the shape, form, pattern, and sequencing of linear movement through time, without particular reference to the force or forces that cause or result from the motion.

Careful kinematic analyses of performance are invaluable for clinicians, physical activity teachers, and coaches. When people learn a new motor skill, a progressive modification of movement kinematics reflects the learning process. Likewise, when a patient rehabilitates an injured joint, the therapist or clinician looks for the gradual return of normal joint kinematics.

The kinematics of a golf swing.

Kinematics spans both qualitative and quantitative forms of analysis. For example, qualitatively describing the kinematics of a soccer kick entails identifying the major joint actions, including hip flexion, knee extension, and possibly plantar flexion at the ankle. A more detailed qualitative kinematic analysis might also describe the precise sequencing and timing of body segment movements, which translates to the degree of skill evident on the part of the kicker. Although most assessments of human movement are carried out qualitatively through visual observation, quantitative analysis is also sometimes appropriate. Physical therapists, for example, often measure the range of motion of an injured joint to help determine the extent to which range of motion exercises may be needed. When a coach measures an athlete's performance in the shot put or long jump, this too is a quantitative assessment.

Sport biomechanists often quantitatively study the kinematic features that characterize an elite performance or the biomechanical factors that may limit the performance of a particular athlete. Sometimes this type of analysis results in constructing a model that details the kinematic characteristics of sound performance for practical use by coaches and athletes.

Most biomechanical studies of human kinematics, however, are performed on nonelite subjects. Kinematic research has shown that infants begin to use stable patterns of coordination in reaching for objects at 12 to 15 months of age, with adult-like reaching movements occurring by about 2 years (13). Scientists have also studied the kinematic characteristics of progressive gait development and throwing ability in young children (12, 16). In collaboration with adapted physical education specialists, biomechanists have documented the characteristic kinematic patterns associated with several relatively common disabling conditions such as cerebral palsy, Down's syndrome, and stroke. Quantitative kinematic screening tests are used to evaluate treatment and progression of a wide variety of motor disorders (21). Kinematic analysis indicates that walking on a treadmill with a grade just greater than 12% may be optimal for minimizing patellofemoral discomfort and potential strain on the anterior cruciate ligament (ACL) in post–ACL reconstruction patients (14).

Biomechanists commonly use high-speed cinematography or videography to perform quantitative kinematic analyses. The process involves taking a carefully planned film or video of a performance, with subsequent computerized or computer-assisted analysis of the performance on a picture-by-picture basis, as described in Chapter 2.

Distance and Displacement

meter
the most common international unit of length, on which the metric system is based

Units of distance and displacement are units of length. In the metric system, the most commonly used unit of distance and displacement is the **meter** (m). A kilometer (km) is 1000 m, a centimeter (cm) is

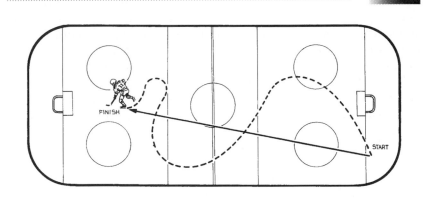

FIGURE 10-1
The distance that a skater travels may be measured from the track on the ice. The skater's displacement is measured in a straight line from initial position to final position.

$\frac{1}{100}$ m, and a millimeter (mm) is $\frac{1}{1000}$ m. In the English system, common units of length are the inch, the foot (0.30 m), the yard (0.91 m), and the mile (1.61 km).

Distance and displacement are assessed differently. Distance is measured along the path of motion. When a runner completes $1\frac{1}{2}$ laps around a 400 m track, the distance that the runner has covered is equal to 600 (400 + 200) m. **Linear displacement** is measured in a straight line from position 1 to position 2, or from initial position to final position. At the end of $1\frac{1}{2}$ laps around the track, the runner's displacement is the length of the straight imaginary line that transverses the field, connecting the runner's initial position to the runner's final position halfway around the track (see Introductory Problem 1). At the completion of 2 laps around the track, the distance run is 800 m. Because initial and final positions are the same, however, the runner's displacement is 0. When a skater moves around an ice rink, the distance the skater travels may be measured along the tracks left by the skates. The skater's displacement is measured along a straight line from initial to final positions on the ice (Figure 10-1).

Another difference is that distance is a scalar quantity and displacement is a vector quantity. Consequently, the displacement includes more than just the length of the line between two positions. Of equal importance is the *direction* in which the displacement occurs. The direction of a displacement relates the final position to the initial position. For example, the displacement of a yacht that has sailed 900 m on a tack due south would be identified as 900 m to the south.

The direction of a displacement may be indicated in several different, equally acceptable ways. Compass directions, such as south or northwest, the terms *left* and *right, up* and *down,* or *positive* and *negative* are all appropriate labels. The positive direction is typically defined as upwards or to the right, with negative regarded as downwards or to the

The metric system is the predominant standard of measurement in every major country in the world except for the United States.

linear displacement
change in location, or the directed distance from initial to final location

left. This enables indication of direction using plus and minus signs. Most important is being consistent in using the system or convention adopted for indicating direction in a given context. It would be confusing to describe a displacement as 500 m north followed by 300 m to the right.

Either distance or displacement may be the more important quantity of interest depending on the situation. Many 5 km and 10 km racecourses are set up so that the finish line is only a block or two from the starting line. Participants in these races are usually interested in the number of kilometers of distance covered or the number of kilometers left to cover as they progress along the racecourse. Knowledge of displacement is not particularly valuable during this type of event. In other situations, however, displacement is more important. For example, triathlon competitions may involve a swim across a lake. Because swimming in a perfectly straight line across a lake is virtually impossible, the actual distance a swimmer covers is always somewhat greater than the width of the lake (Figure 10-2). However, the course is set up so that the identified length of the swim course is the length of the displacement between the entry and exit points on the lake.

Displacement magnitude and distance covered can be identical. When a cross-country skier travels down a straight path through the woods, both distance covered and displacement are equal. However, any time the path of motion is not rectilinear, the distance traveled and the size of the displacement will differ.

Distance covered and displacement may be equal for a given movement. Or, distance may be greater than displacement, but the reverse is never true.

Speed and Velocity

Two quantities that parallel distance and linear displacement are speed and **linear velocity.** These terms are often used synonymously in general conversation, but in mechanics they have precise and different meanings. Speed, a scalar quantity, is defined as the distance covered divided by the time taken to cover it:

linear velocity
the rate of change in location

Displacement and velocity are vector equivalents of the scalar quantities distance and speed.

$$\text{Speed} = \frac{\text{length (or distance)}}{\text{change in time}}$$

Velocity (v) is the change in position or the displacement that occurs during a given period of time:

$$v = \frac{\text{change in position}}{\text{change in time}}$$

$$v = \frac{\text{displacement}}{\text{change in time}}$$

Because the capital Greek letter delta (Δ) is commonly used in mathematical expressions to mean *change in,* a shorthand version of the

FIGURE 10-2

S A M P L E   P R O B L E M   1

A swimmer crosses a lake that is 0.9 km wide in 30 minutes. What was his average velocity? Can his average speed be calculated?

Known

After reading the problem carefully, the next step is to sketch the problem situation, showing all quantities that are known or may be deduced from the problem statement:

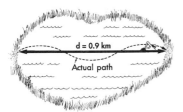

$$t = 30 \text{ min } (0.5 \text{ hr})$$

Solution

In this situation, we know that the swimmer's displacement is 0.9 km. However, we know nothing about the exact path that the swimmer may have followed. The next step is to identify the appropriate formula to use to find the unknown quantity, which is velocity:

$$v = \frac{d}{t}$$

The known quantities can now be filled in to solve for velocity:

$$v = \frac{0.9 \text{ km}}{0.5 \text{ hr}}$$
$$v = 1.8 \text{ km/hr}$$

Speed is calculated as distance divided by time. Although we know the time taken to cross the lake, we do not know, nor can we surmise from the information given, the exact distance covered by the swimmer.

Therefore, the swimmer's speed cannot be calculated.

relationship expressed follows, with t representing the amount of time elapsed during the velocity assessment:

$$v = \frac{\Delta \text{ position}}{\Delta \text{ time}}$$
$$v = \frac{d}{\Delta t}$$

FIGURE 10-3

The velocity of a swimmer in a river is the vector sum of the swimmer's velocity and the velocity of the current.

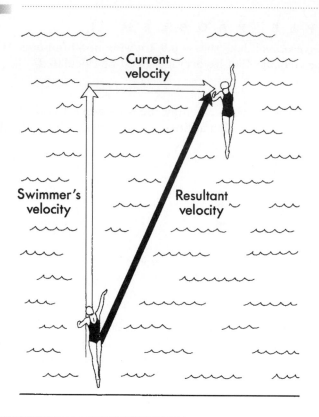

Another way to express change in position is position$_2$ − position$_1$, in which position$_1$ represents the body's position at one point in time and position$_2$ represents the body's position at a later point:

$$velocity = \frac{position_2 - position_1}{time_2 - time_1}$$

Because velocity is based on displacement, it is also a vector quantity. Consequently, description of velocity must include an indication of both the direction and the magnitude of the motion. If the direction of the motion is positive, velocity is positive; if the direction is negative, velocity is a negative quantity. A change in a body's velocity may represent a change in its speed, movement direction, or both.

Whenever two or more velocities act, the laws of vector algebra govern the ultimate speed and direction of the resultant motion. For example, the path actually taken by a swimmer crossing a river is determined by the vector sum of the swimmer's speed in the intended direction and the velocity of the river's current (Figure 10-3). The sample problem shown in Figure 10-4 provides an illustration of this situation.

FIGURE 10-4

S A M P L E   P R O B L E M   2

A swimmer orients herself perpendicular to the parallel banks of a river. If the swimmer's velocity is 2 m/s and the velocity of the current is 0.5 m/s, what will be the swimmer's resultant velocity? How far will the swimmer actually have to swim to get to the other side if the banks of the river are 50 m apart?

Solution

A diagram showing vector representations of the velocities of the swimmer and the current is drawn:

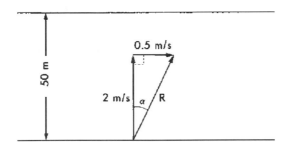

The resultant velocity can be found graphically by measuring the length and the orientation of the vector resultant of the two given velocities:

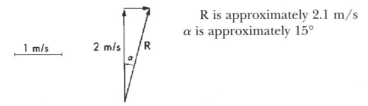

R is approximately 2.1 m/s
α is approximately 15°

The resultant velocity can also be found using trigonometric relationships. The magnitude of the resultant velocity may be calculated using the Pythagorean theorem:

$$R^2 = (2 \text{ m/s})^2 + (0.5 \text{ m/s})^2$$
$$R = \sqrt{(2 \text{ m/s})^2} + \sqrt{(0.5 \text{ m/s})^2}$$
$$R = 2.06 \text{ m/s}$$

The direction of the resultant velocity may be calculated using the cosine relationship:

$$R \cos \alpha = 2 \text{ m/s}$$
$$(2.06 \text{ m/s}) \cos \alpha = 2 \text{ m/s}$$
$$\alpha = \arccos\left(\frac{2 \text{ m/s}}{2.06 \text{ m/s}}\right)$$
$$\alpha = 14°$$

If the swimmer travels in a straight line in the direction of her resultant velocity, the cosine relationship may be used to calculate her resultant displacement:

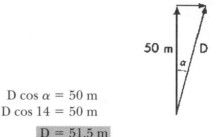

$$D \cos \alpha = 50 \text{ m}$$
$$D \cos 14 = 50 \text{ m}$$
$$\boxed{D = 51.5 \text{ m}}$$

Units of speed and velocity are always units of length divided by units of time.

Units of speed and velocity are units of length divided by units of time. In the metric system, common units for speed and velocity are meters per second (m/s) and kilometers per hour (km/hr). However, any unit of length divided by any unit of time yields an acceptable unit of speed or velocity. For example, a speed of 5 m/s can also be expressed as 5000 mm/s or 18,000 m/hr. It is usually most practical to select units that will result in expression of the quantity in the smallest, most manageable form.

For human gait, speed is the product of stride length and stride frequency. Adults in a hurry tend to walk with both longer stride lengths and faster stride frequency than under more leisurely circumstances. However, for toddlers, women wearing high-heeled shoes, and elderly individuals, it is often difficult to significantly increase stride length without losing balance.

During running, stride length is not simply a function of the runner's body height, but appears to be also influenced by muscle fiber composition, footwear, state of fatigue, injury history, and the inclination (grade) of the running surface (1). Runners traveling at a slow pace tend to increase velocity primarily by increasing stride length. At faster running speeds, recreational runners rely more on increasing stride frequency to increase velocity (Figure 10-5). Overstriding, or using an overly long stride length, should be avoided since it is a risk factor for hamstring strains.

Those who run regularly for exercise usually prefer a given stride frequency over a range of slow to moderate running speeds. One reason for this may be related to running economy—the oxygen consumption required for performing a given task (2). Most runners tend to choose a combination of stride length and stride frequency that minimizes the physiological cost of running (3). As discussed in Chapter 1, many species of animals do the same thing.

Since maximizing speed is the objective of all racing events, sport biomechanists have focused on the kinematic features that appear to ac-

Running speed is the product of stride length and stride frequency.

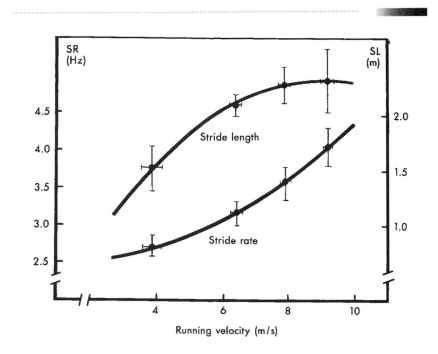

FIGURE 10-5

Changes in stride length and stride rate with running velocity. (From Luhtanen P and Komi PV: Mechanical factors influencing running speed. In Asmussen E and Jorgensen K, eds: *Biomechanics VI-B*, Baltimore, 1978, University Park Press.)

company fast performances in running, skiing, skating, cycling, swimming, and rowing events. Research has shown that the best male and female sprinters are distinguished from their less-skilled counterparts by extremely high stride frequencies and short ground contact times, although their stride lengths are usually only average or slightly greater than average (7). In contrast, the fastest cross-country skiers have longer-than-average cycle lengths, with cycle rates that are only average (22). Research on skating kinematics has shown that better ice skaters appear to excel because of higher stride rates (17), whereas elite roller skaters are distinguished by longer strides (24).

When racing performances are analyzed, comparisons are usually based on pace rather than speed or velocity. Pace is the inverse of speed. Rather than units of distance divided by units of time, pace is presented as units of time divided by units of distance. Pace is the time taken to cover a given distance and is commonly quantified as minutes per km or minutes per mile.

Acceleration

We are well aware that the consequence of pressing down or letting up on the accelerator pedal of an automobile is usually a change in the

linear acceleration
the rate of change in linear velocity

automobile's speed (and velocity). **Linear acceleration** (a) is defined as the rate of change in velocity or the change in velocity occurring over a given time interval (t):

$$a = \frac{\text{change in velocity}}{\text{change in time}}$$

$$a = \frac{\Delta v}{\Delta t}$$

Another way to express change in velocity is $v_2 - v_1$, in which v_1 represents velocity at one point in time and v_2 represents velocity at a later point:

$$a = \frac{v_2 - v_1}{\Delta t}$$

Units of acceleration are units of velocity divided by units of time. If a car increases its velocity by 1 km/hr each second, its acceleration is 1 km/hr/s. If a skier increases velocity by 1 m/s each second, the acceleration is 1 m/s/s. In mathematical terms, it is simpler to express the skier's acceleration as 1 m/s squared (1 m/s^2). A common unit of acceleration in the metric system is m/s^2.

Acceleration is the rate of change in velocity, or the degree with which velocity is changing with respect to time. For example, a body accelerating in a positive direction at a constant rate of 2 m/s^2 is increasing its velocity by 2 m/s each second. If the body's initial velocity was 0, a second later its velocity would be 2 m/s, a second after that its velocity would be 4 m/s, and a second after that its velocity would be 6 m/s.

In general usage, the term *accelerating* means speeding up, or increasing in velocity. If v_2 is greater than v_1, acceleration is a positive number, and the body in motion may have speeded up during the time period in question. However, because it is sometimes appropriate to label the direction of motion as positive or negative, a positive value of acceleration may not mean that the body is speeding up.

If the direction of motion is described in terms other than positive or negative, a positive value of acceleration does indicate that the body being analyzed has speeded up. For example, if a sprinter's velocity is 3 m/s on leaving the blocks and is 5 m/s one second later, calculation of the acceleration that has occurred will yield a positive number. Because $v_1 = 3$ m/s, $v_2 = 5$ m/s, and $t = 1$ s:

$$a = \frac{v_2 - v_1}{\Delta t}$$

$$a = \frac{5 \text{ m/s} - 3 \text{ m/s}}{1 \text{ s}}$$

$$a = 2 \text{ m/s}^2$$

Sliding into a base involves negative acceleration of the base runner.

Whenever the direction of motion is described in terms other than positive or negative, and v_2 is greater than v_1, the value of acceleration will be a positive number and the object in question is speeding up.

Acceleration can also assume a negative value. As long as the direction of motion is described in terms other than positive or negative, negative acceleration indicates that the body in motion is slowing down, or that its velocity is decreasing. For example, when a base runner slides to a stop over home plate, acceleration is negative. If a base runner's velocity is 4 m/s when going into a 0.5 s slide that stops the motion, $v_1 = 4$ m/s, $v_2 = 0$, and $t = 0.5$ s. Acceleration may be calculated as the following:

$$a = \frac{v_2 - v_1}{t}$$
$$a = \frac{0 - 4 \text{ m/s}}{0.5 \text{ s}}$$
$$a = -8 \text{ m/s}^2$$

Whenever v_1 is greater than v_2 in this type of situation, acceleration will be negative. The sample problem in Figure 10-6 provides another example of a situation involving negative acceleration.

Understanding acceleration is more complicated when one direction is designated as positive and the opposite direction is designated as negative. In this situation, a positive value of acceleration can indicate either that the object is speeding up in a positive direction or that it is slowing down in a negative direction (Figure 10-7).

Consider the case of a ball being dropped from a hand. As the ball falls faster and faster because of the influence of gravity, it is gaining speed, for example, 0.3 m/s to 0.5 m/s to 0.8 m/s. Because the downward direction is considered as the negative direction, the

FIGURE 10-6

A soccer ball is rolling down a field. At t = 0, the ball has an instantaneous velocity of 4 m/s. If the acceleration of the ball is constant at -0.3 m/s^2, how long will it take the ball to come to a complete stop?

Known
After reading the problem carefully, the next step is to sketch the problem situation, showing all quantities that are known or given in the problem statement.

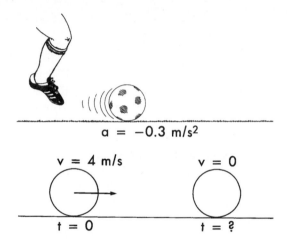

$$a = -0.3 \text{ m/s}^2$$

v = 4 m/s   v = 0

t = 0   t = ?

Solution
The next step is to identify the appropriate formula to use to find the unknown quantity:

$$a = \frac{v_2 - v_1}{t}$$

The known quantities can now be filled in to solve for the unknown variable (time):

$$-0.3 \text{ m/s}^2 = \frac{0 - 4 \text{ m/s}}{t}$$

Rearranging the equation, we have the following:

$$t = \frac{0 - 4 \text{ m/s}}{-0.3 \text{ m/s}^2}$$

Simplifying the expression on the right side of the equation, we have the solution:

$$t = 13.3 \text{ s}$$

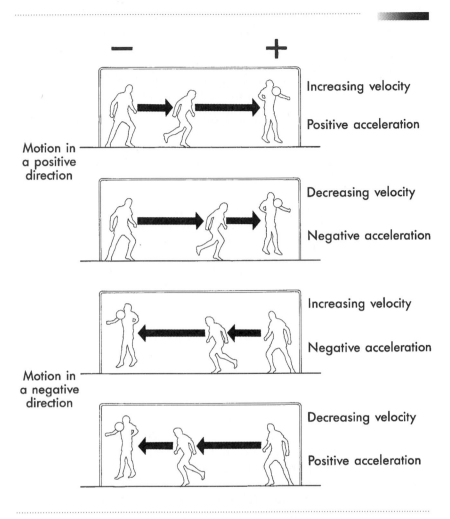

Motion in a positive direction

Motion in a negative direction

FIGURE 10-7
Right is regarded as the positive direction, and *left* is the negative direction. Acceleration may be positive, negative, or equal to zero, based on the direction of the motion and the direction of the change in velocity.

ball's velocity is actually -0.3 m/s to -0.5 m/s to -0.8 m/s. If $v_1 = -0.3$ m/s, $v_2 = -0.5$ m/s, and $t = 0.02$ s, acceleration is calculated as follows:

$$a = \frac{v_2 - v_1}{t}$$

$$a = \frac{-0.5 \text{ m/s} - (-0.3 \text{ m/s})}{0.02 \text{ s}}$$

$$a = -10 \text{ m/s}^2$$

In this situation the ball is speeding up, yet its acceleration is negative because it is speeding up in a negative direction. If acceleration is negative, velocity may be either increasing in a negative direction or decreasing in a positive direction. Alternatively, if acceleration is positive, velocity may be either increasing in a positive direction or decreasing in a negative direction.

The third alternative is for acceleration to be equal to 0. Acceleration is 0 whenever velocity is constant, that is, when v_1 and v_2 are the same. In the middle of a 100 m sprint, a sprinter's acceleration should be close to 0, because at that point the runner should be running at a constant, near maximum velocity.

Acceleration and deceleration (the lay term for negative acceleration) have implications for injury of the human body, since changing velocity results from the application of force (see Chapter 12). The anterior cruciate ligament, which restricts the forward sliding of the femur on the tibial plateaus during knee flexion, is often injured when an athlete who is running decelerates rapidly or changes directions quickly.

It is important to remember that since acceleration is a vector quantity, changing directions, even while maintaining a constant speed, represents a change in acceleration. The concept of angular acceleration, with direction constantly changing, is discussed in Chapter 11. The forces associated with change in acceleration based on change in direction must be compensated for by skiers and velodrome cyclists, in particular. That topic is discussed in Chapter 14.

When acceleration is 0, velocity is constant.

The instantaneous velocity of the shot at the moment of release primarily determines the ultimate horizontal displacement of the shot.

instantaneous
occurring during a small interval of time

average
occurring over a designated time interval

Average and Instantaneous Quantities

It is often of interest to determine the velocity of acceleration of an object or body segment at a particular time. For example, the **instantaneous** velocity of a shot or a discus at the moment the athlete releases it greatly affects the distance that the implement will travel. It is sometimes sufficient to quantify the **average** speed or velocity of the entire performance.

When speed and velocity are calculated, the procedures depend on whether the average or the instantaneous value is the quantity of interest. Average velocity is calculated as the final displacement divided by the total time period. Average acceleration is calculated as the difference in the final and initial velocities divided by the entire time interval. Calculation of instantaneous values can be approximated by dividing differences in velocities over an extremely small time interval. With calculus, velocity can be calculated as the derivative of displacement, and acceleration as the derivative of velocity.

Selection of the time interval over which speed or velocity is quantified is important when analyzing the performance of athletes in racing events. Many athletes can maintain world-record paces for the first one-half or three-fourths of the event, but slow during the last leg because of fatigue. In a study involving female high school sprinters performing the 100 m run, it was found that maximum running speeds of 8.0 to 8.4 m/s were reached 23 to 37 m from the start, and that an average of 7.3% of maximum speed was lost when the runners entered the final 10 m (4). Alternatively, some athletes may intentionally perform at a controlled pace during some segments of a race and then achieve maximum speed at the end. The longer the event is, the more information is potentially lost or concealed when only the final time or average speed is reported.

KINEMATICS OF PROJECTILE MOTION

Bodies projected into the air are **projectiles.** A basketball, a discus, a high jumper, and a sky diver are all projectiles as long as they are moving through the air unassisted. Depending on the projectile, different kinematic quantities are of interest. The resultant horizontal displacement of the projectile determines the winner of the contest in field events such as the shot put, discus throw, and javelin throw. High jumpers and pole vaulters maximize ultimate vertical displacement to win events. Sky divers manipulate both horizontal and vertical components of velocity to land as close as possible to targets on the ground.

However, not all objects that fly through the air are projectiles. A projectile is a body in free fall that is subject only to the forces of gravity and air resistance. Therefore, objects such as airplanes and rockets do not qualify as projectiles, because they are also influenced by the forces generated by their engines.

projectile
a body in free fall that is subject only to the forces of gravity and air resistance

Horizontal and Vertical Components

Just as it is more convenient to analyze general motion in terms of its linear and angular components, it is usually more meaningful to analyze the horizontal and vertical components of projectile motion separately. This is true for two reasons. First, the vertical component is influenced by gravity, whereas no force (neglecting air resistance) affects the horizontal component. Second, the horizontal component of motion relates to the distance the projectile travels, and the vertical component relates to the maximum height achieved by the projectile. Once a body has been projected into the air, its overall (resultant) velocity is constantly changing because of the forces acting on it. When examined separately, however, the horizontal and vertical components of projectile velocity change predictably.

Horizontal and vertical components of projectile motion are independent of each other. In the example shown in Figure 10-8, a baseball is dropped from a height of 1 m at the same instant that a second ball is horizontally struck by a bat at a height of 1 m, resulting in a line drive.

The human body becomes a projectile during the airborne phase of a jump.

FIGURE 10-8

The vertical and horizontal components of projectile motion are independent. A ball hit horizontally has the same vertical component as a ball dropped with no horizontal velocity.

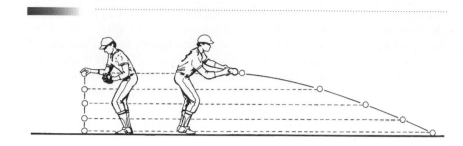

Both balls land on the level field simultaneously because the vertical components of their motions are identical. However, because the line drive also has a horizontal component of motion, it also undergoes some horizontal displacement.

Influence of Gravity

A major factor that influences the vertical but not the horizontal component of projectile motion is the force of gravity, which accelerates bodies in a vertical direction toward the surface of the earth (Figure 10-9). Unlike aerodynamic factors that may vary with the velocity of the wind, gravitational force is a constant, unchanging force that produces a constant downward vertical acceleration. Using the convention that upward is positive and downward is negative, the acceleration of gravity is treated as a negative quantity (-9.81 m/s^2). This acceleration remains constant regardless of the size, shape, or weight of the projectile. The vertical component of the initial projection velocity determines the maximum vertical displacement achieved by a body projected from a given relative projection height.

The force of gravity produces a constant acceleration on bodies near the surface of the earth equal to approximately -9.81 m/s^2.

FIGURE 10-9

Projectile trajectories **A**, without and **B**, with gravitational influence.

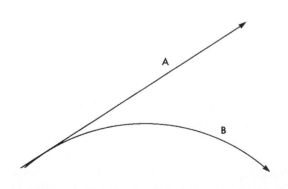

FIGURE 10-10
The pattern of change in the vertical velocity of a projectile is symmetrical about the apex of the trajectory.

Figure 10-10 illustrates the influence of gravity on projectile flight in the case of a ball tossed upward into the air by a juggler. The ball leaves the juggler's hand with a certain vertical velocity. As the ball travels higher and higher, the magnitude of its velocity decreases because it is undergoing a negative acceleration (the acceleration of gravity in a downward direction). At the peak or **apex** of the flight, which is that instant between going up and coming down, vertical velocity is 0. As the ball falls downward, its speed progressively increases, again because of gravitational acceleration. Since the direction of motion is downward, the ball's velocity is becoming progressively more negative. If the ball is caught at the same height from which it was tossed, the ball's speed is exactly the same as its initial speed, although its direction is now reversed. Graphs of the vertical displacement, velocity, and acceleration of a tossed ball are shown in Figure 10-11.

apex
the highest point in the trajectory of a projectile

Influence of Air Resistance

If an object were projected in a vacuum (with no air resistance), the horizontal component of its velocity would remain exactly the same throughout the flight. However, in most real-life situations, air resistance affects the horizontal component of projectile velocity. A ball thrown with a given initial velocity in an outdoor area will travel much farther if it is thrown with a tailwind rather than into a headwind. Because the

Neglecting air resistance, the horizontal speed of a projectile remains constant throughout the trajectory.

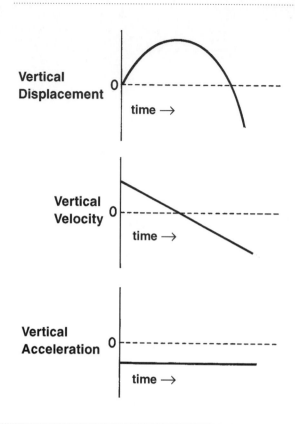

effects of air resistance are variable, however, for purposes of simplification, the horizontal component of a given projectile's velocity will be regarded as an unchanging (constant) quantity in this chapter.

When a projectile drops vertically through the air in a typical real-life situation, its velocity at any point is also related to air resistance. A skydiver's velocity, for example, is much smaller after the opening of the parachute than before its opening.

FACTORS INFLUENCING PROJECTILE TRAJECTORY

trajectory
the flight path of a projectile

The three mechanical factors that determine a projectile's motion are projection angle, projection speed, and relative height of projection.

Three factors influence the **trajectory** (flight path) of a projectile: the angle of projection, the projection speed, and the relative height of projection (Figure 10-12) (Table 10-1). Understanding how these factors interact is useful within the context of sport both for determining how to best project balls and other implements and for predicting how to best catch or strike projected balls.

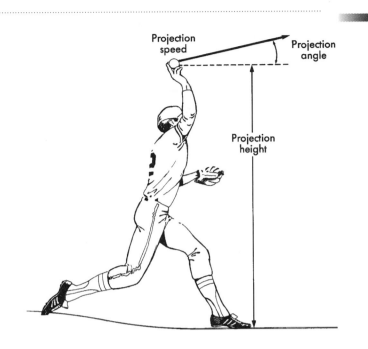

FIGURE 10-12
Factors affecting the trajectory of a projectile include projection angle, projection speed, and relative height of projection.

TABLE 10-1

FACTORS INFLUENCING PROJECTILE MOTION (NEGLECTING AIR RESISTANCE)	
VARIABLE	FACTORS OF INFLUENCE
Flight time	Initial vertical velocity Relative projection height
Horizontal displacement	Horizontal velocity Relative projection height
Vertical displacement	Initial vertical velocity Relative projection height
Trajectory	Initial speed Projection angle Relative projection height

Projection Angle

The **angle of projection** and the effects of air resistance govern the shape of a projectile's trajectory. Changes in projection speed influence the size of the trajectory, but trajectory shape is solely dependent on

angle of projection
the direction at which a body is projected with respect to the horizontal

FIGURE 10-13
The effect of projection angle on projectile trajectory.

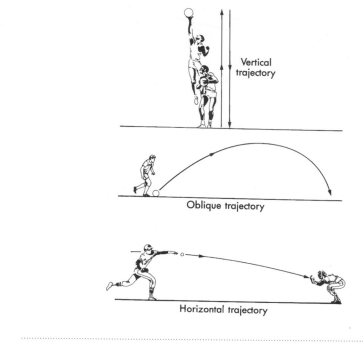

Vertical trajectory

Oblique trajectory

Horizontal trajectory

Projection angle is particularly important in the sport of basketball. A common error among novice players is shooting the ball with too flat a trajectory.

projection angle. In the absence of air resistance, the trajectory of a projectile assumes one of three general shapes, depending on the angle of projection. If the projection angle is perfectly vertical, the trajectory is also perfectly vertical, with the projectile following the same path straight up and then straight down again. If the projection angle is oblique (at some angle between 0° and 90°), the trajectory is *parabolic,* which means shaped like a parabola. A parabola is symmetrical, so its right and left halves are mirror images of each other. A body projected perfectly horizontally (at an angle of 0°) will follow a trajectory resembling one half of a parabola (Figure 10-13). Figure 10-14 displays scaled, theoretical trajectories for an object projected at different angles at a given speed. A ball thrown upward at a projection angle of 80° to the horizontal follows a relatively high and narrow trajectory, achieving more height than horizontal distance. A ball projected upward at a 10° angle to the horizontal follows a trajectory that is flat and long in shape.

Projection angle has direct implications for success in the sport of basketball, since a steep angle of entry into the basket allows a somewhat larger margin of error than a shallow angle of entry. Within 4.57 m of the basket, jump shot release angles are about 52° to 55°, providing a relatively steep angle of entry, whereas shots taken from 6.40 m tend to be released at 48° to 50°, allowing for a minimum release speed, but a less steep angle of entry (19).

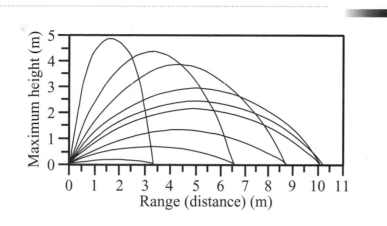

FIGURE 10-14
This scaled diagram shows the size and shape of trajectories for an object projected at 10 m/s.

In projection situations on a field, air resistance may, in reality, create irregularities in the shape of a projectile's trajectory. A typical modification in trajectory caused by air resistance is displayed in Figure 10-15. For purposes of simplification, the effects of aerodynamic forces will be disregarded in the discussion of projectile motion.

Projection Speed

When projection angle and other factors are constant, the **projection speed** determines the length or size of a projectile's trajectory. For example, when a body is projected vertically upward, the projectile's initial speed determines the height of the trajectory's apex. For a body that is projected at an oblique angle, the speed of projection determines both the height and the horizontal length of the trajectory (Figure 10-16). The combined effects of projection speed and projection angle on the horizontal displacement or **range** of a projectile are shown in Table 10-2.

projection speed
the magnitude of projection velocity

A projectile's range is the product of its horizontal speed and flight time.

range
the horizontal displacement of a projectile at landing

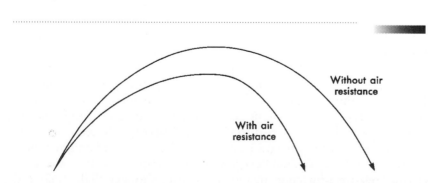

FIGURE 10-15
In real-life situations, air resistance causes a projectile to deviate from its theoretical parabolic trajectory.

Without air resistance

With air resistance

FIGURE 10-16

The effect of projection speed on projectile trajectory with projection angle held constant.

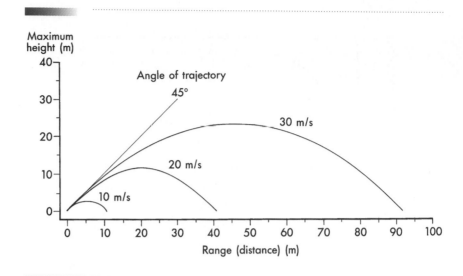

Performance in the execution of a vertical jump on a flat surface is entirely dependent on takeoff speed; that is, the greater the vertical velocity at takeoff, the higher the jump, and the higher the jump, the greater the amount of time the jumper is airborne (see margin). The time required for the performance of a vertical jump can be an important issue for dance choreographers. The incorporation of vertical jumps into a performance must be planned carefully (15). If the tempo of the music necessitates that vertical jumps be executed within one-third of a second, the height of the jumps is restricted to approximately 12 cm. The choreographer must be aware that under these circumstances, most dancers do not have sufficient floor clearance to point their toes during jump execution.

VERTICAL JUMP HEIGHT (CM)	FLIGHT TIME (S)
5	0.2
11	0.3
20	0.4
31	0.5
44	0.6
60	0.7
78	0.8
99	0.9

relative projection height
the difference between projection height and landing height

A projectile's flight time is increased by increasing the vertical component of projection velocity or by increasing the relative projection height.

Relative Projection Height

The third major factor influencing projectile trajectory is the **relative projection height** (Figure 10-17). This is the difference in the height from which the body is initially projected and the height at which it lands or stops. When a discus is released by a thrower from a height of $1\frac{1}{2}$ m above the ground, the relative projection height is $1\frac{1}{2}$ m, because the projection height is $1\frac{1}{2}$ m greater than the height of the field on which the discus lands. If a driven golf ball becomes lodged in a tree, the relative projection height is negative, because the landing height is greater than the projection height. When projection velocity is constant, greater relative projection height translates to longer flight time and greater horizontal displacement of the projectile.

TABLE 10-2

THE EFFECT OF PROJECTION ANGLE ON RANGE (RELATIVE PROJECTION HEIGHT = 0)

PROJECTION SPEED (M/S)	PROJECTION ANGLE (DEGREES)	RANGE (M)
10	10	3.49
10	20	6.55
10	30	8.83
10	40	10.04
10	45	10.19
10	50	10.04
10	60	8.83
10	70	6.55
10	80	3.49
20	10	13.94
20	20	26.21
20	30	35.31
20	40	40.15
20	45	40.77
20	50	40.15
20	60	35.31
20	70	26.21
20	80	13.94
30	10	31.38
30	20	58.97
30	30	79.45
30	40	90.35
30	45	91.74
30	50	90.35
30	60	79.45
30	70	58.97
30	80	31.38

FIGURE 10-17
The relative projection
height.

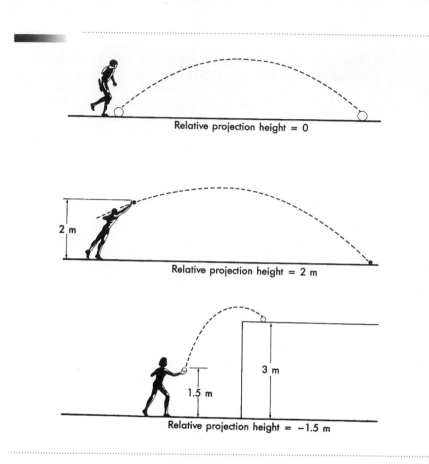

Relative projection height = 0

2 m

Relative projection height = 2 m

3 m

1.5 m

Relative projection height = −1.5 m

In the sport of diving, relative projection height is the height of the springboard or platform above the water. If a diver's center of gravity is elevated 1.5 m above the springboard at the apex of the trajectory, flight time is about 1.2 s from a 1 m board and 1.4 s from a 3 m board. This provides enough time for a skilled diver to complete 3 somersaults from a 1 m board and $3\frac{1}{2}$ somersaults from a 3 m board (25). The implication is that a diver attempting to learn a $3\frac{1}{2}$ somersault dive from the 3 m springboard should first be able to easily execute a $2\frac{1}{2}$ somersault dive from the 1 m board.

Optimum Projection Conditions

In sporting events based on achieving maximum horizontal displacement or maximum vertical displacement of a projectile, the athlete's primary goal is to maximize the speed of projection (10, 20). In the throwing events, another objective is to maximize release height because greater relative projection height produces longer flight time, and con-

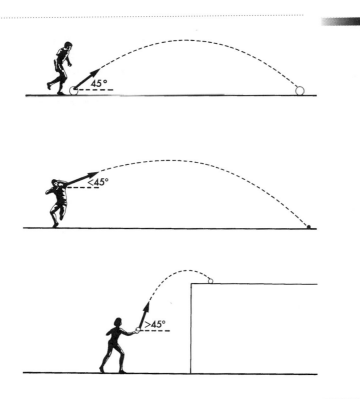

FIGURE 10-18
When projection speed is constant and aerodynamics are not considered, the optimum projection angle is based on the relative height of projection. When the relative projection height is 0, an angle of 45° is optimum. As the relative projection height increases, optimum projection angle decreases. As the relative projection height becomes increasingly negative, the optimum projection angle increases.

sequently greater horizontal displacement of the projectile. However, it is generally not prudent for a thrower to sacrifice release speed for added release height.

The factor that varies the most, both with the event and the performer, is the optimum angle of projection. When relative projection height is zero, the angle of projection that produces maximum horizontal displacement is 45°. As relative projection height increases, the optimum angle of projection decreases, and as relative projection height decreases, the optimum angle increases (Figure 10-18).

When the human body is the projectile during a jump, determining the optimum projection angle is further complicated by the fact that takeoff angle can affect projection speed. In the performance of the long jump, for example, because takeoff and landing heights are the same, the theoretically optimum angle of takeoff is 45° with respect to the horizontal. However, it has been estimated by Hay (8) that to obtain this theoretically optimum takeoff angle, long jumpers would decrease the horizontal velocity they could otherwise obtain by approximately 50%. Researchers have shown that success in the long jump, high jump, and pole vault are all related to the athlete's ability to maximize horizontal velocity going into takeoff (6, 9). The actual takeoff angles employed by elite long jumpers range from approximately 18° to 27° (8). Takeoff angles during all three phases of the triple jump are even smaller elite performers than those used in the long jump (18). In the ski jump, where athletes have the advantage of a large relative height between takeoff and landing, takeoff angles are as small as 4.6° to 6.2° (23). In an event such as the high jump, in which the goal is to maximize vertical displacement, takeoff angles among skilled Fosbury Flop style jumpers range from 40° to 48° (5).

In the throwing events, the aerodynamic characteristics of the projected implements also influence the trajectory (see Chapter 15). In these events (shot, discus, javelin, and hammer), only the trajectory of the shot is not appreciably affected by aerodynamic forces. The concept that the optimum angle of release must not restrict release speed is still a paramount consideration for performance in the shot put. The release angles reported among elite throwers in the shot are approximately 36° to 37° (11).

ANALYZING PROJECTILE MOTION

initial velocity
vector quantity incorporating both angle and speed of projection

Because velocity is a vector quantity, the **initial velocity** of a projectile incorporates both the initial speed (magnitude) and the angle of projection (direction) into a single quantity. When the initial velocity of a projectile is resolved into horizontal and vertical components, the horizontal component has a certain speed or magnitude in a horizontal direction, and the vertical component has a speed or magnitude in a vertical direction (Figure 10-19). The magnitudes of the horizontal and vertical components are always quantified so that if they were added together through the process of vector composition, the resultant velocity vector would be equal in magnitude and direction to the original initial velocity vector. The horizontal and vertical components of initial velocity may be quantified both graphically and trigonometrically (Figure 10-20).

The vertical speed of a projectile is constantly changing because of gravitational acceleration.

The horizontal acceleration of a projectile is always 0.

For purposes of analyzing the motion of projectiles, it will be assumed that the horizontal component of projectile velocity is constant throughout the trajectory and that the vertical component of projectile velocity is constantly changing because of the influence of gravity (Figure 10-21). Since horizontal projectile velocity is constant, horizontal acceleration is equal to the constant of 0 throughout the trajectory. The vertical acceleration of a projectile is equal to the constant -9.81 m/s^2.

Equations of Constant Acceleration

When a body is moving with a constant acceleration (positive, negative, or equal to 0), certain interrelationships are present among the kinematic quantities associated with the motion of the body. These inter-

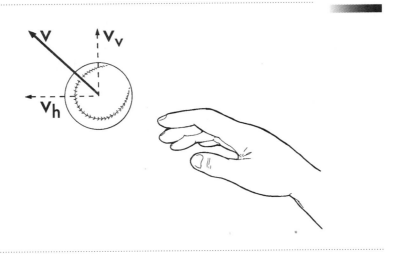

FIGURE 10-19
The vertical and
horizontal components of
projection velocity.

relationships may be expressed using three mathematical equations originally derived by Galileo, which are known as the **laws of constant acceleration** or the laws of uniformly accelerated motion. Using the variable symbols d, v, a, and t (representing displacement, velocity,
acceleration, and time, respectively) and with the subscripts 1 and 2
(representing first or initial and second or final points in time), the
equations are the following:

$$v_2 = v_1 + at \qquad (1)$$
$$d = v_1 t + \left(\tfrac{1}{2}\right)at^2 \qquad (2)$$
$$v_2^2 = v_1^2 + 2ad \qquad (3)$$

Notice that each of the equations contains a unique combination of
three of the four kinematic quantities: displacement, velocity, acceleration, and time. This provides considerable flexibility for solving problems in which two of the quantities are known and the objective is to
solve for a third.

It is instructive to examine these relationships as applied to the horizontal component of projectile motion in which a = 0. In this case, each
term containing acceleration may be removed from the equation. The
equations then appear as the following:

$$v_2 = v_1 \qquad (1H)$$
$$d = v_1 t \qquad (2H)$$
$$v_2^2 = v_1^2 \qquad (3H)$$

Equations 1H and 3H reaffirm that the horizontal component of projectile velocity is a constant. Equation 2H indicates that horizontal

**laws of constant
acceleration**
formulas relating
displacement, velocity,
acceleration, and time when
acceleration is unchanging

FIGURE 10-20

S A M P L E   P R O B L E M   4

A basketball is released with an initial speed of 8 m/s at an angle of 60°. Find the horizontal and vertical components of the ball's initial velocity, both graphically and trigonometrically.

Known

A diagram showing a vector representation of the initial velocity is drawn using a scale of 1 cm = 2 m/s:

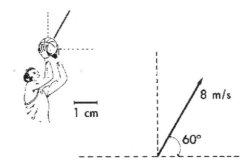

Solution

The horizontal component is drawn in along the horizontal line to a length that is equal to the length that the original velocity vector extends in the horizontal direction. The vertical component is then drawn in the same fashion in a direction perpendicular to the horizontal line:

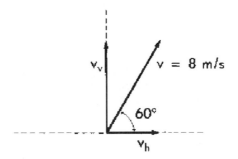

The lengths of the horizontal and vertical components are then measured:

$$\text{length of horizontal component} = 2 \text{ cm}$$
$$\text{length of vertical component} = 3.5 \text{ cm}$$

To calculate the magnitudes of the horizontal and vertical components, use the scale factor of 2 m/s/cm:

Magnitude of horizontal component:

$$v_h = 2 \text{ cm} \times 2 \text{ m/s/cm}$$

$$\boxed{v_h = 4 \text{ m/s}}$$

Magnitude of vertical component:

$$v_h = 3.5 \text{ cm} \times 2 \text{ m/s/cm}$$

$$\boxed{v_h = 7 \text{ m/s}}$$

To solve for v_h and v_v trigonometrically, construct a right triangle with the sides being the horizontal and vertical components of initial velocity and the initial velocity represented as the hypotenuse:

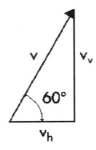

The sine and cosine relationships may be used to quantify the horizontal and vertical components:

$$v_h = (8 \text{ m/s})(\cos 60)$$

$$\boxed{v_h = 4 \text{ m/s}}$$

$$v_v = (8 \text{ m/s})(\sin 60)$$

$$\boxed{v_v = 6.9 \text{ m/s}}$$

Note that the magnitude of the horizontal component is *always* equal to the magnitude of the initial velocity multiplied by the cosine of the projection angle. Similarly, the magnitude of the initial vertical component is *always* equal to the magnitude of the initial velocity multiplied by the sine of the projection angle.

FIGURE 10-21

The horizontal and vertical components of projectile velocity. Notice that the horizontal component is constant, and the vertical component is constantly changing.

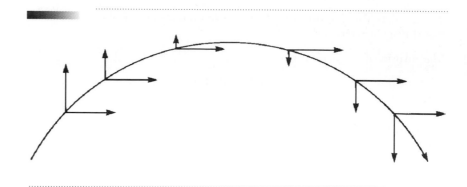

displacement is equal to the product of horizontal velocity and time (Figure 10-22).

When the constant acceleration relationships are applied to the vertical component of projectile motion, acceleration is equal to -9.81 m/s^2, and the equations cannot be simplified by the deletion of the acceleration term. However, in analysis of the vertical component of projectile motion, the initial velocity (v_1) is equal to 0 in certain cases. For example, when an object is dropped from a stationary position, the initial velocity of the object is 0. When this is the case, the equations of constant acceleration may be expressed as the following:

$$v_2 = at \qquad (1V)$$
$$d = (\tfrac{1}{2})at^2 \qquad (2V)$$
$$v_2^2 = 2ad \qquad (3V)$$

When an object is dropped, equation 1V relates that the object's velocity at any instant is the product of gravitational acceleration and the amount of time the object has been in free fall. Equation 2V indicates that the vertical distance through which the object has fallen can be calculated from gravitational acceleration and the amount of time the object has been falling. Equation 3V expresses the relationship between the object's velocity and vertical displacement at a certain time and gravitational acceleration.

It is useful in analyzing projectile motion to remember that at the apex of a projectile's trajectory the vertical component of velocity is 0. If the goal is to determine the maximum height achieved by a projectile, v_2 in equation 3 may be set equal to 0:

$$0 = v_1^2 + 2ad \qquad (3A)$$

An example of this use of equation 3A is shown in Figure 10-23. If the problem is to determine the total flight time, one approach is to calculate the time it takes to reach the apex, which is one-half of the total flight time if the projection and landing heights are equal. In this case,

FIGURE 10-22

S A M P L E   P R O B L E M   5

The score was tied at 20 to 20 in the final 1987 AFC playoff game between the Denver Broncos and the Cleveland Browns. During the first overtime period, Denver had the opportunity to kick a field goal, with the ball placed on the tee at a distance of 29 m from the goal posts. If the ball was kicked with the horizontal component of initial velocity being 18 m/s and a flight time of 2 seconds, was the kick long enough to make the field goal?

Known

$$v_h = 18 \text{ m/s}$$
$$t = 2 \text{ s}$$

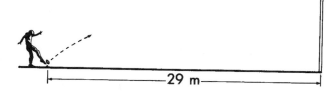

Solution

The formula 2H is selected to solve the problem since two of the variables contained in the formula (v_h and t) are known quantities and since the unknown variable (d) is the quantity we wish to find:

$$d_h = v_h t$$
$$d = (18 \text{ m/s})(2 \text{ s})$$
$$d = 36 \text{ m}$$

The ball did travel a sufficient distance for the field goal to be good, and Denver won the game, advancing to Super Bowl XXI.

---

v_2 in equation 1 for the vertical component of the motion may be set equal to 0 because vertical velocity is 0 at the apex:

$$0 = v_1 + at \qquad (1A)$$

The sample problem in Figure 10-24 illustrates this use of equation 1A.

When using the equations of constant acceleration, it is important to remember that they may be applied to the horizontal component of projectile motion or to the vertical component of projectile motion, but not to the resultant motion of the projectile. If the horizontal component of motion is being analyzed, a = 0, but if the vertical

FIGURE 10-23

A volleyball is deflected vertically by a player in a game housed in a high school gymnasium where the ceiling clearance is 10 m. If the initial velocity of the ball is 15 m/s, will the ball contact the ceiling?

Known

$$v_1 = 15 \text{ m/s}$$
$$a = -9.81 \text{ m/s}^2$$

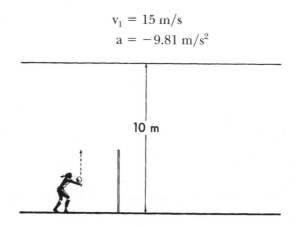

10 m

Solution

The equation selected for use in solving this problem must contain the variable d for vertical displacement. Equation 2 contains d but also contains the variable t, which is an unknown quantity in this problem. Equation 3 contains the variable d and, recalling that vertical velocity is zero at the apex of the trajectory, Equation 3A can be used to find d:

$$v_2^2 = v_1^2 + 2ad \qquad\qquad (3)$$
$$0 = v_1^2 + 2ad \qquad\qquad (3A)$$
$$0 = (15 \text{ m/s})^2 + (2)(-9.81 \text{ m/s}^2)d$$
$$(19.62 \text{ m/s}^2)d = 225 \text{ m}^2/\text{s}^2$$
$$\boxed{d = 11.47 \text{ m}}$$

Therefore, the ball has sufficient velocity to contact the 10 m ceiling.

FIGURE 10-24

S A M P L E   P R O B L E M   7

A ball is kicked at a 35° angle, with an initial speed of 12 m/s. How high and how far does the ball go?

Solution

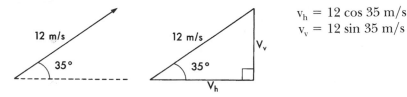

$v_h = 12 \cos 35$ m/s
$v_v = 12 \sin 35$ m/s

How high does the ball go?
Equation 1 cannot be used because it does not contain d. Equation 2 cannot be used unless t is known. Since vertical velocity is 0 at the apex of the ball's trajectory, Equation 3A is selected:

$$0 = v_1^2 + 2ad \qquad\qquad (3A)$$
$$0 = (12 \sin 35 \text{ m/s})^2 + (2)(-9.81 \text{ m/s}^2)d$$
$$(19.62 \text{ m/s}^2)d = 47.37 \text{ m}^2/\text{s}^2$$
$$\boxed{d = 2.41 \text{ m}}$$

How far does the ball go?
Equation 2H for horizontal motion cannot be used because t for which the ball was in the air is not known. Equation 1A can be used to solve for the time it took the ball to reach its apex:

$$0 = v_1 + at \qquad\qquad (1A)$$
$$0 = 12 \sin 35 \text{ m/s} + (-9.81 \text{ m/s}^2)t$$
$$t = \frac{6.88 \text{ m/s}}{9.81 \text{ m/s}^2}$$
$$t = 0.70 \text{ s}$$

Recalling that the time to reach the apex is one-half of the total flight time, total time is the following:

$$t = (0.70 \text{ s})(2)$$
$$t = 1.40 \text{ s}$$

Equation 2H can then be used to solve for the horizontal distance the ball traveled:

$$d_h = v_h t \qquad\qquad (2H)$$
$$d_h = (12 \cos 35 \text{ m/s})(1.40 \text{ s})$$
$$\boxed{d_h = 13.76 \text{ m}}$$

FIGURE 10-25
Summary table of
equations pertaining to
projectile motion.

FORMULAS RELATING TO PROJECTILE MOTION

The Equations of Constant Acceleration

These equations may be used to relate linear kinematic quantities whenever acceleration (a) is a constant, unchanging value.

$$v_2 = v_1 + at \tag{1}$$

$$d = v_1 t + (\tfrac{1}{2})at^2 \tag{2}$$

$$v_2^2 = v_1^2 + 2ad \tag{3}$$

Special Case Applications of the Equations of Constant Acceleration

For the horizontal component of projectile motion, with a = 0:

$$d_h = v_h t \tag{2H}$$

For the vertical component of projectile motion, with $v_1 = 0$, as when the projectile is dropped from a static position:

$$v_2 = at \tag{1V}$$

$$d = (\tfrac{1}{2})at^2 \tag{2V}$$

$$v_2^2 = 2ad \tag{3V}$$

For the vertical component of projectile motion, with $v_2 = 0$, as when the projectile is at its apex:

$$0 = v_1 + at \tag{1A}$$

$$0 = v_1^2 + 2ad \tag{2A}$$

component is being analyzed, $a = -9.81$ m/s^2. The equations of constant acceleration and their special variations are summarized in the box in Figure 10-25.

SUMMARY

Linear kinematics is the study of the form or sequencing of linear motion with respect to time. Linear kinematic quantities include the scalar quantities of distance and speed, and the vector quantities of displacement, velocity, and acceleration. Depending on the motion being analyzed, either a vector quantity or its scalar equivalent and either an instantaneous or an average quantity may be of interest.

A projectile is a body in free fall that is affected only by gravity and air resistance. Projectile motion is analyzed in terms of its horizontal and vertical components. The two components are independent of each other, and only the vertical component is influenced by gravitational force. Factors that determine the height and distance the projectile achieves are projection angle, projection speed, and relative projection height. The equations of constant acceleration can be used to quantitatively analyze projectile motion, with vertical acceleration being -9.81 m/s^2 and horizontal acceleration being 0.

INTRODUCTORY PROBLEMS

Note: Some problems require vector algebra (see Chapter 3).

1. A runner completes $6\frac{1}{2}$ laps around a 400 m track during a 12 minute (720 s) run test. Calculate the following quantities:
   a. The distance the runner covered
   b. The runner's displacement at the end of 12 minutes
   c. The runner's average speed
   d. The runner's average velocity
   e. The runner's average pace
   (Answer: a. 2.6 km; b. 160 m; c. 3.6 m/s; d. 0.22 m/s; e. 4.6 min/km)

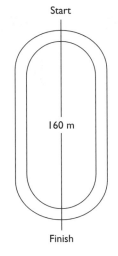

Start

160 m

Finish

2. A ball rolls with an acceleration of -0.5 m/s^2. If it stops after 7 seconds, what was its initial speed? (Answer: 3.5 m/s)
3. A wheelchair marathoner has a speed of 5 m/s after rolling down a small hill in 1.5 s. If the wheelchair underwent a constant acceleration of 3 m/s^2 during the descent, what was the marathoner's speed at the top of the hill? (Answer: 0.5 m/s)
4. An orienteer runs 400 m directly east and then 500 m to the northeast (at a 45° angle from due east and from due north). Provide a graphic solution to show final displacement with respect to the starting position.
5. An orienteer runs north at 5 m/s for 120 seconds, and then west at 4 m/s for 180 seconds. Provide a graphic solution to show the orienteer's resultant displacement.
6. Why are the horizontal and vertical components of projectile motion analyzed separately?
7. A soccer ball is kicked with an initial horizontal speed of 5 m/s and an initial vertical speed of 3 m/s. Assuming that projection and landing heights are the same and neglecting air resistance, identify the following quantities:
   a. The ball's horizontal speed 0.5 seconds into its flight
   b. The ball's horizontal speed midway through its flight
   c. The ball's horizontal speed immediately before contact with the ground
   d. The ball's vertical speed at the apex of the flight
   e. The ball's vertical speed midway through its flight
   f. The ball's vertical speed immediately before contact with the ground

8. If a baseball, a basketball, and a 71.2 N shot were dropped simultaneously from the top of the Empire State Building (and air resistance was not a factor), which would hit the ground first? Why?

9. A tennis ball leaves a racket during the execution of a perfectly horizontal ground stroke with a speed of 22 m/s. If the ball is in the air for 0.7 seconds, what horizontal distance does it travel? (Answer: 15.4 m)

10. A trampolinist springs vertically upward with an initial speed of 9.2 m/s. How high above the trampoline will the trampolinist go? (Answer: 4.31 m)

ADDITIONAL PROBLEMS

1. Explain what specific factors (joint positions and ranges of motion, muscle group torque production, etc.) could be contributing to performance errors resulting in (a) too little horizontal speed of a projectile, and (b) too little initial vertical speed of a projectile, for one of the following motor skills:
   a. standing long jump
   b. basketball jump shot
   c. ground stroke in tennis
   d. punt in football
   e. triple axel in dance or ice-skating

2. Provide a trigonometric solution for Introductory Problem 4. (Answer: D = 832 m; $\angle$ = 25° north of due east)

3. Provide a trigonometric solution for Introductory Problem 5. (Answer: D = 937 m; $\angle$ = 50° west of due north)

4. A buoy marking the turn in the ocean swim leg of a triathlon becomes unanchored. If the current carries the buoy southward at 0.5 m/s, and the wind blows the buoy westward at 0.7 m/s, what is the resultant displacement of the buoy after 5 minutes? (Answer: 258 m; $\angle$ = 54.5° west of due south)

5. A sailboat is being propelled westerly by the wind at a speed of 4 m/s. If the current is flowing at 2 m/s to the northeast, where will the boat be in 10 minutes with respect to its starting position? (Answer: D = 1.8 km; $\angle$ = 29° north of due west)

6. A Dallas Cowboy carrying the ball straight down the near sideline with a velocity of 8 m/s crosses the 50-yard line at the same time that the last Buffalo Bill who can possibly hope to catch him starts running from the 50-yard line at a point that is 13.7 m from the near sideline. What must the Bill's velocity be if he is to catch the Cowboy just short of the goal line? (Answer: 8.35 m/s)

7. A soccer ball is kicked from the playing field at a 45° angle. If the ball is in the air for 3 seconds, what is the maximum height achieved? (Answer: 11.0 m)

8. A ball is kicked a horizontal distance of 45.8 m. If it reaches a maximum height of 24.2 m with a flight time of 4.4 seconds, was the ball kicked at a projection angle less than, greater than, or equal to 45°? Provide a rationale for your answer based on the appropriate calculations. (Answer: >45°)

9. A badminton shuttlecock is struck by a racquet at a 35° angle, giving it an initial speed of 10 m/s. How high will it go? How far will it travel horizontally before being contacted by the opponent's racquet at the same height from which it was projected? (Answer: d_v = 1.68 m; d_h = 9.58 m)

10. An archery arrow is shot with a speed of 45 m/s at an angle of 10°. How far horizontally can the arrow travel before hitting a target at the same height from which it was released? (Answer: 70.6 m)

LABORATORY EXPERIENCES

1. On a track of known length, run two laps, one at a slow speed and one at a fast speed, counting the number of steps you take during each lap and timing each lap. Calculate your average step length, your speed (m/s), and your pace (min/mi) during each lap and write an explanation of the results. How do your results compare to the graph in Figure 10-5?

2. Answer the following questions pertaining to the split times (in seconds) presented below for Ben Johnson and Carl Lewis during the 100 m sprint in the 1988 Olympic Games.

	Johnson	Lewis
10m	1.86	1.88
20m	2.87	2.96
30m	3.80	3.88
40m	4.66	4.77
50m	5.55	5.61
60m	6.38	6.45
70m	7.21	7.29
80m	8.11	8.12
90m	8.98	8.99
100m	9.83	9.86

a. Plot velocity and acceleration curves for both sprinters. In what ways are the curves similar and different?

b. What general conclusions can you draw about performance in elite sprinters?

3. Videotape the trajectory of a ball tossed vertically into the air and caught at the same height from which it was tossed. Tape a piece of tracing paper over the video display and using single-frame advance,

trace the sequential images of the ball. What do the tracings reveal about the nature of projectile motion?

4. Simultaneously drop and horizontally project two balls of the same type. Which ball lands first? Why?

5. Construct a "treasure map" by using a compass and pacing off distances between a starting point and the treasure. Include at least 3 changes of direction in the path to the treasure. Indicate approximate distances on the map in meters. (It may be helpful to adjust your step length to the length of 1 m in pacing off distances.)

REFERENCES

1. Cavanagh PR and Kram R: Stride length in distance running: Velocity, body dimensions, and added mass effects. In Cavanagh PR, ed: *Biomechanics of distance running,* Champaign, 1990, Human Kinetics.
2. Cavanagh PR and Kram R: The efficiency of human movement—a statement of the problem, Med Sci Sports Exerc 17:304, 1985.
3. Cavanagh PR and Williams KR: The effect of stride length variation on oxygen uptake during distance running, Med Sci Sports Exerc 14:30, 1982.
4. Chow JW: Maximum speed of female high school runners, Int J Sport Biomech 3:110, 1987.
5. Dapena J: Mechanics of translation in the fosbury flop, Med Sci Sports Exerc 12:37, 1980.
6. Dapena J and Chung CS: Vertical and radial motions of the body during the take-off phase of high jumping, Med Sci Sports Exerc 20:290, 1988.
7. Dillman CJ: Overview of the United States Olympic Committee sports medicine biomechanics program. In Butts NK, Gushiken TT, and Zarins BT, eds: *The elite athlete,* New York, 1985, Spectrum Publications, Inc.
8. Hay JG: The biomechanics of the long jump, Exerc Sport Sci Rev 14:401, 1986.
9. Hay JG and Nohara H: The techniques used by elite long jumpers in preparation for take-off, J Biomech 23:229, 1990.
10. Hay JG and Yu B: Critical characteristics of technique in throwing the discus, J Sports Sci 13:125, 1996.
11. Hubbard M: The throwing events in track and field. In Vaughan CL, ed: *Biomechanics of sport,* Boca Raton, Fla, 1989, CRC Press, Inc.
12. Jeng SF, Liao HF, Lai JS, and Hou JW: Optimization of walking in children: Med Sci Sports Exerc 29:370, 1997.
13. Konczak J and Dichgans J: The development toward stereotypic arm kinematics during reaching in the first 3 years of life, Exp Brain Res 117:346, 1997.
14. Lange GW, Hintermeister RA, Schlegel T, Dillman CJ, and Steadman JR: Electromyographic and kinematic analysis of graded treadmill walking and the implications for knee rehabilitation, J Orthop Sports Phys Ther 23:294, 1996.
15. Laws K: *The physics of dance,* New York, 1984, Schirmer Books.
16. Marques Bruna P and Grimshaw PN: 3-dimensional kinematics of overarm throwing action of children age 15 to 30 months, Percept Mot Skills 84:1267, 1997.

17. McCaw ST and Hoshizaki TB: A kinematic comparison of novice, intermediate, and elite ice skaters. In Jonsson B, ed: *Biomechanics X-B*, Champaign, Ill, 1987, Human Kinetics Publishers, Inc.

18. Miller JA and Hay JG: Kinematics of a world record and other world-class performances in the triple jump, Int J Sport Biomech 2:272, 1986.

19. Miller S and Bartlett R: The relationship between basketball shooting kinematics, distance and playing position, J Sports Sci 14:243, 1996.

20. Morriss C and Bartlett R: Biomechanical factors critical for performance in the men's javelin throw, Sports Med 21:438, 1996.

21. Ramos E, Latash MP, Hurvitz EA, and Brown SH: Quantification of upper extremity function using kinematic analysis, Arch Phys Med Rehabil 78:491, 1997.

22. Rundell KW and McCarthy JR: Effect of kinematic variables on performance in women during a cross-country ski race, Med Sci Sports Exerc 28:1413, 1996.

23. Watanabe K: Ski-jumping, alpine-, cross-country-, and nordic combination skiing. In Vaughan CL, ed: *Biomechanics of sport*, Boca Raton, Fla, 1989, CRC Press, Inc.

24. Wilson BD, McDonald M, and Neal RJ: Roller skating sprint technique. In Jonsson B, ed: *Biomechanics X-B*, Champaign, Ill, 1987, Human Kinetics Publishers, Inc.

25. Yeadon MR: Theoretical models and their application to aerial movement. In Van Gheluwe B and Atha J, eds: *Current research in sports biomechanics*, Basel, 1987, Karger.

ANNOTATED READINGS

Anderson T: Biomechanics and running economy, Sports Med 22:76, 1996.
  Reviews current knowledge about kinematic and other factors that may contribute to economy during running.
Hay JG: Citius, altius, longius (faster, higher, longer): The biomechanics of jumping for distance, J Biomech 26:7, 1993.
  Discusses current research related to performance in the long jump and triple jump.
Maugh TH: Physics of basketball: Those golden arches, Science 1:106, 1981.
  Describes practical considerations for shooting a basketball to maximize the probability of its going through the hoop.
Townend MS: Throwing. In Townend MS: *Mathematics in sport*, New York, 1984, John Wiley & Sons, Inc.
  Covers an in-depth mathematical analysis of the mechanical principles of relevance in projecting the shot, hammer, discus, javelin, and basketball.

RELATED WEB SITES

The Catapult Museum Online
http://www.nzp.com
  Provides links to 3-D animations, RealAudio narration, 3-D models, and chat sessions dedicated to catapults.
The History of Projectile Motion
http://tqd.advanced.org/2779/History.html
  Provides textual information and historic drawings of the first accurate description of projectile motion by Galileo, plus links to projectile animations, including a projectile water balloon game with sound effects.

Illustration of Projectile Motion

http://nucphy.ph.msstate.edu/~tm/1013/tutor/projectile.html

Includes a demonstration of a projectile trajectory that provides the user the ability to click on buttons that bring up the horizontal and vertical velocities and accelerations along with explanations.

Multimedia Physics Lab—Projectile Motion

http://134.153.162.27/mpl/Physics1054/Mechanics/moduleM3/moduleM3P5.html

Includes links to an entire laboratory assignment on projectile motion, including a projectile video, VideoPoint analysis, determining the scale factor and trajectory, and plotting horizontal and vertical displacements.

Programming Example: Projectile Motion

http://www.cs.mtu.edu/~shene/COURSES/cs201/NOTES/chap02/projectile.html

Provides a documented, downloadable computer program for calculating horizontal and vertical displacements, resultant velocity, and direction of a projectile.

Projectile Motion

http://www.polar.sunynassau.edu/~schoenf/projectl.htm

Provides step-by-step instructions for solving projectile problems involving both horizontal and vertical motion components.

Projectile Motion

http://suhep.phy.syr.edu/courses/PHY211.96Fall/asg7/trajplot.html

Generates a scaled trajectory for a projectile with user-provided launch angle and damping coefficient.

Projectile Motion: Howitzer and Tunnel

http://storm.ph.utexas.edu/~phy-demo/demo-txt/1d60-10.html

Shows photograph and explanation of a demonstration of a small, rolling cart that shoots a ball upwards before entering a tunnel. The ball falls onto the cart on the other side of the tunnel.

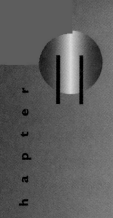

Chapter

ANGULAR KINEMATICS
OF HUMAN MOVEMENT

After completing this chapter, the reader will be able to:

Distinguish angular motion from rectilinear and curvilinear motion.

Discuss the relationships among angular kinematic variables.

Correctly associate angular kinematic quantities with their units of measure.

Explain the relationships between angular and linear displacement, angular and linear velocity, and angular and linear acceleration.

Solve quantitative problems involving angular kinematic quantities and the relationships between angular and linear kinematic quantities.

Why is a driver longer than a 9-iron? Why do batters slide their hands up the handle of the bat to execute a bunt but not to execute a power hit? How does the angular motion of the discus or hammer during the windup relate to the linear motion of the implement after release?

These questions relate to angular motion, or rotational motion around an axis. The axis of rotation is a line, real or imaginary, oriented perpendicular to the plane in which the rotation occurs, like the axle for the wheels of a cart. Like linear motion, angular motion is a basic component of general motion.

OBSERVING THE ANGULAR KINEMATICS
OF HUMAN MOVEMENT

Understanding angular motion is particularly important for the student of human movement, because most volitional human movement in-

368

volves rotation of one or more body segments around the joints at which they articulate. Translation of the body as a whole during gait occurs by virtue of rotational motions taking place at the hip, knee, and ankle around imaginary transverse (frontal) axes of rotation. During the performance of jumping jacks, both the arms and the legs rotate around imaginary anteroposterior axes passing through the shoulder and hip joints. The angular motion of sport implements such as golf clubs, baseball bats, and hockey sticks, as well as household and garden tools, is also often of interest.

As discussed in Chapter 2, clinicians, coaches, and teachers of physical activities routinely analyze human movement based on visual observation. What is actually observed in such situations is the angular kinematics of human movement. Based on observation of the timing and range of motion of joint actions, the experienced analyst can make inferences about the coordination of muscle activity producing the joint actions and the forces resulting from those joint actions.

Much of the reported description of the developmental stages of motor skills is based on analysis of angular kinematics. For example, three developmental stages for kicking among children aged 2–6 years have been identified (3). In stage 1, the child kicks using a small range of motion of hip flexion, with no coordinated motion apparent at any other joint. In stage 2, knee extension is coordinated with hip flexion and the arms are abducted at the shoulders to promote balance. Stage 3 is characterized by increased hip flexion and knee extension, and elbow flexion is present in addition to shoulder abduction to improve balance. The knowledgeable analyst can obtain a great deal of information about the relative developmental and skill levels of the performer through careful observation of angular kinematics.

MEASURING ANGLES

As reviewed in Appendix A, an angle is composed of two sides that intersect at a vertex. Quantitative kinematic analysis can be achieved by projecting filmed images of the human body onto a piece of paper, with joint centers then marked with dots and the dots connected with lines representing the longitudinal axes of the body segments (Figure 11-1). A protractor can be used to make hand measurements of angles of interest from this representation, with the joint centers forming the vertices of the angles between adjacent body segments. (The procedure for measuring angles with a protractor is reviewed in Appendix A.) Videotapes and films of human movement can also be analyzed using this same basic procedure

FIGURE 11-1
For the human body, joint centers form the vertices of body segment angles.

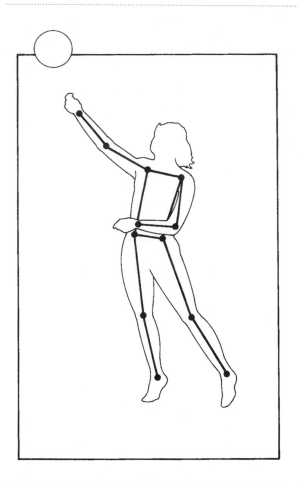

to evaluate the angles present at the joints of the human body and the angular orientations of the body segments. The angle assessments are usually done with computer software from stick-figure representations of the human body constructed in computer memory.

Relative versus Absolute Angles

Assessing the angle at a joint involves measuring the angle of one body segment relative to the other body segment articulating at the joint. The **relative angle** at the knee is the angle formed between the longitudinal axis of the thigh and the longitudinal axis of the lower leg (Figure 11-2). When joint range of motion is quantified, it is the relative joint angle that is measured.

The convention used for measuring relative joint angles is that in anatomical reference position, all joint angles are at 0°. As discussed in

relative angle
angle at a joint formed between the longitudinal axes of adjacent body segments

Relative angles should consistently be measured on the same side of a given joint.

The straight, fully extended position at a joint is regarded as 0 degrees.

FIGURE 11-2
Angles measured at joints
are the angles between
adjacent body segments.

Chapter 5, joint motion is then measured directionally. For example, when the extended arm is elevated 30° in front of the body in the sagittal plane, the arm is in 30° of flexion at the shoulder. When the leg is abducted at the hip, the range of motion in abduction is likewise measured from 0° in anatomical reference position.

The relative angle at the knee (measured between adjacent body segments) and the absolute angle of the trunk (measured with respect to the right horizontal).

FIGURE 11-3

Angles of orientation of
individual body segments
are measured with respect
to an absolute (fixed) line
of reference.

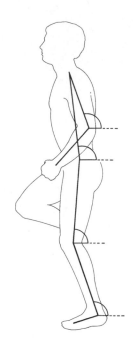

absolute angle
angular orientation of a
body segment with respect
to a fixed line of reference

*Absolute angles should
consistently be measured in
the same direction from a
single reference—either
horizontal or vertical.*

Other angles of interest are often the orientations of the body segments themselves. As discussed in Chapter 9, when the trunk is in flexion, the angle of inclination of the trunk directly affects the amount of force that must be generated by the trunk extensor muscles to support the trunk in the position assumed. The angle of inclination of a body segment, referred to as its **absolute angle,** is measured with respect to an absolute reference line, usually either horizontal or vertical. Figure 11-3 shows quantification of segment angles with respect to the right horizontal.

Tools for Measuring Body Angles

Goniometers are commonly used by clinicians for direct measurement of relative joint angles on a live human subject. A goniometer is essentially a protractor with two long arms attached. One arm is fixed so that it extends from the protractor at an angle of 0°. The other arm extends from the center of the protractor and is free to rotate. The center of the protractor is aligned over the joint center, and the two arms are aligned over the longitudinal axes of the two body segments that connect at the joint. The angle at the joint is then read at the intersection of the freely rotating arm and the protractor scale. The accuracy of the reading depends on the accuracy of the positioning of the goniometer. Knowledge of the underlying joint anatomy is essential for proper lo-

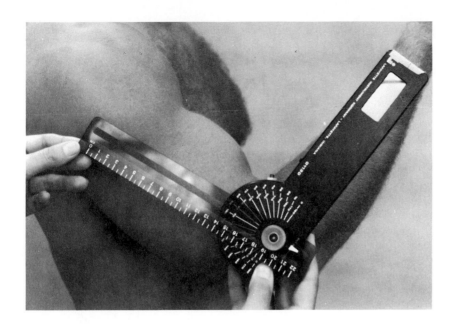

A goniometer is used to measure joint angles.

cation of the joint center of rotation. Placing marks on the skin to identify the location of the center of rotation at the joint and the longitudinal axes of the body segments before aligning the goniometer is sometimes helpful, particularly if repeated measurements are being taken at the same joint.

Other instruments available for quantifying angles relative to the human body are the electrogoniometer and various inclinometers. The electrogoniometer (referred to as an *elgon*) was developed by Karpovich in the late 1950s. An elgon is simply a goniometer with an electrical potentiometer at its vertex. When the arms of the elgon are attached with tape or velcro straps over a joint center, changes in the relative angle at the joint cause proportional changes in the electrical current emitted by the elgon. These changes are usually recorded on a moving strip chart, which marks the angle present at the joint with respect to time during a movement of interest. Inclinometers are other devices used for direct assessment of human body segment angles. These are usually gravitationally based instruments that identify the absolute angle of orientation of a body segment. Because accurate positioning of inclinometers is critical to the accuracy of the readings obtained, the measurement validity and reliability of these devices is controversial, particularly for measuring spinal orientation (2, 10, 13, 14).

Instant Center of Rotation

Quantification of joint angles is complicated by the fact that joint motion is often accompanied by displacement of one bone with respect to

FIGURE 11-4
The path of the instant center at the knee during knee extension.

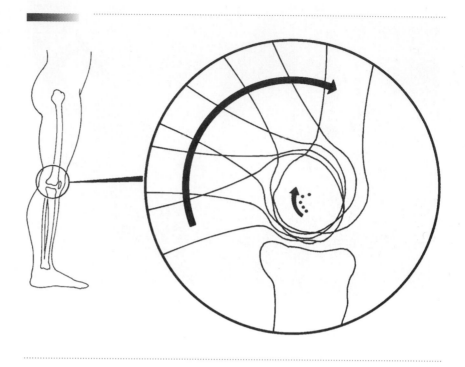

instant center
precisely located center of rotation at a joint at a given instant in time

the articulating bone at the joint. This phenomenon is caused by normal asymmetries in the shapes of the articulating bone surfaces. One example is the tibiofemoral joint, at which medial rotation and anterior displacement of the femur on the tibial plateau accompany flexion (Figure 11-4). As a result, the location of the exact center of rotation at the joint changes slightly when joint angle changes. The center of rotation at a given joint angle, or at a given instant in time during a dynamic movement, is called the **instant center.** The exact location of the instant center for a given joint may be determined through measurements taken from roentgenograms (X rays), which are usually taken at 10° intervals throughout the range of motion at the joint (11). An instrumented spatial linkage involving intracortical pin fixation has also been used to quantify relative linear and angular movements at the knee (7). The instant center at the knee shifts during angular movement due to accompanying linear displacements between the femur and tibia in all three planes (7).

ANGULAR KINEMATIC RELATIONSHIPS

The interrelationships among angular kinematic quantities are similar to those discussed in Chapter 10 for linear kinematic quantities. Although the units of measure associated with the angular kinematic quan-

tities are different from those used with their linear counterparts, the relationships among angular units also parallel those present among linear units.

Angular Distance and Displacement

Consider a pendulum swinging back and forth from a point of support. The pendulum is rotating around an axis passing through its point of support perpendicular to the plane of motion. If the pendulum swings through an arc of 60°, it has swung through an angular distance of 60°. If the pendulum then swings back through 60° to its original position, it has traveled an angular distance totaling 120° (60° + 60°). Angular distance is measured as the sum of all angular changes undergone by a rotating body.

The same procedure may be used for quantifying the angular distances through which the segments of the human body move. If the angle at the elbow joint changes from 90° to 160° during the flexion phase of a forearm curl exercise, the angular distance covered is 70°. If the extension phase of the curl returns the elbow to its original position of 90°, an additional 70° have been covered, resulting in a total angular distance of 140° for the complete curl. If 10 curls are performed, the angular distance transcribed at the elbow is 1400° (10 × 140°).

Just as with its linear counterpart, **angular displacement** is assessed as the difference in the initial and final positions of the moving body. If the angle at the knee of the support leg changes from 5° to 12° during

angular displacement
change in the angular position or orientation of a line segment

FIGURE 11-5
A, The path of motion of a swinging pendulum. **B,** The angular distance is the sum of all angular changes that have occurred; the angular displacement is the angle between the initial and final positions.

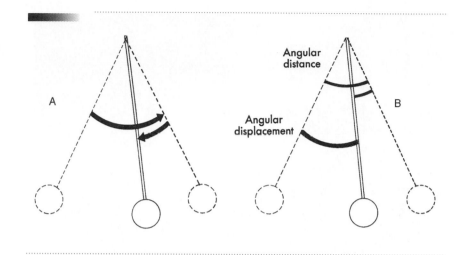

the initial support phase of a running stride, the angular distance and the angular displacement at the knee are 7°. If extension occurs at the knee, returning the joint to its original 5° position, angular distance totals 14° (7° + 7°), but angular displacement is 0 because the final position of the joint is the same as its original position. The relationship between angular distance and angular displacement is represented in Figure 11-5.

Like linear displacement, angular displacement is defined by both magnitude and direction. Since rotation observed from a side view occurs in either a clockwise or a counterclockwise direction, the direction of angular displacement may be indicated using these terms. The counterclockwise direction is conventionally designated as positive (+), and the clockwise direction as negative (−) (Figure 11-6). With the human body, it is also appropriate to indicate the direction of angular displacement with joint-related terminology such as flexion or abduction.

The counterclockwise direction is regarded as positive, and the clockwise direction is regarded as negative.

Three units of measure are commonly used to represent angular distance and angular displacement. The most familiar of these units is the degree. A complete circle of rotation transcribes an arc of 360°, an arc of 180° subtends a straight line, and 90° forms a right angle between perpendicular lines (Figure 11-7).

Another unit of angular measure sometimes used in biomechanical analyses is the **radian.** A line connecting the center of a circle to any point on the circumference of the circle is a radius. A radian is defined as the size of the angle subtended at the center of a circle by an arc equal in length to the radius of the circle (Figure 11-8). One radian is equivalent to 57.3°. Because a radian is much larger than a degree, it is

radian
unit of angular measure used in angular-linear kinematic quantity conversions; equal to 57.3°

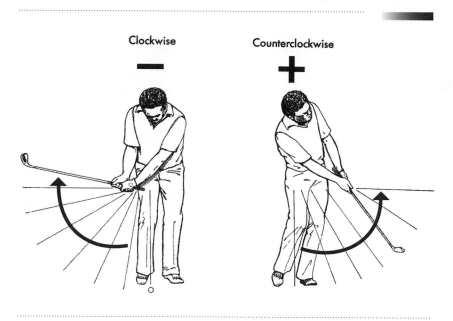

Clockwise   Counterclockwise

−   +

FIGURE 11-6
The direction of rotational
motion is commonly
identified as
counterclockwise (or
positive) versus clockwise
(or negative).

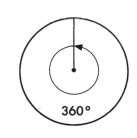

360°

180°

90°

FIGURE 11-7
Angles measured in
degrees.

FIGURE 11-8

A radian is defined as the size of the angle subtended at the center of a circle by an arc equal in length to the radius of the circle.

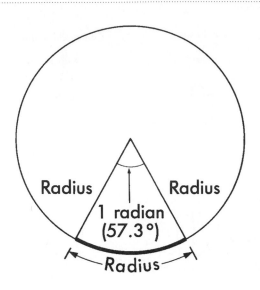

Pi (π) is a mathematical constant equal to approximately 3.14, which is the ratio of the circumference to the diameter of a circle.

a more convenient unit for the representation of extremely large angular distances or displacements. Radians are often quantified in multiples of pi (π). One complete circle is an arc of 2π radians.

The third unit sometimes used to quantify angular distance or displacement is the revolution. One revolution transcribes an arc equal to a circle. Dives and some gymnastic skills are often described by the number of revolutions the human body undergoes during their execution. The one and a half forward somersault dive is a descriptive example. Figure 11-9 illustrates the way in which degrees, radians, and revolutions compare as units of angular measure.

Angular Speed and Velocity

Angular speed is a scalar quantity and is defined as the angular distance covered divided by the time interval over which the motion occurred:

$$\text{angular speed} = \frac{\text{angular distance}}{\text{change in time}}$$

$$\sigma = \frac{\phi}{\Delta t}$$

The small Greek letter sigma (σ) represents angular speed, the small Greek letter phi (ϕ) represents angular distance, and t represents time.

Angular velocity is calculated as the change in angular position or the angular displacement that occurs during a given period of time:

angular velocity
rate of change in the angular position or orientation of a line segment

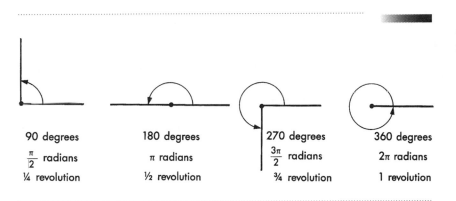

FIGURE 11-9
Comparison of degrees, radians, and revolutions.

$$\text{angular velocity} = \frac{\text{change in angular position}}{\text{change in time}}$$

$$\omega = \frac{\Delta \text{ angular position}}{\Delta \text{ time}}$$

$$\text{angular velocity} = \frac{\text{angular displacement}}{\text{change in time}}$$

$$\omega = \frac{\theta}{\Delta t}$$

The small Greek letter omega (ω) represents angular velocity, the capital Greek letter theta (θ) represents angular displacement, and t represents the time elapsed during the velocity assessment. Another way to express change in angular position is angular position$_2$ − angular position$_1$, in which angular position$_1$ represents the body's position at one point in time and angular position$_2$ represents the body's position at a later point:

$$\omega = \frac{\text{angular position}_2 - \text{angular position}_1}{\text{time}_2 - \text{time}_1}$$

Because angular velocity is based on angular displacement, it must include an identification of the direction (clockwise or counterclockwise, negative or positive) in which the angular displacement on which it is based occurred.

Units of angular speed and angular displacement are units of angular distance or angular displacement divided by units of time. The unit of time most commonly used is the second. Units of angular speed and angular velocity are degrees per second (deg/s), radians per second (rad/s), revolutions per second (rev/s), and revolutions per minute (rpm).

Moving the body segments at a high rate of angular velocity is a characteristic of skilled performance in many sports. Angular velocities at the joints of the throwing arm in major league baseball pitchers have been reported to reach 6180 deg/s (107.9 rad/s) of internal rotation at the shoulder and approximately 4595 deg/s (80.2 rad/s) of elbow extension (12). A comparison of the fast ball and curve ball pitches among members of the Australian National pitching squad showed average angular velocities at the wrist of 3.1 rad/s and 5.8 rad/s for the fast ball and curve ball pitches respectively (5). Angular velocity of the racket during serves executed by professional male tennis players has been found to range from 1900 to 2200 deg/s (33.2 to 38.4 rad/s) just before ball impact (4).

As discussed in Chapter 10, when the human body becomes a projectile during the execution of a jump, the height of the jump determines the amount of time the body is in the air. When figure skaters perform a triple or quadruple axel, as compared to a single axel or a double axel, this means that either jump height or rotational velocity of the body must be greater. Measurements of these two variables indicate that it is the skater's angular velocity that increases, with skilled skaters rotating their bodies in excess of 5 rev/s while airborne during the triple axel (9). Higher rotational velocities of the body with increasing skill difficulty have also been documented in gymnastics, with representative values of 6.80 rad/s for the handspring, 7.77 rad/s for the handspring incorporating a somersault and one-half twist, and 10.2 rad/s for the backward somersault layout with two twists (1).

Angular Acceleration

angular acceleration
rate of change in angular velocity

Angular acceleration is the rate of change in angular velocity, or the change in angular velocity occurring over a given time. The conventional symbol for angular acceleration is the small Greek letter alpha (α):

$$\text{angular acceleration} = \frac{\text{change in angular velocity}}{\text{change in time}}$$

$$\alpha = \frac{\Delta\omega}{\Delta t}$$

The calculation formula for angular acceleration is therefore the following:

$$\alpha = \frac{\omega_2 - \omega_1}{t_2 - t_1}$$

In this formula, ω_1 represents angular velocity at an initial point in time, ω_2 represents angular velocity at a second or final point in time, and t_1 and t_2 are the times at which velocity was assessed. Use of this formula is illustrated in the sample problem in Figure 11-10.

FIGURE 11-10

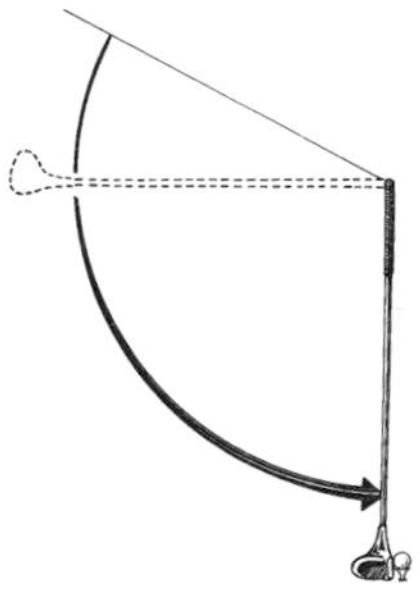

S A M P L E   P R O B L E M   1

A golf club is swung with an average angular acceleration of 1.5 rad/s². What is the angular velocity of the club when it strikes the ball at the end of a 0.8 second swing? (Provide an answer in both radian and degree-based units.)

Known

$$\alpha = 1.5 \text{ rad/s}^2$$

$$t = 0.8 \text{ s}$$

Solution

The formula to be used is the equation relating angular acceleration, angular velocity, and time:

$$\alpha = \frac{\omega_2 - \omega_1}{t}$$

Substituting in the known quantities yields the following:

$$1.5 \text{ rad/s}^2 = \frac{\omega_2 - \omega_1}{0.8 \text{ s}}$$

It may also be deduced that the angular velocity of the club at the beginning of the swing was 0:

$$1.5 \text{ rad/s}^2 = \frac{\omega_2 - 0}{0.8 \text{ s}}$$

$$(1.5 \text{ rad/s}^2)(0.8 \text{ s}) = \omega_2 - 0$$

$$\boxed{\omega_2 = 1.2 \text{ rad/s}}$$

In degree-based units:

$$\omega_2 = (1.2 \text{ rad/s})(57.3 \text{ deg/rad})$$

$$\boxed{\omega_2 = 68.8 \text{ deg/s}}$$

Just as with linear acceleration, angular acceleration may be positive, negative, or 0. When angular acceleration is 0, angular velocity is constant. Just as with linear acceleration, positive angular acceleration may indicate either increasing angular velocity in the positive direction or decreasing angular velocity in the negative direction. Similarly, a negative value of angular acceleration may represent either decreasing angular velocity in the positive direction or increasing angular velocity in the negative direction.

Human movement rarely involves constant velocity or constant acceleration.

Table 11-1

COMMON UNITS OF MEASURE			
	DISPLACEMENT	VELOCITY	ACCELERATION
Linear	meters	meters/second	meters/second^2
Angular	radians	radians/second	radians/second^2

Units of angular acceleration are units of angular velocity divided by units of time. Common examples are degrees per second squared (deg/s^2), radians per second squared (rad/s^2), and revolutions per second squared (rev/s^2). Units of angular and linear kinematic quantities are compared in Table 11-1.

Angular Motion Vectors

right hand rule
procedure for identifying the direction of an angular motion vector

Because representing angular quantities using symbols such as curved arrows would be impractical, angular quantities are represented with conventional straight vectors using what is called the **right hand rule.** According to this rule, when the fingers of the right hand are curled in the direction of an angular motion, the vector used to represent the motion is oriented perpendicular to the plane of rotation in the direction the extended thumb points in (Figure 11-11). The magnitude of the quantity may be indicated through proportionality to the vector's length.

Average versus Instantaneous Angular Quantities

Angular speed, velocity, and acceleration may be calculated as instantaneous or average values, depending on the length of the time interval selected. The instantaneous angular velocity of a baseball bat at the instant of contact with a ball is typically of greater interest than the average angular velocity of the swing, because the former directly affects the resultant velocity of the ball.

RELATIONSHIPS BETWEEN LINEAR AND ANGULAR MOTION

Linear and Angular Displacement

Timing is important in the execution of a groundstroke in tennis. If the ball is contacted too soon or too late, it may be hit out of bounds.

The greater the radius is between a given point on a rotating body and the axis of rotation, the greater is the linear distance undergone by that point during an angular motion (Figure 11-12). This observation is expressed in the form of a simple equation:

$$s = r\phi$$

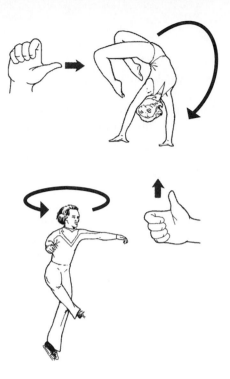

FIGURE 11-11
An angular motion vector
is oriented perpendicular
to the linear displacement
(d) of a point on a
rotating body.

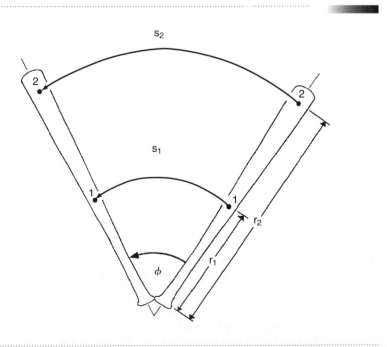

FIGURE 11-12
The larger the radius of
rotation (r), the greater
the linear distance (s)
traveled by a point on a
rotating body.

radius of rotation
distance from the axis of
rotation to a point of
interest on a rotating body

The curvilinear distance traveled by the point of interest (s) is the product of r, the point's **radius of rotation,** and ϕ, the angular distance through which the rotating body moves, which is quantified in radians.

For this relationship to be valid, two conditions must be met: the linear distance and the radius of rotation must be quantified in the same units of length, and angular distance must be expressed in radians. Although units of measure are normally balanced on opposite sides of an equal sign when a valid relationship is expressed, this is not the case here. When the radius of rotation (expressed in meters) is multiplied by angular displacement in radians, the result is linear displacement in meters. Radians disappear on the right side of the equation in this case because, as may be observed from the definition of the radian, the radian serves as a conversion factor between linear and angular measurements.

Linear and Angular Velocity

The same type of relationship exists between the angular velocity of a rotating body and the linear velocity of a point on that body at a given instant in time. The relationship is expressed as the following:

$$v = r\omega$$

The linear (tangential) velocity of the point of interest is v, r is the radius of rotation for that point, and ω is the angular velocity of the rotating body. For the equation to be valid, angular velocity must be expressed in radian-based units (typically rad/s) and velocity must be expressed in the units of the radius of rotation divided by the appropriate units of time. Radians are again used as a linear-angular conversion factor and are not balanced on opposite sides of the equals sign:

$$m/s = (m)(rad/s)$$

The use of radian-based units for conversions between linear and angular velocities is shown in Figure 11-13.

During several sport activities an immediate performance goal is to direct an object such as a ball, shuttlecock, or hockey puck accurately, while imparting a relatively large amount of velocity to it with a bat, club, racket, or stick. In baseball batting, the initiation of the bat swing and the angular velocity of the swing must be timed precisely to make contact with the ball and direct it into fair territory. A 40 m/s pitch reaches the batter 0.41 seconds after leaving the pitcher's hand. It has been estimated that a difference of 0.001 seconds in the time of initiation of the swing can determine whether the ball is directed to center field or down the foul line, and that a swing initiated 0.003 seconds too early or too late will result in no contact with the ball (6).

With all other factors held constant, the greater the radius of rotation at which a swinging implement hits a ball, the greater the linear ve-

Skilled performances of high-velocity movements are characterized by precisely coordinated timing of body segment rotations.

FIGURE 11-13

S A M P L E   P R O B L E M   2

Two baseballs are consecutively hit by a bat. The first ball is hit 20 cm from the bat's axis of rotation and the second ball is hit 40 cm from the bat's axis of rotation. If the angular velocity of the bat was 30 rad/s at the instant that both balls were contacted, what was the linear velocity of the bat at the two contact points?

Known

$$r_1 = 20 \text{ cm}$$
$$r_2 = 40 \text{ cm}$$
$$\omega_1 = \omega_2 = 30 \text{ rad/s}$$

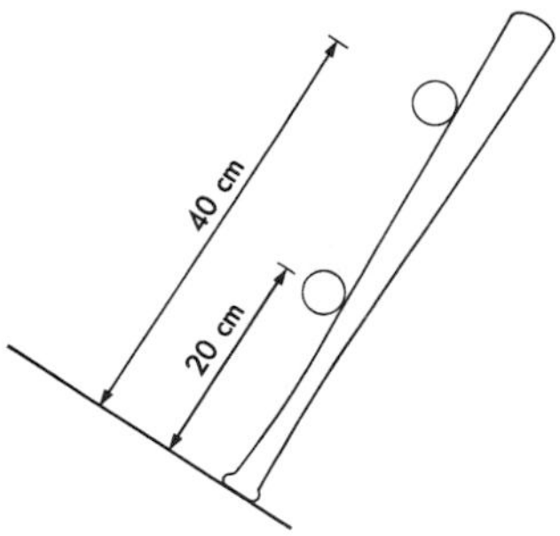

Solution

The formula to be used is the equation relating linear and angular velocities:

$$v = r\omega$$

For ball 1:

$$v_1 = (0.20 \text{ m})(30 \text{ rad/s})$$
$$v_1 = 6 \text{ m/s}$$

For ball 2:

$$v_2 = (0.40 \text{ m})(30 \text{ rad/s})$$
$$v_2 = 12 \text{ m/s}$$

locity imparted to the ball. In golf, longer clubs are selected for longer shots, and shorter clubs are selected for shorter shots. However, the magnitude of the angular velocity figures as heavily as the length of the radius of rotation in determining the linear velocity of a point on a swinging implement. Little Leaguers often select long bats, which increase the potential radius of rotation if a ball is contacted, but are also too heavy for the young players to swing as quickly as shorter, lighter bats. The relationship between the radius of rotation of the contact point between a striking implement and a ball and the subsequent velocity of the ball is shown in Figure 11-14.

It is important to recognize that the linear velocity of a ball struck by a bat, racket, or club is *not* identical to the linear velocity of the contact point on the swinging implement. Other factors, such as the directness of the hit and the elasticity of the impact, also influence ball velocity.

The greater the angular velocity of a baseball bat, the farther a struck ball will travel, other conditions being equal.

FIGURE 11-14

A simple experiment in which a rotating stick strikes three balls demonstrates the significance of the radius of rotation.

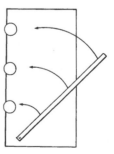

Top view

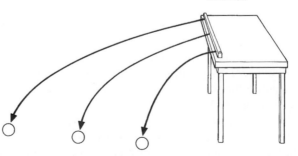

Side view

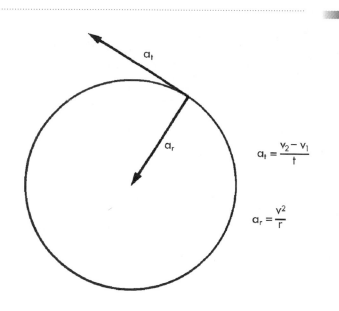

$$a_t = \frac{v_2 - v_1}{t}$$

$$a_r = \frac{v^2}{r}$$

FIGURE 11-15
Tangential and radial
acceleration vectors shown
relative to a circular path
of motion.

Linear and Angular Acceleration

The acceleration of a body in angular motion may be resolved into two perpendicular linear acceleration components. These components are directed along and perpendicular to the path of angular motion at any point in time (Figure 11-15).

The component directed along the path of angular motion takes its name from the term *tangent*. A tangent is a line that touches, but does not cross, a curve at a single point. The tangential component, known as **tangential acceleration,** represents the change in linear speed for a body traveling on a curved path. The formula for tangential acceleration is the following:

$$a_t = \frac{v_2 - v_1}{t}$$

Tangential acceleration is a_t, v_1 is the tangential linear velocity of the moving body at an initial time, v_2 is the tangential linear velocity of the moving body at a second time, and t is the time interval over which the velocities are assessed.

When a ball is thrown, the ball follows a curved path as it is accelerated by the muscles of the shoulder, elbow, and wrist. The tangential component of ball acceleration represents the rate of change in the linear speed of the ball. Because the speed of projection greatly affects a projectile's range, tangential velocity should be maximum just before ball release if the objective is to throw the ball fast or for distance. Once

tangential acceleration
component of acceleration of a body in angular motion directed along a tangent to the path of motion; represents change in linear speed

At the instant that a thrown ball is released, its tangential and radial accelerations become equal to 0, because a thrower is no longer applying force.

ball release occurs, tangential acceleration is 0 because the thrower is no longer applying a force.

The relationship between tangential acceleration and angular acceleration is expressed as follows:

$$a_t = r\alpha$$

Linear acceleration is a, r is the radius of rotation, and α is angular acceleration. The units of linear acceleration and the radius of rotation must be compatible, and angular acceleration must be expressed in radian-based units for the relationship to be accurate.

Although the linear speed of an object traveling along a curved path may not change, its direction of motion is constantly changing. The second component of angular acceleration represents the rate of change in direction of a body in angular motion. This component is called **radial acceleration,** and it is always directed toward the center of curvature. Radial acceleration may be quantified by using the following formula:

radial acceleration
component of acceleration of a body in angular motion directed toward the center of curvature; represents change in direction

$$a_r = \frac{v^2}{r}$$

Radial acceleration is a_r, v is the tangential linear velocity of the moving body, and r is the radius of rotation. An increase in linear velocity or a decrease in the radius of curvature increases radial acceleration. Thus, the smaller the radius of curvature (the tighter the curve), the more difficult it is for a cyclist to negotiate the curve at a high velocity (see Chapter 13).

During execution of a ball throw, the ball follows a curved path because the thrower's arm and hand restrain it. This restraining force causes radial acceleration toward the center of curvature throughout the motion. When the thrower releases the ball, radial acceleration no longer exists and the implement follows the path of the tangent to the curve at that instant. The timing of release is therefore critical: if release occurs too soon or too late, the ball will be directed to the left or the right rather than straight ahead. The sample problem in Figure 11-16 demonstrates the effects of the tangential and radial components of acceleration.

Both tangential and radial components of motion can contribute to the resultant linear velocity of a projectile at release. For example, during somersault dismounts from the high bar in gymnastics routines, although the primary contribution to linear velocity of the body's center of gravity is generally from tangential acceleration, the radial component can contribute up to 50% of the resultant velocity (8). The size of the contribution from the radial component, and whether the contribution is positive or negative, varies with the performer's technique.

FIGURE 11-16

SAMPLE PROBLEM 3

A windmill-style softball pitcher executes a pitch in 0.65 seconds. If her pitching arm is 0.7 m long, what are the magnitudes of the tangential and radial accelerations on the ball just before ball release when tangential ball speed is 20 m/s? What is the magnitude of the total acceleration on the ball at this point?

Known

$$t = 0.65 \text{ s}$$
$$r = 0.7 \text{ m}$$
$$v_2 = 20 \text{ m/s}$$

Solution

To solve for tangential acceleration, use the following formula:

$$a_t = \frac{v_2 - v_1}{t}$$

Substitute in what is known and assume that $v_1 = 0$:

$$a_t = \frac{20 \text{ m/s} - 0}{0.65 \text{ s}}$$

$$\boxed{a_t = 30.8 \text{ m/s}^2}$$

To solve for radial acceleration, use the following formula:

$$a_r = \frac{v^2}{r}$$

Substitute in what is known:

$$a_r = \frac{(20 \text{ m/s})^2}{0.7 \text{ m}}$$

$$\boxed{a_r = 571.4 \text{ m/s}^2}$$

To solve for total acceleration, perform vector composition of tangential and radial acceleration. Since tangential and radial acceleration are oriented perpendicular to each other, the Pythagorean theorem can be used to calculate the magnitude of total acceleration.

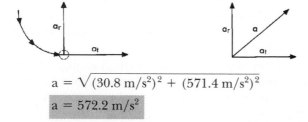

$$a = \sqrt{(30.8 \text{ m/s}^2)^2 + (571.4 \text{ m/s}^2)^2}$$

$$\boxed{a = 572.2 \text{ m/s}^2}$$

SUMMARY

An understanding of angular motion is an important part of the study of biomechanics because most volitional motion of the human body involves the rotation of bones around imaginary axes of rotation passing through the joint centers at which the bones articulate. The angular kinematic quantities—angular displacement, angular velocity, and angular acceleration—possess the same interrelationships as their linear counterparts, with angular displacement representing change in angular position, angular velocity defined as the rate of change in angular position, and angular acceleration indicating the rate of change in angular velocity during a given time. Depending on the selection of the time interval, either average or instantaneous values of angular velocity and angular acceleration may be quantified.

Angular kinematic variables may be quantified for the relative angle formed by the longitudinal axes of two body segments articulating at a joint or for the absolute angular orientation of a single body segment with respect to a fixed reference line. Different instruments are available for direct measurement of angles on a human subject.

INTRODUCTORY PROBLEMS

1. The relative angle at the knee changes from $0°$ to $85°$ during the knee flexion phase of a squat exercise. If 10 complete squats are performed, what is the total angular distance and the total angular displacement undergone at the knee? (Provide answers in both degrees and radians.) (Answer: $\phi = 1700$ deg, 29.7 rad; $\theta = 0$)

2. Identify the angular displacement, the angular velocity, and the angular acceleration of the second hand on a clock over the time interval in which it moves from the number 12 to the number 6. Provide answers in both degree and radian-based units. (Answer: $\theta = -180$ deg, $-\pi$ rad; $\omega = -6$ deg/s, $-\pi/30$ rad/s; $\alpha = 0$)

3. How many revolutions are completed by a top spinning with a constant angular velocity of $3\,\pi$ rad/s during a 20 second time interval? (Answer: 30 rev)

4. A kicker's extended leg is swung for 0.4 seconds in a counterclockwise direction while accelerating at 200 deg/s^2. What is the angular velocity of the leg at the instant of contact with the ball? (Answer: 80 deg/s, 1.4 rad/s)

5. The angular velocity of a runner's thigh changes from 3 rad/s to 2.7 rad/s during a 0.5 second time period. What has been the average angular acceleration of the thigh? (Answer: -0.6 rad/s^2, -34.4 deg/s^2)

6. Identify three movements during which the instantaneous angular velocity at a particular time is the quantity of interest. Explain your choices.

7. Fill in the missing corresponding values of angular measure in the table below.

DEGREES	RADIANS	REVOLUTIONS
90	?	?
?	1	?
180	?	?
?	?	1

8. Measure and record the following angles for the drawing shown below:

   a. The relative angle at the shoulder
   b. The relative angle at the elbow
   c. The absolute angle of the upper arm
   d. The absolute angle of the forearm
   Use the right horizontal as your reference for the absolute angles.

9. Calculate the following quantities for the diagram shown below:
   a. The angular velocity at the hip over each time interval
   b. The angular velocity at the knee over each time interval
   Would it provide meaningful information to calculate the average angular velocities at the hip and knee for the movement shown? Provide a rationale for your answer.

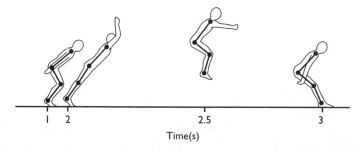

Time(s)

10. A tennis racket swung with an angular velocity of 12 rad/s strikes a motionless ball at a distance of 0.5 m from the axis of rotation. What is the linear velocity of the racket at the point of contact with the ball? (Answer: 6 m/s)

ADDITIONAL PROBLEMS

1. A 1.2 m golf club is swung in a planar motion by a right-handed golfer with an arm length of 0.76 m. If the initial velocity of the golf ball is 35 m/s, what was the angular velocity of the left shoulder at the point of ball contact? (Assume that the left arm and the club form a straight line and that the initial velocity of the ball is the same as the linear velocity of the club head at impact.) (Answer: 17.86 rad/s)

2. David is fighting Goliath. If David's 0.75 m sling is accelerated for 1.5 seconds at 20 rad/s^2, what will be the initial velocity of the projected stone? (Answer: 22.5 m/s)

3. A baseball is struck by a bat 46 cm from the axis of rotation when the angular velocity of the bat is 70 rad/s. If the ball is hit at a height of 1.2 m at a 45° angle, will the ball clear a 1.2 m fence 110 m away? (Assume that the initial linear velocity of the ball is the same as the linear velocity of the bat at the point at which it is struck.) (Answer: No, the ball will fall through a height of 1.2 m at a distance of 105.7 m.)

4. A polo player's arm and stick form a 2.5 m rigid segment. If the arm and stick are swung with an angular speed of 1.0 rad/s as the player's horse gallops at 5 m/s, what is the resultant velocity of a motionless ball that is struck head-on? (Assume that ball velocity is the same as the linear velocity of the end of the stick.) (Answer: 7.5 m/s)

5. Explain how the velocity of the ball in Problem 4 would differ if the stick were swung at a 30° angle to the direction of motion of the horse.

6. List three movements for which a relative angle at a particular joint is important and three movements for which the absolute angle of a body segment is important. Explain your choices.

7. A majorette in the Rose Bowl Parade tosses a baton into the air with an initial angular velocity of 2.5 rev/s. If the baton undergoes a constant acceleration while airborne of −0.2 rev/s^2 and its angular velocity is 0.8 rev/s when the majorette catches it, how many revolutions did it make in the air? (Answer: 14 rev)

8. A cyclist enters a curve of 30 m radius at a speed of 12 m/s. As the brakes are applied, speed is decreased at a constant rate of 0.5 m/s^2. What are the magnitudes of the cyclist's radial and tangential accelerations when his speed is 10 m/s? (Answer: a_r = 3.33 m/s^2; a_t = −0.5 m/s^2)

9. A hammer is being accelerated at 15 rad/s^2. Given a radius of rotation of 1.7 m, what are the magnitudes of the radial and tangential components of acceleration when tangential hammer speed is 25 m/s? (Answer: a_r = 367.6 m/s^2; a_t = 25.5 m/s^2)

10. An ice skater increases her speed from 10 m/s to 12.5 m/s over a period of 3 seconds while coming out of a curve of 20 m radius. What are the magnitudes of her radial, tangential, and total accel-

erations as she leaves the curve? (Remember that a_r and a_t are the vector components of total acceleration.) (Answer: $a_r = 7.81$ m/s^2; $a_t = 0.83$ m/s^2; $a = 7.85$ m/s^2)

1. Perform the experiment shown in Figure 11-14 on p. 386. Record the distances traveled by the three balls and write a brief explanation.
2. With a partner, use a goniometer to measure range of motion for wrist flexion and hyperextension and for hip flexion and hyperextension. Provide an explanation for differences in these ranges of motion between your partner and yourself.
3. Observe a young child executing a kick or a throw. Write a brief description of the angular kinematics of the major joint actions. What features distinguish the performance from that of a reasonably skilled adult?
4. Working in a small group, observe 2 volunteers performing simultaneous maximal vertical jumps from a side view. Either videotape (and replay) the jumps or have the subjects repeat the jumps several times. Write a comparative description of the angular kinematics of the jumps, including both relative and absolute angles of importance. Does your description suggest a reason for one jump being higher than the other?
5. Tape a piece of tracing paper over the monitor of a video cassette recorder and using the single frame advance button, draw at least 3 sequential stick-figure representations of a person performing a movement of interest. (If the movement is slow, you may need to skip a consistent number of frames in between tracings.) Use a protractor to measure the angle present at one major joint of interest on each figure. Given $\frac{1}{30}$ s between adjacent video pictures, calculate the angular velocity at the joint between pictures 1 and 2 and between pictures 2 and 3. Record your answers in both degree- and radian-based units.

REFERENCES

1. Bruggemann G-P: Biomechanics in gymnastics. In Van Gheluwe B and Atha J: *Current research in sports biomechanics,* Basel, 1987, Karger.
2. Chen SP, et al: Reliability of three lumbar sagittal motion measurement methods: Surface inclinometers, J Occup Environ Med 39:217, 1997.
3. Deach D: Genetic development of motor skills in children two through six years of age. Unpublished doctoral dissertation, University of Michigan, 1950.
4. Elliott BC: Tennis strokes and equipment. In Vaughan CL, ed: *Biomechanics of sport,* Boca Raton, Fla, 1989, CRC Press, Inc.
5. Elliott B, et al: A three-dimensional cinematographic analysis of the fastball and curveball pitches in baseball, Int J Sport Biomech 2:20, 1986.

6. Gutman D: The physics of foul play, Discover, p 70, Apr 1988.
7. Ishii Y, Terajima S, and Koga Y: Three-dimensional kinematics of the human knee with intracortical pin fixation, Clin Orthop 343:144, 1997.
8. Kerwin DG, Yeadon MR, and Harwood MJ: High bar release in triple somersault dismounts, J Appl Biomech 9:279, 1993.
9. King D, Arnold A, and Smith S: A biomechanical comparison of single, double, and triple axels. In Hamill J, Derrick TR, and Elliott EH, eds: *Biomechanics in sports XI*, Amherst, MA, 1993, The International Society of Biomechanics in Sports.
10. Mayer TG, Kondraske G, Beals SB, and Gatchel RJ: Spinal range of motion. Accuracy and sources of error with inclinometric measurement, Spine 22: 1976, 1997.
11. Nordin M and Frankel VH: Biomechanics of the knee. In Nordin M and Frankel VH, eds: *Biomechanics of the musculoskeletal system*, 2nd ed, Philadelphia, 1989, Lea & Febiger.
12. Pappas AM, Zawacki RM and Sullivan TJ: Biomechanics of baseball pitching, Am J Sports Med 13:216, 1985.
13. Samo DG, et al.: Validity of three lumbar sagittal motion measurement methods: Surface inclinometers compared with radiographs, J Occup Environ Med 39L: 209, 1997.
14. Sauer PM, Ensink FB, Frese K, Seeger D, Hildebrandt J: Lumbar range of motion: Reliability and validity of the inclinometer technique in the clinical measurement of trunk flexibility, Spine 21:1332, 1996.

ANNOTATED READINGS

Bruggemann G-P: Biomechanics in gymnastics. In Van Gheluwe B and Atha J: *Current research in sports biomechanics*, Basel, Karger, 1987.
   Presents research on kinematic and other aspects of human body rotation in gymnastics.
Dainty DA and Norman RW: *Standardizing biomechanical testing in sport*, Champaign Human Kinetics Publishers, 1987.
   Describes procedures for quantifying angular and linear kinematic quantities.
Pons DJ and Vaughan CL: Mechanics of cycling. In Vaughan CL, ed: *Biomechanics of sport*, Boca Raton, Fla, CRC Press, Inc., 1989.
   Includes a detailed description on kinematic and other aspects of cycling.
Zatsiorsky VN: *Kinematics of human motion*. Champaign, IL: Human Kinetics, 1988.
   Contains chapters on joint kinematics and the kinematics of individual joints.

RELATED WEB SITES

The Exploratorium's Science of Hockey
http://www.exploratorium.edu/hockey
   Explains scientific concepts related to hockey, including how to translate rotational motion of the arms into linear motion of the puck.
Physics Lecture Notes
http://sutherland.monroe.edu/pavone/HTMLnotes/lecturenotes.html
   Includes links to pages on a number of topics from mechanics, including rotational kinematics.
Rotational Kinematics
http://dps.phys.psu.edu/Act/Mech/Rotkin/rotkin1/rotkin1.htm
   Presents descriptions, diagrams, and formulae related to rotational kinematics.

12

LINEAR KINETICS OF HUMAN MOVEMENT

After completing this chapter, the reader will be able to:

Identify Newton's laws of motion and gravitation and describe practical illustrations of the laws.

Explain what factors affect friction and discuss the role of friction in daily activities and sports.

Define impulse and momentum and explain the relationship between them.

Explain what factors govern the outcome of a collision between two bodies.

Discuss the interrelationships among mechanical work, power, and energy.

Solve quantitative problems related to kinetic concepts.

What can people do to improve traction when walking on icy streets? Why do some balls bounce higher on one surface than on another? How can football linemen push larger opponents backward? In this chapter, the topic of kinetics will be introduced with a discussion of some important basic concepts and principles relating to linear kinetics.

NEWTON'S LAWS

Sir Isaac Newton (1642–1727) discovered many of the fundamental relationships that form the foundation for the field of modern mechanics. These principles highlight the interrelationships among the basic kinetic quantities introduced in Chapter 2.

Law of Inertia

Newton's first law of motion is known as the *law of inertia*. This law states the following:

> A body will maintain a state of rest or constant velocity unless acted on by an external force that changes the state.

In other words, a motionless object will remain motionless unless there is a net force (a force not counteracted by another force) acting on it. Similarly, a body traveling with a constant speed along a straight path will continue its motion unless acted on by a net force that alters either the speed or the direction of the motion.

It seems intuitively obvious that an object in a static (motionless) situation will remain motionless barring the action of some external force. We assume that a piece of furniture such as a chair will maintain a fixed position unless pushed or pulled by a person exerting a net force to cause its motion. When a body is traveling with a constant velocity, however, the enactment of the law of inertia is not so obvious because, in most situations, external forces do act to reduce velocity. For example, the law of inertia implies that a skater gliding on ice will continue gliding with the same speed and in the same direction, barring the action of an external force. But in reality, friction and air resistance are two forces normally present that act to slow skaters and other moving bodies.

A skater has a tendency to continue gliding with constant speed and direction because of inertia.

Law of Acceleration

Newton's second law of motion is an expression of the interrelationships among force, mass, and acceleration. This law, known as the *law of acceleration,* may be stated as follows for a body with constant mass:

> A force applied to a body causes an acceleration of that body of a magnitude proportional to the force, in the direction of the force, and inversely proportional to the body's mass.

When a ball is thrown, kicked, or struck with an implement, it tends to travel in the direction of the line of action of the applied force. Similarly, the greater the amount of force applied, the greater the speed the ball has. The algebraic expression of the law is a well-known formula that expresses the quantitative relationships among an applied force, a body's mass, and the resulting acceleration of the body:

$$F = ma$$

Thus, if a 1 kg ball is struck with a force of 10 N, the resulting acceleration of the ball is 10 m/s². If the ball has a mass of 2 kg, the application of the same 10 N force results in an acceleration of only 5 m/s².

Newton's second law also applies to a moving body. When a defensive football player running down the field is blocked by an opposing player, the velocity of the defensive player following contact is a function of the player's original direction and speed and the direction and magnitude of the force exerted by the offensive player.

Law of Reaction

The third of Newton's laws of motion states that every applied force is accompanied by a reaction force:

> For every action, there is an equal and opposite reaction.

In terms of forces, the law may be stated as follows:

> When one body exerts a force on a second, the second body exerts a reaction force that is equal in magnitude and opposite in direction on the first body.

The law of reaction holds true even when the bodies in contact are of significantly different masses.

When a person leans with a hand against a rigid wall, the wall pushes back on the hand with a force that is equal and opposite to that exerted by the hand on the wall. The harder the hand pushes against the wall, the greater is the amount of pressure felt across the surface of the hand where it contacts the wall. Another illustration of Newton's third law of motion is found in Figure 12-1.

During gait, every contact of a foot with the floor or ground generates an upward reaction force. Researchers and clinicians measure and study these ground reaction forces in analyzing differences in gait patterns across the lifespan and among individuals with handicapping conditions. Recent research in this area shows that young children achieve an adultlike gait pattern with a significant heel strike between two and three years of age (21).

In accordance with Newton's third law of motion, ground reaction forces are sustained with every footfall during running.

Researchers have studied the ground reaction forces that are sustained with every footfall during running to investigate factors related to both performance and running-related injuries. The magnitude of the vertical component of the ground reaction force (GRF) during running is generally two to three times the runner's body weight, with the pattern of force sustained during ground contact varying with running style. Runners are classified as rearfoot, midfoot, or forefoot strikers, according to the portion of the shoe first making contact with the ground. Typical vertical GRF patterns for rearfoot strikers and others are shown in Figure 12-2. Factors influencing GRF patterns include running speed, footwear, ground surface, and grade (4). The running shoes worn and the use of orthotics are other factors that can affect GRF patterns (23).

FIGURE 12-1

SAMPLE PROBLEM 1

A 90 kg ice hockey player collides head-on with an 80 kg ice hockey player. If the first player exerts a force of 450 N on the second player, how much force is exerted by the second player on the first?

Known

$m_1 = 90 \text{ kg}$
$m_2 = 80 \text{ kg}$
$F_1 = 450 \text{ N}$

Solution

This problem does not require computation. According to Newton's third law of motion, for every action there is an equal and opposite reaction. If the force exerted by the first player on the second has a magnitude of 450 N and a positive direction, then the force exerted by the second player on the first has a magnitude of 450 N and a negative direction.

-450 N

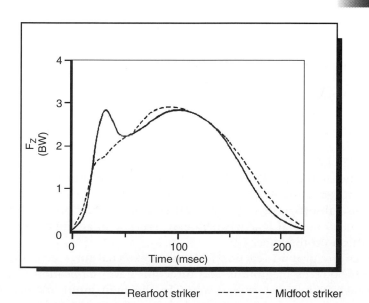

Rearfoot striker   - - - - - - - Midfoot striker

FIGURE 12-2
Typical ground reaction force patterns for rearfoot strikers and others. Runners may be classified as rearfoot, midfoot, or forefoot strikers according to the portion of the shoe that usually contacts the ground first.

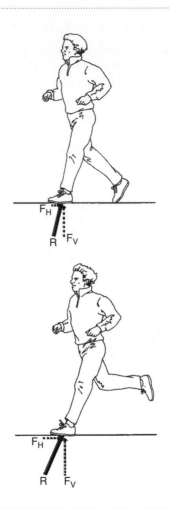

*Since the ground reaction force
is an external force acting on
the human body, its magnitude
and direction affect the body's
velocity.*

As discussed in Chapter 10, runners generally increase stride length
as running speed increases over the slow to moderate speed range.
Longer strides tend to generate GRFs with larger retarding horizontal
components (Figure 12-3). This is one reason that overstriding can be
counterproductive. With longer stride lengths, muscles crossing the
knee also absorb more of the shock that is transmitted upward through
the musculoskeletal system, which may translate to additional stress being placed on the knees (8).

Since the ground reaction force is an external force acting on the
human body, its magnitude and direction have implications for performance in many sporting events. In the high jump, for example,
skilled performers are moving with a large horizontal velocity and a
slight downwardly directed vertical velocity at the beginning of the

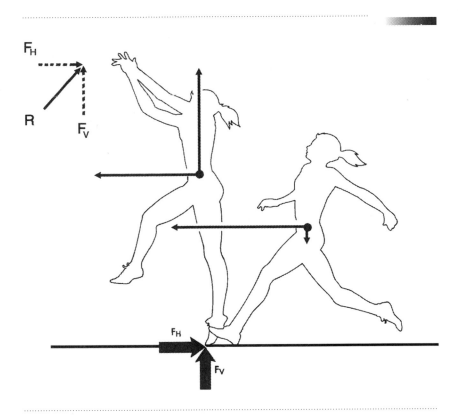

FIGURE 12-4
During the high jump takeoff, the horizontal component (F_H) of the ground reaction force (R) decreases the performer's horizontal velocity and the vertical component (F_v) can contribute to upward vertical velocity.

stride before takeoff. The GRF reduces the jumper's horizontal velocity and creates an upwardly directed vertical velocity (Figure 12-4). Better jumpers not only enter the takeoff phase of the jump with high horizontal velocities but also effectively use the GRF to convert horizontal velocity to upward vertical velocity (6). Skilled performance in baseball pitching is characterized by ground reaction shear forces of 35% of body weight in the direction of the pitch with the push-off leg, which emphasizes the importance of the lower extremity contribution to the throwing motion (17).

Aerobic dance exercise is a popular exercise mode that results in a large number of lower extremity overuse injuries. Shoes for aerobic dance exercise are specifically designed to cushion the metatarsal region of the foot, where much of the ground reaction force is sustained. Study of aerobic step exercise indicates a significantly smaller ground reaction force with a 6-inch, as compared to an 8- or 10-inch, step height, but little difference between the 8- and 10-inch step heights (18). This research evidence suggests that participants are well advised to use a lower step height to avoid injury.

Law of Gravitation

Newton's discovery of the law of universal gravitation was one of the most significant contributions to the Scientific Revolution and is considered by many to mark the beginning of modern science (5). According to legend, Newton's thoughts on gravitation were provoked either by his observation of a falling apple or by his actually being struck on the head by a falling apple. In his writings on the subject, Newton used the example of the falling apple to illustrate the principle that every body attracts every other body (3). Newton's law of gravitation states the following:

> All bodies are attracted to one another with a force proportional to the product of their masses and inversely proportional to the distance between them.

Stated algebraically, the law is the following:

$$F_g = G \frac{m_1 m_2}{d^2}$$

The force of gravitational attraction is F_g, G is a numerical constant, m_1 and m_2 are the masses of the bodies, and d is the distance between the mass centers of the bodies.

For the example of the falling apple, Newton's law of gravitation indicates that just as the earth attracts the apple, the apple attracts the earth, although to a much smaller extent. As the formula for gravitational force shows, the greater the mass of either body, the greater the attractive force between the two. Similarly, the greater the distance between the bodies, the smaller the attractive force between them.

For biomechanical applications, the only gravitational attraction of consequence is that generated by the earth because of its extremely large mass. The rate of gravitational acceleration at which bodies are attracted toward the surface of the earth (9.81 m/s^2) is based on the earth's mass and the distance to the center of the earth.

MECHANICAL BEHAVIOR OF BODIES IN CONTACT

According to Newton's third law of motion, for every action there is an equal and opposite reaction. However, consider the case of a horse hitched to a cart. According to Newton's third law, when the horse exerts a force on the cart to cause forward motion, the cart exerts a backward force of equal magnitude on the horse (Figure 12-5). Considering the horse and the cart as a single mechanical system, if the two forces are equal in magnitude and opposite in direction, their vector sum is 0. How does the horse and cart system achieve forward motion? The answer relates to the presence of another force that acts with a different magnitude on the cart than on the horse: the force of friction.

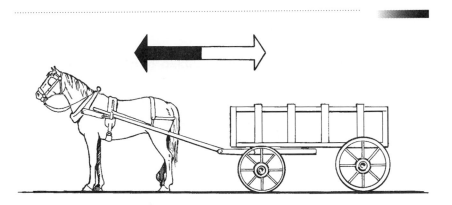

FIGURE 12-5
When a horse attempts to pull a cart forward, the cart exerts an equal and opposite force on the horse, in accordance with Newton's third law.

Friction

Friction is a force that acts at the interface of surfaces in contact in the direction opposite the direction of motion or impending motion. Because friction is a force, it is quantified in units of force (N). The magnitude of the generated friction force determines the relative ease or difficulty of motion for two objects in contact.

Consider the example of a box sitting on a level table top (Figure 12-6). The two forces acting on the undisturbed box are its own weight and a reaction force (R) applied by the table. In this situation the reaction force is equal in magnitude and opposite in direction to the box's weight.

When an extremely small horizontal force is applied to this box, it remains motionless. The box can maintain its static position because the applied force causes the generation of a friction force at the box/table interface that is equal in magnitude and opposite in direction to the small applied force. As the magnitude of the applied force becomes greater and greater, the magnitude of the opposing friction force also increases to a certain critical point. At that point the friction force present is termed **maximum static friction** (F_m). If the magnitude of the applied force is increased beyond this value, motion will occur (the box will slide).

Once the box is in motion, an opposing friction force continues to act. The friction force present during motion is referred to as **kinetic friction** (F_k). Unlike static friction, the magnitude of kinetic friction remains at a constant value that is *less than* the magnitude of maximum static friction. Regardless of the amount of the applied force or the speed of the occurring motion, the kinetic friction force remains the same. Figure 12-7 illustrates the relationship between friction and an applied external force.

friction
force acting over the area of contact between two surfaces in the direction opposite that of motion or motion tendency

maximum static friction
maximum amount of friction that can be generated between two static surfaces

kinetic friction
constant-magnitude friction generated between two surfaces in contact during motion

FIGURE 12-6

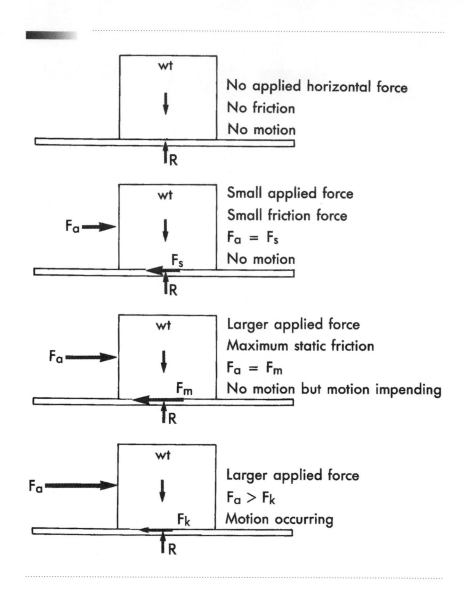

wt

No applied horizontal force
No friction
No motion

R

F_a

wt

Small applied force
Small friction force
$F_a = F_s$
No motion

F_s

R

F_a

wt

Larger applied force
Maximum static friction
$F_a = F_m$
No motion but motion impending

F_m

R

F_a

wt

Larger applied force
$F_a > F_k$
Motion occurring

F_k

R

coefficient of friction
number that serves as an index of the interaction between two surfaces in contact

normal reaction force
force acting perpendicular to two surfaces in contact

What factors determine the amount of applied force needed to move an object? More force is required to move a refrigerator than to move the empty box in which the refrigerator was delivered. More force is also needed to slide the refrigerator across a carpeted floor than to do so across a smooth linoleum floor. Two factors govern the magnitude of the force of maximum static friction or kinetic friction in any situation: the **coefficient of friction,** represented by the small Greek letter mu (μ), and the **normal** (perpendicular) **reaction force** (R):

$$F = \mu R$$

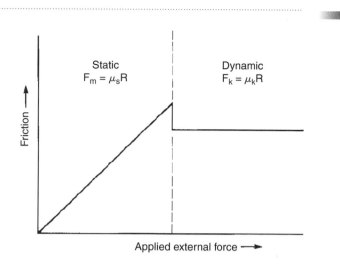

FIGURE 12-7
As long as a body is static, the magnitude of the friction force developed is equal to that of an applied external force. Once motion is initiated, the magnitude of the friction force remains at a constant level below that of maximum static friction.

The coefficient of friction is a unitless number indicating the relative ease of sliding, or the amount of mechanical and molecular interaction between two surfaces in contact. Factors influencing the value of μ are the relative roughness and hardness of the surfaces in contact and the type of molecular interaction between the surfaces. The greater the mechanical and molecular interaction, the greater the value of μ. For example, the coefficient of friction between two blocks covered with rough sandpaper is larger than the coefficient of friction between a skate and a smooth surface of ice. The coefficient of friction describes the interaction between two surfaces in contact and is not descriptive of either surface alone. The coefficient of friction for the blade of an ice skate in contact with ice is different from that for the blade of the same skate in contact with concrete or wood.

The coefficient of friction between two surfaces assumes one or two different values, depending on whether the bodies in contact are motionless (static) or in motion (kinetic). The two coefficients are known as the *coefficient of static friction* (μ_s) and the *coefficient of kinetic friction* (μ_k). The magnitude of maximum static friction is based on the coefficient of static friction:

$$F_m = \mu_s R$$

The magnitude of the kinetic friction force is based on the coefficient of kinetic friction:

$$F_k = \mu_k R$$

For any two bodies in contact, μ_k is always smaller than μ_s. Kinetic friction coefficients as low as 0.003 have been reported between the blade

Because μ_k is always smaller than μ_s, the magnitude of kinetic friction is always less than the magnitude of maximum static friction.

FIGURE 12-8

The coefficient of static friction between a sled and the snow is 0.18, with a coefficient of kinetic friction of 0.15. A 250 N boy sits on the 200 N sled. How much force directed parallel to the horizontal surface is required to start the sled in motion? How much force is required to keep the sled in motion?

Known

$\mu_s = 0.18$
$\mu_k = 0.15$
wt $= 250$ N $+ 200$ N

Solution

To start the sled in motion, the applied force must exceed the force of maximum static friction:

$$F_m = \mu_s R$$
$$F_m = (0.18)(250 \text{ N} + 200 \text{ N})$$
$$F_m = 81 \text{ N}$$

The applied force must be greater than 81 N.

To maintain motion, the applied force must equal the force of kinetic friction:

$$F_k = \mu_k R$$
$$F_k = (0.15)(250 \text{ N} + 200 \text{ N})$$
$$F_k = 67.5 \text{ N}$$

The applied force must be at least 67.5 N.

of a racing skate and a properly treated ice rink under optimal conditions (29). Use of the coefficients of static and kinetic friction is illustrated in Figure 12-8.

The other factor affecting the magnitude of the friction force generated is the normal reaction force. If weight is the only vertical force acting on a body sitting on a horizontal surface, R is equal in magnitude to the weight. If the object is a tackling sled with a 100 kg coach standing on it, R is equal to the weight of the sled plus the weight of the coach. Other vertically directed forces such as pushes or pulls can also affect the magnitude of R, which is always equal to the vector sum of all forces or force components acting normal to the surfaces in contact (Figure 12-9).

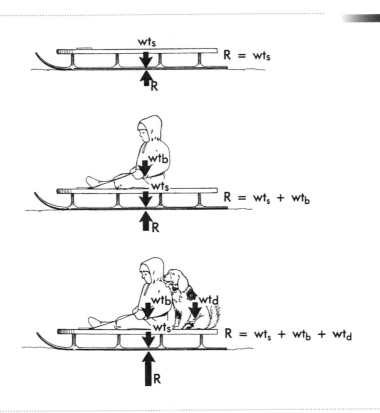

FIGURE 12-9
As weight increases, the normal reaction force increases.

The magnitude of R can be intentionally altered to increase or decrease the amount of friction present in a particular situation. When a football coach stands on the back of a tackling sled, the normal reaction force exerted by the ground on the sled is increased, with a concurrent increase in the amount of friction generated, making it more difficult for a tackler to move the sled. Alternatively, if the magnitude of R is decreased, friction is decreased and it is easier to initiate motion.

How can the normal reaction force be decreased? Suppose you need to rearrange the furniture in a room. Is it easier to push or pull an object such as a desk to move it? When a desk is pushed, the force exerted is typically directed diagonally downward. In contrast, force is usually directed diagonally upward when a desk is pulled. The vertical component of the push or pull either adds to or subtracts from the magnitude of the normal reaction force, thus influencing the magnitude of the friction force generated and the relative ease of moving the desk (Figure 12-10).

The amount of friction present between two surfaces can also be changed by altering the coefficient of friction between the surfaces. For example, the use of gloves in sports such as golf and racquetball

It is advantageous to pull with a line of force that is directed slightly upward when moving a heavy object.

Racquetball and golf gloves are designed to increase the friction between the hand and the racquet or club, as are the grips on the handles of the rackets and clubs themselves.

FIGURE 12-10
From a mechanical
perspective, it is easier to
pull than to push an
object such as a desk,
since pulling tends to
decrease the magnitude of
R and F, whereas pushing
tends to increase R and F.

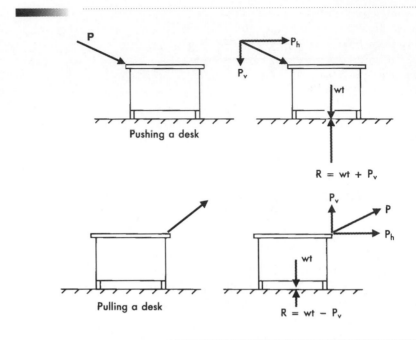

Pushing a desk

$R = wt + P_v$

Pulling a desk

$R = wt - P_v$

increases the coefficient of friction between the hand and the grip of the club or racquet. Similarly, lumps of wax applied to a surfboard increase the roughness of the board's surface, thereby increasing the coefficient of friction between the board and the surfer's feet. The application of a thin, smooth coat of wax to the bottom of cross-country skis is designed to decrease the coefficient of friction between the skis and the snow, with different waxes used for various snow conditions.

A widespread misconception about friction is that greater contact surface area generates more friction. Advertisements often imply that wide-track automobile tires provide better traction (friction) against the road than tires of normal width. However, the only factors known to affect friction are the coefficient of friction and the normal reaction force. Because wide-track tires typically weigh more than normal tires, they do increase friction to the extent that they increase R. However, the same effect can be achieved by carrying bricks or cinder blocks in the trunk of the car, a practice often followed by people who regularly drive on icy roads. Wide-track tires do tend to provide the advantages of increased lateral stability and increased wear, since larger surface area reduces the stress on a properly inflated tire.

Friction exerts an important influence during many daily activities. Walking depends on a proper coefficient of friction between a person's shoes and the supporting surface. If the coefficient of friction is too low, as when a person with smooth-soled shoes walks on a patch of ice, slippage will occur. The bottom of a wet bathtub or shower stall should provide a coefficient of friction with the soles of bare feet that is sufficiently large to prevent slippage.

The amount of friction present between ballet shoes and the dance studio floor must be controlled so that movements involving some amount of sliding or pivoting—such as *glissades, assembles,* and *pirouettes*—can be executed smoothly but without slippage. Rosin is often applied to dance floors because it provides a large coefficient of static friction but a significantly smaller coefficient of dynamic friction (15). This helps to prevent slippage in static situations and allows desired movements to occur freely.

The amount of friction present during sport situations has engendered heated controversies. The National Football League Players Association has attempted to have artificial turf declared a "hazardous substance" partly because the high coefficient of friction between artificial turf and a football shoe often does not allow rotation of a planted foot. Many knee injuries have been attributed to the immobility of a foot planted on artificial turf when a player is tackled. This possibility is heightened with increased turf temperature, since the coefficient of friction between football shoes and artificial turf increases with temperature (28). Football shoes that are not designed for use on artificial turf also tend to generate more friction on artificial turf than those shoes that are designed for such use (12). However, playing on artificial turf continues, because the Consumer Products Safety Commission has concluded that there is insufficient evidence supporting the NFL players' claim (25).

Another controversial disagreement occurred between Glenn Allison, a retired professional bowler and member of the American Bowling Congress Hall of Fame, and the American Bowling Congress. The dispute arose over the amount of friction present between Allison's ball and the lanes on which he bowled a perfect score of 300 in three consecutive games. According to the congress, his scores could not be recognized because the lanes he used did not conform to congress standards for the amount of conditioning oil present (14).

The magnitude of the rolling friction present between a rolling object, such as a bowling ball or an automobile tire, and a flat surface is approximately one-hundredth to one-thousandth of that present between sliding surfaces. Rolling friction occurs because both the curved and the flat surfaces are slightly deformed during contact. The coefficient of friction between the surfaces in contact, the normal reaction force, and the size of the radius of curvature of the rolling body all influence the magnitude of rolling friction. For bicycle tires, rolling friction is inversely proportional to the wheel diameter (30). It decreases with bicycle tire width and increases with reduced tire pressure (20).

The amount of friction present in a sliding or rolling situation is dramatically reduced when a layer of fluid, such as oil or water, intervenes

The coefficient of friction between a dancer's shoes and the floor must be small enough to allow freedom of motion but large enough to prevent slippage.

Rolling friction is influenced by the weight, radius, and deformability of the rolling object, as well as by the coefficient of friction between the two surfaces.

FIGURE 12-11

A horse can pull a cart if the horse's hooves generate more friction than the wheels of the cart.

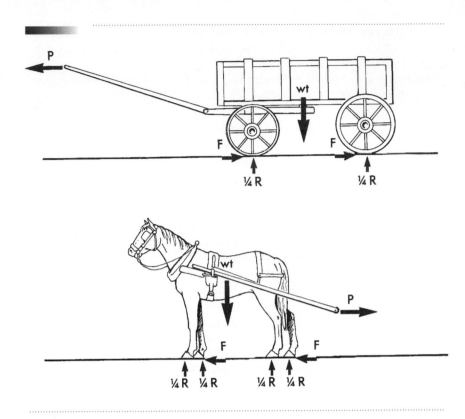

The synovial fluid present at many of the joints of the human body greatly reduces the friction between the articulating bones.

between two surfaces in contact. The presence of synovial fluid serves to reduce the friction, and subsequently the mechanical wear, on the diarthrodial joints of the human body. The coefficient of friction in a total hip prosthesis is approximately 0.01 (26). Researchers attribute the extremely low coefficients of friction between speed skates and the ice to a liquid-like film layer on the surface of the ice (7). The amount of friction between a bowling ball and a properly oiled lane is also extremely small, and according to the American Bowling Congress, an insufficient amount of oil on the lanes gave Allison the unfair advantage of added ball traction (14).

Revisiting the question presented earlier about the horse and cart, the force of friction is the determining factor for movement. The system moves forward if the magnitude of the friction force generated by the horse's hooves against the ground exceeds that produced by the wheels of the cart against the ground (Figure 12-11). Because most horses are shod to increase the amount of friction between their hooves and the ground, and most cart wheels are round and smooth to minimize the amount of friction they generate, the horse is usually at an advantage. However, if the horse stands on a slippery surface or if the cart rests in deep sand or is heavily loaded, motion may not be possible.

Momentum

Another factor that affects the outcome of interactions between two bodies is momentum, a mechanical quantity that is particularly important in situations involving collisions. Momentum may be defined generally as the quantity of motion that an object possesses. More specifically, **linear momentum** is the product of an object's mass and its velocity:

$$M = mv$$

A static object (with 0 velocity) has no momentum; that is, its momentum equals 0. A change in a body's momentum may be caused by either a change in the body's mass or a change in its velocity. In most human movement situations, changes in momentum result from changes in velocity. Units of momentum are units of mass multiplied by units of velocity, expressed in terms of kg · m/s. Because velocity is a vector quantity, momentum is also a vector quantity and is subject to the rules of vector composition and resolution.

When a head-on collision between two objects occurs, there is a tendency for both objects to continue moving in the direction of motion originally possessed by the object with the greatest momentum. If a 90 kg hockey player traveling at 6 m/s to the right collides head-on with an 80 kg player traveling at 7 m/s to the left, the momentum of the first player is the following:

$$M = mv$$
$$= (90 \text{ kg})(6 \text{ m/s})$$
$$= 540 \text{ kg} \cdot \text{m/s}$$

The momentum of the second player is expressed as follows:

$$M = mv$$
$$= (80 \text{ kg})(7 \text{ m/s})$$
$$= 560 \text{ kg} \cdot \text{m/s}$$

Since the second player's momentum is greater, both players would tend to continue moving in the direction of the second player's original velocity after the collision. Actual collisions are also affected by the extent to which the players become entangled, by whether one or both players remain on their feet, and by the elasticity of the collision.

Neglecting these other factors that may influence the outcome of the collision, it is possible to calculate the magnitude of the combined velocity of the two hockey players after the collision using a modified statement of Newton's first law of motion. Newton's first law may be restated as the *principle of conservation of momentum:*

> In the absence of external forces, the total momentum of a given system remains constant.

linear momentum quantity of motion, measured as the product of a body's mass and its velocity

Momentum is a vector quantity.

In the absence of external forces, momentum is conserved. However, friction and air resistance are forces that normally act to reduce momentum.

The principle is expressed in equation format as the following:

$$M_1 = M_2$$
$$(mv)_1 = (mv)_2$$

Subscript 1 designates an initial point in time and subscript 2 represents a later time.

Applying this principle to the hypothetical example of the colliding hockey players, the vector sum of the two players' momenta before the collision is equal to their single, combined momentum following the collision (Figure 12-12). In reality, friction and air resistance are external forces that typically act to reduce the total amount of momentum present.

Impulse

When external forces do act, they change the momentum present in a system predictably. Changes in momentum depend not only on the magnitude of the acting external forces but also on the length of time over which each force acts. The product of force and time is known as **impulse:**

impulse
product of a force and the time interval over which the force acts

$$Impulse = Ft$$

When an impulse acts on a system, the result is a change in the system's total momentum. The relationship between impulse and momentum is derived from Newton's second law:

$$F = ma$$
$$F = m\frac{(v_2 - v_1)}{t}$$
$$Ft = (mv)_2 - (mv)_1$$
$$Ft = \Delta M$$

Subscript 1 designates an initial time and subscript 2 represents a later time. An application of this relationship is presented in Figure 12-13.

Significant changes in an object's momentum may result from a small force acting over a large time interval or from a large force acting over a small time interval. A golf ball rolling across a green gradually loses momentum because its motion is constantly opposed by the force of rolling friction. The momentum of a baseball struck vigorously by a bat also changes because of the large force exerted by the bat during the fraction of a second it is in contact with the ball.

The amount of impulse generated by the human body is often intentionally manipulated. When a vertical jump is performed on a force platform, a graphical display of the vertical ground reaction force across

FIGURE 12-12

SAMPLE PROBLEM 3

A 90 kg hockey player traveling with a velocity of 6 m/s collides head-on with an 80 kg player traveling at 7 m/s. If the two players entangle and continue traveling together as a unit following the collision, what is their combined velocity?

Known

$m_1 = 90$ kg
$v_1 = 6$ m/s
$m_2 = 80$ kg
$v_2 = -7$ m/s

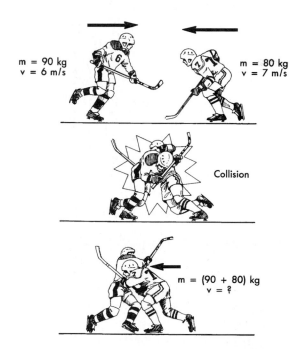

m = 90 kg
v = 6 m/s

m = 80 kg
v = 7 m/s

Collision

m = (90 + 80) kg
v = ?

Solution

The law of conservation of momentum may be used to solve the problem, with the two players considered as the total system.

$$\text{Before collision} \quad \text{After collision}$$
$$m_1v_1 + m_2v_2 = (m_1 + m_2)(v)$$
$$(90 \text{ kg})(6 \text{ m/s}) + (80 \text{ kg})(-7 \text{ m/s}) = (90 \text{ kg} + 80 \text{ kg})(v)$$
$$540 \text{ kg} \cdot \text{m/s} - 560 \text{ kg} \cdot \text{m/s} = (170 \text{ kg})(v)$$
$$-20 \text{ kg} \cdot \text{m/s} = (170 \text{ kg})(v)$$

$v = 0.12$ m/s in the 80 kg player's original direction of travel.

FIGURE 12-13

A toboggan race begins with the two crew members pushing the toboggan to get it moving as quickly as possible before they climb in. If crew members apply an average force of 100 N in the direction of motion of the 90 kg toboggan for a period of 7 seconds before jumping in, what is the toboggan's speed (neglecting friction) at that point?

Known

F = 100 N
t = 7 s
m = 90 kg

Solution

The crew members are applying an impulse to the toboggan to change the toboggan's momentum from 0 to a maximum amount. The impulse-momentum relationship may be used to solve the problem.

$$Ft = (mv)_2 - (mv)_1$$
$$(100\ N)(7\ s) = (90\ kg)(v) - (90\ kg)(0)$$

$v = 7.78$ m/s in the direction of force application.

---

time can be generated (Figure 12-14). Since impulse is the product of force and time, the impulse is the area under the force-time curve. The larger the impulse generated against the floor, the greater the change in the performer's momentum and the higher the resulting jump. Theoretically, impulse can be increased by increasing either the magnitude of applied force or the time interval over which the force acts. Practically, however, when time of force application against the ground is prolonged during vertical jump execution, the magnitude of the force that can be generated is dramatically reduced, with the ultimate result being a smaller impulse. For performing a maximal vertical jump, the performer must maximize impulse by optimizing the trade-off between applied force magnitude and force duration. The motion of the arms during a counter-movement jump contributes 12%–13% of the total upward momentum, indicating the importance of arm motion when jumping for maximum height (16).

Impulse can also be intentionally manipulated during a landing from a jump (Figure 12-15). A performer who lands rigidly will experience a relatively large ground reaction force sustained over a relatively short time interval. Alternatively, allowing the hip, knee, and ankle joints to

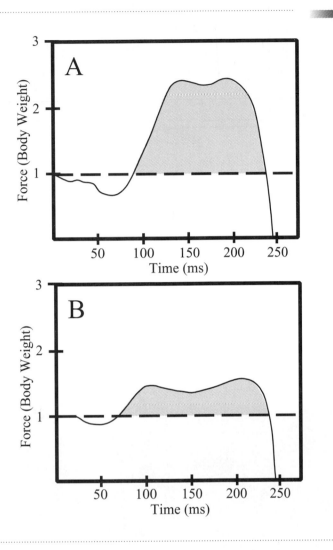

FIGURE 12-14
Force-time histories for **A**, high, and **B**, low vertical jumps by the same performer. The shaded area represents the impulse generated against the floor during the jump.

undergo flexion during the landing increases the time interval over which the landing force is absorbed, thereby reducing the magnitude of the force sustained.

It is also useful to manipulate impulse when catching a hard-thrown ball. "Giving" with the ball after it initially contacts the hands or the glove before bringing the ball to a complete stop will prevent the force of the ball from causing the hands to sting. The greater the time period is between initial hand contact with the ball and bringing the ball to a complete stop, the less is the magnitude of the force exerted by the ball against the hand and the smaller is the likelihood of experiencing a sting.

FIGURE 12-15

Representations of ground
reaction forces during
vertical jump
performances. **A**, A rigid
landing. **B**, A landing with
hip, knee, and ankle
flexion occurring. Note
the differences in the
magnitudes and times of
the landing impulses.

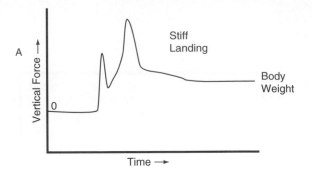

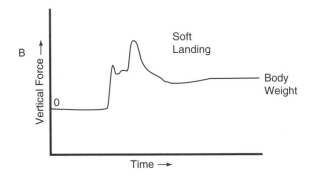

*"Giving" with the ball
during a catch serves to
lessen the magnitude of the
impact force sustained by
the catcher.*

Impact

The type of collision that occurs between a struck baseball and a bat is known as an **impact.** An impact involves the collision of two bodies over an extremely small time interval during which the two bodies exert relatively large forces on each other. The behavior of two objects following an impact depends not only on their collective momentum but also on the nature of the impact.

For the hypothetical case of a **perfectly elastic impact,** the relative velocities of the two bodies after impact are the same as their relative velocities before impact. The impact of a superball with a hard surface approaches perfect elasticity, because the ball's speed diminishes little during its collision with the surface. At the other end of the range is the **perfectly plastic impact,** during which at least one of the bodies in contact deforms and does not regain its original shape, and the bodies do not separate. This occurs when modeling clay is dropped on a surface.

Most impacts are neither perfectly elastic nor perfectly plastic, but are somewhere between the two. The **coefficient of restitution** describes the relative elasticity of an impact. It is a unitless number between 0 and 1. The closer the coefficient of restitution is to 1, the more elastic is the impact; and the closer the coefficient is to 0, the more plastic is the impact.

The coefficient of restitution governs the relationship between the relative velocities of two bodies before and after an impact. This relationship, which was originally formulated by Newton, may be stated as follows:

> When two bodies undergo a direct collision, the difference in their velocities immediately after impact is proportional to the difference in their velocities immediately before impact.

This relationship can also be expressed algebraically as the following:

$$-e = \frac{\text{relative velocity after impact}}{\text{relative velocity before impact}}$$

$$-e = \frac{v_1 - v_2}{u_1 - u_2}$$

In this formula, e is the coefficient of restitution, u_1 and u_2 are the velocities of the bodies just before impact, and v_1 and v_2 are the velocities of the bodies immediately after impact (Figure 12-16).

In tennis, the nature of the game depends on the type of impacts between ball and racket and between ball and court. All other conditions being equal, a tighter grip on the racket increases the apparent coefficient of restitution between ball and racket (11). Other factors of influence are racket size, shape, balance, flexibility, string type and

impact
collision characterized by the exchange of a large force during a small time interval

perfectly elastic impact
impact during which the velocity of the system is conserved

perfectly plastic impact
impact resulting in the total loss of system velocity

coefficient of restitution
number that serves as an index of elasticity for colliding bodies

FIGURE 12-16

The differences in two ball's velocities before impact is proportional to the difference in their velocities after impact. The factor of proportionality is the coefficient of restitution.

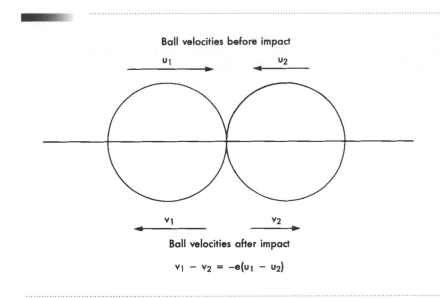

Ball velocities before impact

u_1 u_2

v_1 v_2

Ball velocities after impact

$$v_1 - v_2 = -e(u_1 - u_2)$$

Ball/racket and ball/court impacts determine the nature of the game in tennis.

tension, and swing kinematics (9, 10). The condition of the ball is also significant. Tests have shown that rebound from a surface is higher after 800 impacts than when a ball is new, because the loss of nap increases the coefficient of restitution between ball and surface and decreases the ball's aerodynamic drag. However, the major factor affecting ball rebound is the amount of time the ball has been outside a pressurized can. A loss of rebound height for both used and unused balls occurs after 5 days out of a can (24). The surface of the court also influences ball rebound during play, with differences in the coefficients of restitution and friction between ball and surface making some courts "fast" and others "slow."

In the case of an impact between a moving body and a stationary one, Newton's law of impact can be simplified because the velocity of the stationary body remains 0. The coefficient of restitution between a ball and a flat, stationary surface onto which the ball is dropped may be approximated using the following formula:

$$e = \sqrt{\frac{h_b}{h_d}}$$

In this equation, e is the coefficient of restitution, h_d is the height from which the ball is dropped, and h_b is the height to which the ball bounces (Figure 12-17). The coefficient of restitution describes the interaction between two bodies during an impact; it is *not* descriptive of any single object or surface. Dropping a basketball, a golf ball, a racquetball, and a baseball onto several different surfaces demonstrates that some balls bounce higher on certain types of surfaces (Figure 12-18).

FIGURE 12-17

S A M P L E   P R O B L E M   5

A basketball is dropped from a height of 2 m onto a gymnasium floor. If the coefficient of restitution between ball and floor is 0.9, how high will the ball bounce?

Known:

$h_d = 2$ m
$e = 0.9$

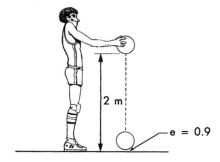

Solution

$$e = \sqrt{\frac{h_b}{h_d}}$$

$$0.9 = \sqrt{\frac{h_b}{2 \text{ m}}}$$

$$0.81 = \frac{h_b}{2 \text{ m}}$$

$$h_b = 1.6 \text{ m}$$

FIGURE 12-18

Bounce heights of a basketball, golf ball, racquetball, and baseball all dropped onto the same surface from a height of 1 m.

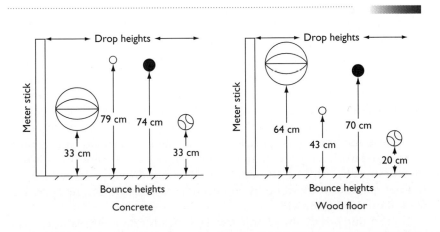

Increases in impact velocity and temperature increase the coefficient of restitution.

The coefficient of restitution is increased by increases in both impact velocity and temperature. In sports such as baseball and tennis, increases in both incoming ball velocity and bat or racquet velocity increase the coefficient of restitution between bat or racquet and ball, and contribute to a livelier ball rebound from the striking instrument. In racquetball and squash, where the ball is constantly being deformed against the wall, the ball's thermal energy (temperature) is increased over the course of play. As ball temperature increases, its rebound from both racquet and wall becomes more lively.

WORK, POWER, AND ENERGY RELATIONSHIPS

Work

The word *work* is commonly used in a variety of contexts. A person can speak of "working out" in the weight room, doing "yard work," or "working hard" to prepare for an exam. However, from a mechanical standpoint, **work** is defined as force applied against a resistance, multiplied by the displacement of the resistance in the direction of the force:

$$W = Fd$$

work
calculated as force multiplied by the displacement of the resistance in the direction of the force

When a body is moved a given distance as the result of the action of an applied external force, the body has had work performed on it, with the quantity of work equal to the product of the magnitude of the applied force and the distance through which the body was moved. When a force is applied to a body but no net force results because of opposing forces such as friction or the body's own weight, no mechanical work has been done, since there has been no movement of the body.

When the muscles of the human body produce tension resulting in the motion of a body segment, the muscles perform work on the body segment and the mechanical work performed may be characterized as either positive or negative work, according to the type of muscle action that predominates. When both the net muscle torque and the direction of angular motion at a joint are in the same direction, the work done by the muscles is said to be *positive*. Alternatively, when the net muscle torque and the direction of angular motion at a joint are in opposite directions, the work done by the muscles is considered to be *negative*. Although many movements of the human body involve co-contraction of agonist and antagonist muscle groups, when concentric contraction prevails the work is positive, and when eccentric contraction prevails the work is negative. During an activity such as running on a level surface, the net negative work done by the muscles is equal to the net positive work done by the muscles.

Performing positive mechanical work typically requires greater caloric expenditure than performing the same amount of negative mechanical work. However, no simple relationship between the caloric energy re-

Mechanical work should not be confused with caloric expenditure.

quired for performing equal amounts of positive and negative mechanical work has been discovered. When individuals are monitored
during performances of positive and negative mechanical work, energy
expenditures vary considerably both across and within individual performances (1). Energy expenditure with the use of elbow crutches is
2–3 times higher than in normal walking, although the mechanical work
is only 1.3–1.5 times greater (27). This indicates that use of elbow
crutches is also accompanied by a decrease in the efficiency of mechanical work production.

Units of work are units of force multiplied by units of distance. In
the metric system, the common unit of force (N) multiplied by a common unit of distance (m) is termed the *joule* (J).

$$1\,J = 1\,Nm$$

Power

Another term used in different contexts is **power.** In mechanics, power
refers to the amount of mechanical work performed in a given time:

$$Power = \frac{Work}{Change\ in\ time}$$

$$P = \frac{W}{\Delta t}$$

power
rate of work production
that is calculated as work
divided by the time during
which the work was done

Using the relationships previously described, power can also be defined
as the following:

$$Power = \frac{force \times distance}{change\ in\ time}$$

$$P = \frac{Fd}{\Delta t}$$

Because velocity equals the directed distance divided by the change in
time, the equation can also be expressed as the following:

$$P = Fv$$

Units of power are units of work divided by units of time. In the metric system, joules divided by seconds are termed *watts* (W).

$$1\,W = 1\,J/s$$

In sports such as throwing, jumping, and sprinting and in Olympic
weight lifting, the athlete's ability to exert mechanical power or the
combination of force and velocity is critical to successful performance.
A sample problem involving mechanical work and power is shown in
Figure 12-19.

*The ability to produce
mechanical power is critically
important for athletes
competing in explosive track
and field events.*

FIGURE 12-19

A 580 N person runs up a flight of 30 stairs of riser (height) of 25 cm during a 15-second period. How much mechanical work is done? How much mechanical power is generated?

Known

wt (F) = 580 N
h = 30 × 25 cm
t = 15 s

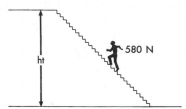

Solution

For mechanical work:

$$W = Fd$$
$$W = (580 \text{ N})(30 \times 0.25 \text{ m})$$
$$\boxed{W = 4350 \text{ J}}$$

For mechanical power:

$$P = \frac{W}{t}$$
$$P = \frac{4350 \text{ J}}{15 \text{ s}}$$
$$\boxed{P = 290 \text{ watts}}$$

Energy

Energy is defined generally as the capacity to do work. Mechanical energy is therefore the capacity to do mechanical work. Units of mechanical energy are the same as units of mechanical work (joules, in the metric system). There are two forms of mechanical energy: **kinetic energy** and **potential energy.**

Kinetic energy (KE) is the energy of motion. A body possesses kinetic energy only when in motion. Formally, the kinetic energy of linear motion is defined as one-half of a body's mass multiplied by the square of its velocity:

$$KE = \tfrac{1}{2} mv^2$$

If a body is motionless (v equals 0), its kinetic energy is also 0. Because velocity is squared in the expression for kinetic energy, increases in a

kinetic energy
energy of motion, calculated as $\tfrac{1}{2} mv^2$

potential energy
energy by virtue of a body's position or configuration, calculated as the product of weight and height

body's velocity create dramatic increases in its kinetic energy. For example, a 2 kg ball rolling with a velocity of 1 m/s has a kinetic energy of 1 J:

$$KE = \tfrac{1}{2}mv^2$$
$$- (0.5)(2\text{ kg})(1\text{ m/s})^2$$
$$= (1\text{ kg})(1\text{ m}^2/\text{s}^2)$$
$$= 1\text{ J}$$

If the velocity of the ball is increased to 3 m/s, kinetic energy is significantly increased:

$$KE = \tfrac{1}{2}mv^2$$
$$= (0.5)(2\text{ kg})(3\text{ m/s})^2$$
$$= (1\text{ kg})(9\text{ m}^2/\text{s}^2)$$
$$= 9\text{ J}$$

The other major category of mechanical energy is potential energy (PE), which is the energy of position. More specifically, potential energy is a body's weight multiplied by its height above a reference surface:

$$PE = wt \cdot h$$
$$PE = ma_g h$$

In the second formula, m represents mass, a_g is the acceleration of gravity, and h is the body's height. The reference surface is usually the floor or the ground, but in special circumstances it may be defined as another surface.

Because in biomechanical applications the weight of a body is typically fixed, changes in potential energy are usually based on changes in the body's height. For example, when a 50 kg bar is elevated to a height of 1 m, its potential energy at that point is 490.5 J:

$$PE = ma_g h$$
$$= (50\text{ kg})(9.81\text{ m/s}^2)(1\text{ m})$$
$$= 490.5\text{ J}$$

Potential energy may also be thought of as stored energy. The term *potential* implies potential for conversion to kinetic energy. A special form of potential energy is called **strain energy** (SE) or elastic energy. Strain energy may be defined as follows:

$$SE = \tfrac{1}{2}kx^2$$

In this formula, k is a spring constant, representing a material's relative stiffness or ability to store energy on deformation, and x is the distance over which the material is deformed. When an object is stretched, bent, or otherwise deformed, it stores this particular form

strain energy
capacity to do work by virtue of a deformed body's return to its original shape

of potential energy for later use. For example, when the muscles of the human body are stretched, they store strain energy that is released to increase the force of subsequent contraction, as discussed in Chapter 6. During an activity such as a maximal-effort throw, stored energy in stretched muscles can contribute significantly to the force and power generated and to the resulting velocity of the throw (19). Likewise, when the end of a diving board or a trampoline surface is depressed, strain energy is created. Subsequent conversion of the stored energy to kinetic energy enables the surface to return to its original shape and position. The poles used by vaulters store strain energy as they bend, and then release kinetic energy as they straighten during the performance of the vault. In 1963, the increase of approximately 23 cm in the world record for the pole vault was attributed largely to the advent of vaulting poles made of fiberglass, a material capable of storing more strain energy than the bamboo, steel, or aluminum of which earlier poles were constructed (13).

Conservation of Mechanical Energy

Consider the changes that occur in the mechanical energy of a ball tossed vertically into the air (Figure 12-20). As the ball gains height, it also gains potential energy (ma_gh). However, since the ball is losing velocity with increasing height because of gravitational acceleration, it is also losing kinetic energy ($\frac{1}{2}mv^2$). At the apex of the ball's trajectory (the instant between rising and falling), its height and potential energy are at a maximum value and its velocity and kinetic energy are 0. As the ball starts to fall, it progressively gains kinetic energy while losing potential energy.

The correlation between the kinetic and potential energies of the vertically tossed ball illustrates a concept that applies to all bodies when the only external force acting is gravity. The concept is known as the *law of conservation of mechanical energy*, which may be stated as follows:

> When gravity is the only acting external force, a body's mechanical energy remains constant.

Since the mechanical energy a body possesses is the sum of its potential and kinetic energies, the relationship may also be expressed as the following:

$$(PE + KE) = C$$

In this formula, C is a constant; that is, it is a number that remains constant throughout the period of time during which gravity is the only external force acting. Figure 12-21 quantitatively illustrates this principle.

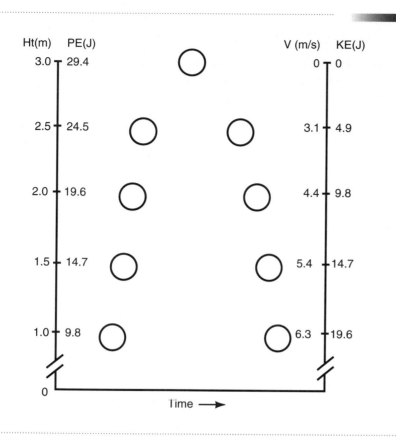

FIGURE 12-20
Height, velocity, potential energy, and kinetic energy changes for a 1 kg ball tossed upward from a height of 1 m. Note that PE + KE = C (a constant) throughout the trajectory.

Principle of Work and Energy

There is a special relationship between the quantities of mechanical work and mechanical energy. This relationship is described as the *principle of work and energy*, which may be stated as follows:

> The work of a force is equal to the change in energy that it produces in the object acted on.

Algebraically, the principle may be represented as the following:

$$W = \Delta KE + \Delta PE + \Delta TE$$

In this formula, KE is kinetic energy, PE is potential energy, and TE is thermal energy (heat). The algebraic statement of the principle of work and energy indicates that the change in the sum of the forms of energy produced by a force is quantitatively equal to the mechanical work done by that force. When a tennis ball is projected into the air by a ball-throwing machine, the mechanical work performed on the ball by the

FIGURE 12-21

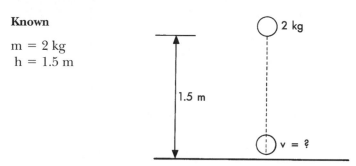

S A M P L E   P R O B L E M   7

A 2 kg ball is dropped from a height of 1.5 m. What is its velocity immediately before impact with the floor?

Known

m = 2 kg
h = 1.5 m

Solution

The principle of the conservation of mechanical energy may be used to solve the problem. The total energy possessed by the ball when it is held at a height of 1.5 m is its potential energy. Immediately before impact, the ball's height (and potential energy) may be assumed to be 0, and 100% of its energy at that point is kinetic.

Total (constant) mechanical energy possessed by the ball:

$$PE + KE = C$$
$$(wt)(h) + \tfrac{1}{2}mv^2 = C$$
$$(2 \text{ kg})(9.81 \text{ m/s}^2)(1.5 \text{ m}) + 0 = C$$
$$29.43 \text{ J} = C$$

Velocity of the ball before impact:

$$PE + KE = 29.43 \text{ J}$$
$$(wt)(h) + \tfrac{1}{2}mv^2 = 29.43 \text{ J}$$
$$(2 \text{ kg})(9.81 \text{ m/s}^2)(0) + \tfrac{1}{2}(2 \text{ kg}) v^2 = 29.43 \text{ J}$$
$$v^2 = 29.43 \text{ J/kg}$$

$$v = 5.42 \text{ m/s}$$

machine results in an increase in the ball's mechanical energy. Prior to projection, the ball's potential energy is based on its weight and height, and its kinetic energy is 0. The ball-throwing machine increases the ball's total mechanical energy by imparting kinetic energy to it. In this situation, the change in the ball's thermal energy is negligible. Figure 12-22 provides a quantitative illustration of the principle of work and energy.

FIGURE 12-22

S A M P L E   P R O B L E M   8

How much mechanical work is required to catch a 1.3 kg ball traveling at a velocity of 40 m/s?

Known

m = 1.3 kg
v = 40 m/s

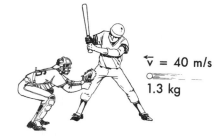

$\overleftarrow{v}$ = 40 m/s

1.3 kg

Solution

The principle of work and energy may be used to calculate the mechanical work required to change the ball's kinetic energy to 0. Assume that the potential energy and thermal energy of the ball do not change:

$$W = \Delta KE$$
$$W = (\tfrac{1}{2} mv^2)_2 - (\tfrac{1}{2} mv^2)_1$$
$$W = 0 - (\tfrac{1}{2})(1.3 \text{ kg})(40 \text{ m/s})^2$$

$$\boxed{W = 1040 \text{ J}}$$

The work-energy relationship is also evident during movements of the human body. For example, the arches in runners' feet act as a mechanical spring to store, and subsequently return, strain energy as they cyclically deform and then regain their resting shapes. For a 70 kg man running at 4.5 m/s, each arch stores approximately 17 J of energy at midstance. Combined with the estimated 35 J stored by each of the Achilles tendons, this equals a storage and partial return of approximately one-half of the mechanical energy expended, or one-half of the mechanical work required of the muscles during the stance phase (2). The ability of the arches to function as a spring reduces the amount of mechanical work that would otherwise be required during running.

Two-joint muscles in the human body also serve to transfer mechanical energy from one joint to another, thereby reducing the mechanical work required of the muscles crossing the second joint during a given movement. For example, during takeoff for a vertical jump, when the hip extensors work concentrically to produce hip extension, if the rectus femoris remains isometrically contracted, a secondary effect is an extensor torque exerted at the knee. In this case, it is the hip extensors that produce the knee extensor torque, since the length of the rectus femoris does not change. Research indicates that during both a vertical jump and the push-off phase of running, the rectus femoris and gastrocnemius help to extend the distal joints they cross by transferring mechanical energy from the proximal joints of the leg to the distal ones (22). During landings and the shock-absorbing phase of running, the process is reversed, with these two-joint muscles transferring energy from the distal joints to the proximal ones, thereby promoting the dissipation of mechanical energy (22).

It is important not to confuse the production of mechanical energy or mechanical work by the muscles of the human body with the consumption of chemical energy or caloric expenditure. Factors such as concentric versus eccentric muscular contractions, the transfer of energy between body segments, elastic storage and reuse of energy, and limitations in joint ranges of motion complicate direct quantitative calculation of the relationship between mechanical and physiological energy estimates (31). Approximately 25% of the energy consumed by the muscles is converted into work, with the remainder changed to heat or used in the body's chemical processes.

FORMULA SUMMARY

DESCRIPTION	FORMULA
Force = (mass)(acceleration)	$F = ma$
Friction = (coefficient of friction)(normal reaction force)	$F = \mu R$
Linear momentum = (mass)(velocity)	$M = mv$
Coefficient of restitution = $\dfrac{\text{relative velocity after impact}}{\text{relative velocity before impact}}$	$-e = \dfrac{v_1 - v_2}{u_1 - u_2}$
Work = (force) (displacement of resistance)	$W = Fd$
Power = $\dfrac{\text{Work}}{\text{time}}$	$P = \dfrac{W}{t}$
Power = (force)(velocity)	$P = Fv$
Kinetic energy = $\frac{1}{2}$(mass) (velocity squared)	$KE = \frac{1}{2}mv^2$
Potential energy = (weight)(height)	$PE = ma_g h$
Strain energy = $\frac{1}{2}$(spring constant) (deformation squared)	$SE = \frac{1}{2}kx^2$
Potential energy + kinetic energy = constant	$PE + KE = C$
Work = change in energy	$W = \Delta KE + \Delta PE + \Delta TE$

SUMMARY

Linear kinetics is the study of the forces associated with linear motion. The interrelationships among many basic kinetic quantities are identified in the physical laws formulated by Sir Isaac Newton.

Friction is a force generated at the interface of two surfaces in contact when there is motion or a tendency for motion of one surface with respect to the other. The magnitudes of maximum static friction and kinetic friction are determined by the coefficient of friction between the two surfaces and by the normal reaction force pressing the two surfaces together. The direction of friction force always opposes the direction of motion or motion tendency.

Other factors that affect the behavior of two bodies in contact when a collision is involved are momentum and elasticity. Linear momentum is the product of an object's mass and its velocity. The total momentum present in a given system remains constant barring the action of external forces. Changes in momentum result from impulses, external forces acting over a time interval. The elasticity of an impact governs the amount of velocity present in the system following the impact. The relative elasticity of two impacting bodies is represented by the coefficient of restitution.

Mechanical work is the product of force and the distance through which the force acts. Mechanical power is the mechanical work done over a time interval. Mechanical energy has two major forms: kinetic and potential. When gravity is the only acting external force, the sum of the kinetic and potential energies possessed by a given body remains constant. Changes in a body's energy are equal to the mechanical work done by an external force.

INTRODUCTORY PROBLEMS

1. How much force must be applied by a kicker to give a stationary 2.5 kg ball an acceleration of 40 m/s^2? (Answer: 100 N)
2. A high jumper with a body weight of 712 N exerts a force of 3 kN against the ground during takeoff. How much force is exerted by the ground on the high jumper? (Answer: 3 kN)
3. What factors affect the magnitude of friction?
4. If m_s between a basketball shoe and a court is 0.56, and the normal reaction force acting on the shoe is 350 N, how much horizontal force is required to cause the shoe to slide? (Answer: >196 N)
5. A football player pushes a 670 N tackling sled. The coefficient of static friction between sled and grass is 0.73 and the coefficient of dynamic friction between sled and grass is 0.68.
   a. How much force must the player exert to start the sled in motion?

b. How much force is required to keep the sled in motion?

c. Answer the same two questions with a 100 kg coach standing on the back of the sled.

(Answer: a. >489.1 N; b. 455.6 N; c. >1205.2 N, 1122.7 N)

6. Lineman A has a mass of 100 kg and is traveling with a velocity of 4 m/s when he collides head-on with Lineman B, who has a mass of 90 kg and is traveling at 4.5 m/s. If both players remain on their feet, what will happen? (Answer: Lineman B will push Lineman A backward with a velocity of 0.03 m/s)

7. Two skaters gliding on ice run into each other head-on. If the two skaters hold onto each other and continue to move as a unit after the collision, what will be their resultant velocity? Skater A has a velocity of 5 m/s and a mass of 65 kg. Skater B has a velocity of 6 m/s and a mass of 60 kg. (Answer: v = 0.28 m/s in the direction originally taken by Skater B)

8. A ball dropped on a surface from a 2 m height bounces to a height of 0.98 m. What is the coefficient of restitution between ball and surface? (Answer: 0.7)

9. A set of 20 stairs, each of 20 cm height, is ascended by a 700 N man in a period of 1.25 seconds. Calculate the mechanical work, power, and change in potential energy during the ascent. (Answer: W = 2800 J, P = 2240 W, PE = 2800 J)

10. A pitched ball with a mass of 1 kg reaches a catcher's glove traveling at a velocity of 28 m/s.

a. How much momentum does the ball have?

b. How much impulse is required to stop the ball?

c. If the ball is in contact with the catcher's glove for 0.5 seconds during the catch, how much average force is applied by the glove?

(Answer: a. 28 kg · m/s; b. 28 N s; c. 56 N)

ADDITIONAL PROBLEMS

1. Identify three practical examples of each of Newton's laws of motion and clearly explain how each example illustrates the law.

2. Select one sport or daily activity and identify the ways in which the amount of friction present between surfaces in contact affects performance outcome.

3. A 2 kg block sitting on a horizontal surface is subjected to a horizontal force of 7.5 N. If the resulting acceleration of the block is 3 m/s^2, what is the magnitude of the friction force opposing the motion of the block? (Answer: 1.5 N)

4. Explain the interrelationships among mechanical work, power, and energy within the context of a specific human motor skill.

5. Explain in what ways mechanical work is and is not related to caloric expenditure. Include in your answer the distinction between positive and negative work and the influence of anthropometric factors.

6. A 108 cm, 0.73 kg golf club is swung for 0.5 seconds with a constant acceleration of 10 rad/s^2. What is the linear momentum of the club head when it impacts the ball? (Answer: 3.9 kg · m/s)

7. A 6.5 N ball is thrown with an initial velocity of 20 m/s at a 35° angle from a height of 1.5 m.
   a. What is the velocity of the ball if it is caught at a height of 1.5 m?
   b. If the ball is caught at a height of 1.5 m, how much mechanical work is required?
   (Answer: a. 20 m/s; b. 132.5 J)

8. A 50 kg person performs a maximum vertical jump with an initial velocity of 2 m/s.
   a. What is the performer's maximum kinetic energy during the jump?
   b. What is the performer's maximum potential energy during the jump?
   c. What is the performer's minimum kinetic energy during the jump?
   d. How much is the performer's center of mass elevated during the jump?
   (Answer: a. 100 J; b. 100 J; c. 0; d. 20 cm)

9. Using the principle of conservation of mechanical energy, calculate the maximum height achieved by a 7 N ball tossed vertically upward with an initial velocity of 10 m/s. (Answer: 5.1 m)

10. Select one of the following sport activities and speculate about the changes that take place between kinetic and potential forms of mechanical energy.
   a. A single leg support during running
   b. A tennis serve
   c. A pole vault performance
   d. A springboard dive

LABORATORY EXPERIENCES

1. Have each member of your lab group remove one shoe. Use a spring scale to determine the magnitude of maximum static friction for each shoe on two different surfaces. (Depending on the sensitivity of the spring scale, you may need to load the shoe with weight.) Present your results in a table and write a paragraph explaining the results.

2. Line up three or four blocks of wood of different sizes and weights on a smooth, inclined surface with the blocks separated by 8–10 cm. Using a long stick such as a meter stick held on both ends, apply an equal force to each block simultaneously. Observe the different distances the blocks travel. Write a descriptive explanation of your results.

3. Perform a maximal vertical jump on a force platform. Use a planimeter to measure the impulses generated against the platform during takeoff and landing on the force-time record for the jump. Explain the similarities and differences in the takeoff and landing impulses.

4. Drop 5 different balls from a height of 2 m on 2 different surfaces and carefully observe and record the bounce heights. Calculate the coefficient of restitution for each ball on each surface and write a paragraph explaining your results.

5. Using a stopwatch, time each member of your lab group running up a flight of stairs. Use a ruler to measure the height of one stair, then multiply by the number of stairs to calculate the total change in height. Calculate work, power, and change in potential energy for your own ascent.

REFERENCES

1. Aura O and Komi PV: Mechanical efficiency of pure positive and pure negative work with special reference to the work intensity, Int J Sports Med 7:44, 1986.
2. Bennett MS et al: Elastic properties of the human foot and their significance for running. In Bennett MS et al: *Biomechanics in sport,* London, 1988, Mechanical Engineering Publications, Ltd.
3. Burke J: *The day the universe changed,* Boston, 1985, Little, Brown & Co, Inc.
4. Cavanagh PR and Lafortune MA: Ground reaction forces in distance running, J Biomech 13:397, 1980.
5. Cohen BI: Newton's discovery of gravity, Sci Am 244:166, 1981.
6. Dapena J: Biomechanics of elite high jumpers. In Terauds J et al, eds: *Sports biomechanics,* Del Mar, Calif, 1984, Academic Publishers.
7. de Koning JJ, de Groot G, and van Ingen Schenau GJ: Ice friction during speed skating, J Biomech 25:565, 1992.
8. Derrick TR, Hamill J, and Caldwell GE: Energy absorption of impacts during running at various stride lengths, Med Sci Sports Exerc 30:128, 1998.
9. Groppel JL et al: Effects of different string tension patterns and racket motion on tennis racket-ball impact, Int J Sport Biomech 3:142, 1987.
10. Groppel JL et al: The effects of string type and tension on impact in mid-sized and oversized tennis racquets, Int J Sport Biomech 3:40, 1987.
11. Hatze H: The relationship between the coefficient of restitution and energy losses in tennis rackets, J Appl Biomech 9:124, 1993.
12. Heidt RS et al: Differences in friction and torsional resistance in athletic shoe−turf surface interfaces, Am J Sports Med 24:834, 1996.
13. Jerome J: Pole vaulting: Biomechanics at the bar. In Schrier EW and Allman WF, eds: *Newton at the bat,* New York, 1984, Charles Scribner's Sons.
14. Kiefer J: Bowling: The great oil debate. In Schrier EW and Allman WF, eds: *Newton at the bat,* New York, 1984, Macmillan−Charles Scribner's Sons, 1984.
15. Laws K: *The physics of dance,* New York, 1984, Schirmer Books.
16. Lees A and Barton G: The interpretation of relative momentum data to assess the contribution of the free limbs to the generation of vertical velocity in sports activities, J Sports Sci 14:503, 1996.

17. MacWilliams BA, Choi T, Perezous MK, Chao EY, and McFarland EG: Characteristic ground-reaction forces in baseball pitching, Am J Sports Med 26:66, 1998.

18. Maybury MC and Waterfield J: An investigation into the relation between step height and ground reaction forces in step exercise: A pilot study, Br J Sports Med 31:109, 1997.

19. Newton RU et al: Influence on load and stretch shortening cycle on the kinematics, kinetics and muscle activation that occurs during explosive upper-body movements, Eur J Appl Physiol 75:333, 1997.

20. Pons DJ and Vaughan CL: Mechanics of cycling. In Vaughan CL, ed: *Biomechanics of sport,* Boca Raton, Fla, 1989, CRC Press, Inc.

21. Preis S, Klemms A, and Müller K: Gait analysis by measuring ground reaction forces in children: Changes to an adaptive gait pattern between the ages of one and five years, Dev Med Child Neurol 39:228, 1997.

22. Prilutsky BI and Zatsiorsky VM: Tendon action of two-joint muscles: Transfer of mechanical energy between joints during jumping, landing, and running, J Biomech 27:25, 1994.

23. Putnam CA and Kozey JW: Substantive issues in running. In Vaughan CL, ed: *Biomechanics of sport,* Boca Raton, Fla, 1989, CRC Press, Inc.

24. Rand KT, Hyer MW, and Williams MH: A dynamic test for comparison rebound characteristics of three brands of tennis balls. In Groppell JL, ed: *Proceedings of the national symposium on racquet sports,* Champaign, Ill, 1979.

25. Rapoport R: Artificial turf: Is the grass greener? In Schrier EW and Allman WF, eds: *Newton at the bat,* New York, 1984, MacMillan–Charles Scribner's Sons.

26. Saikko VO: A three-axis hip joint simulator for wear and friction studies on total hip prostheses, Proc Inst Mech Eng [H] 210:175, 1996.

27. Thys H, Willems PA, and Saels P: Energy cost, mechanical work and muscular efficiency in swing-through gait with elbow crutches, J Biomech 29:1473, 1996.

28. Torg JS, Stilwell G, and Rogers K: The effect of ambient temperature on the shoe-surface interface release coefficient, Am J Sports Med 24:79, 1996.

29. van Ingen Schenau GJ, DeBoer RW, and DeGroot G: Biomechanics of speed skating. In Vaughan CL, ed: *Biomechanics of sport,* Boca Raton, Fla, 1989, CRC Press, Inc.

30. Whitt FR and Wilson DG: *Bicycling science: Ergonomics and mechanics,* Cambridge, Mass, 1974, The MIT Press.

31. Williams KR: The relationship between mechanical and physiological energy estimates, Med Sci Sports Exerc 17:317, 1985.

ANNOTATED READINGS

de Koning JJ, de Groot G, and van Ingen Schenau GJ: Ice friction during speed skating, J Biomech 25:565, 1992.
   Presents an interesting discussion on theories as to why the coefficient of friction between a racing skate and the ice is so small.

McMahon TA and Greene PR: Fast running tracks, Sci Am 239:148, 1978.
   Includes a fascinating discussion of the thought processes behind the degree of elasticity engineered into Harvard's indoor track. Using this track, runners can better their best times by nearly 3%.

Prilutsky BI and Zatsiorsky VM: Tendon action of two-joint muscles: Transfer of mechanical energy between joints during jumping, landing, and running, J Biomech 27:25, 1994.
Provides an excellent explanation of research on mechanical energy transfer between joints.

Townend MS: *Mathematics in sport,* New York, 1984, John Wiley & Sons, Inc.
The chapter on jumping provides in-depth analyses of changes in quantities such as momentum, mechanical work, and mechanical energy during performance of the strad-dle and Fosbury flop high jump techniques, the pole vault, the long jump, and the triple jump.

RELATED WEB SITES

Advanced Medical Technology, Inc.
http://www.amtiweb.com
Provides information on the AMTI force platforms, with reference to ground reaction forces in gait analysis, balance and posture, and other topics.

The Exploratorium's Science of Hockey
http://www.exploratorium.edu/hockey
Explains scientific concepts related to hockey, including the friction between ice and skate and the mechanics of skating.

Explore Science Home Page
http://www.explorescience.com
Provides a link to a page on 2-D elastic and inelastic collisions on a flat or tilted table.

Kistler
http://www.kistler.com
Describes a series of force platforms for measuring ground reaction forces.

Physics Lecture Notes
http://sutherland.monroe.edu/pavone/HTMLnotes/lecturenotes.html
Includes links to pages on a number of topics from mechanics, including Newton's Laws, momentum, and work, power, and energy.

EQUILIBRIUM AND HUMAN MOVEMENT

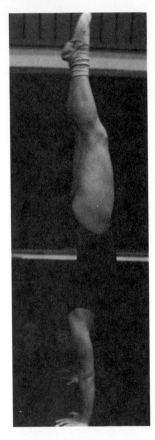

Many athletic skills require mechanical stability.

After completing this chapter, the reader will be able to:

Define torque, quantify resultant torques, and identify the factors that affect resultant joint torques.

Identify the mechanical advantages associated with the different classes of levers and explain the concept of leverage within the human body.

Solve basic quantitative problems using the equations of static equilibrium.

Define center of gravity and explain the significance of center of gravity location in the human body.

Explain how mechanical factors affect a body's stability.

W hy do long jumpers and high jumpers lower their centers of gravity before takeoff? What mechanical factors enable a wheelchair to remain stationary on a graded ramp or a Sumo wrestler to resist the attack of his opponent? A body's mechanical stability is based on its resistance to both linear and angular motion. This chapter introduces the kinetics of angular motion, along with the factors that affect mechanical stability.

EQUILIBRIUM

Torque

As discussed in Chapter 3, the rotary effect created by an applied force is known as **torque** or *moment of force*. Torque, which may be thought of as *rotary force,* is the angular equivalent of linear force. Algebraically,

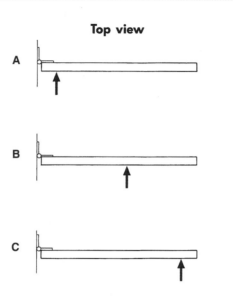

Top view

A

B

C

FIGURE 13-1
Which position of force
application is best for
opening the swinging
door? Experience should
verify that position **C** is
best.

torque is the product of force and the force's **moment arm,** or the perpendicular distance from the force's line of action to the axis of rotation.

$$T = Fd_\perp$$

Thus, both the magnitude of a force and the length of its moment arm equally affect the amount of torque generated (Figure 13-1).

As may be observed in Figure 13-2, the moment arm is the shortest distance between the force's line of action and the axis of rotation. A force directed through an axis of rotation produces no torque because the force's moment arm is 0.

Within the human body, the moment arm for a muscle with respect to a joint center is the perpendicular distance between the muscle's line of action and the joint center (Figure 13-3). As a joint moves through a range of motion, there are changes in the moment arms of the muscles crossing the joint. For any given muscle, the moment arm is largest when the angle of pull on the bone is closest to 90°. At the elbow as the angle of pull moves away from 90° in either direction, the moment arm for the elbow flexors is progressively diminished. Changes in moment

torque
the rotary effect of a force
about an axis of rotation,
measured as the product of
the force and the
perpendicular distance
between the force's line of
action and the axis

moment arm
shortest (perpendicular)
distance between a force's
line of action and an axis of
rotation

*It is easiest to initiate rotation
when force is applied
perpendicularly and as far
away as possible from the axis
of rotation.*

FIGURE 13-2

The moment arm of a force is the perpendicular distance from the force's line of action to the axis of rotation (the door hinge).

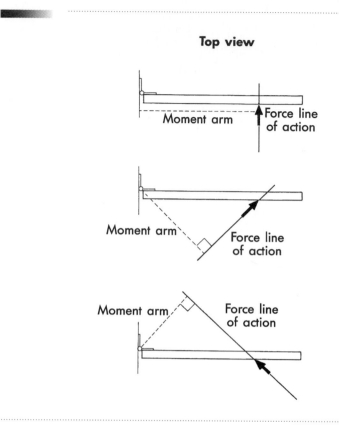

Top view

Moment arm — Force line of action

Moment arm — Force line of action

Moment arm — Force line of action

arm directly affect the joint torque that a muscle generates. For a muscle to generate a constant joint torque during an exercise, it must produce more force as its moment arm decreases.

In the sport of rowing where adjacent crew members traditionally row on opposite sides of the hull, the moment arm between the force applied by the oar and the stern of the boat is a factor affecting performance (Figure 13-4). With the traditional arrangement, the rowers on one side of the boat are positioned farther from the stern than their counterparts on the other side, thus causing a net torque and a resulting lateral oscillation about the stern during rowing (16). The Italian rig eliminates this problem by positioning rowers so that no net torque is produced, assuming that the force produced by each rower with each stroke is nearly the same (Figure 13-4). Italian and German rowers have similarly developed alternative positionings for the eight-member crew (Figure 13-5).

Another example of the significance of moment arm length is provided by a dancer's foot placement during preparation for execution of a total body rotation around the vertical axis. When a dancer initiates

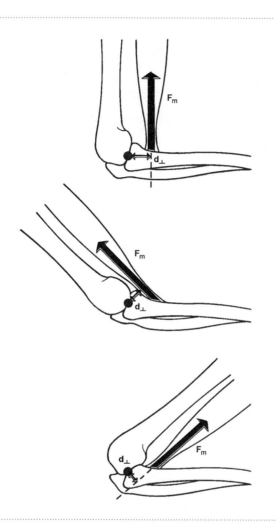

FIGURE 13-3
A muscle's moment arm is maximal at a 90° angle of pull. As the line of pull moves away from 90° in either direction, the moment arm becomes progressively smaller.

a turn, the torque producing the turn is provided by equal and oppositely directed forces exerted by the feet against the floor. A pair of equal and opposite forces is known as a force **couple.** Because the forces in a couple are positioned on opposite sides of the axis of rotation, they produce torque in the same direction. The torque generated by a couple is therefore the sum of the products of each force and its moment arm. Turning from fifth position, with a small distance between the feet, requires greater force production by a dancer than turning at the same rate from fourth position, in which the moment arms of the forces in the couple are longer (Figure 13-6). Significantly more force is required when the torque is generated by a single support foot for which the moment arm is reduced to the distance between the metatarsals and the calcaneus (12).

couple
pair of equal, oppositely directed forces that act on opposite sides of an axis of rotation to produce torque

FIGURE 13-4

A, This crew arrangement creates a net torque about the stern of the boat because the sum of the top side oar moment arms $(d_1 + d_2)$ is less than the sum of the bottom side moment arms $(d_3 + d_4)$. **B**, This arrangement eliminates the problem, assuming that all rowers stroke simultaneously and produce equal force because $(d_1 + d_2) = (d_3 + d_4)$.

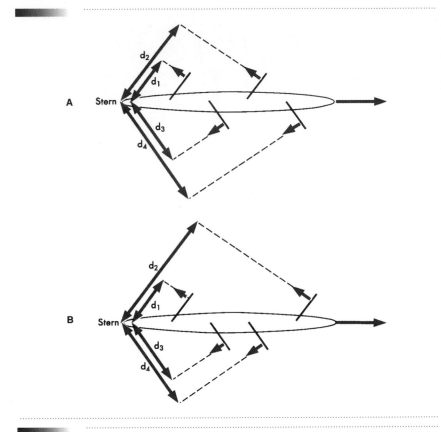

FIGURE 13-5

The Italians and Germans have used alternative positionings for eight-member crews. The torques produced by the oar forces with respect to the stern are balanced in arrangements **B** and **C**, but not in the traditional arrangement shown in **A**.

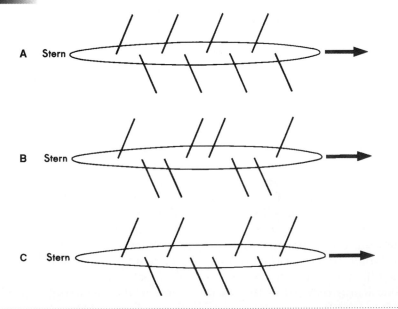

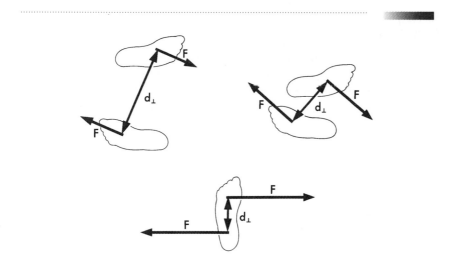

FIGURE 13-6
The wider a dancer's stance, the greater the moment arm for the force couple generated by the feet when a turn is executed. When rotation is initiated from a single foot stance, the moment arm becomes the distance between the support points of the foot.

Torque is a vector quantity and is therefore characterized by both magnitude and direction. The magnitude of the torque created by a given force is equal to $Fd_\perp$, and the direction of a torque may be described as clockwise or counterclockwise. As discussed in Chapter 11, the counterclockwise direction is conventionally referred to as the positive (+) direction, and the clockwise direction is regarded as negative (−). The magnitudes of two or more torques acting at a given axis of rotation can be added using the rules of vector composition (Figure 13-7).

Resultant Joint Torques

The concept of torque is important in the study of human movement because torque produces movement of the body segments. As discussed in Chapter 6, when a muscle crossing a joint develops tension, it produces a force pulling on the bone to which it attaches, thereby creating torque at the joint the muscle crosses.

Much human movement involves simultaneous tension development in agonist and antagonist muscle groups. The tension in the antagonists controls the velocity of the movement and enhances the stability of the joint at which the movement is occurring. Since antagonist tension development creates torque in the direction opposite that of the torque produced by the agonist, the resulting movement at the joint is a function of the net torque. When net torque and joint movement occur in the same direction, the torque is termed *concentric,* and torque in the direction opposite joint motion is considered to be *eccentric.* Although these terms are generally useful descriptors in analysis of muscular function,

The product of muscle tension and muscle moment arm produces a torque at the joint crossed by the muscle.

FIGURE 13-7

S A M P L E   P R O B L E M   I

Two children sit on opposite sides of a playground seesaw. If Joey, weighing 200 N, is 1.5 m from the seesaw's axis of rotation and Susie, weighing 190 N, is 1.6 m from the axis of rotation, which end of the seesaw will drop?

Known

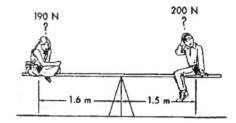

Joey: $\text{wt}(F_J) = 200 \text{ N}$

$d_{\perp J} = 1.5 \text{ m}$

Susie: $\text{wt}(F_S) = 190 \text{ N}$

$d_{\perp S} = 1.6 \text{ m}$

Solution

The seesaw will rotate in the direction of the resultant torque at its axis of rotation. To find the resultant torque, the torques created by both children are summed according to the rules of vector composition. The torque produced by Susie's body weight is in a counterclockwise (positive) direction, and the torque produced by Joey's body weight is in a clockwise (negative) direction.

$$T_a = (F_S)(d_{\perp S}) - (F_J)(d_{\perp J})$$
$$T_a = (190 \text{ N})(1.6 \text{ m}) - (200 \text{ N})(1.5 \text{ m})$$
$$T_a = 304 \text{ N-m} - 300 \text{ N-m}$$
$$T_a = 4 \text{ N-m}$$

The resultant torque is in a positive direction, and Susie's end of the seesaw will fall.

their application is complicated when two-joint or multijoint muscles are considered, since there may be concentric torque at one joint and eccentric torque at a second joint crossed by the same muscle.

Because directly measuring the forces produced by muscles during the execution of most movement skills is not practical, measurements or estimates of resultant joint torques (joint moments) are often studied to investigate the patterns of muscle contributions. A number of factors, including the weight of body segments, the motion of the body segments, and the action of external forces, may contribute to net joint torques. Young infants generate irregular patterns of joint torques, possibly because of their inexperience in predicting the magnitude and direction of external forces (11). Among adults, however, joint torque profiles are typically matched to the requirements of the task

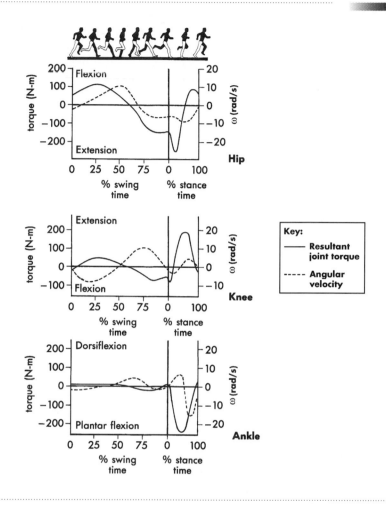

FIGURE 13-8
Representative resultant
joint torques and joint
angular velocity curves for
the lower extremity during
running. (Modified from
Putnam CA and Kozey JW:
Substantive issues in
running. In Vaughan CL,
ed: *Biomechanics of sport,*
Boca Raton, Fla. 1989,
CRC Press, Inc.)

at hand and provide at least general estimates of muscle group contribution levels.

To better understand muscle function during running, a number of
investigators have studied resultant joint torques at the hip, knee, and
ankle throughout the running stride. Figure 13-8 displays representative resultant joint torques and angular velocities for the hip, knee, and
ankle during a running stride as calculated from film and force platform data. In Figure 13-8, when the resultant joint torque curve and the
angular velocity curves are on the same side of the 0 line, the torque is
concentric; the torque is eccentric when the reverse is true. As may be
observed from Figure 13-8, both concentric and eccentric torques are
present at the lower extremity joints during running.

FIGURE 13-9
Absolute average joint
torques for the hip, knee,
and ankle versus pedaling
rate during cycling.
(Modified from Redfield R
and Hull ML: On the
relation between joint
moments and pedalling at
constant power in
bicycling, J Biomech
19:317, 1986.)

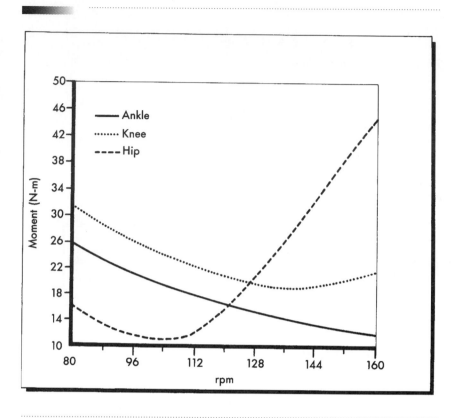

*The torques required at the
hip, knee, and ankle
during cycling at a given
power are influenced by
body position and cycle
dimensions.*

During both running and walking, individuals with anterior cruciate ligament (ACL) injury use greater extensor torques at the hip and ankle and lower extensor torques at the knee as compared to uninjured people (5). Research indicates that one contributing factor to this pattern may be the functional knee braces often worn by ACL-deficient individuals (5).

Lower extremity joint torques during cycling at a given power are affected by pedaling rate, seat height, length of the pedal crank arm, and the distance from the pedal spindle to the ankle joint. Average hip and knee torques during cycling under cruising conditions have been reported to be minimum at approximately 105 rotations per minute (15). Figure 13-9 shows the changes in average resultant torque at the hip, knee, and ankle joints with changes in pedaling rate at a constant power.

It is widely assumed that the muscular force (and subsequently, joint torque) requirements of resistance exercise increase as the amount of resistance increases. However, this is true only as long as movement kinematics remain constant. It has been shown, for example, that during the squat exercise, subtle changes in kinematic variables such as absolute trunk angle and knee flexion may reflect dramatically altered joint torques at the hip, knee, and ankle. As the load for the squat increased from 60% to 80% of maximum, one subject in a study displayed more than twice the expected increase in torque at the knee, while another subject actually decreased torque at the knee (7).

Another factor influencing joint torques during exercise is movement speed. When other factors remain constant, increased movement speed is associated with increased resultant joint torques during exercises such as the squat (1, 9). However, increased movement speed during weight training is generally undesirable because increased speed increases not only the muscle tension required but also the likelihood of incorrect technique and subsequent injury. Acceleration of the load early in the performance of a resistance exercise also generates momentum, which means that the involved muscles need not work as hard throughout the range of motion as would otherwise be the case. For these reasons it is both safer and more effective to perform exercises at slow, controlled movement speeds.

Levers

When muscles develop tension, pulling on bones to support or move the resistance created by the weight of the body segment(s) and possibly the weight of an added load, the muscle and bone are functioning mechanically as a **lever.** A lever is a rigid bar that rotates about an axis or **fulcrum.** Force applied to the lever moves a resistance. In the human body, the bone acts as the rigid bar, the joint is the axis or fulcrum, and the muscles apply force. The three relative arrangements of the applied force, resistance, and axis of rotation for a lever are shown in Figure 13-10.

With a **first class lever** the applied force and resistance are located on opposite sides of the axis. The playground seesaw is an example of a first class lever, as are a number of commonly used tools, including scissors, pliers, and crowbars (Figure 13-11). Within the human body the simultaneous action of agonist and antagonist muscle groups on opposite sides of a joint axis is analogous to the functioning of a first class lever, with the agonists providing the applied force and the antagonists supplying a resistance force. With a first class lever, the applied force and resistance may be at equal distances from the axis, or one may be farther away from the axis than the other.

In a **second class lever** the applied force and resistance are on the same side of the axis, with the resistance closer to the axis. A wheelbarrow, lug nut wrench, and nutcracker are examples of second class levers, although there are no completely analogous examples in the human body (Figure 13-11).

With a **third class lever** the force and the resistance are on the same side of the axis, but the applied force is closer to the axis. A canoe

lever
simple machine consisting of a relatively rigid barlike body that may be made to rotate about an axis

fulcrum
point of support or axis about which a lever may be made to rotate

first class lever
lever positioned with the applied force and the resistance on opposite sides of the axis of rotation

second class lever
lever positioned with the resistance between the applied force and the fulcrum

third class lever
lever positioned with the applied force between the fulcrum and the resistance

FIGURE 13-10

Relative locations of the applied force, the resistance, and the fulcrum or axis of rotation determine lever classifications.

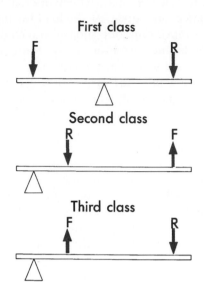

FIGURE 13-11

A, First class levers.
B, Second class levers.
C, Third class levers. Note that the paddle and shovel function as third class levers only when the top hand does not apply force but serves as a fixed axis of rotation.

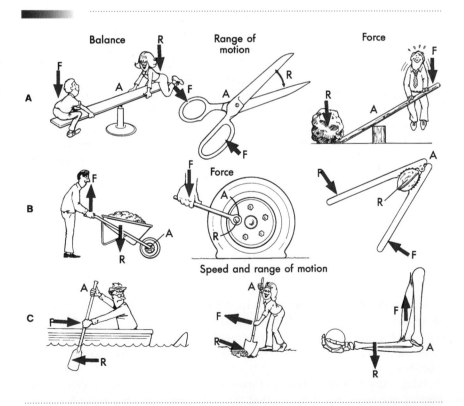

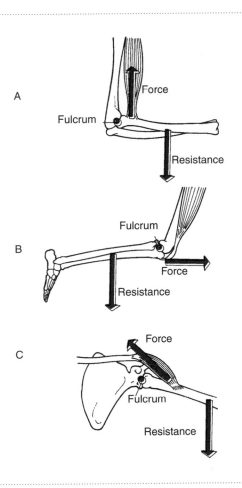

A

Force

Fulcrum

Resistance

Fulcrum

B

Force

Resistance

Force

C

Fulcrum

Resistance

FIGURE 13-12
Most levers within the human body are third class. **A**, The biceps at the elbow. **B**, The patellar tendon at the knee. **C**, The medial deltoid at the shoulder.

paddle and a shovel can serve as third class levers (Figure 13-11). Most muscle-bone lever systems of the human body are also of the third class for concentric contractions, with the muscle supplying the applied force and attaching to the bone at a short distance from the joint center compared to the distance at which the resistance supplied by the weight of the body segment or that of a more distal body segment acts (Figure 13-12). As shown in Figure 13-13, however, during eccentric contractions it is the muscle that supplies the resistance against the applied external force. During eccentric contractions muscle and bone function as a second class lever.

A lever system can serve one of two purposes (Figure 13-14). Whenever the moment arm of the applied force is greater than the moment arm of the resistance, the magnitude of the applied force needed to move a given resistance is less than the magnitude of the resistance.

The elbow flexors contract eccentrically to apply a braking resistance and control movement speed during the down phase of a curl exercise. In this case the muscle-bone lever system is second class.

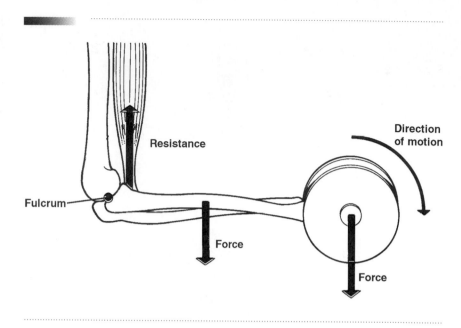

mechanical advantage
ratio of force arm to resistance arm for a given lever

The moment arm of an applied force can also be referred to as the force arm, and the moment arm of a resistance can be referred to as the resistance arm.

Whenever the resistance arm is longer than the force arm, the resistance may be moved through a relatively large distance. The mechanical effectiveness of a lever for moving a resistance may be expressed quantitatively as its **mechanical advantage,** which is the ratio of the moment arm of the force to the moment arm of the resistance:

$$\text{Mechanical advantage} = \frac{\text{Moment arm (force)}}{\text{Moment arm (resistance)}}$$

Whenever the moment arm of the force is longer than the moment arm of the resistance, the mechanical advantage ratio reduces to a number that is greater than one, and the magnitude of the applied force required to move the resistance is less than the magnitude of the resistance. The ability to move a resistance with a force that is smaller than the resistance offers a clear advantage when a heavy load must be moved. As shown in Figure 13-11, a wheelbarrow combines second class leverage with rolling friction to facilitate transporting a load. When removing a lug nut from an automobile wheel, it is helpful to use as long an extension as is practical on the wrench to increase mechanical advantage.

Alternatively, when the mechanical advantage ratio is less than one, a force that is larger than the resistance must be applied to cause motion of the lever. Although this arrangement is less effective in the sense that more force is required, a small movement of the lever at the point

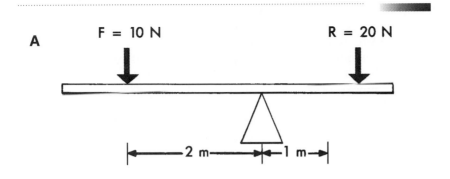

A

F = 10 N   R = 20 N

|←——2 m——→|←1 m→|

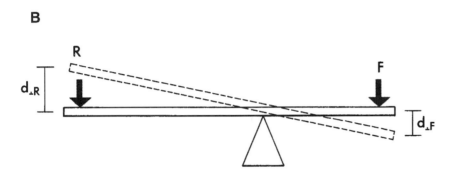

B

R   F

$d_{\perp R}$

$d_{\perp F}$

Figure 13-14
A, A force can balance a larger resistance when its moment arm is longer than the moment arm of the resistance. **B**, A force can move a resistance through a larger range of motion when the moment arm of the force is shorter than the moment arm of the resistance.

of force application moves the resistance through a larger range of motion (Figure 13-14).

During wheelchair propulsion, mechanical advantage is the ratio of handrim radius to wheel radius. Since handrim radius is always smaller than wheel radius, the mechanical advantage for wheelchair propulsion is always less than one. This is advantageous because movements of the handrims translate to larger movements of the wheels and the resistance, which is the force of rolling friction, is relatively low. For a given applied force on the pushrim, wheelchair velocity is proportional to mechanical advantage. Researchers have found that a mechanical advantage of 0.43 is more mechanically efficient than mechanical advantages ranging up to 0.87 for wheelchair propulsion, because at lower mechanical advantage (and lower velocity), the wheelchair occupant is able to apply force more directly in line with the path of handrim rotation (19). During wheelchair propulsion, mechanical advantage has been shown to have a significant effect on oxygen uptake, energy cost, mechanical efficiency, and stroke frequency (17).

Skilled pitchers often maximize the length of the moment arm between the ball hand and the total body axis of rotation during the delivery of a pitch to maximize the effect of the torque produced by the muscles.

Anatomical Levers

Skilled athletes in many sports intentionally maximize the length of the effective moment arm for force application to maximize the effect of the torque produced by muscles about a joint. During execution of the serve in tennis, expert players not only strike the ball with the arm fully extended but also vigorously rotate the body in the transverse plane, making the axis of rotation the spine and maximizing the length of the anatomical lever delivering the force. The same strategy is employed by accomplished baseball pitchers. As discussed in Chapter 11, the longer the radius of rotation, the greater the linear velocity of the racket head or hand delivering the pitch and the greater the resultant velocity of the struck or thrown ball.

In the human body, most muscle-bone lever systems are of the third class and therefore have a mechanical advantage of less than one. Although this arrangement promotes range of motion and angular speed of the body segments, the muscle forces generated must be in excess of the resistance force or forces if positive mechanical work is to be done.

The angle at which a muscle pulls on a bone also affects the mechanical effectiveness of the muscle-bone lever system. The force of muscular tension is resolved into two force components—one perpendicular to the attached bone and one parallel to the bone (Figure 13-15). As discussed in Chapter 6, only the component of muscle force acting perpendicular to the bone—the rotary component—actually causes the bone to rotate about the joint center. The component of muscle force directed parallel to the bone pulls the bone either away from the joint center (a dislocating component) or toward the joint center (a stabilizing component), depending on whether the angle between the bone and the attached muscle is less than or greater than 90°. The angle of maximum mechanical advantage for any muscle is the angle at which the most rotary force can be produced. At a joint such as the elbow, the relative angle present at the joint is close to the angles of attachment of the elbow flexors. The maximum mechanical advantages for the brachialis, biceps, and brachioradialis occur between angles at the elbow of approximately 75 and 90 degrees (Figure 13-16).

As joint angle and mechanical advantage change, muscle length also changes. Alterations in the lengths of the elbow flexors associated with changes in angle at the elbow are shown in Figure 13-17. These changes affect the amount of tension a muscle can generate, as discussed in Chapter 6. The angle at the elbow at which maximum flexion torque is produced is approximately 80°, with torque capability progressively diminishing as the angle at the elbow changes in either direction (18).

The varying mechanical effectiveness of muscle groups for producing joint rotation with changes in joint angle is the underlying basis for the design of modern variable-resistance strength training devices. These machines are designed to match the changing torque-generating capa-

The force-generating capability of a muscle is affected by muscle length, cross-sectional area, moment arm, angle of attachment, shortening velocity, and state of training.

Variable resistance training devices are designed to match the resistance offered to the torque-generating capability of the muscle group as it varies throughout a range of motion.

The term isokinetic implies constant angular velocity at a joint when applied to exercise machinery.

FIGURE 13-15

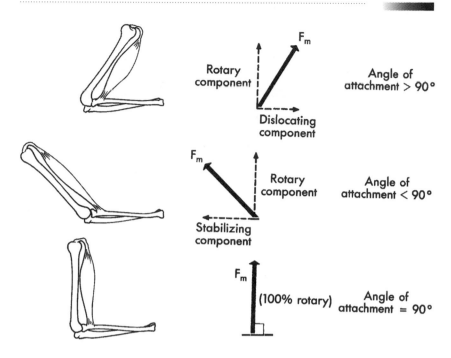

bility of a muscle group throughout the range of motion at a joint. Machines manufactured by Universal (the Centurion) and Nautilus are examples. Although these machines offer more relative resistance through the extremes of joint range of motion than free weights, the resistance patterns incorporated are not an exact match for average human strength curves (6).

Isokinetic machines represent another approach to matching torque-generating capability with resistance. These devices are generally designed so that an individual applies force to a lever arm that rotates at a constant angular velocity. If the joint center is aligned with the center of rotation of the lever arm, the body segment rotates with the same (constant) angular velocity of the lever arm. If volitional torque production by the involved muscle group is maximum throughout the range of motion, a maximum matched resistance is theoretically achieved. However, when force is initially applied to the lever arm of isokinetic machines, acceleration occurs and the angular velocity of the arm fluctuates until the set rotational speed is reached (6). Because optimal use of isokinetic resistance machines requires that the user be focused on exerting maximal effort throughout the range of motion, some individuals prefer other modes of resistance training.

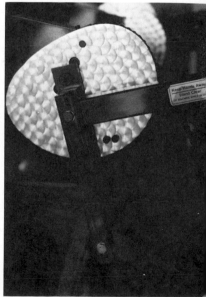

A cam in a variable resistance training machine is designed to match the resistance offered to the mechanical advantage of the muscle.

FIGURE 13-16

Mechanical advantage of the brachialis (●), biceps (□), and brachioradialis (▽) as a function of elbow angle. (Modified from van Zuylen EJ, van Zelzen A, and van der Gon JJD: A biomechanical model for flexion torques of human arm muscles as a function of elbow angle, J Biomech 21:183, 1988.)

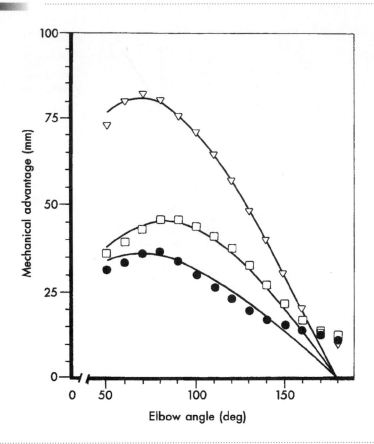

Equations of Static Equilibrium

Equilibrium is a state characterized by balanced forces and torques (no net forces and torques). In keeping with Newton's first law, a body in equilibrium is either motionless or moving with a constant velocity. Whenever a body is completely motionless, it is in **static equilibrium.** Three conditions must be met for a body to be in a state of static equilibrium:

static equilibrium

motionless state characterized by $\Sigma F_v = 0$, $\Sigma F_h = 0$, and $\Sigma T = 0$

1. The sum of all vertical forces (or force components) acting on the body must be 0,

2. the sum of all horizontal forces (or force components) acting on the body must be 0,

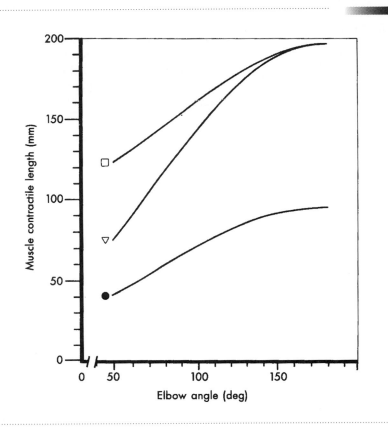

FIGURE 13-17
Contractile length of the brachialis (●), biceps (□), and brachioradialis (▽) as a function of elbow angle. (Modified from van Zuylen EJ, van Zelzen A, and van der Gon JJD: A biomechanical model for flexion torques of human arm muscles as a function of elbow angle, J Biomech 21:183, 1988.)

3. the sum of all torques must be zero:

$$\Sigma F_v = 0$$
$$\Sigma F_h = 0$$
$$\Sigma T = 0$$

The capital Greek letter sigma (Σ) means *the sum of,* F_v represents vertical forces, F_h represents horizontal forces, and T is torque. Whenever an object is in a static state, it may be inferred that all three conditions are in effect, since the violation of any one of the three conditions would result in motion of the body. The conditions of static equilibrium are valuable tools for solving problems relating to human movement (Figure 13-18 to Figure 13-20).

The presence of a net force acting on a body results in acceleration of the body.

FIGURE 13-18

SAMPLE PROBLEM 2

How much force must be produced by the biceps brachii, attaching at 90° to the radius at 3 cm from the center of rotation at the elbow joint, to support a weight of 70 N held in the hand at a distance of 30 cm from the elbow joint? (Neglect the weight of the forearm and hand, and neglect any action of other muscles.)

Known

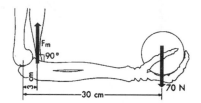

d_m = 3 cm

wt = 70 N

d_{wt} = 30 cm

Solution

Since the situation described is static, the sum of the torques acting at the elbow must be equal to 0.

$$\Sigma T_e = 0$$
$$\Sigma T_e = (F_m)(d_m) - (wt)(d_{wt})$$
$$0 = (F_m)(0.03 \text{ m}) - (70 \text{ N})(0.30 \text{ m})$$
$$F_m = \frac{(70 \text{ N})(0.30 \text{ m})}{0.03 \text{ m}}$$

$$\boxed{F_m = 700 \text{ N}}$$

....

Equations of Dynamic Equilibrium

dynamic equilibrium (D'Alembert's principle)
concept indicating a balance between applied forces and inertial forces for a body in motion

Bodies in motion are considered to be in a state of **dynamic equilibrium,** with all acting forces resulting in equal and oppositely directed inertial forces. This general concept was first identified by the French mathematician D'Alembert and is known as *D'Alembert's principle.* Modified versions of the equations of static equilibrium, which incorporate factors known as *inertia vectors,* describe the conditions of dynamic equilibrium. The equations of dynamic equilibrium may be stated as follows:

$$\Sigma F_x - m\bar{a}_x = 0$$
$$\Sigma F_y - m\bar{a}_y = 0$$
$$\Sigma T_G - \bar{I}\alpha = 0$$

The sums of the horizontal and vertical forces acting on a body are ΣF_x and ΣF_y; $m\bar{a}_x$ and $m\bar{a}_y$ are the products of the body's mass and the horizontal and vertical accelerations of the body's center of mass; ΣT_G is

FIGURE 13-19

S A M P L E   P R O B L E M   3

Two individuals apply force to opposite sides of a frictionless swinging door. If A applies a 30 N force at a 40° angle 45 cm from the door's hinge and B applies force at a 90° angle 38 cm from the door's hinge, what amount of force is applied by B if the door remains in a static position?

Known

$F_A = 30$ N
$d_{\perp A} = (0.45$ m$)(\sin 40)$
$d_{\perp B} = 0.38$ m

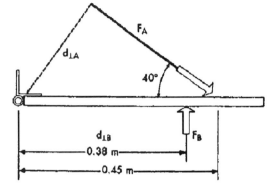

Solution

The equations of static equilibrium are used to solve for F_B. The solution may be found by summing the torques created at the hinge by both forces.

$\Sigma T_h = 0$
$\Sigma T_h = (F_A)(d_{\perp A}) - (F_B)(d_{\perp B})$
$0 = (30$ N$)(0.45$ m$)(\sin 40) - (F_B)(0.38$ m$)$

$F_B = 22.8$ N

the sum of torques about the body's center of mass, and $\bar{I}\alpha$ is the product of the body's moment of inertia about the center of mass and the body's angular acceleration (Figure 13-21). (The concept of moment of inertia is discussed in Chapter 14.)

A familiar example of the effect of D'Alembert's principle is the change in vertical force experienced when riding in an elevator. As the elevator accelerates upward, an inertial force in the opposite direction is created and body weight as measured on a scale in the elevator increases. As the elevator accelerates downward, an upwardly directed inertial force decreases body weight as measured on a scale in the elevator. Although body weight remains constant, the vertical inertial force changes the magnitude of the reaction force measured on the scale.

FIGURE 13-20

S A M P L E   P R O B L E M   4

The quadriceps tendon attaches to the tibia at a $30°$ angle 4 cm from the joint center at the knee. When an 80 N weight is attached to the ankle 28 cm from the knee joint, how much force is required of the quadriceps to maintain the leg in a horizontal position? What is the magnitude and direction of the reaction force exerted by the femur on the tibia? (Neglect the weight of the leg and the action of other muscles.)

Known

wt = 80 N
d_{wt} = 0.28 m
d_F = 0.04 m

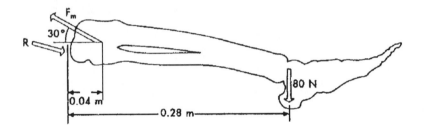

Solution

The equations of static equilibrium can be used to solve for the unknown quantities:

$$\Sigma T_k = 0$$
$$\Sigma T_k = (F_m \sin 30)(d_F) - (wt)(d_{wt})$$
$$0 = (F_m \sin 30)(0.04 \text{ m}) - (80 \text{ N})(0.28 \text{ m})$$

$$F_m = 1120 \text{ N}$$

The equations of static equilibrium can be used to solve for the vertical and horizontal components of the reaction force exerted by the femur on the tibia. Summation of vertical forces yields the following:

$$\Sigma F_v = 0$$
$$\Sigma F_v - R_v \mid (F_m \sin 30) - wt$$
$$0 = R_v + 1120 \sin 30 \text{ N} - 80 \text{ N}$$
$$R_v = -480 \text{ N}$$

Summation of horizontal forces yields the following:

$$\Sigma F_h = 0$$
$$\Sigma F_h = R_h - (F_m \cos 30)$$
$$0 = R_h - 1120 \cos 30 \text{ N}$$
$$R_h = 970 \text{ N}$$

The Pythagorean theorem can now be used to find the magnitude of the resultant reaction force:

$$R = \sqrt{(-480 \text{ N})^2 + (970 \text{ N})^2}$$
$$R = 1082 \text{ N}$$

The tangent relationship can be used to find the angle of orientation of the resultant reaction force:

$$\tan \alpha = \frac{480 \text{ N}}{970 \text{ N}}$$
$$\alpha = 26.3$$

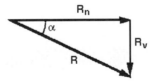

R = 1082 N, α = 26.3 degrees

FIGURE 13-21

A 580 N skydiver in free fall is accelerating at -8.8 m/s^2 rather than -9.81 m/s^2 because of the force of air resistance. How much drag force is acting on the skydiver?

Known

$$\text{wt} = -580 \text{ N}$$
$$a = -8.8 \text{ m/s}^2$$
$$\text{Mass} = \frac{580 \text{ N}}{9.81 \text{ m/s}^2} = 59.12 \text{ kg}$$

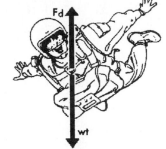

Solution

Since the skydiver is considered to be in dynamic equilibrium, D'Alembert's principle may be used. All identified forces acting are vertical forces, so the equation of dynamic equilibrium summing the vertical forces to zero is used:

$$\Sigma F_y - \overline{m}a_y = 0$$

Given that $\Sigma F_y = -580 \text{ N} + F_d$, substitute the known information into the equation:

$$-580 \text{ N} + F_d - (59.12 \text{ kg})(-8.8 \text{ m/s}^2) = 0$$

$$\boxed{F_d = 59.7 \text{ N}}$$

CENTER OF GRAVITY

center of mass
mass centroid
center of gravity
the point around which the mass and weight of a body are balanced in all directions

A body's mass is the matter of which it is composed. A unique point is associated with every body, around which the body's mass is equally distributed in all directions. This point is known as the **center of mass** or the **mass centroid** of the body. In the analysis of bodies subject to gravitational force, the center of mass may also be referred to as the **center of gravity** (CG), the point about which a body's weight is equally balanced in all directions or the point about which the sum of torques produced by the weights of the body segments is equal to 0. This definition implies not that the weights positioned on opposite sides of the CG are equal, but that the torques created by the weights on opposite sides of the CG are equal. As illustrated in Figure 13-22, equal weight and equal torque generation on opposite sides of a point can be quite different. The terms *center of mass* and *center of gravity* are more com-

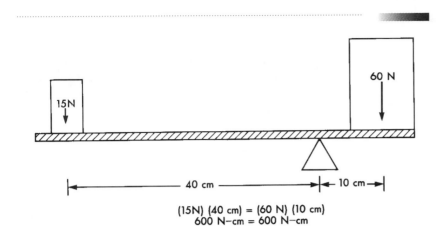

$$(15N) (40\ cm) = (60\ N) (10\ cm)$$
$$600\ N\text{-}cm = 600\ N\text{-}cm$$

FIGURE 13-22
The presence of equal torques on opposite sides of an axis of rotation does not necessitate the presence of equal weights on opposite sides of the axis.

monly used for biomechanics applications than *mass centroid,* although all three terms refer to exactly the same point. Because the masses of bodies on the earth are subject to gravitational force, the center of gravity is probably the most accurately descriptive of the three to use for biomechanical applications.

The CG of a perfectly symmetrical object of homogeneous density, and therefore homogeneous mass and weight distribution, is at the exact center of the object. For example, the CG of a spherical shot or a solid rubber ball is at its geometric center. If the object is a homogeneous ring, the CG is located in the hollow center of the ring. However, when mass distribution within an object is not constant, the CG shifts in the direction of greater mass. It is also possible for an object's CG to be located physically outside of the object (Figure 13-23).

Locating the Center of Gravity

The location of the CG for a one-segment object, such as a baseball bat, a broom, or a shovel, can be approximately determined using two approaches. The first involves the use of a fulcrum to determine the location of a balance point for the object in three different planes. Because the CG is the point around which the mass of a body is equally distributed, it is also the point around which the body is balanced in all directions. Since balancing some objects in certain planes may be difficult, suspending a plumb line from the object in question in three different positions is an alternative procedure. The point of intersection of the three paths of the plumb line approximates the location of the CG (Figure 13-24).

FIGURE 13-23

The center of gravity is the single point associated with a body around which the body's weight is equally balanced in all directions.

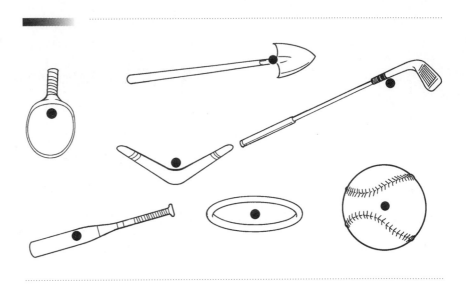

The running kinematics of a young child include noticeable vertical oscillations of the CG.

The location of a body's CG is of interest because, mechanically, a body behaves as though all of its mass were concentrated at the CG. For example, when the human body acts as a projectile, the body's CG follows a parabolic trajectory, regardless of any changes in the configurations of the body segments while in the air. Another implication is that when a weight vector is drawn for an object displayed in a free body diagram, the weight vector acts at the CG. Because the body's mechanical behavior can be traced by following the path of the total body CG, this factor has been studied as a possible indicator of performance proficiency in several sports.

For example, it has been hypothesized that skilled runners display less vertical oscillation of the CG during performance. Although this has not been well documented for adult runners (20), in children of 4 to 7 years of age, vertical oscillation of the CG decreases with age and concomitant maturation of running gait (13).

The path of the CG during takeoff in several of the jumping events is one factor believed to distinguish skilled from less-skilled performance. Research indicates that better Fosbury flop style high jumpers employ both body lean and body flexion (especially of the support leg) just before takeoff to lower the CG and prolong support foot contact time, thus resulting in increased takeoff impulse (3). In the long jump, better athletes maintain a normal sprinting stride, with CG height relatively constant, through the second-last step (8). During the last step, however, they markedly lower CG height, then increase CG height going into the jump step (8). Among better pole vaulters there is progressive elevation of the CG from the third-last step through takeoff. This is partially due to the elevation of the arms as the vaulter prepares

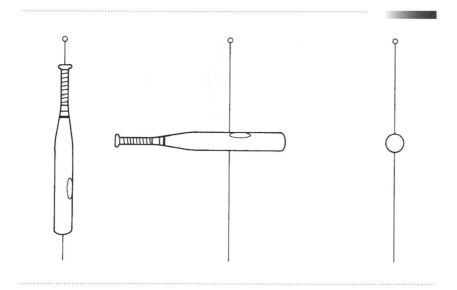

FIGURE 13-24
The suspension method for determining CG location involves suspending an object from a given point about which it is free to rotate. When the object comes to rest, a plumb line suspended from the same point can be used to mark a vertical line on the object about which it is balanced. When the procedure is repeated with the object aligned in two other planes, the intersection of the three lines approximates the location of the CG.

to plant the pole. However, research indicates that better vaulters lower their hips during the second-last step, then progressively elevate the hips (and the CG) through takeoff.

The strategy of lowering the CG prior to takeoff enables the athlete to lengthen the vertical path over which the body is accelerated during takeoff, thus facilitating a high vertical velocity at takeoff (Figure 13-25). The speed and angle of takeoff primarily determine the trajectory of the performer's CG during the jump. The only other influencing factor is air resistance, which exerts an extremely small effect on performance in the jumping events.

Locating the Human Body Center of Gravity

Locating the CG for a body containing two or more movable, interconnected segments is more difficult than doing so for a nonsegmented body, because every time the body changes configuration, its weight distribution and CG location are changed. Every time an arm, leg, or finger moves, the CG location as a whole is shifted at least slightly in the direction in which the weight is moved.

Some relatively simple procedures exist for determining the location of the CG of the human body. In the seventeenth century, the Italian mathematician Borelli used a simple balancing procedure for CG location that involved positioning a person on a wooden board (Figure 13-26). A more sophisticated version of this procedure enables calculation

The speed and projection angle of an athlete's total body center of mass largely determine performance outcome in the high jump.

FIGURE 13-25

Height of the athlete's CG during preparation for takeoff in the long jump. (Modified from Nixdorf E and Bruggemann P: Zur Absprungvorbereitung beim Weitsprung—Eine biomechanische Untersuchung zum problem der Korperschwerpunktsenkung, Lehre Leichtathlet p. 1539, 1983.)

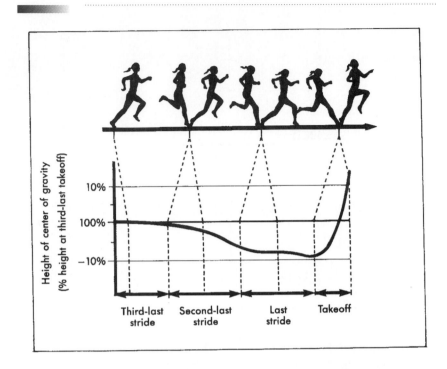

reaction board
specially constructed board for determining the center of gravity location of a body positioned on top of it

Location of the CG of the human body is complicated by the fact that its constituents (such as bone, muscle, and fat) have different densities and are unequally distributed throughout the body.

segmental method
procedure for determining total body center of mass location based on the masses and center of mass locations of the individual body segments

of the location of the plane passing through the CG of a person positioned on a **reaction board.** This procedure requires the use of a scale, a platform of the same height as the weighing surface of the scale, and a rigid board with sharp supports on either end (Figure 13-27). The calculation of the location of the plane containing the CG involves the summation of torques acting about the platform support. Forces creating torques at the support include the person's body weight, the weight of the board, and the reaction force of the scale on the platform (indicated by the reading on the scale). Although the platform also exerts a reaction force on the board, it creates no torque, because the distance of that force from the platform support is 0. Since the reaction board and subject are in static equilibrium, the sum of the three torques acting at the platform support must be 0, and the distance of the subject's CG plane to the platform may be calculated (Figure 13-28).

A commonly used procedure for estimating the location of the total body CG from projected film images of the human body is known as the **segmental method.** This procedure is based on the concept that since the body is composed of individual segments (each with an individual CG), the location of the total body CG is a function of the locations of the respective segmental CGs. Some body segments, however,

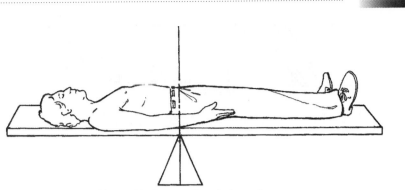

Top and bottom portions balanced

FIGURE 13-26
The relatively crude procedure devised by seventeenth century mathematician Borelli for approximating the CG location of the human body.

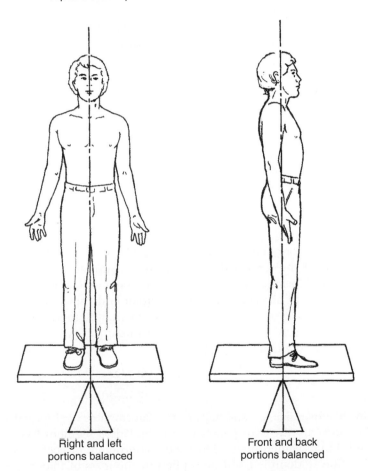

Right and left
portions balanced

Front and back
portions balanced

FIGURE 13-27
By summing torques at
point a, d (the distance
from a to the subject's
CG) may be calculated.

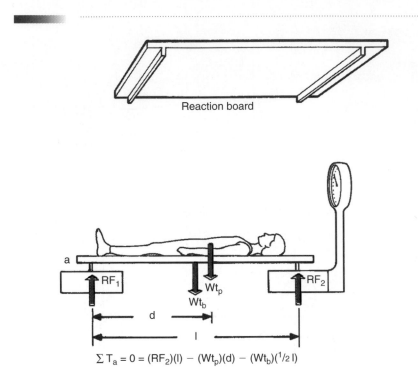

Reaction board

$$\Sigma T_a = 0 = (RF_2)(l) - (Wt_p)(d) - (Wt_b)(^1\!/_2\,l)$$

*The location of the CG of a
multisegmented object is more
influenced by the positions of
the heavier segments than by
those of the lighter segments.*

*The segmental method is most
commonly implemented
through a computer program
that reads x,y coordinates of
joint centers from a file
created by a digitizer.*

are much more massive than others and have a larger influence on the
location of the total body CG. When the products of each body segment's CG location and its mass are summed and subsequently divided
by the sum of all segmental masses (total body mass), the result is the
location of the total body CG. The segmental method uses data for average locations of individual body segment CGs as related to a percentage of segment length (2, 4):

$$X_{cg} = \Sigma(x_s)(m_s)/\Sigma m_s$$
$$Y_{cg} = \Sigma(y_s)(m_s)/\Sigma m_s$$

In this formula, X_{cg} and Y_{cg} are the coordinates of the total body CG,
x_s and y_s are the coordinates of the individual segment CGs, and m_s is
individual segment mass. Thus the x coordinate of each segment's CG
location is identified and multiplied by the mass of that respective segment. The $(x_s)(m_s)$ products for all of the body segments are then
summed and subsequently divided by total body mass to yield the x coordinate of the total body CG location. The same procedure is followed
to calculate the y coordinate for total body CG location (Figure 13-29).

FIGURE 13-28

S A M P L E   P R O B L E M   6

Find the distance from the platform support to the subject's CG, given the following information for the diagram in Figure 13-27:

Known

$$\text{Mass (subject)} = 73 \text{ kg}$$
$$\text{Mass (board alone)} = 28 \text{ kg}$$
$$\text{Scale reading} = 44 \text{ kg}$$
$$l_b = 2 \text{ m}$$

Solution

$$Wt_p = (73 \text{ kg})(9.81 \text{ m/s}^2)$$
$$= 716.13 \text{ N}$$
$$Wt_b = (44 \text{ kg})(9.81 \text{ m/s}^2)$$
$$= 431.64 \text{ N}$$
$$RF_2 = (66 \text{ kg})(9.81 \text{ m/s}^2)$$
$$= 647.46 \text{ N}$$

Use an equation of static equilibrium:

$$\Sigma T_a = 0 = (RF_2)(l) - (Wt_p)(d) - (Wt_b)(\tfrac{1}{2}l)$$
$$0 = (647.46 \text{ N})(2 \text{ m}) - (716.13 \text{ N})(d) - (431.64 \text{ N}\tfrac{1}{2})(2 \text{ m})$$
$$d = 1.2 \text{ m}$$

STABILITY AND BALANCE

A concept closely related to the principles of equilibrium is **stability.** Stability is defined mechanically as resistance to both linear and angular acceleration or resistance to disruption of equilibrium. In some circumstances, such as a Sumo wrestling contest or the defense of a quarterback by an offensive lineman, maximizing stability is desirable. In other situations, an athlete's best strategy is to intentionally minimize stability. Sprinters and swimmers in the preparatory stance before the start of a race intentionally assume a body position allowing them to accelerate quickly and easily at the sound of the starter's pistol. An individual's ability to control equilibrium is known as **balance.**

Different mechanical factors affect a body's stability. According to Newton's second law of motion (F = ma), the more massive an object is, the greater is the force required to produce a given acceleration. Football linemen who are expected to maintain their positions despite the forces exerted on them by opposing linemen are therefore more

stability
resistance to disruption of equilibrium

balance
ability to control equilibrium

FIGURE 13-29

S A M P L E   P R O B L E M   7

The x,y coordinates of the CGs of the upper arm, forearm, and hand segments are provided on the diagram below. Use the segmental method to find the CG for the entire arm, using the data provided for segment masses.

Known

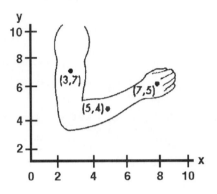

SEGMENT	MASS%	x	(x)(MASS %)	y	(y)(MASS %)
Upper arm	0.45				
Forearm	0.43				
Hand	0.12				
Σ					

Solution

First list the x and y coordinates in their respective columns, and then calculate and insert the product of each coordinate and the mass percentage for each segment into the appropriate columns. Sum the product columns, which yield the x,y coordinates of the total arm CG.

SEGMENT	MASS%	x	(x)(MASS %)	y	(y)(MASS%)
Upper arm	0.45	3	1.35	7	3.15
Forearm	0.43	5	2.15	4	1.72
Hand	0.12	7	0.84	5	0.60
Σ			4.34		5.47

x = 4.34

y = 5.47

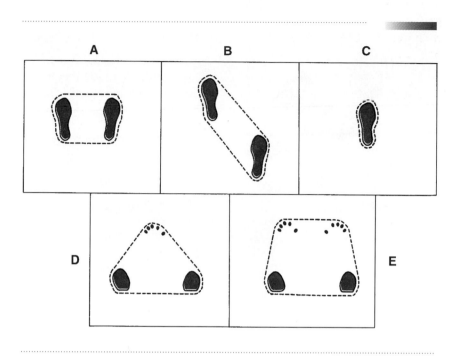

FIGURE 13-30
The base of support for **A**, a square stance, **B**, an angled stance, **C**, a one foot stance, **D**, a three point stance, and **E**, a four point stance. Areas of contact between body parts and the support surface are shaded. The base of support is the area enclosed by the dashed line.

mechanically stable if they are more massive. In contrast, gymnasts are at a disadvantage with greater body mass because execution of most gymnastic skills involves disruption of stability.

The greater the amount of friction is between an object and the surface or surfaces it contacts, the greater is the force requirement for initiating or maintaining motion. Toboggans and racing skates are designed so that the friction they generate against the ice will be minimal, enabling a quick disruption of stability at the beginning of a run or race. However, racquetball, golf, and batting gloves are designed to increase the stability of the player's grip on the implement.

Another factor affecting stability is the size of the **base of support.** This consists of the area enclosed by the outermost edges of the body in contact with the supporting surface or surfaces (Figure 13-30). When the line of action of a body's weight (directed from the CG) moves outside the base of support, a torque is created that tends to cause angular motion of the body, thereby disrupting stability, with the CG falling toward the ground. The larger the base of support is, the lower is the likelihood that this will occur. Martial artists typically assume a wide stance during defensive situations to increase stability. Alternatively, sprinters in the starting blocks maintain a relatively small base of support so that they can quickly disrupt stability at the start of the race. Maintaining balance during an *arabesque on pointe*, in which the dancer

base of support
area bound by the outermost regions of contact between a body and support surface or surfaces

Performing an arabesque en pointe *requires excellent balance because lateral movement of the dancer's line of gravity outside the small base of support will result in loss of balance.*

A swimmer on the blocks positions her center of gravity close to the front boundary of her base of support to prepare for forward acceleration.

is balanced on the toes of one foot, requires continual adjustment of CG location through subtle body movements (12).

The horizontal location of the CG relative to the base of support can also influence stability. The closer the horizontal location of the CG is to the boundary of the base of support, the smaller is the force required to push it outside the base of support, thereby disrupting equilibrium. Athletes in the starting position for a race consequently assume stances that position the CG close to the forward edge of the base of support. Alternatively, if a horizontal force must be sustained, stability is enhanced if the CG is positioned closer to the oncoming force, since the CG can be displaced farther before being moved outside the base of support. Sumo wrestlers lean toward their opponents when being pushed.

The height of the CG relative to the base of support can also affect stability. The higher the positioning of the CG, the greater the potentially disruptive torque created if the body undergoes an angular dis-

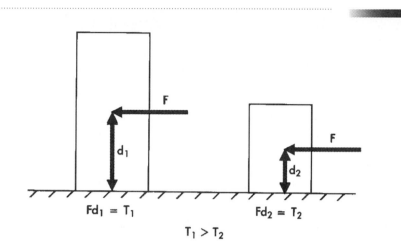

$$Fd_1 = T_1 \qquad Fd_2 = T_2$$

$$T_1 > T_2$$

FIGURE 13-31
The higher the CG location, the greater the amount of torque its motion creates about the intersection of the line of gravity and the support surface.

placement (Figure 13-31). During an earthquake when the motion of the ground causes a building to oscillate horizontally, taller buildings (with higher CGs) undergo more horizontal motion than one-story buildings, and there is more horizontal motion with each successive floor up. Athletes often crouch in sport situations when added stability is desirable. A common instructional cue for beginners in many sports is "Bend your knees!"

Although these principles of stability (summarized in the shaded box) are generally true, their application to the human body should be made only with the recognition that neuromuscular factors are also influential. Changes in foot position have been found to affect two measures of standing balance: the location of the line of gravity and postural sway. Research has shown that frontal plane sway decreases when the base of support is widened to 15 cm, but that increasing the width of the base of support beyond 15 cm does not further reduce body sway in the frontal plane. Researchers have also found that when people stand with one foot 30 cm in front of the other, frontal plane sway increases, but a surprising increase in sagittal plane sway occurs as well. The angle of foot position during normal stance does not affect balance; however, an extreme toe-in position causes more body sway (10).

Although under normal conditions the size of the base of support is a primary determiner of stability, recent research shows that a variety of other factors can also limit control of balance. Friction coefficient levels of less than 0.82, reduction in dorsiflexion range of motion greater than 51%, and reduction in plantar flexion strength of more than 35% are all risk factors for falling (14). More research is needed to clarify the application of the principles of stability to human balance.

PRINCIPLES OF MECHANICAL STABILITY

When other factors are held constant, a body's ability to maintain equilibrium is increased by the following:

1. Increasing body mass
2. Increasing friction between the body and the surface or surfaces contacted
3. Increasing the size of the base of support in the direction of the line of action of an external force
4. Horizontally positioning the center of gravity near the edge of the base of support on the oncoming external force
5. Vertically positioning the center of gravity as low as possible

SUMMARY

Rotary motion is caused by torque, a vector quantity with magnitude and direction. When a muscle develops tension, it produces torque at the joint or joints that it crosses. Rotation of the body segment occurs in the direction of the resultant joint torque.

Mechanically, muscles and bones function as levers. Most joints function as third class lever systems, well structured for maximizing range of motion and movement speed, but requiring muscle force of greater magnitude than that of the resistance to be overcome. The angle at which a muscle pulls on a bone also affects its mechanical effectiveness because only the rotary component of muscle force produces joint torque.

When a body is motionless, it is in static equilibrium. The three conditions of static equilibrium are $\Sigma F_v = 0$, $\Sigma F_h = 0$, and $\Sigma T = 0$. A body in motion is in dynamic equilibrium when inertial factors are considered.

The mechanical behavior of a body subject to force or forces is greatly influenced by the location of its center of gravity—the point around which the body's weight is equally balanced in all directions. Different procedures are available for determining center of gravity location.

A body's mechanical stability is its resistance to both linear and angular acceleration. A number of factors influence a body's stability, including mass, friction, center of gravity location, and base of support.

INTRODUCTORY PROBLEMS

1. Why does a force directed through an axis of rotation not cause rotation at the axis?
2. Why does the orientation of a force acting on a body affect the amount of torque it generates at an axis of rotation within the body?
3. A 23 kg boy sits 1.5 m from the axis of rotation of a seesaw. At what distance from the axis of rotation must a 21 kg boy be positioned on the other side of the axis to balance the seesaw? (Answer: 1.6 m)
4. How much force must be produced by the biceps brachii at a perpendicular distance of 3 cm from the axis of rotation at the elbow to support a weight of 200 N at a perpendicular distance of 25 cm from the elbow? (Answer: 1667 N)
5. Two people push on opposite sides of a swinging door. If A exerts a force of 40 N at a perpendicular distance of 20 cm from the hinge and B exerts a force of 30 N at a perpendicular distance of 25 cm from the hinge, what is the resultant torque acting at the hinge and which way will the door swing? (Answer: $T_h = 0.5$ N-m; in the direction that A pushes)
6. To which lever classes do a golf club, a swinging door, and a broom belong? Explain your answers, including free body diagrams.
7. Is the mechanical advantage of a first class lever greater than, less than, or equal to one? Explain.

8. Using a diagram, identify the magnitudes of the rotary and stabilizing components of a 100 N muscle force that acts at an angle of 20° to a bone. (Answer: rotary component = 34 N, stabilizing component = 94 N)
9. A 10 kg block sits motionless on a table in spite of an applied horizontal force of 2 N. What are the magnitudes of the reaction force and friction force acting on the block? (Answer: R = 98.1 N, F = 2 N)
10. Given the following data for the reaction board procedure, calculate the distance from the platform support to the subject's CG: RF_2 = 400 N, 1 = 2.5 m, wt = 600 N. (Answer: 1.67 m)

1. For one joint of the lower extremity, explain why eccentric torque occurs during gait.
2. Select one human motor skill with which you are familiar and sketch a graph showing how you would expect center of gravity height to change during that skill. Use Figure 13-29 as a model.
3. A 35 N hand and forearm are held at a 45° angle to the vertically oriented humerus. The CG of the forearm and hand is located at a distance of 15 cm from the joint center at the elbow, and the elbow flexor muscles attach at an average distance of 3 cm from the joint center. (Assume that the muscles attach at an angle of 45° to the bones.)
   a. How much force must be exerted by the forearm flexors to maintain this position?
   b. How much force must the forearm flexors exert if a 50 N weight is held in the hand at a distance along the arm of 25 cm? (Answer: a. 175 N; b. 591.7 N)
4. A hand exerts a force of 90 N on a scale at 32 cm from the joint center at the elbow. If the triceps attach to the ulna at a 90° angle and at a distance of 3 cm from the elbow joint center and if the weight of the forearm and hand is 40 N with the hand/forearm CG located 17 cm from the elbow joint center, how much force is being exerted by the triceps? (Answer: 733.3 N)

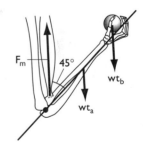

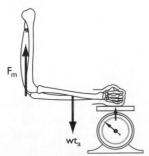

5. A patient rehabilitating a knee injury performs knee extension exercises wearing a 15 N weight boot. Calculate the amount of torque generated at the knee by the weight boot for the four positions shown, given a distance of 0.4 m between the weight boot's CG and the joint center at the knee. (Answer: a. 0; b. 3 N-m; c. 5.2 N-m; d. 6 N-m)

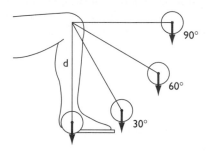

6. A 600 N person picks up a 180 N suitcase positioned so that the suitcase's CG is 20 cm lateral to the location of the person's CG before picking up the suitcase. If the person does not lean to compensate for the added load in any way, where is the combined CG location for the person and suitcase with respect to the person's original CG location? (Answer: Shifted 4.6 cm toward the suitcase)

7. A worker leans over and picks up a 90 N box at a distance of 0.7 m from the axis of rotation in her spine. Neglecting the effect of body weight, how much added force is required of the low back muscles with an average moment arm of 6 cm to stabilize the box in the position shown? (Answer: 1050 N)

8. A man carries a 3 m, 32 N board over his shoulder. If the board extends 1.8 m behind the shoulder and 1.2 m in front of the shoulder, how much force must the man apply vertically downward with his hand that rests on the board 0.2 m in front of the shoulder to stabilize the board in this position? (Assume that the weight of the board is evenly distributed throughout its length.) (Answer: 48 N)

9. A therapist applies a lateral force of 80 N to the forearm at a distance of 25 cm from the axis of rotation at the elbow. The biceps attaches to the radius at a 90° angle and at a distance of 3 cm from the elbow joint center.

a. How much force is required of the biceps to stabilize the arm in this position?

b. What is the magnitude of the reaction force exerted by the humerus on the ulna? (Answer: a. 666.7 N; b. 586.7 N)

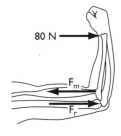

10. Tendon forces T_a and T_b are exerted on the patella. The femur exerts force F on the patella. If the magnitude of T_b is 80 N, what are the magnitudes of T_a and F, if no motion is occurring at the joint? (Answer: T_a = 44.8 N, F = 86.1 N)

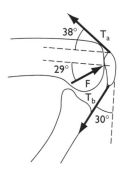

1. Experiment by removing lug nuts from an automobile tire with a lug wrench, a lug wrench with a short extension on the handle, and a lug wrench with a longer extension on the handle. Write a paragraph explaining your findings and draw a free body diagram showing the applied force, resistance, and axis of rotation. What provides the resistance?

2. Position a pole across the back of a chair (serving as a fulcrum) and hang a 2 lb weight on one end of the pole. Position a 5 lb weight on the other side of the pole such that the weights are balanced. Measure and record the distances of the two weights from the fulcrum and write an explanation of your results.

3. Perform curl-up exercises under the following conditions: 1) arms folded across the chest, 2) hands behind the neck, 3) holding a 5 lb weight above the head. Write a paragraph explaining your findings and draw a free body diagram showing the applied force, resistance, and axis of rotation.

4. Use the reaction board procedure to calculate the sagittal, frontal, and transverse plane positions of the center of gravity of a subject in anatomical position. Repeat the calculations with the subject 1) extending both arms overhead, and 2) extending one arm to the right. Present your results in a table and write a paragraph of explanation.

5. Using a picture of a person from a magazine or photograph and the anthropometric data from Appendix D, calculate and mark the location of total body center of gravity using the segmental method. First draw and scale x and y axes around the picture. Next, mark the approximate locations of segmental centers of gravity on the picture, using the data from Appendix D. Finally, construct a table using the table in Sample Problem 7 as a model.

REFERENCES

1. Cappozzo A et al: Lumbar spine loading during half-squat exercises, Med Sci Sports Exerc 17:613, 1985.
2. Clauser CE, McConville JT, and Young JW: *Weight, volume, and center of mass of segments of the human body,* AMRL Tech Rep, 1969, Wright-Patterson Air Force Base.
3. Dapena J, McDonald C, and Cappaert J: A regression analysis of high jumping technique, Int J Sport Biomech, 6:246, 1990.
4. Dempster WT: *Space requirements of the seated operator,* WADC Tech Rep 55–159, 1955, Wright-Patterson Air Force Base.
5. DeVita P, Torry M, Glover KL, and Speroni DL: A functional knee brace alters joint torque and power patterns during walking and running, J Biomech 29:583, 1996.
6. Garhammer J: Weight lifting and training. In Vaughan CL, ed: *Biomechanics of sport,* Boca Raton, FL 1989, CRC Press, Inc.
7. Hay JG, Andrews JG, and Vaughan CL: The influence of external load on joint torques exerted in a squat exercise, Proceedings of the biomechanics symposium, Bloomington, Ind, 1980.
8. Hay, JG and Nohara, H: The techniques used by elite long jumpers in preparation for take-off, J Biomech, 23:229, 1990.
9. Hay JG et al: Load, speed, and equipment effects in strength training exercises. In Matsui H and Kobayashi K, eds: *Biomechanics VIII-B,* Champaign, Ill, 1983, Human Kinetic Publishers, Inc. Press.
10. Kirby RL, Price NA, and MacLeod DA: The influence of foot position on standing balance, J Biomech 20:423, 1987.
11. Konczak J, Borutta M, and Dichgans J: The development of goal-directed reaching in infants. II. Learning to produce task-adequate patterns of joint torque, Exp Brain Res 113:465, 1997.
12. Laws K: *The physics of dance,* New York, 1984, Schirmer Books.
13. Miyamaru M et al: Path of the whole body center of gravity for young children in running. In Jonsson B, ed: *Biomechanics X-B,* Champaign, Ill, 1987, Human Kinetics Publishers, Inc.
14. Pai YC and Patton J: Center of mass velocity-position predictions for balance control, J Biomech 30:347, 1997.
15. Redfield R and Hull ML: On the relation between joint moments and pedalling rates at constant power in bicycling, J Biomech 19:317, 1986.
16. Townend MS: *Mathematics in sport,* New York, 1984, John Wiley & Sons, Inc.
17. van der Woude LH, Botden E, Vriend I, and Veeger D: Mechanical advantage in wheelchair lever propulsion: Effect on physical strain and efficiency, J Rehabil Res Dev 34:286, 1997.

18. van Zuylen EJ, van Zelzen A, and van der Gon JJD: A biomechanical model for flexion torques of human arm muscles as a function of elbow angle, J Biomech, 21:183, 1988.

19. Veeger HEJ, Van der Woude LHV, and Rozendal RH: Effect of handrim velocity on mechanical efficiency in wheelchair propulsion, Med Sci Sports Exerc, 24:100, 1992.

20. Williams KR: Biomechanics of running, Exerc Sport Sci Rev 13:389, 1985.

ADDITIONAL READINGS

Garhammer J: Weight lifting and training. In Vaughan CL, ed: *Biomechanics of sport,* Boca Raton, Fla, 1989, CRC Press, Inc.
Critically reviews the scientific literature on weight lifts and resistance training.

Hay JF: Citius, altius, longius (faster, higher, longer): The biomechanics of jumping for distance, J Biomech, 26:7, 1993.
Reviews current research on biomechanical factors affecting performance in the long jump and triple jump.

Laws K: *The physics of dance,* New York, 1984, Schirmer Books.
Relates principles of torque production and balance to the performance of various ballet movements.

Winter DA: *Biomechanics and motor control of human movement,* 2nd ed, New York, 1990, John Wiley & Sons, Inc.
Chapter on kinetics includes useful discussions on interpreting moment of force (torque) curves and the difference between the center of gravity and center of pressure.

RELATED WEB SITES

Advanced Medical Technology, Inc.
http://www.amtiweb.com
Provides information on the AMTI force platforms, with reference to ground reaction forces in gait analysis, balance and posture, and other topics.

Interactive Physics Problem Set
http://info.itp.berkeley.edu/Vol1/Contents.html
Provides links to interactive problems related to use of the equations of static equilibrium.

14

Chapter

ANGULAR KINETICS OF HUMAN MOVEMENT

After completing this chapter, the reader will be able to:

Identify the angular analogues of mass, force, momentum, and impulse.

Explain why changes in the configuration of a rotating airborne body can produce changes in the body's angular velocity.

Identify and provide examples of the angular analogues of Newton's laws of motion.

Define centripetal force and explain where and how it acts.

Solve quantitative problems relating to the factors that cause or modify angular motion.

W hy do sprinters run with more swing phase flexion at the knee than distance runners? Why do dancers and ice skaters spin more rapidly when their arms are brought in close to the body? How do cats always land on their feet? In this chapter more concepts pertaining to angular kinetics are explored from the perspective of the similarities and differences between linear and angular kinetic quantities.

RESISTANCE TO ANGULAR ACCELERATION

Moment of Inertia

Inertia is a body's tendency to resist acceleration (see Chapter 3). Although inertia itself is a concept rather than a quantity that can be measured in units, a body's inertia is directly proportional to its mass (Figure 14-1). According to Newton's second law, the greater a body's mass, the greater its resistance to linear acceleration. Therefore,

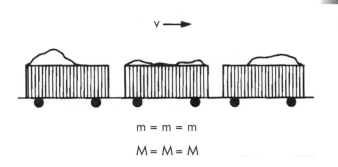

FIGURE 14-1
The distribution of mass in
a system does not affect its
linear momentum.

mass is a body's inertial characteristic for considerations relative to linear motion.

Resistance to angular acceleration is also a function of a body's mass. The greater the mass, the greater the resistance to angular acceleration. However, the relative ease or difficulty of initiating or halting angular motion depends on an additional factor: the distribution of mass with respect to the axis of rotation.

Consider the baseball bats shown in Figure 14-2. Suppose a player warming up in the batter's box adds a weight ring to the bat he is swinging. Will the relative ease of swinging the bat be greater with the weight positioned near the striking end of the bat or near the bat's grip? Similarly, is it easier to swing a bat held by the grip (the normal hand position) or with the bat turned around and held by the barrel?

Experimentation with a baseball bat or some similar object makes it apparent that the more closely concentrated the mass is to the axis of rotation, the easier it is to swing the object. Conversely, the more mass is positioned away from the axis of rotation, the more difficult it is to initiate (or stop) angular motion. Resistance to angular acceleration, therefore, depends not only on the amount of mass possessed by an object but also on the distribution of that mass with respect to the axis of rotation. The inertial property for angular motion must therefore incorporate both factors.

The inertial property for angular motion is **moment of inertia,** represented as [I.] Every body is composed of particles of mass, each with its own particular distance from a given axis of rotation. The

The more closely mass is distributed to the axis of rotation, the easier it is to initiate or stop angular motion.

moment of inertia
inertial property for rotating bodies representing resistance to angular acceleration; based on both mass and the distance the mass is distributed from the axis of rotation

FIGURE 14-2
Although both bats have
the same mass, Bat A is
harder to swing than Bat B
because the weight ring on
it is positioned farther
from the axis of rotation.

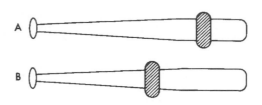

moment of inertia for a single particle of mass may be represented as
the following:

$$I = mr^2$$

In this formula, m is the particle's mass and r is the particle's radius
of rotation. The moment of inertia of an entire body is the sum of
the moments of inertia of all the mass particles the object contains
(Figure 14-3):

$$I = \Sigma mr^2$$

The distribution of mass with respect to the axis of rotation is more important than the total amount of body mass in determining resistance
to angular acceleration because r is squared. Since r is the distance between a given particle and an axis of rotation, values of r change as the
axis of rotation changes. Thus, when a player grips a baseball bat, "chok-
ing up" on the bat reduces the bat's moment of inertia with respect to
the axis of rotation at the player's wrists and concomitantly increases
the relative ease of swinging the bat. Little League baseball players often unknowingly make use of this concept when swinging bats that are
longer and heavier than they can effectively handle.

Within the human body, the distribution of mass with respect to an
axis of rotation can dramatically influence the relative ease or difficulty
of moving the body limbs. For example, during gait, the distribution of
a given leg's mass, and therefore its moment of inertia with respect to
the primary axis of rotation at the hip, depends largely on the angle
present at the knee. In sprinting, maximum angular acceleration of the
legs is desired, and considerably more flexion at the knee is present during the swing phase than during running at slower speeds. This greatly
reduces the moment of inertia of the leg with respect to the hip, thus
reducing resistance to hip flexion. Runners who have leg morphology
involving mass distribution closer to the hip with more massive thighs
and slimmer lower legs than others have a smaller moment of inertia
of the leg with respect to the hip. This is an anthropometric characteristic that contributes to running economy (2). During walking, in which

*During sprinting, extreme
flexion at the knee reduces
the moment of inertia of
the swinging leg.*

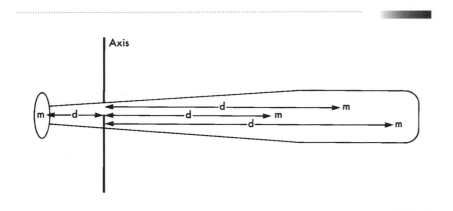

FIGURE 14-3
Moment of inertia is the
sum of the products of
each particle's mass and
radius of rotation squared.

minimal angular acceleration of the legs is required, flexion at the knee
during the swing phase remains relatively small, and the leg's moment
of inertia with respect to the hip is relatively large.

Modern-day golf irons are commonly constructed with the heads bottom weighted, perimeter weighted, or heel and toe weighted. These manipulations of the amount of mass and the distribution of mass within
the head of the club are designed to increase club head inertia, thus reducing the tendency of the club to rotate about the shaft during an off-
center hit. Research findings indicate that perimeter weighted club
heads perform best for eccentric ball contacts outside the club head
CG, with a simple blade head club superior for contacts below the club
head CG (15). The most consistent performance, however, was exhibited by a toe and bottom weighted club head, which was second best for
all eccentric hits (15). A goffer's individual preference, feel, and experience should ultimately determine selection of club type.

Determining Moment of Inertia

Assessing moment of inertia for a body with respect to an axis by measuring the distance of each particle of body mass from an axis of rotation and then applying the formula is obviously impractical. In practice,
mathematical procedures are used to calculate moment of inertia for
bodies of regular geometric shapes and known dimensions. Because the
human body is composed of segments that are of irregular shapes and
heterogeneous mass distributions, either experimental procedures or
mathematical models are used to approximate moment of inertia values for individual body segments and for the body as a whole in different positions. Moment of inertia for the human body and its segments
has been approximated by using average measurements from cadaver

*The fact that bone, muscle,
and fat have different densities
and are distributed dissimilarly
in individuals complicates
efforts to calculate human
body segment moments of
inertia.*

FIGURE 14-4

Knee angle affects the moment of inertia of the swinging leg with respect to the hip because of changes in the radius of gyration for the lower leg (k_2) and foot (k_3).

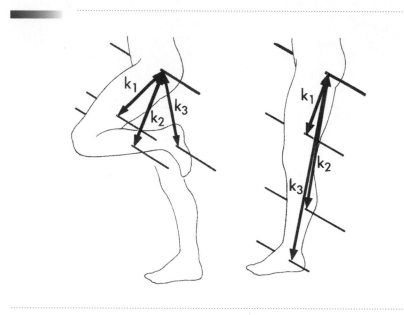

Because there are formulas available for calculating the moment of inertia of regularly shaped solids, some investigators have modeled the human body as a composite of various geometric shapes.

radius of gyration
distance from the axis of rotation to a point where the body's mass could be concentrated without altering its rotational characteristics

studies, measuring the acceleration of a swinging limb, employing photogrammetric methods, and applying mathematical modeling (9).

Once moment of inertia for a body of known mass has been assessed, the value may be characterized using the following formula:

$$I = mk^2$$

In this formula, I is moment of inertia with respect to an axis, m is total body mass, and k is a distance known as the **radius of gyration.** The radius of gyration represents the object's mass distribution with respect to a given axis of rotation. It is the distance from the axis of rotation to a point at which the mass of the body can theoretically be concentrated without altering the inertial characteristics of the rotating body. This point is *not* the same as the segmental CG (Figure 14-4). Since the radius of gyration is based on r^2 for individual particles, it is always longer than the radius of rotation, the distance to the segmental CG.

The length of the radius of gyration changes as the axis of rotation changes. As mentioned earlier, it is easier to swing a baseball bat when the bat is grasped by the barrel end rather than by the bat's grip. When the bat is held by the barrel, k is much shorter than when the bat is held properly, since more mass is positioned close to the axis of rotation. Likewise, the radius of gyration for a body segment such as the forearm is greater with respect to the wrist than with respect to the elbow.

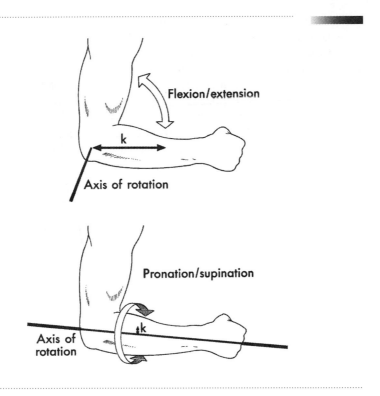

FIGURE 14-5
The radius of gyration (k) of the forearm for flexion/extension movements is much larger than for pronation/supination.

The radius of gyration is a useful index of moment of inertia when a given body's resistance to rotation with respect to different axes is discussed. Units of moment of inertia parallel the formula definition of the quantity, and therefore consist of units of mass multiplied by units of length squared ($kg \cdot m^2$).

Human Body Moment of Inertia

Moment of inertia can only be defined with respect to a specific axis of rotation. The axis of rotation for a body segment in sagittal and frontal plane motions is typically an axis passing through the center of a body segment's proximal joint. When a segment rotates around its own longitudinal axis, its moment of inertia is quite different from its moment of inertia during flexion and extension or abduction and adduction, because its mass distribution and therefore its moment of inertia are markedly different with respect to this axis of rotation. Figure 14-5 illustrates the difference in the lengths of the radii of gyration for the forearm with respect to the transverse and longitudinal axes of rotation.

principal axes
three mutually perpendicular axes passing through the total body center of gravity

principal moment of inertia
total body moment of inertia relative to one of the principal axes

The ratio of muscular strength (the ability of a muscle group to produce torque about a joint) to segmental moments of inertia (resistance to rotation at a joint) is an important contributor to performance capability in gymnastic events.

angular momentum
quantity of angular motion possessed by a body; measured as the product of moment of inertia and angular velocity

The moment of inertia of the human body as a whole is also different with respect to different axes. When the entire human body rotates free of support, it moves around one of three **principal axes:** the transverse (or frontal), the anteroposterior (or sagittal), or the longitudinal (or vertical) axis, each of which passes through the total body CG. Moment of inertia with respect to one of these axes is known as a **principal moment of inertia.** Figure 14-6 shows quantitative estimates of principal moments of inertia for the human body in several different positions. When the body assumes a tucked position during a somersault, its principal moment of inertia (and resistance to angular motion) about the transverse axis is clearly less than when the body is in anatomical position. Divers performing a somersaulting dive undergo changes in principal moment of inertia about the transverse axis on the order of 15 kg · m^2 to 6.5 kg · m^2 as the body goes from a layout position to a pike position (7).

As children grow from childhood through adolescence and into adulthood, developmental changes result in changing proportions of body segment lengths, masses, and radii of gyration, all affecting segment moments of inertia (8). Segment moments of inertia affect resistance to angular rotation and therefore performance capability in sports such as gymnastics and diving. Several prominent female gymnasts who achieved world-class status during early adolescence faded from the public view before reaching the age of 20 because of declines in their performance capabilities generally attributed to changes in body proportions with growth. Substantial changes in principal moments of inertia for the body segments occur with age, and large interindividual differences exist in the growth patterns of these principal moments of inertia (9). According to Jensen (8), the best predictor of moment of inertia values among children is the product of body mass and body height squared, $(m)(ht)^2$, rather than age.

ANGULAR MOMENTUM

Since moment of inertia is the inertial property for rotational movement, it is an important component of other angular kinetic quantities. As discussed in Chapter 12, the quantity of motion that an object possesses is referred to as its *momentum.* Linear momentum is the product of the linear inertial property (mass) and linear velocity. The quantity of angular motion that a body possesses is likewise known as **angular momentum.** Angular momentum, represented as [H], is the product of the angular inertial property (moment of inertia) and angular velocity:

For linear motion: $M = mv$
For angular motion: $H = I\omega$
Or: $H = mk^2\omega$

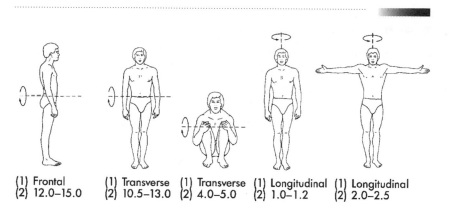

(1) Frontal
(2) 12.0–15.0

(1) Transverse
(2) 10.5–13.0

(1) Transverse
(2) 4.0–5.0

(1) Longitudinal
(2) 1.0–1.2

(1) Longitudinal
(2) 2.0–2.5

FIGURE 14-6
Principal moments of inertia of the human body in different positions with respect to different principal axes. (1) Principal axis. (2) Moment of inertia $(kg \cdot m^2)$. (Modified from Hochmuth G: Biomechanik sportlicher bewegungen, Frankfurt, Germany, 1967, Wilhelm Limpert, Verlag.)

Three factors affect the magnitude of a body's angular momentum: its mass (m), the distribution of that mass with respect to the axis of rotation (k), and the angular velocity of the body (ω). If a body has no angular velocity, it has no angular momentum. As mass or angular velocity increases, angular momentum increases proportionally. The factor that most dramatically influences angular momentum is the distribution of mass with respect to the axis of rotation, because angular momentum is proportional to the square of the radius of gyration (Figure 14-7). Units of angular momentum result from multiplying units of mass, units of length squared, and units of angular velocity, which yields $kg \cdot m^2/s$.

For a multisegmented object such as the human body, angular momentum about a given axis of rotation is the sum of the angular momenta of the individual body segments. During an airborne somersault, the angular momentum of a single segment such as the lower leg with respect to the principal axis of rotation passing through the total body CG consists of two components: the local term and the remote term. The local term is based on the segment's angular momentum about its own segmental CG, and the remote term represents the segment's angular momentum about the total body CG. Angular momentum for this segment about a principal axis is the sum of the local term and the remote term:

$$H = I_s \omega_s + mr^2 \omega_g$$

In the local term, I_s is the segment's moment of inertia and ω_s is the segment's angular velocity, both with respect to a transverse axis through the segment's own CG. In the remote term, m is the segment's mass, r is the distance between the total body and segmental CGs, and ω_g is the angular velocity of the segmental CG about the principal transverse axis (Figure 14-8). The sum of the angular momenta of all the body segments about a principal axis yields the total body angular momentum about that axis.

FIGURE 14-7

SAMPLE PROBLEM 1

Consider a rotating 10 kg body for which k = 0.2 m and ω = 3 rad/s. What is the effect on the body's angular momentum if the mass doubles? The radius of gyration doubles? The angular velocity doubles?

Solution
The body's original angular momentum is the following:

$$H = mk^2\omega$$
$$H = (10 \text{ kg}) \ (0.2 \text{ m})^2 \ (3 \text{ rad/s})$$
$$H = 1.2 \text{ kg} \cdot \text{m}^2/\text{s}$$

With mass doubled:

$$H = mk^2\omega$$
$$H = (20 \text{ kg}) \ (0.2 \text{ m})^2 \ (3 \text{ rad/s})$$
$$H = 2.4 \text{ kg} \cdot \text{m}^2/\text{s}$$

H is doubled.

With k doubled:

$$H = mk^2\omega$$
$$H = (10 \text{ kg}) \ (0.4 \text{ m})^2 \ (3 \text{ rad/s})$$
$$H = 4.8 \text{ kg} \cdot \text{m}^2/\text{s}$$

H is quadrupled.

With ω doubled:

$$H = mk^2\omega$$
$$H = (10 \text{ kg}) \ (0.2 \text{ m})^2 \ (6 \text{ rad/s})$$
$$H = 2.4 \text{ kg} \cdot \text{m}^2/\text{s}$$

H is doubled.

During takeoff from a springboard or platform, a competitive diver must attain sufficient linear momentum to reach the necessary height (and safe distance from the board or platform) and sufficient angular momentum to perform the required number of rotations. For multiple-rotation, nontwisting platform dives, the angular momentum generated at takeoff increases as the rotational requirements of the dive increase (6). Angular momentum values as high as 66 kg · m^2/s and 70 kg · m^2/s have been reported for multiple gold medalist Greg Louganis during his back two and a half and forward three and a half springboard dives, respectively (13). When a twist is also incorporated into a somersaulting dive, the angular momentum required is further increased. Inclusion of

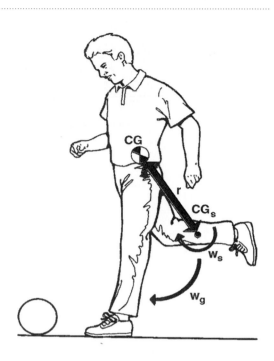

FIGURE 14-8
The angular momentum
of the swinging leg is the
sum of its local term, $I_s\omega_s$
and its remote term,
$mr^2\omega_g$.

a twist during forward one and a half springboard dives is associated with increased angular momentum at takeoff of 6% to 19% (17).

Conservation of Angular Momentum

Whenever gravity is the only acting external force, angular momentum is conserved. For angular motion, the principle of conservation of momentum may be stated as follows:

> The total angular momentum of a given system remains constant in the absence of external torques.

Gravitational force acting at a body's CG produces no torque because $d_\perp$ equals 0 and therefore it creates no change in angular momentum.

The principle of conservation of angular momentum is particularly useful in the mechanical analysis of diving, trampolining, and gymnastics events in which the human body undergoes controlled rotations while airborne. In a one and a half front somersault dive the diver leaves the springboard with a fixed amount of angular momentum. According to the principle of conservation of angular momentum, the amount of angular momentum present at the instant of takeoff remains constant throughout the dive. As the diver goes from an extended layout position

The magnitude and direction of the angular momentum vector for an airborne performer are established at the instant of takeoff.

FIGURE 14-9

When angular momentum is conserved, changes in body configuration produce a tradeoff between moment of inertia and angular velocity, with a tuck position producing greater angular velocity.

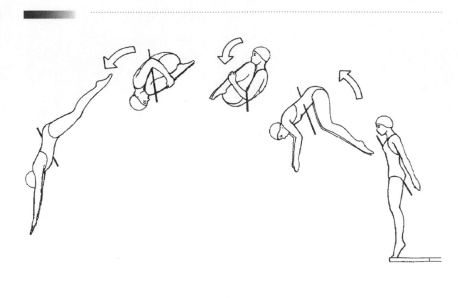

into a tuck, the radius of gyration is decreased, thus reducing the body's principal moment of inertia about the transverse axis. Because angular momentum remains constant, a compensatory increase in angular velocity must accompany the decrease in moment of inertia (Figure 14-9). The tighter the diver's tuck, the greater the angular velocity. Once the somersault is completed, the diver extends to a full layout position, thereby increasing total body moment of inertia with respect to the axis of rotation. Again, because angular momentum remains constant, an equivalent decrease in angular velocity occurs. For the diver to appear to enter the water perfectly vertically, minimal angular velocity is desirable. The sample problem in Figure 14-10 quantitatively illustrates this example.

Other examples of conservation of angular momentum occur when an airborne performer has a total body angular momentum of 0 and a forceful movement, such as a jump pass or volleyball spike, is executed. When a volleyball player performs a spike, moving the hitting arm with a high angular velocity and a large angular momentum, there is a compensatory rotation of the lower body producing an equal amount of angular momentum in the opposite direction (Figure 14-11). The moment of inertia of the two legs with respect to the hips is much greater than that of the spiking arm with respect to the shoulder. The angular velocity of the legs generated to counter the angular momentum of the swinging arm is therefore much less than the angular velocity of the spiking arm.

FIGURE 14-10

S A M P L E   P R O B L E M   2

A 60 kg diver is positioned so that his radius of gyration is 0.5 m as he leaves the board with an angular velocity of 4 rad/s. What is the diver's angular velocity when he assumes a tuck position, altering his radius of gyration to 0.25 m?

Known
m = 60 kg
 k = 0.5 m
ω = 4 rad/s

Position 1

Position 2

m = 60 kg
 k = 0.25 m

Solution
To find ω, calculate the amount of angular momentum that the diver possesses when he leaves the board, since angular momentum remains constant during the airborne phase of the dive:

Position 1:

$$H = mk^2\omega$$
$$H = (60 \text{ kg}) \ (0.5 \text{ m})^2 \ (4 \text{ rad/s})$$
$$H = 60 \text{ kg} \cdot \text{m}^2/\text{s}$$

Use this constant value for angular momentum to determine ω when k = 0.25 m:

Position 2:

$$H = mk^2\omega$$
$$60 \text{ kg} \cdot \text{m}^2/\text{s} = (60 \text{ kg})(0.25 \text{ m})^2(\omega)$$

$$\omega = 16 \text{ rad/s}$$

Transfer of Angular Momentum

Although angular momentum remains constant in the absence of external torques, transferring angular velocity at least partially from one principal axis of rotation to another is possible. This occurs when a diver changes from a primarily somersaulting rotation to one that is primarily twisting, and vice versa. An airborne performer's angular velocity vector does not necessarily occur in the same direction as the angular momentum vector. It is possible for a body's somersaulting angular momentum and its twisting angular momentum to be altered in midair, though the vector sum of the two (the total angular momentum) remains constant in magnitude and direction.

Researchers have observed several procedures for changing the total body axis of rotation. Asymmetrical arm movements and rotation of the hips (termed *hula movement*) can tilt the axis of rotation out of the original plane of motion (Figure 14-12). The lesser-used hula movement can produce tilting of the principal axis of rotation when the body is som-

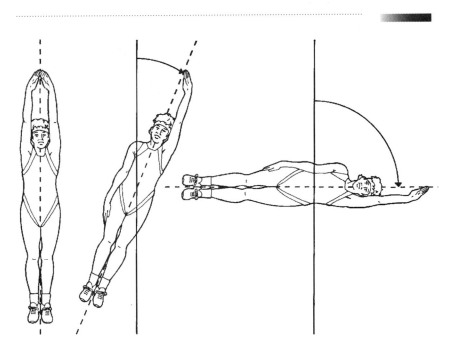

FIGURE 14-12
Asymmetrical positioning of the arms with respect to the axis of angular momentum can shift the axis of rotation.

ersaulting in a piked position. These asymmetrical movements can be used to generate twist and to eliminate twist (18). The results of a study on twisting somersault performances executed from a trampoline indicate that less-skilled performers tend to rush their asymmetrical movements, detracting from performance quality (16).

Even when total body angular momentum is 0, generating a twist in midair is possible using skillful manipulation of a body composed of at least two segments. Prompted by the observation that a domestic cat seems to always land on its feet no matter what position it falls from, scientists have studied this apparent contradiction of the principle of conservation of angular momentum (5). Gymnasts and divers can use this procedure, referred to as *cat rotation,* without violating the conservation of angular momentum.

Cat rotation is basically a two-phase process. It is accomplished most effectively when the two body segments are in a 90° pike position, so that the radius of gyration of one segment is maximum with respect to the longitudinal axis of the other segment (Figure 14-13). The first phase consists of the internally generated rotation of Segment 1 around its longitudinal axis. Because angular momentum is conserved, there is a compensatory rotation of Segment 2 in the opposite direction around the longitudinal axis of Segment 1. However, the resulting rotation is of a relatively small velocity, because k for Segment 2 is relatively large with respect to Axis 1. The second phase of the process consists of rotation

FIGURE 14-13

A skillful human performer can rotate 180° or more in the air with zero angular momentum because in a piked position there is a large discrepancy between the radii of gyration for the upper and lower extremities with respect to the longitudinal axes of these two major body segments, as explained in the text.

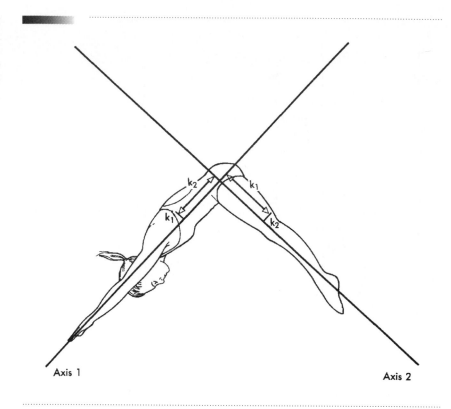

Axis 1   Axis 2

of Segment 2 around its longitudinal axis in the same direction originally taken by Segment 1. Accompanying this motion is a compensatory rotation of Segment 1 in the opposite direction around Axis 2. Again, angular velocity is relatively small, because k for Segment 1 is relatively large with respect to Axis 2. Using this procedure, a skilled diver can initiate a twist in midair and turn through as much as 450° (5). Cat rotation is performed around the longitudinal axes of the two major body segments. It is easier to initiate rotation about the longitudinal principal axis than about either the transverse or the anteroposterior principal axes, because total body moment of inertia with respect to the longitudinal axis is much smaller than the total body moments of inertia with respect to the other two axes.

Change in Angular Momentum

When an external torque does act, it changes the amount of angular momentum present in a system predictably. Just as with changes in linear momentum, changes in angular momentum depend not only on the magnitude and direction of acting external torques but also on the length of the time interval over which each torque acts:

$$\text{Linear impulse} = Ft$$
$$\text{Angular impulse} = Tt$$

When an **angular impulse** acts on a system, the result is a change in the total angular momentum of the system. The impulse-momentum relationship for angular quantities may be expressed as the following:

$$Tt = \Delta H$$
$$Tt = (I\omega)_2 - (I\omega)_1$$

As before, the symbols T, t, H, I, and ω represent torque, time, angular momentum, moment of inertia, and angular velocity respectively, and subscripts 1 and 2 denote initial and second or final points in time. Because angular impulse is the product of torque and time, significant changes in an object's angular momentum may result from the action of a large torque over a small time interval or from the action of a small torque over a large time interval. Since torque is the product of a force's magnitude and the perpendicular distance to the axis of rotation, both of these factors affect angular impulse. The effect of angular impulse on angular momentum is shown in Figure 14-14.

In the throwing events in track and field, the object is to maximize the angular impulse exerted on an implement before release to maximize its momentum and the ultimate horizontal displacement following release. As discussed in Chapter 11, linear velocity is directly related to angular velocity, with the radius of rotation serving as the factor of proportionality. As long as the moment of inertia (mk^2) of a rotating body remains constant, increased angular momentum translates directly to increased linear momentum when the body is projected. This concept is particularly evident in the hammer throw, in which the athlete first swings the hammer two or three times around the body with the feet planted, and then executes the next three or four whole body turns while facing the hammer before release. Some hammer throwers perform the first one or two of the whole body turns with the trunk in slight flexion (called *countering with the hips*), thereby enabling a farther reach with the hands (Figure 14-15). This tactic increases the radius of rotation, and thus the moment of inertia of the hammer with respect to the axis of rotation, so that if angular velocity is not reduced, the angular momentum of the thrower/hammer system is increased. For this strategy, the final turns are completed with the entire body leaning away from the hammer, or *countering with the shoulders*. Researchers have suggested that although the ability to lean forward throughout the turns should increase the angular momentum imparted to the hammer, a natural tendency to protect against excessive spinal stresses or shoulder strength limitations may prevent the thrower from accomplishing this technique modification (4).

angular impulse
change in angular momentum equal to the product of torque and the time interval over which the torque acts

FIGURE 14-14

SAMPLE PROBLEM 3

What average amount of force must be applied by the elbow flexors inserting at an average perpendicular distance of 1.5 cm from the axis of rotation at the elbow over a period of 0.3 seconds to stop the motion of the 3.5 kg arm swinging with an angular velocity of 5 rad/s when k = 20 cm?

Known

$d = 0.015$ m
$t = 0.3$ s
$m = 3.5$ kg
$k = 0.20$ m
$\omega = 5$ rad/s

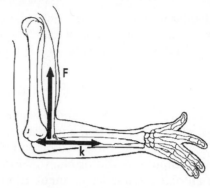

Solution

The impulse-momentum relationship for angular motion can be used.

$$Tt = \Delta H$$
$$Fdt = (mk^2\omega)_2 - (mk^2\omega)_1$$
$$F(0.015 \text{ m}) (0.3 \text{ s}) = 0 - (3.5 \text{ kg}) (0.20 \text{ m})^2 (5 \text{ rad/s})$$

$$F = 155.56 \text{ N}$$

FIGURE 14-15

A hammer thrower must counter the centrifugal force of the hammer to avoid being pulled out of the throwing ring. **A**, countering with the shoulders results in a smaller radius of rotation for the hammer than **B**, countering with the hips.

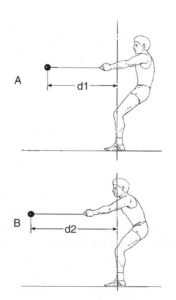

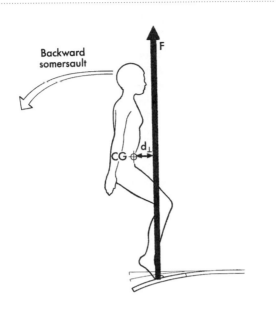

FIGURE 14-16
The product of the springboard reaction force (F) and its moment arm with respect to the diver's center of gravity (d$_\perp$) creates a torque, which generates the angular impulse that produces the diver's angular momentum at takeoff.

When a support surface reaction force is directed through the performer's center of gravity, linear but not angular impulse is generated.

The angular momentum required for the total body rotations executed during aerial skills is primarily derived from the angular impulse created by the reaction force of the support surface during takeoff. During back dives performed from a platform, the major angular impulse is produced during the final weighting of the platform, when the diver comes out of a crouched position through extension at the hip, knee, and ankle joints and executes a vigorous arm swing simultaneously (14). The vertical component of the platform reaction force, acting in front of the diver's CG, creates most of the backward angular momentum required (Figure 14-16).

On a springboard, the position of the fulcrum with respect to the tip of the board can usually be adjusted and can influence performance. Setting the fulcrum farther back from the tip of the board results in greater downward board tip vertical velocity at the beginning of takeoff, which allows the diver more time in contact with the board to generate angular momentum and increased vertical velocity going into the dive (10). Concomitant disadvantages, however, include the requirement of increased hurdle flight duration and the necessity of reversing downward motion from a position of greater flexion at the knees (10).

The motions of the body segments during takeoff determine the magnitude and direction of the reaction force generating linear and angular impulses. During both platform and springboard dives, the rotation of the arms at takeoff generally contributes more to angular

The arm swing during takeoff contributes significantly to the diver's angular momentum.

momentum than the motion of any other segment (6, 12). Highly skilled divers perform the arm swing with the arms fully extended, thus maximizing the moment of inertia of the arms and the angular momentum generated. Less-skilled divers often must use flexion at the elbow to reduce the moment of inertia of the arms about the shoulders so that arm swing can be completed during the time available (12). In contrast to the takeoff during a dive, during the takeoff for aerial somersaults performed from the floor in gymnastics, it is forceful extension of the legs that contributes the most to angular momentum (3).

Angular impulse produced through the support surface reaction force is also essential for performance of the *tour jeté*, a dance movement that consists of a jump accompanied by a 180° turn, with the dancer landing on the foot opposite the takeoff foot. When the movement is performed properly, the dancer appears to rise straight up and then rotate about the principal vertical axis in the air. In reality, the jump must be executed so that a reaction torque around the dancer's vertical axis is generated by the floor. The extended leg at the initiation of the jump creates a relatively large moment of inertia relative to the axis of rotation, thereby resulting in a relatively low total body angular velocity. At the peak of the jump the dancer's legs simultaneously cross the axis of rotation and the arms simultaneously come together overhead, close to the axis of rotation. These movements dramatically reduce moment of inertia, thus increasing angular velocity (11).

Similarly, when a skater performs a double or triple axel jump in figure skating, angular momentum is generated by the skater's movements and changes in total body moment of inertia prior to takeoff. Over half of the angular momentum for a double axel is generated during the preparatory glide on one skate going into the jump (1). Most of this angular momentum is contributed by motion of the free leg, which is extended somewhat horizontally to increase total body moment of inertia around the skater's vertical axis (1). As the skater becomes airborne, both legs are extended vertically and the arms are tightly crossed to minimize moment of inertia around the vertical axis and thereby maximize rotational velocity.

The surface reaction force is used by the dancer to generate angular momentum during the takeoff of the tour jete.

ANGULAR ANALOGUES OF NEWTON'S LAWS OF MOTION

Table 14-1 presents linear and angular kinetic quantities in a parallel format. With the many parallels between linear and angular motion, it is not surprising that Newton's laws of motion may also be stated in terms of angular motion. It is necessary to remember that torque and moment of inertia are the angular equivalents of force and mass in substituting terms.

TABLE 14-1

LINEAR AND ANGULAR KINETIC QUANTITIES	
LINEAR	ANGULAR
mass (m)	moment of inertia (I)
force (F)	torque (T)
momentum (M)	angular momentum (H)
impulse (Ft)	angular impulse (Tt)

Newton's First Law

The angular version of the first law of motion may be stated as follows:

A rotating body will maintain a state of constant rotational motion unless acted on by an external torque.

In the analysis of human movement in which mass remains constant throughout, this angular analogue forms the underlying basis for the principle of conservation of angular momentum. Because angular velocity may change to compensate for changes in moment of inertia resulting from alterations in the radius of gyration, the quantity that remains constant in the absence of external torque is angular momentum.

Newton's Second Law

In angular terms, Newton's second law may be stated algebraically and in words as the following:

$$T = I\alpha$$

A net torque produces angular acceleration of a body that is directly proportional to the magnitude of the torque, in the same direction as the torque, and inversely proportional to the body's moment of inertia.

In accordance with Newton's second law for angular motion, the angular acceleration of the forearm is directly proportional to the magnitude of the net torque at the elbow and in the direction (flexion) of the net torque at the elbow. The greater the moment of inertia is with respect to the axis of rotation at the elbow, the smaller is the resulting angular acceleration (Figure 14-17).

FIGURE 14-17

S A M P L E   P R O B L E M   4

The knee extensors insert on the tibia at an angle of 30° at a distance of 3 cm from the axis of rotation at the knee. How much force must the knee extensors exert to produce an angular acceleration at the knee of 1 rad/s², given a mass of the lower leg and foot of 4.5 kg and k = 23 cm?

Known

$d = 0.03$ m
$\alpha = 1$ rad/s²
$m = 4.5$ kg
$k = 0.23$ m

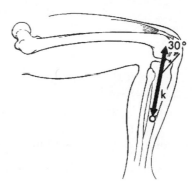

Solution

The angular analogue of Newton's second law of motion may be used to solve the problem:

$$T = I\alpha$$
$$Fd = mk^2\alpha$$
$$(F \sin 30 \text{ N}) (0.03 \text{ m}) = (4.5 \text{ kg}) (0.23 \text{ m})^2(1 \text{ rad/s}^2)$$

$$\boxed{F = 15.9 \text{ N}}$$

Newton's Third Law

The law of reaction may be stated in angular form as the following:

> For every torque exerted by one body on another, there is an equal and opposite torque exerted by the second body on the first.

When a baseball player forcefully swings a bat, rotating the mass of the upper body, a torque is created around the player's longitudinal axis. If the batter's feet are not firmly planted, the lower body tends to rotate around the longitudinal axis in the opposite direction. However, since the feet usually are planted, the torque generated by the upper body is translated to the ground, where the earth generates a torque of equal magnitude and opposite direction on the cleats of the batter's shoes.

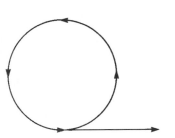

FIGURE 14-18
An object swung in a circle and then released will follow a linear path tangential to the curve at the point of release, since this is the direction of motion at the point of release.

CENTRIPETAL FORCE

Bodies undergoing rotary motion around a fixed axis are also subject to a linear force. When an object attached to a line is whirled around in a circular path and then released, the object flies off on a path that forms a tangent to the circular path it was following at the point at which it was released, since this is the direction it was traveling in at the point of release (Figure 14-18). **Centripetal force** prevents the rotating body from leaving its circular path while rotation occurs around a fixed axis. The direction of a centripetal force is always toward the center of rotation; this is the reason it is also known as *center-seeking force*. Centripetal force produces the radial component of the acceleration of a body traveling on a curved path (see Chapter 11). The following formula quantifies the magnitude of a centripetal force in terms of the tangential linear velocity of the rotating body:

centripetal force
force directed toward the center of rotation for a body in rotational motion

$$F_c = \frac{mv^2}{r}$$

In this formula, F_c is centripetal force, m is mass, v is the tangential linear velocity of the rotating body at a given point in time, and r is the radius of rotation. Centripetal force may also be defined in terms of angular velocity:

$$F_c = mr\omega^2$$

As is evident from both equations, the speed of rotation is the most influential factor on the magnitude of centripetal force, because centripetal force is proportional to the square of velocity or angular velocity.

When a cyclist rounds a curve, the ground exerts centripetal force on the tires of the cycle. The forces acting on the cycle/cyclist system are weight and the ground reaction force (Figure 14-19). The horizontal component of the ground reaction force (centripetal force) creates a torque about the cycle/cyclist CG. To prevent rotation toward the

Cyclists and runners lean into a curve to offset the torque created by centripetal force acting on the base of support.

FIGURE 14-19
Free body diagram of a
cyclist on a curve. R_H is
centripetal force. When
the cyclist is balanced,
$(wt)(d_{wt}) = (R_H)(D_{RH})$.

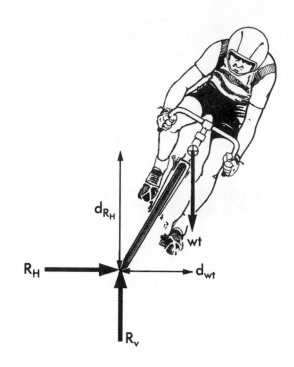

outside of the curve, the cyclist must lean to the inside of the curve so
that the moment arm of the system's weight relative to the contact point
with the ground is large enough to produce an oppositely directed
torque of equal magnitude.

SUMMARY

Whereas a body's resistance to linear acceleration is proportional to its
mass, resistance to angular acceleration is related to both mass and the
distribution of mass with respect to the axis of rotation. Resistance to
angular acceleration is known as moment of inertia, a quantity that incorporates both the amount of mass and its distribution relative to the
center of rotation.

Just as linear momentum is the product of the linear inertial property (mass) and linear velocity, angular momentum is the product of
moment of inertia and angular velocity. In the absence of external
torques, angular momentum is conserved. An airborne human performer can alter total body angular velocity by manipulating moment

FORMULA SUMMARY

DESCRIPTION	FORMULA
Moment of inertia = (mass)(radius of gyration squared)	$I = mk^2$
Angular momentum = (moment of inertia)(angular velocity)	$H = I\omega$
Local term for angular momentum = (segment moment of inertia about segment CG)(segment angular velocity about segment CG)	$H_l = I_s\omega_s$
Remote term for angular momentum = (segment mass) (distance between total body and segment CGs, squared) (angular velocity of segment about the principal axis)	$H_r = mr^2\omega_g$
Angular impulse = change in angular momentum	$Tt = \Delta H$ $Tt = (mk^2\omega)_2 - (mk^2\omega)_1$
Newton's Second Law (Rotational Version)	$T = I\alpha$
Centripetal force = $\dfrac{(mass)(velocity\ squared)}{radius\ of\ rotation}$	$F_c = \dfrac{mv^2}{r}$
Centripetal force = (mass)(radius of rotation)(angular velocity squared)	$F_c = mr\omega^2$

of inertia through changes in body configuration relative to the principal axis around which rotation is occurring. Skilled performers can also alter the axis of rotation and initiate rotation when no angular momentum is present while airborne. The principle of conservation of angular momentum is based on the angular version of Newton's first law of motion. The second and third laws of motion may also be expressed in angular terms by substituting moment of inertia for mass, torque for force, and angular acceleration for linear acceleration.

A linear force that acts on all rotating bodies is centripetal (or a center-seeking) force, which is always directed toward the center of rotation. The magnitude of centripetal force depends on the mass, speed, and radius of rotation of the rotating body.

INTRODUCTORY PROBLEMS

1. If you had to design a model of the human body composed entirely of regular geometric solids, which solid shapes would you choose? Using a straight edge, sketch a model of the human body that incorporates the solid shapes you have selected.

2. Construct a table displaying common units of measure for both linear and angular quantities of the inertial property, momentum, and impulse.
3. Skilled performance of a number of sport skills is characterized by "follow-through." Explain the value of "follow-through" in terms of the concepts discussed in this chapter.
4. Explain the reason the product of body mass and body height squared is a good predictor of body moment of inertia in children.
5. A 1.1 kg racquet has a moment of inertia about a grip axis of rotation of 0.4 kg $\cdot$ m^2. What is its radius of gyration? (Answer: 0.6 m)
6. How much angular impulse must be supplied by the hamstrings to bring a leg swinging at 8 rad/s to a stop, given that the leg's moment of inertia is 0.7 kg $\cdot$ m^2? (Answer: 5.6 mg $\cdot$ m^2/s).
7. Given the following principal transverse axis moments of inertia and angular velocities, calculate the angular momentum of each of the following gymnasts. What body configurations do these moments of inertia represent?

	I_{cg}(kg $\cdot$ m^2)	(rad/s)
A	3.5	20.00
B	7.0	10.00
C	15.0	4.67

(Answer: A = 70 kg $\cdot$ m^2/s; B = 70 kg $\cdot$ m^2/s; C = 70 kg $\cdot$ m^2/s)
8. A volleyball player's 3.7 kg arm moves at an average angular velocity of 15 rad/s during execution of a spike. If the average moment of inertia of the extending arm is 0.45 kg $\cdot$ m^2, what is the average radius of gyration for the arm during the spike? (Answer: 0.35 m)
9. A 50 kg diver in a full layout position, with a total body radius of gyration with respect to her transverse principal axis equal to 0.45 m, leaves a springboard with an angular velocity of 6 rad/s. What is the diver's angular velocity when she assumes a tuck position, reducing her radius of gyration to 0.25 m? (Answer: 19.4 rad/s)
10. If the centripetal force exerted on a swinging tennis racket by a player's hand is 40 N, how much reaction force is exerted on the player by the racket? (Answer: 40 N)

ADDITIONAL PROBLEMS

1. The radius of gyration of the thigh with respect to the transverse axis at the hip is 54% of the segment length. The mass of the thigh is 10.5% of total body mass, and the length of the thigh is 23.2% of total body height. What is the moment of inertia of the thigh with

respect to the hip for males of the following body masses and heights?

	MASS (kg)	Height (m)
A	60	1.6
B	60	1.8
C	70	1.6
D	70	1.8

(Answer: A = 0.25 kg $\cdot$ m^2, B = 0.32 kg $\cdot$ m^2, C = 0.30 kg $\cdot$ m^2, D = 0.37 kg $\cdot$ m^2)

2. Select three sport or daily living implements and explain the ways in which you might modify each implement's moment of inertia with respect to the axis of rotation to adapt it for a person of impaired strength.

3. A 0.68 kg tennis ball is given an angular momentum of 2.72×10^{-3} kg $\cdot$ m^2/s when struck by a racquet. If its radius of gyration is 2 cm, what is its angular velocity? (Answer: 10 rad/s)

4. A 7.27 kg shot makes seven complete revolutions during its 2.5 second flight. If its radius of gyration is 2.54 cm, what is its angular momentum? (Answer: 0.0817 kg $\cdot$ m^2/s)

5. What is the resulting angular acceleration of a 1.7 kg forearm and hand when the forearm flexors, attaching 3 cm from the center of rotation at the elbow, produce 10 N of tension, given a 90° angle at the elbow and a forearm and hand radius of gyration of 20 cm? (Answer: 4.41 rad/s^2)

6. The patellar tendon attaches to the tibia at a 20° angle 3 cm from the axis of rotation at the knee. If the tension in the tendon is 400 N, what is the resulting acceleration of the 4.2 kg lower leg and foot given a radius of gyration of 25 cm for the lower leg/foot with respect to the axis of rotation at the knee? (Answer: 15.6 rad/s^2)

7. A cave woman swings a 0.75 m sling of negligible weight around her head with a centripetal force of 220 N. What is the initial velocity of a 9 N stone released from the sling? (Answer: 13.4 m/s)

8. A 7.27 kg hammer on a 1 m wire is released with a linear velocity of 28 m/s. What reaction force is exerted on the thrower by the hammer at the instant before release? (Answer: 5.7 kN)

9. Discuss the effect of banking a curve on a racetrack. Construct a free body diagram to assist with your analysis.

10. Using the data in Appendix D, calculate the locations of the radii of gyration of all body segments with respect to the proximal joint center for a 1.7 m tall woman.

LABORATORY EXPERIENCES

1. View either a video or a live performance of pitches executed by a skilled performer and by a relatively unskilled performer. Explain the differences you observe in terms of the concepts presented in this chapter.
2. View either a video or a live performance of a long jump from the side view. Explain the motions of the jumper's arms and legs in terms of the concepts presented in this chapter.
3. View either a video or a live performance of dives incorporating pikes or somersaults from the side view. Explain the motions of the diver's arms and legs in terms of the concepts presented in this chapter.
4. Stand on a rotating platform with both arms abducted 90°and have a partner spin you at a moderate angular velocity. Once the partner has let go, quickly fold your arms across your chest, being careful not to lose balance. Write a paragraph explaining the change in angular velocity. Is angular momentum in this situation constant?
5. Stand on a rotating platform and use a hula-hooping motion of the hips to generate rotation. Explain how total body rotation results.

REFERENCES

1. Albert WJ and Miller DI: Takeoff characteristics of single and double axel figure skating jumps, J Appl Biomech 12:72, 1996.
2. Anderson T: Biomechanics and running economy, Sports Med 22:76, 1996.
3. Brüggemann G-P: Biomechanics in gymnastics. In Van Gheluwe B and Atha J: *Current research in sports biomechanics*, Basel, 1987, Karger.
4. Dapena J and McDonald C: A three-dimensional analysis of angular momentum in the hammer throw, Med Sci Sports Exerc 21:206, 1989.
5. Frohlich C: The physics of somersaulting and twisting, Sci Am 242:154, 1980.
6. Hamill J, Ricard MD, and Golden DM: Angular momentum in multiple rotation nontwisting platform dives, Int J Sport Biomech 2:78, 1986.
7. Hay JG: *The biomechanics of sports techniques,* 3rd ed, Englewood Cliffs, NJ, 1985, Prentice-Hall, Inc.
8. Jensen RK: The growth of children's moment of inertia, Med Sci Sports Exerc 18:440, 1987.
9. Jensen RK and Nassas G: Growth of segment principal moments of inertia between four and twenty years, Med Sci Sports Exerc 20:594, 1988.
10. Jones IC and Miller DI: Influence of fulcrum position on springboard response and takeoff performance in the running approach, J Appl Biomech 12: 383, 1996.
11. Laws K: *The physics of dance,* New York, 1984, Schirmer Books.
12. Miller DI, Jones IC, and Pizzimenti MA: Taking off: Greg Louganis' diving style, Soma 2:20, 1988.
13. Miller Di and Munro CF: Body segment contributions to height achieved during the flight of a springboard dive, Med Sci Sports Exerc 16:234, 1984.
14. Miller DI et al: Kinetic and kinematic characteristics of 10-m platform performances of elite divers: I. Back takeoffs, Int J Sport Biomech 5:60, 1989.

15. Nesbit SM et al: A discussion of iron golf club head inertia tensors and their effects on the golfer, J Appl Biomech 12:449, 1996.

16. Sanders RH: Effect of ability on twisting techniques in forward somersaults on the trampoline, J Appl Biomech 11:267, 1995.

17. Sanders RH and Wilson BD: Angular momentum requirements of the twisting and nontwisting forward $1\frac{1}{2}$ somersault dive, Int J Sport Biomech 3:47, 1988.

18. Yeadon MR: The biomechanics of the human in flight, Am J Sports Med 25:575, 1997.

ANNOTATED READINGS

Frohlich C: The physics of somersaulting and twisting, Sci Am 242:154, 1980.
   Describes the way in which divers, gymnasts, astronauts, and cats perform rotational maneuvers in midair that seem to violate the conservation of angular momentum.

Miller DI, Jones IC, and Pizzimenti MA: Taking off: Greg Louganis' diving style, Soma 2:20, 1988.
   Discusses the linear and angular momentum requirements for performing total body rotations at the world class level and describes the methods by which these factors may be studied.

Miller DI et al: Kinetic and kinematic characteristics of 10-m platform performances of elite divers: I. Back takeoffs, Int J Sport Biomech 5:60, 1989.

Miller DI et al: Kinetic and kinematic characteristics of 10-m platform performances of elite divers: II. Reverse takeoffs, Int J Sport Biomech 6:283, 1990.
   Presents a detailed description of a film, video, and force platform study of participants in the Fifth World Diving Championships.

Townend MS: *Mathematics in sport,* New York, 1984, John Wiley & Sons, Inc.
   A chapter on running, a chapter on jumping, and sections on gymnastics and high-board diving discuss the principle of conservation of angular momentum with respect to various applications in the identified sports.

RELATED WEB SITES

Centripetal Force
http://hyperphysics.phy-astr.gsu.edu/cf.html
   Provides descriptive text, diagrams, and interactive calculations for centripetal force and centripetal acceleration.

Centripetal Force
http://www.science.urch.edu/~rubin/pedagogy/131/131labs/centripetal.html
   Provides detailed description of a laboratory exercise on centripetal force.

Centripetal Force and Acceleration
http://rowlf.cc.wwu.edu:8080/~vawter/PhysicsNet/Pages/CentripetalForce.html
   Includes text, diagrams, and formulas.

Conceptest on Centripetal Force
http://motor1.physics.wayne.edu/~cinabro/cinabro/education/conceptest15.html
   Poses several questions related to centripetal force with a link to an answer page.

Conservation of Angular Momentum
http://schutz.ucsc.edu/~josh/5A/book/torque/node15.html
   Describes a demonstration, with diagrams and formulas, for conservation of angular momentum.

Conservation of Angular Momentum: Rotating Stool and Bicycle
http://storm.ph.utexas.edu/~phy-demo/demo-txt/1q40-30.html
   Shows and describes experimental demonstration of conservation of angular momentum.
Exploratorium: Angular Momentum
http://www.exploratorium.edu/xref/phenomena/momentum_-_angular.html
   Provides links to pages on illustrations of related phenomena.
Exploratorium: Centripetal Force
http://www.exploratorium.edu/xref/phenomena/centripetal_force.html
   Provides links to pages on illustrations of related phenomena.
The Interactive Physics Problem Set
http://info.itp.berkeley.edu/Vol1/Contents.html
   Provides links to interactive problem sets on torques, rotational inertia, and angular momentum.
Physics 131: Notes—Centripetal Force
http://www.science.urich.edu/~rubin/pedagogy/131/131notes/131notes_90.html
   Includes detailed discussion of centripetal and centrifugal force.

Chapter 15

HUMAN MOVEMENT IN A FLUID MEDIUM

After completing this chapter, the reader will be able to:

Explain the ways in which the composition and flow characteristics of a fluid affect fluid forces.

Define buoyancy and explain the variables that determine whether a human body will float.

Define drag, identify the components of drag, and identify the factors that affect the magnitude of each component.

Define lift and explain the ways in which it can be generated.

Discuss the theories regarding propulsion of the human body in swimming.

The ability to control the action of fluid forces differentiates elite from average swimmers.

Why are there dimples in a golf ball? Why are some people able to float while others cannot? Why are cyclists, swimmers, downhill skiers, and speed skaters concerned with streamlining their bodies during competition?

Both air and water are fluid mediums that exert forces on bodies moving through them. Some of these forces slow the progress of a moving body; others provide support or propulsion. A general understanding of the actions of fluid forces on human movement activities is an important component of the study of the biomechanics of human movement. This chapter introduces the effects of fluid forces on both human and projectile motion.

THE NATURE OF FLUIDS

Although in general conversation the term *fluid* is often used interchangeably with the term *liquid,* from a mechanical perspective, a **fluid** is any substance that tends to flow or continuously deform when acted on by a shear force (33). Both gases and liquids are fluids with similar mechanical behaviors.

fluid
substance that flows when subjected to a shear stress

Air and water are fluids that exert forces on the human body.

Relative Motion

Because a fluid is a medium capable of flow, the influence of the fluid on a body moving through it depends not only on the body's velocity but also on the velocity of the fluid. Consider the case of waders standing in the shallow portion of a river with a moderately strong current. If they stand still, they feel the force of the current against their legs. If they walk upstream against the current, the current's force against their legs is even stronger. If they walk downstream, the current's force is reduced and perhaps even imperceptible.

The velocity of a body relative to a fluid influences the magnitude of the forces exerted by the fluid on the body.

When a body moves through a fluid, the **relative velocity** of the body with respect to the fluid influences the magnitude of the acting forces. If the direction of motion is directly opposite the direction of the fluid flow, the magnitude of the velocity of the moving body relative to the fluid is the algebraic sum of the speeds of the moving body and the fluid (Figure 15-1). If the body moves in the same direction as the surrounding fluid, the magnitude of the body's velocity relative to the fluid is the difference in the speeds of the object and the fluid. In other words, the relative velocity of a body with respect to a fluid is the vector subtraction of the velocity of the fluid from the velocity of the body (Figure 15-2). Likewise, the relative velocity of a fluid with respect to a body moving through it is the vector subtraction of the velocity of the body from the velocity of the fluid.

relative velocity
velocity of a body with respect to the velocity of something else, such as the surrounding fluid

Laminar versus Turbulent Flow

When an object such as a human hand or a canoe paddle moves through water, there is little apparent disturbance of the immediately surrounding water if the relative velocity of the object with respect to the water

FIGURE 15-1

The relative velocity of a moving body with respect to a fluid is equal to the vector subtraction of the velocity of the wind from the velocity of the body.

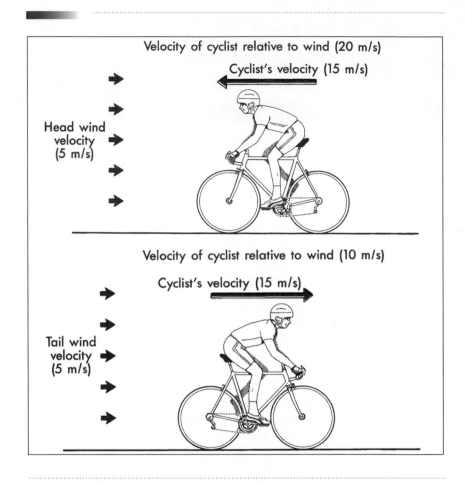

laminar flow
flow characterized by smooth, parallel layers of fluid

is low. However, if the relative velocity of motion through the water is sufficiently high, waves and eddies appear.

When an object moves with sufficiently low velocity relative to any fluid medium, the flow of the adjacent fluid is termed **laminar flow.** Laminar flow is characterized by smooth layers of fluid molecules flowing parallel to one another (Figure 15-3).

When an object moves with sufficiently high velocity relative to a surrounding fluid, the layers of fluid near the surface of the object mix and the flow is termed *turbulent*. The rougher the surface of the body, the lower the relative velocity at which turbulence is caused. Laminar flow and **turbulent flow** are distinct categories. If any turbulence is present, the flow is nonlaminar. The nature of the fluid flow surrounding an object can dramatically affect the fluid forces exerted on the object.

turbulent flow
flow characterized by mixing of adjacent fluid layers

FIGURE 15-2

S A M P L E   P R O B L E M   I

A sailboat is traveling at an absolute speed of 3 m/s against a 0.5 m/s current and with a 6 m/s tailwind. What is the velocity of the current with respect to the boat? What is the velocity of the wind with respect to the boat?

Known

$v_b = 3$ m/s $\rightarrow$
$v_c = 0.5$ m/s $\leftarrow$
$v_w = 6$ m/s $\rightarrow$

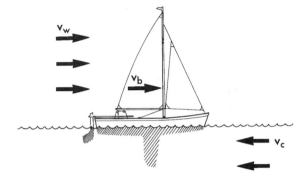

Solution

The velocity of the current with respect to the boat is equal to the vector subtraction of the absolute velocity of the boat from the absolute velocity of the current.

$$v_{c/b} = v_c - v_b$$
$$v_{c/b} = (0.5 \text{ m/s} \leftarrow) - (3 \text{ m/s} \rightarrow)$$
$$v_{c/b} = (3.5 \text{ m/s} \leftarrow)$$

The velocity of the current with respect to the boat is 3.5 m/s in the direction opposite that of the boat.

The velocity of the wind with respect to the boat is equal to the vector subtraction of the absolute velocity of the boat from the absolute velocity of the wind.

$$v_{w/b} = v_w - v_b$$
$$v_{w/b} = (6 \text{ m/s} \rightarrow) - (3 \text{ m/s} \rightarrow)$$
$$v_{w/b} = (3 \text{ m/s} \rightarrow)$$

The velocity of the wind with respect to the boat is 3 m/s in the direction in which the boat is sailing.

Fluid Properties

Other factors that influence the magnitude of the forces a fluid generates are the fluid's density, specific weight, and viscosity. As discussed in

FIGURE 15-3
Laminar flow is
characterized by smooth,
parallel layers of fluid.

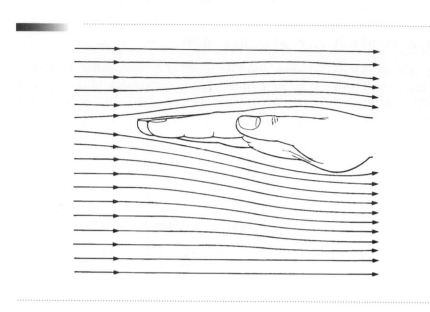

Chapter 3, density ρ, is defined as mass/volume, and the ratio of weight to volume is known as specific weight γ. The denser and heavier the fluid medium surrounding a body, the greater the magnitude of the forces the fluid exerts on the body. The property of fluid viscosity involves the internal resistance of a fluid to flow. The greater the extent to which a fluid resists flow under an applied force, the more viscous the fluid is. A thick molasses, for example, is more viscous than a liquid honey, which is more viscous than water. Increased fluid viscosity results in increased forces exerted on bodies exposed to the fluid.

Atmospheric pressure and temperature influence a fluid's density, specific weight, and viscosity.

Atmospheric pressure and temperature influence a fluid's density, specific weight, and viscosity, with more mass concentrated in a given unit of fluid volume at higher atmospheric pressures and lower temperatures. Because molecular motion in gases increases with temperature, the viscosity of gases also increases. The viscosity of liquids decreases with increased temperature because of a reduction in the cohesive forces among the molecules. The densities, specific weights, and viscosities of common fluids are shown in Table 15-1.

BUOYANCY

Characteristics of the Buoyant Force

Buoyancy is a fluid force that always acts vertically upward. The factors that determine the magnitude of the buoyant force were originally ex-

APPROXIMATE PHYSICAL PROPERTIES OF COMMON FLUIDS			
FLUID	DENSITY (KG/M^3)	SPECIFIC WEIGHT (N/M^3)	VISCOSITY (NS/M^2)
Air	1.20	11.8	.000018
Water	998	9,790	.0010
Sea water*	1,026	10,070	.0014
Ethyl alcohol	799	7,850	.0012
Mercury	13,550.20	133,000.0	.0015
*(10°C, 3.3% salinity)			

Fluids are measured at 20°C and standard atmospheric pressure.

*(10°C, 3.3% salinity)

plained by the ancient Greek mathematician Archimedes. **Archimedes' principle** states that the magnitude of the buoyant force acting on a given body is equal to the weight of the fluid displaced by the body. The latter factor is calculated by multiplying the specific weight of the fluid by the volume of the portion of the body that is surrounded by the fluid. Buoyancy (F_b) is calculated as the product of the displaced volume (V_d) and the fluid's specific weight γ:

$$F_b = V_d \gamma$$

For example, if a water polo ball with a volume of 0.2 m^3 is completely submerged in water at 20°C, the buoyant force acting on the ball is equal to the ball's volume multiplied by the specific weight of water at 20°C:

$$F_b = V_d \gamma$$
$$F_b = (0.2 \text{ m}^3)\ (9790 \text{ N/m}^3)$$
$$F_b = 1958 \text{ N}$$

The more dense the surrounding fluid, the greater the magnitude of the buoyant force. Since sea water is more dense than fresh water, a given object's buoyancy is greater in sea water than in fresh water. Because the magnitude of the buoyant force is directly related to the volume of the submerged object, the point at which the buoyant force acts is the object's **center of volume,** which is also known as the *center of buoyancy.* The center of volume is the point around which a body's volume is equally distributed in all directions.

Archimedes' principle
physical law stating that the buoyant force acting on a body is equal to the weight of the fluid displaced by the body

center of volume
point around which a body's volume is equally distributed and at which the buoyant force acts

Flotation

The ability of a body to float in a fluid medium depends on the relationship between the body's buoyancy and its weight. When weight and the buoyant force are the only two forces acting on a body and their magnitudes are equal, the body floats in a motionless state, in accordance with the principles of static equilibrium. If the magnitude of the weight is greater than that of the buoyant force, the body sinks, moving downward in the direction of the net force.

Most objects float statically in a partially submerged position. The volume of a freely floating object needed to generate a buoyant force equal to the object's weight is the volume that is submerged.

Flotation of the Human Body

In the study of biomechanics, buoyancy is most commonly of interest relative to the flotation of the human body in water. Some individuals cannot float in a motionless position, and others float with little effort. This difference in floatability is a function of body density. Since the density of bone and muscle is greater than the density of fat, individuals who are extremely muscular and have little body fat have higher average body densities than individuals with less muscle, less dense bones, or more body fat. If two individuals have an identical body volume, the one with the higher body density weighs more. Alternatively, if two people have the same body weight, the person with the higher body density has a smaller body volume. For flotation to occur, the body volume must be large enough to create a buoyant force greater than or equal to body weight (Figure 15-4). Many individuals can float only when holding a large volume of inspired air in the lungs, a tactic that increases body volume without practically altering body weight.

The orientation of the human body as it floats in water is determined by the relative position of the total body center of gravity relative to the total body center of volume. The exact locations of the center of gravity and center of volume vary with anthropometric dimensions and body composition. Typically, the center of gravity is inferior to the center of volume due to the relatively large volume and relatively small weight of the lungs. Because weight acts at the center of gravity and buoyancy acts at the center of volume, a torque is created that rotates the body until it is positioned so that these two acting forces are vertically aligned and the torque ceases to exist (Figure 15-5).

When beginning swimmers try to float on their backs, they typically assume a horizontal body position. Once the swimmer relaxes, the lower end of the body sinks because of the acting torque. An experienced teacher instructs beginning swimmers to assume a more diagonal position in the water before relaxing into the back float. This position min-

In order for a body to float, the buoyant force it generates must equal or exceed its weight.

People who cannot float in swimming pools may float in Utah's Great Salt Lake, in which the density of the water surpasses even that of sea water.

FIGURE 15-4

SAMPLE PROBLEM 2

When holding a large quantity of inspired air in her lungs, a 22 kg girl has a body volume of 0.025 m^3. Can she float in fresh water if γ equals 9810 N/m^3? Given her body volume, how much could she weigh and still be able to float?

Known

$m = 22 \text{ kg}$

$V = 0.025 \text{ m}^3$

$\gamma = 9810 \text{ N/m}^3$

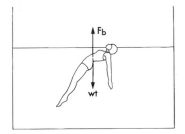

Solution

The free body diagram shows that two forces are acting on the girl—her weight and the buoyant force. According to the conditions of static equilibrium, the sum of the vertical forces must be equal to zero for the girl to float in a motionless position. If the buoyant force is less than her weight, she will sink, and if the buoyant force is equal to her weight, she will float completely submerged. If the buoyant force is greater than her weight, she will float partly submerged. The magnitude of the buoyant force acting on her total body volume is the product of the volume of displaced fluid (her body volume) and the specific weight of the fluid:

$$F_b = V\gamma$$
$$F_b = (0.025 \text{ m}^3)(9810 \text{ N/m}^3)$$
$$F_b = 245.52 \text{ N}$$

Her body weight is equal to her body mass multiplied by the acceleration of gravity:

$$wt = (22 \text{ kg})(9.81 \text{ m/s}^2)$$
$$wt = 215.82 \text{ N}$$

Since the buoyant force is greater than her body weight, the girl will float partly submerged in fresh water.

Yes, she will float.

To calculate the maximum weight that the girl's body volume can support in fresh water, multiply the body volume by the specific weight of water.

$$wt_{max} = (0.025 \text{ m}^3)(9810 \text{ N/m}^3)$$

$$wt_{max} = 245.25 \text{ N}$$

FIGURE 15-5
A, A torque is created on a swimmer by body weight (acting at the center of gravity) and the buoyant force (acting at the center of volume). **B,** When the center of gravity and the center of volume are vertically aligned, this torque is eliminated.

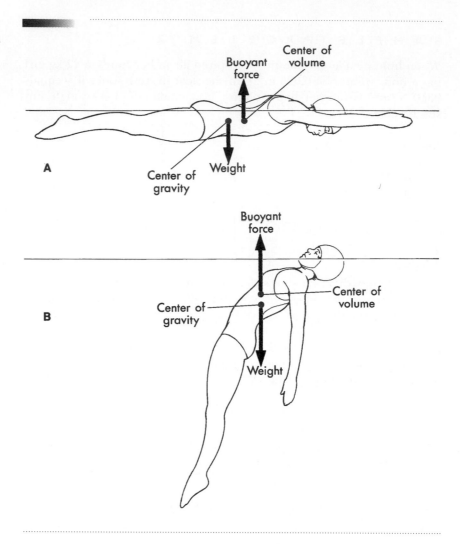

imizes torque and the concomitant sinking of the lower extremity. Other strategies that a swimmer can use to reduce torque on the body when entering a back float position include extending the arms backward in the water above the head and flexing the knees. Both tactics elevate the location of the center of gravity, positioning it closer to the center of volume.

DRAG

Drag is a force caused by the dynamic action of a fluid that acts in the direction of the freestream fluid flow. Generally, a drag is a *resistance* force—a force that slows the motion of a body moving through a fluid.

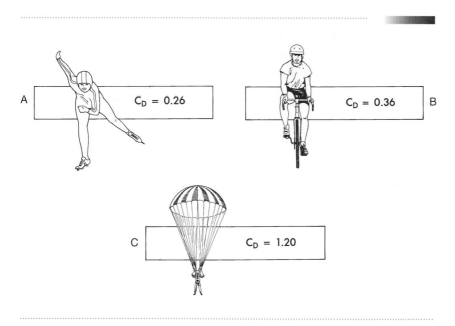

FIGURE 15-6
Approximate coefficients of drag for the human body. **A**, Frontal drag on a speed skater. **B**, Frontal drag on a cyclist in touring position. **C**, Vertical drag on a parachutist falling with the parachute fully opened. (Modified from Roberson JA and Crowe CT: *Engineering fluid mechanics*, ed 2, Boston, 1980, Houghton Mifflin Co.)

The drag force acting on a body in relative motion with respect to a fluid is defined by the following formula:

$$F_D = \tfrac{1}{2}C_D\rho\, A_p v^2$$

In this formula, F_D is drag force, C_D is the **coefficient of drag,** ρ is the fluid density, A_p is the projected area of the body or the surface area of the body oriented perpendicular to the fluid flow, and v is the relative velocity of the body with respect to the fluid. The coefficient of drag is a unitless number that serves as an index of the amount of drag an object can generate. Its size depends on the shape and orientation of a body relative to the fluid flow, with long, streamlined bodies generally having lower coefficients of drag than blunt or irregularly shaped objects. Approximate coefficients of drag for the human body in positions commonly assumed during participation in several sports are shown in Figure 15-6.

The formula for the total drag force demonstrates the exact way in which each of the identified factors affects drag. If the coefficient of drag, the fluid density, and the projected area of the body remain constant, drag increases with the square of the relative velocity of motion. This relationship is referred to as the **theoretical square law.** According to this law, if cyclists double their speed and other factors remain constant, the drag force opposing them increases fourfold. The effect of drag is more consequential when a body is moving with a high velocity, which occurs in sports such as cycling, speed skating, downhill skiing, the bobsled, and the luge.

coefficient of drag
unitless number that is an index of a body's ability to generate fluid resistance

theoretical square law
drag increases approximately with the square of velocity when relative velocity is low

Increase or decrease in the fluid density also results in a proportional change in the drag force. Because air density decreases with increasing altitude, many world records set at the 1968 Olympic Games in Mexico City where the elevation is 2250 m may have been partially attributable to the reduced air resistance acting on the competitors. Mathematical model–based estimates indicate that the reduction in drag attributable to the lesser air density in Mexico City accounts for 0.08 seconds of performance time in the 100 m sprint and 0.16 seconds of race time in the 200 m event (25). Bob Beamon's noteworthy long jump performance of 8.9 m during the games was 2.4 cm longer than if the same jump had been performed at sea level (43). Theoretical calculations indicate that because air density decreases more than VO_2max with increasing altitude, the ideal altitude for cycling performance should be 4000 m (6).

In swimming, research indicates that drag varies with the anthropometric characteristics of the individual swimmer, as well as with the stroke used. Researchers distinguish between passive drag, which is generated by the swimmer's body size, shape, and position in the water, and active drag, which is associated with the swimming motion. Elite swimmers generate much less active drag than average swimmers through superior stroke technique (23). Older swimmers tend to generate more drag than younger swimmers, and sprinters tend to incur more drag than long distance swimmers (22). Furthermore, better swimmers appear to be able to increase swimming speed while simultaneously reducing drag or showing only a small increase (22).

Three forms of resistance contribute to the total drag force. The component of resistance that predominates depends on the nature of the fluid flow immediately adjacent to the body.

Skin Friction

skin friction
surface drag
viscous drag
resistance derived from friction between adjacent layers of fluid near a body moving through the fluid

One component of the total drag is known as **skin friction, surface drag,** or **viscous drag.** This drag is similar to the friction force described in Chapter 12. Skin friction is derived from the sliding contacts between successive layers of fluid close to the surface of a moving body (Figure 15-7). The layer of fluid particles immediately adjacent to the moving body is slowed because of the shear stress the body exerts on the fluid. The next adjacent layer of fluid particles moves with slightly less speed because of friction between the adjacent molecules, and the next layer is affected in turn. The number of layers of affected fluid becomes progressively larger as the flow moves in the downstream direction along the body. The entire region within which fluid velocity is diminished because of the shearing resistance caused by the boundary of the moving body is the **boundary layer.** The force of the body exerts on the fluid in creating the boundary layer results in an oppositely directed reaction force exerted by the fluid on the body. This reaction force is known as *skin friction.*

boundary layer
layer of fluid immediately adjacent to a body

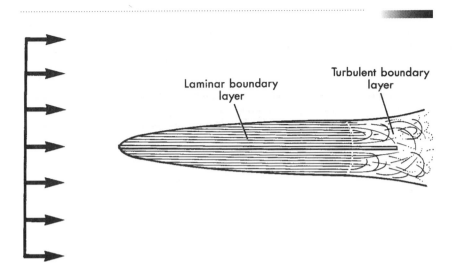

FIGURE 15-7
The fluid boundary layer for a thin, flat plate, shown from the side view. The laminar boundary layer gradually becomes thicker as flow progresses along the plate.

Wearing smooth, snug apparel helps to minimize skin friction.

Several factors affect the magnitude of skin friction drag. It increases proportionally with increases in the relative velocity of fluid flow, the surface area of the body over which the flow occurs, the roughness of the body surface, and the viscosity of the fluid. Skin friction is always one component of the total drag force acting on a body moving relative to a fluid, and it is the major form of drag present when the flow is primarily laminar.

Among these factors, the one that a competitive athlete can readily alter is the relative roughness of the body surface. Athletes can wear tight-fitting clothing composed of a smooth fabric rather than loose-fitting clothing or clothing made of a rough fabric. A 10% reduction of drag occurs when a speed skater wears a smooth spandex suit as opposed to the traditional wool outfit (41). A 6% decrease in air resistance results from cyclists using appropriate clothing, including sleeves, tights, and smooth covers over the laces of the shoes (24). The long cotton socks and loose shorts and singlets commonly worn by runners are particularly aerodynamically inefficient. Smoothing the running shoe and laces and either covering or shaving body hair reduces drag on runners by as much as 10% (25). Competitive male swimmers and cyclists often shave body hair to reduce skin friction.

The other factor affecting skin friction that athletes can alter in some circumstances is the amount of surface area in contact with the fluid. Carrying an extra passenger such as a cox in a rowing event results in a larger wetted surface area of the hull because of the added weight and, as a result, skin friction drag is increased.

Common running attire is aerodynamically inefficient.

FIGURE 15-8
Form drag results from
the suction-like force
created between the
positive pressure zone on a
body's leading edge and
the negative pressure zone
on the trailing edge when
turbulence is present.

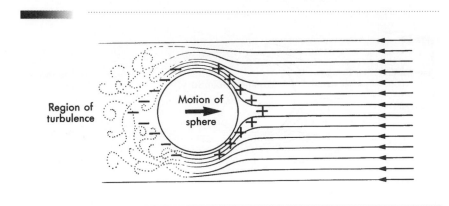

form drag
profile drag
pressure drag
resistance created by a
pressure differential between
the lead and rear sides of a
body moving through a fluid

Form Drag

A second component of the total drag acting on a body moving through
a fluid is **form drag,** which is also known as **profile drag** or **pressure
drag.** Form drag is always one component of the drag on a body moving relative to a fluid. When the boundary layer of fluid molecules next
to the surface of the moving body is primarily turbulent, form drag predominates. Form drag is the major contributor to overall drag during
most human and projectile motion.

When a body moves through a fluid medium with sufficient velocity
to create a pocket of turbulence behind the body, an imbalance in the
pressure surrounding the body—a *pressure differential*—is created (Figure 15-8). At the upstream end of the body where fluid particles meet
the body head-on, a zone of relative high pressure is formed. At the
downstream end of the body where turbulence is present, a zone of relative low pressure is created. Whenever a pressure differential exists, a
force is directed from the region of high pressure to the region of low
pressure. For example, a vacuum cleaner creates a suction force because
a region of relative low pressure (the relative vacuum) exists inside the
machine housing. This force, directed from front to rear of the body in
relative motion through a fluid, constitutes form drag.

Several factors affect the magnitude of form drag, including the relative velocity of the body with respect to the fluid, the magnitude of the
pressure gradient between the front and rear ends of the body, and the
size of the surface area that is aligned perpendicular to the flow. Both
the size of the pressure gradient and the amount of surface area perpendicular to the fluid flow can be reduced to minimize the effect of
form drag on the human body. For example, streamlining the overall
shape of the body reduces the magnitude of the pressure gradient.
Streamlining minimizes the amount of turbulence created and hence
minimizes the negative pressure that is created at the object's rear

*Streamlining helps to minimize
form drag.*

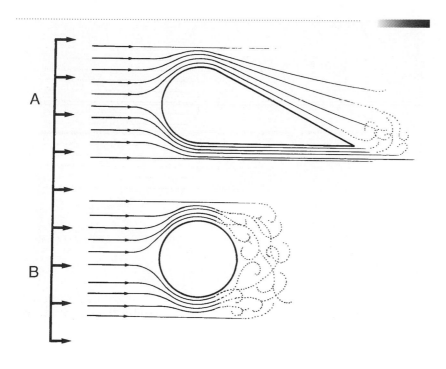

FIGURE 15-9
The effect of streamlining
is a reduction in the
turbulence created at the
trailing edge of a body in
a fluid. **A**, A streamlined
shape. **B**, A sphere.

(Figure 15-9). Assuming a more crouched body position also reduces the
body's projected surface area oriented perpendicular to the fluid flow.

Competitive cyclists, skaters, and skiers assume a streamlined body
position with the smallest possible area of the body oriented perpendicular to the oncoming airstream. Even though the low-crouched aero-
position assumed by competitive cyclists increases the cyclist's metabolic
cost as compared to riding in an upright position, the aerodynamic benefit is an over tenfold reduction in drag (16). Similarly, race cars, yacht
hulls, and some cycling helmets are designed with streamlined shapes.
The aerodynamic frame and handlebar designs for racing cycles also reduce drag (7, 32).

Streamlining is also an effective way to reduce form drag in the water. The ability to streamline body position during freestyle swimming
is a characteristic that distinguishes elite from sub-elite performers (8).
Using a triathlon wet suit can reduce the drag on a competitor swimming at a typical triathlon race pace of 1.25 m/sec by as much as 14%
because the buoyant effect of the wet suit results in reduced form drag
on the swimmer (38).

The nature of the boundary layer at the surface of a body moving
through a fluid can also influence form drag by affecting the pressure
gradient between the front and rear ends of the body. When the boundary layer is primarily laminar, the fluid separates from the boundary

A streamlined cycling helmet.

close to the front end of the body, creating a large turbulent pocket with a large negative pressure and thereby a large form drag (Figure 15-10). In contrast, when the boundary layer is turbulent, the point of flow separation is closer to the rear end of the body, the turbulent pocket created is smaller, and the resulting form drag is smaller.

The nature of the boundary layer depends on the roughness of the body's surface and the body's velocity relative to the flow. As the relative velocity of motion for an object such as a golf ball increases, changes in the acting drag occur (Figure 15-11). As relative velocity increases up to a certain critical point, the theoretical square law is in effect, with drag increasing with the square of velocity. After this critical velocity is reached, the boundary layer becomes more turbulent than laminar, and form drag diminishes because the pocket of reduced pressure on the trailing edge of the ball becomes smaller. As velocity increases further, the effects of skin friction and form drag grow, increasing the total drag. The dimples in a golf ball are carefully engineered to produce a turbulent boundary layer at the ball's surface that reduces form drag on the ball over the range of velocities at which a golf ball travels.

Another way in which form drag can be manipulated is through drafting, the process of following closely behind another participant in speed-based sports such as cycling and automobile racing. Drafting provides the advantage of reducing form drag on the follower, since the leader partially shelters the follower's leading edge from increased pressure against the fluid. Depending on the size of the pocket of reduced pres-

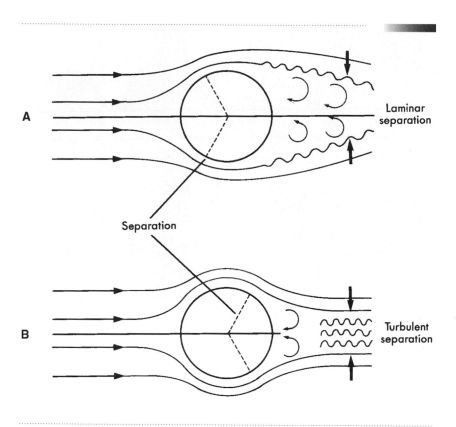

FIGURE 15-10
A, Laminar flow results in an early separation of flow from the boundary and a larger drag producing wake as compared to **B**, turbulent boundary flow.

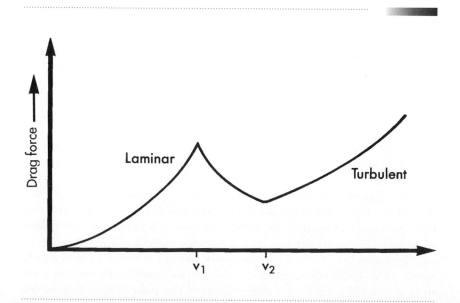

FIGURE 15-11
Drag increases approximately with the square of velocity until there is sufficient relative velocity (v_1) to generate a turbulent boundary layer. As velocity increases beyond this point, form drag decreases. After a second critical relative velocity (v_2) is reached, the drag again increases.

Cyclists drafting to minimize form drag.

wave drag
resistance created by the generation of waves at the interface between two different fluids, such as air and water

sure behind the leader, a suctionlike force may also help to propel the follower forward.

Wave Drag

The third type of drag acts at the interface of two different fluids, for example, at the interface between water and air. Although bodies that are completely submerged in a fluid are not affected by **wave drag,** this form of drag can be a major contributor to the overall drag acting on a human swimmer, particularly when the swim is done in open water. When a swimmer moves a body segment along, near, or across the air and water interface, a wave is created in the more dense fluid (the water). The reaction force the water exerts on the swimmer constitutes wave drag.

The magnitude of wave drag increases with greater up-and-down motion of the body and increased swimming speed. The height of the bow wave generated in front of a swimmer increases proportionally with swimming velocity, although at a given velocity, skilled swimmers produce smaller waves than less-skilled swimmers, presumably due to better technique (less up-and-down motion) (37). At fast swimming speeds, wave drag is generally the largest component of the total drag acting on the

The bow wave generated by a competitive swimmer.

swimmer. For this reason, competitive swimmers typically propel themselves underwater to eliminate wave drag for a small portion of the race in events in which the rules permit it. One underwater stroke is allowed following the dive or a turn in the breaststroke, and a distance of up to 15 m is allowed underwater after a turn in the backstroke. In most swimming pools the lane lines are designed to minimize wave action by dissipating moving surface water.

LIFT FORCE

While drag forces act in the direction of the freestream fluid flow, another force, known as **lift,** is generated perpendicular to the fluid flow. Although the name *lift* suggests that this force is directed vertically upward, it may assume any direction, as determined by the direction of the fluid flow and the orientation of the body. The factors affecting the magnitude of lift are basically the same factors that affect the magnitude of drag:

$$F_L = \tfrac{1}{2} C_L \rho A_p v^2$$

In this equation, F_L represents lift force, C_L is the **coefficient of lift,** ρ is the fluid density, A_p is the surface area against which lift is generated,

lift
force acting on a body in a fluid in a direction perpendicular to the fluid flow

coefficient of lift
unitless number that is an index of a body's ability to generate lift

The lane lines in modern swimming pools are designed to minimize wave action, enabling faster racing times.

and v is the relative velocity of a body with respect to a fluid. The factors affecting the magnitudes of the fluid forces discussed are summarized in Table 15-2.

Foil Shape

foil
shape capable of generating lift in the presence of a fluid flow

One way in which lift force may be created is for the shape of the moving body to resemble that of a **foil** (Figure 15-12). When the fluid stream encounters a foil, the fluid separates, with some flowing over the curved surface and some flowing straight back along the flat surface on the opposite side. The fluid that flows over the curved surface is positively accelerated relative to the fluid flow, creating a region of relative high velocity flow. The difference in the velocity of flow on the curved side of the foil as opposed to the flat side of the foil creates a pressure difference in the fluid, in accordance with a relationship derived by the Italian scientist Bernoulli. According to the **Bernoulli principle,** regions of relative high-velocity fluid flow are associated with regions of relative low pressure, and regions of relative low-velocity flow are associated with regions of relative high pressure. When these regions of relative low and high pressure are created on opposite sides of the foil, the result is a lift force directed perpendicular to the foil from the zone of high pressure toward the low pressure zone.

Bernoulli principle
an expression of the inverse relationship between relative velocity and relative pressure in a fluid flow

Different factors affect the magnitude of the lift force acting on a foil. The greater the velocity of the foil relative to the fluid, the greater the pressure differential and the lift force generated. Other contribut-

TABLE 15-2

FACTORS AFFECTING THE MAGNITUDES OF FLUID FORCES	
FORCE	FACTORS
Buoyant force	specific weight of the fluid volume of fluid displaced
Skin friction	relative velocity of the fluid amount of body surface area exposed to the flow roughness of the body surface viscosity of the fluid
Form drag	relative velocity of the fluid pressure differential between leading and rear edges of the body amount of body surface area perpendicular to the flow
Wave drag	relative velocity of the wave amount of surface area perpendicular to the wave viscosity of the fluid
Lift force	relative velocity of the fluid density of the fluid size, shape, and orientation of the body

ing factors are the fluid density and the surface area of the flat side of the foil. As both of these variables increase, lift increases. An additional factor of influence is the *coefficient of lift,* which indicates a body's ability to generate lift based on its shape.

The human hand resembles a foil shape when viewed from a lateral perspective. When a swimmer slices a hand through the water, it

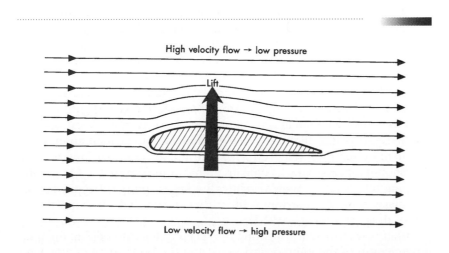

FIGURE 15-12
Lift force generated by a foil shape is directed from the region of relative high pressure on the flat side of the foil toward the region of relative low pressure on the curved side of the foil.

generates lift force directed perpendicular to the palm. Synchronized swimmers use a sculling motion, rapidly slicing their hands back and forth, to maneuver their bodies through various positions in the water. The lift force generated by rapid sculling motions enables elite synchronized swimmers to support their bodies in an inverted position with both legs extended completely out of the water.

The semi-foil shapes of projectiles such as the discus, javelin, football, boomerang, and frisbee generate some lift force when oriented at appropriate angles with respect to the direction of the fluid flow. Spherical projectiles such as a shot or a ball, however, do not sufficiently resemble a foil and cannot generate lift by virtue of shape.

The angle of orientation of the projectile with respect to the fluid flow—the **angle of attack**—is an important factor in launching a lift-producing projectile for maximum range (horizontal displacement). A positive angle of attack is necessary to generate a lift force (Figure 15-13). As the angle of attack increases, the amount of surface area exposed perpendicularly to the fluid flow also increases, thereby increasing the amount of form drag acting. With too steep an attack angle, the fluid cannot flow along the curved side of the foil to create lift. Airplanes that assume too steep an ascent can stall and lose altitude until pilots reduce the attack angle of the wings to enable lift (27).

To maximize the flight distance of a projectile such as the discus or javelin, it is advantageous to maximize lift and minimize drag. Form drag, however, is minimum at an angle of attack of 0, which is a poor angle for generating lift. The optimum angle of attack for maximizing range is the angle at which the **lift/drag ratio** is maximum. The largest lift/drag ratio for a discus traveling at a relative velocity of 24 m/s is generated at an angle of attack of 10° (15). For both the discus and the javelin, however, the single most important factor related to distance achieved is release speed (2, 17).

When the projectile is the human body during the performance of a jump, maximizing the effects of lift while minimizing the effects of drag is more complicated. In the ski jump, because of the relatively long period of time during which the body is airborne, the lift/drag ratio for the human body is particularly important. Research on ski jumping indicates that for optimal performance, ski jumpers should have a flattened body with a large frontal area (for generating lift) and a small body weight (for enabling greater acceleration) during takeoff. During the first part of the flight, jumpers should assume a small angle of attack to minimize drag (Figure 15-14). During the latter part of the flight, they should increase attack angle up to that of maximum lift. On smaller jumping hills where takeoff velocities are lower, jumpers should assume the attack angle for maximum lift earlier in the flight because the effect of drag is not so great (31).

Other researchers investigating the ski jump have noted that the maximum length of the jump occurs neither at the body angle of attack for

angle of attack
angle between the longitudinal axis of a body and the direction of the fluid flow

lift/drag ratio
the magnitude of the lift force divided by the magnitude of the total drag force acting on a body at a given time

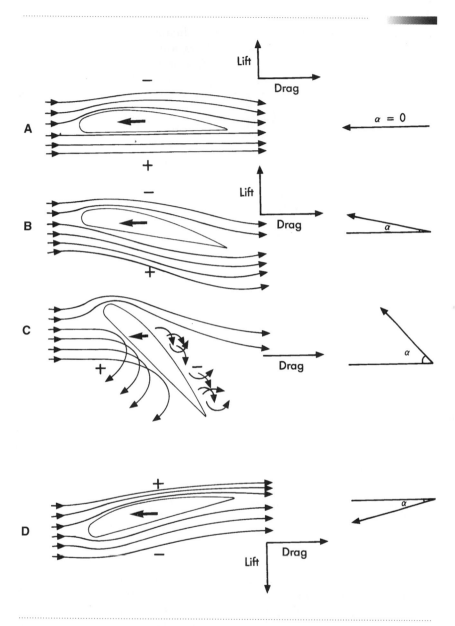

FIGURE 15-13
A, Drag and lift are small because the angle of attack (α) does not create a sufficiently high pressure differential across the top and bottom surfaces of the foil. **B**, An angle of attack that promotes lift. **C**, When the angle of attack is too large, the fluid cannot flow over the curved surface of the foil and no lift is generated. **D**, When the angle of attack is below the horizontal, lift is created in a downward direction. (Modified from Maglischo E: *Swimming faster: A comprehensive guide to the science of swimming*, Palo Alto, Calif, 1982, Mayfield Publishing Co.)

maximum lift nor at the attack angle for which the lift/drag is maximum, but somewhere between the two (Figure 15-15) (12). The ski jump consists of takeoff, flight, and landing, with body position during each phase influencing body position during the subsequent phase or phases. Research indicates that better performance is achieved when ski position is changed during the flight. Ski positions include the classic style,

FIGURE 15-14

The angle of attack is the angle formed between the primary axis of a body and the direction of the fluid flow.

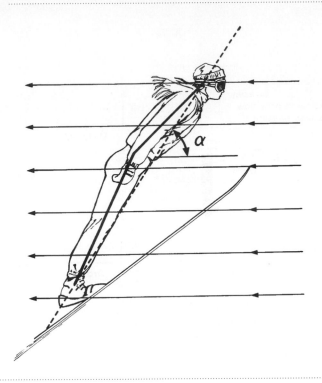

FIGURE 15-15

The relationship between ski jump length and the performer's angle of attack. (Modified from Denoth J, Luethi SM, and Gasser HH: Methodological problems in optimization of the flight phase in ski jumping, Int J Sport Biomech 3:404, 1987.)

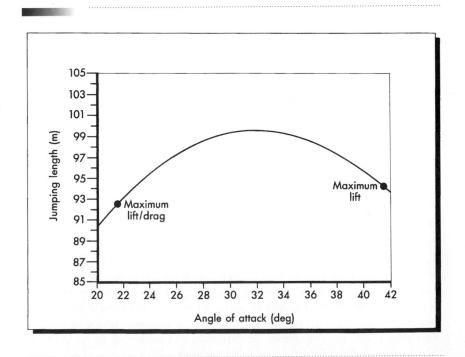

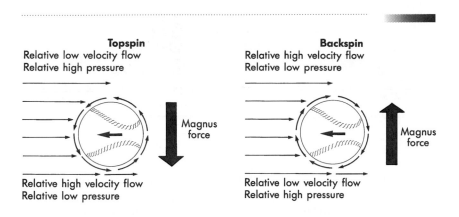

FIGURE 15-16
Magnus force results from
the pressure differential
created by a spinning
body.

in which the skis are parallel, the V style, in which the skis are in a her-
ringbone position in the frontal plane, and the flat V style, in which the
skis are more flat in the sagittal plane than with the V style. Optimal
performance results from initiating the jump in the classic style or the
flat V style, and changing to the V style 1.3–1.6 seconds into the jump
(19). This is true because the classic and flat V styles produce less drag
than the V style in the first part of the jump, and the V style maximizes
lift in the later part of the jump.

Magnus Effect

Spinning objects also generate lift. When an object in a fluid medium
spins, the boundary layer of fluid molecules adjacent to the object spins
with it. When this happens, the fluid molecules on one side of the spinning body collide head-on with the molecules in the fluid freestream
(Figure 15-16). This creates a region of relative low velocity and high
pressure. On the opposite side of the spinning object the boundary layer
moves in the same direction as the fluid flow, thereby creating a zone
of relative high velocity and low pressure. The pressure differential creates what is called the **Magnus force,** a lift force directed from the high-
pressure region to the low-pressure region.

Magnus force affects the flight path of a spinning projectile as it travels through the air, causing the path to deviate progressively in the direction of the spin, a deviation known as the **Magnus effect.** When a tennis ball or table tennis ball is hit with topspin, the ball drops more rapidly
than it would without spin and the ball tends to rebound low and fast, often making it more difficult for the opponent to return the shot. The nap
on a tennis ball traps a relatively large boundary layer of air with it as it
spins, thereby accentuating the Magnus effect. The Magnus effect can also
result from sidespin, as when a pitcher throws a curve ball (Figure 15-17).
The modern-day version of the curve ball is a ball that is intentionally

Magnus force
lift force created by spin

Magnus effect
deviation in the trajectory of
a spinning object toward the
direction of spin, resulting
from the Magnus force

FIGURE 15-17

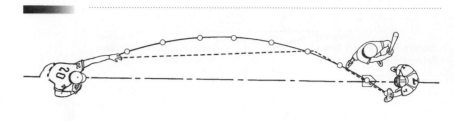

FIGURE 15-17
The trajectory of a ball thrown with sidespin follows a regular curve due to the Magnus effect. The dashed line shows the illusion seen by the players on the field.

A ball projected with spin follows a trajectory that curves in the direction of the spin.

pitched with spin, causing it to follow a curved path in the direction of the spin throughout its flight path. Curve balls thrown by major league pitchers spin as quickly as 27 revolutions per second and deviate horizontally as much as 40 cm over the pitcher-to-batter distance (39).

In 1982 a controversy between professional baseball players and scientists arose over the behavior of a pitched curve ball. The players claimed that the path taken by a curve ball was along a straight line until a certain critical point at which the ball "broke" and suddenly curved. This appearance is enhanced when topspin is imparted to the ball, since the Magnus effect of topspin accentuates the effect of gravity. However, the actual path of a curve ball is a smooth arc, which has been documented with high-speed movie film (1).

Soccer players also use the Magnus effect when it is advantageous for a kicked ball to follow a curved path, as may be the case when a player executing a free kick attempts to score. The "banana shot" consists of a kick executed so that the kicker places a lateral spin on the ball, curving it around the wall of defensive players in front of the goal (Figure 15-18).

The Magnus effect is maximal when the axis of spin is perpendicular to the direction of relative fluid velocity. Golf clubs are designed to impart some backspin to the struck ball, thereby creating an upwardly directed Magnus force that increases flight time and flight distance (Figure 15-19). When a golf ball is hit laterally off-center, a spin about a vertical axis is also produced, causing a laterally deviated Magnus force that causes the ball to deviate from a straight path. When backspin and sidespin have been imparted to the ball, the resultant effect of the Magnus force on the path of the ball depends on the orientation of the ball's resultant axis of rotation to the airstream and on the velocity with which the ball was struck. When a golf ball is struck laterally off-center, commonly known as a hook (to the left) or a slice (to the right), the ball unfortunately follows a curved path to one side.

PROPULSION IN A FLUID MEDIUM

Whereas a headwind slows a runner or cyclist by increasing the acting drag force, a tailwind can actually contribute to forward propulsion. Theoretical calculations indicate that a tailwind of 2 m/s improves run-

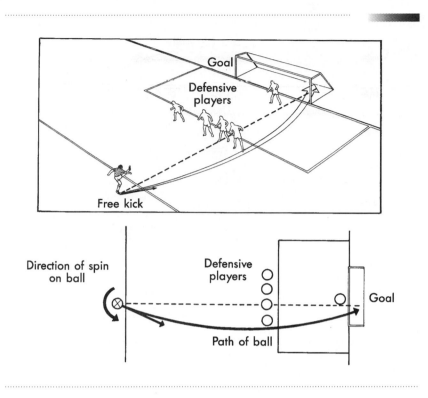

FIGURE 15-18
A banana shot in soccer results from imparting sidespin to the ball.

ning time during a 100 m sprint by approximately 0.18 seconds (42). A tailwind affects the relative velocity of a body with respect to the air, thereby modifying the resistive drag acting on the body. Thus, a tailwind of a velocity greater than the velocity of the moving body produces a drag force in the direction of motion (Figure 15-20). This force has been termed **propulsive drag.**

propulsive drag
force acting in the direction of a body's motion

FIGURE 15-19
The loft on a golf club is designed to produce backspin on the ball. A properly hit ball rises because of the Magnus effect.

FIGURE 15-20
Drag force acting in the same direction as the body's motion may be thought of as propulsive drag because it contributes to the forward velocity of the body.

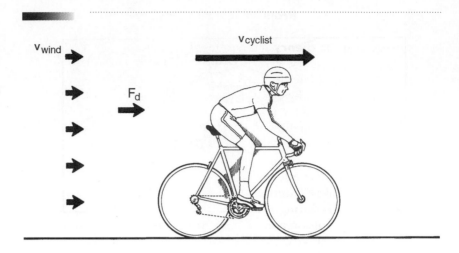

FIGURE 15-20
Drag force acting in the same direction as the body's motion may be thought of as propulsive drag because it contributes to the forward velocity of the body.

Analyzing the fluid forces acting on a swimmer is more complicated. Resistive drag acts on a swimmer, yet the propulsive forces exerted by the water in reaction to the swimmer's movements are responsible for the swimmer's forward motion through the water. The motions of the body segments during swimming produce a complex combination of drag and lift forces throughout each stroke cycle, and even among elite swimmers a wide range of kinetic patterns during stroking have been observed (36). As a result, researchers have proposed several theories regarding the ways in which swimmers propel themselves through the water.

Propulsive Drag Theory

propulsive drag theory
theory attributing propulsion in swimming to propulsive drag on the swimmer

The oldest theory of swimming propulsion is the **propulsive drag theory,** which was proposed by Counsilman and Silvia (9) and is based on Newton's third law of motion. According to this theory, as a swimmer's hands and arms move backward through the water, the forwardly directed reaction force generated by the water produces propulsion. The theory also suggests that the horizontal components of the downward and backward motion of the foot and the upward and backward motion of the opposite foot generate a forwardly directed reaction force from the water.

When high-speed movie films of skilled swimmers revealed that swimmer's hands and feet followed a zigzag rather than a straight-back path through the water, the theory was modified. It was suggested that this type of movement pattern enabled the body segments to push against still or slowly moving water instead of water already accelerated backward, thereby creating more propulsive drag. However, propulsive drag may not be the major contributor to propulsion in swimming.

Propulsive Lift Theory

The **propulsive lift theory** was proposed by Counsilman in 1971 (5). According to this theory, swimmers use the foil-like shape of the hand by employing rapid lateral movements through the water to generate lift. The lift is resisted by downward movement of the hand and by stabilization of the shoulder joint, which translates the forward directed force to the body, propelling it past the hand. The theory was modified by Firby (14) in 1975, with the suggestion that swimmers use their hands and feet as propellers, constantly changing the pitches of the body segments to use the most effective angle of attack.

A number of investigators have since studied the forces generated by the body segments during swimming. It has been shown that lift does contribute to propulsion and that a combination of lift and drag forces acts throughout a stroke cycle. The relative contributions of lift and drag vary with the stroke performed, the phase within the stroke, and the individual swimmer. For example, lift is the primary force acting during the breaststroke, whereas lift and drag contribute differently to various phases of the front crawl stroke (35). Drag generated by the swimmer's hand is maximal when hand orientation is nearly perpendicular to the flow, and lift is maximal when the hand moves in the direction of either the thumb or the little finger (3).

propulsive lift theory
theory attributing propulsion in swimming at least partially to lift acting on the swimmer

Vortex Generation

Researchers have found a poor correlation between physiological and mechanical approaches to calculating propelling efficiency in swimming (4). This has led to the speculation that some unknown processes may play a role in swimming propulsion, with one possibility being the generation of vortices in the water by the swimmer. Vortex shedding has been found to play a role in the propulsion of both flying and swimming vertebrates and insects (13, 30). The generation of thrust in racing canoe and kayak paddling has also been described in terms of the mechanics of vortex-ring wakes (18). The observation that a swimmer performing the dolphin kick leaves behind a series of bound vortices, or columns of rotating water, has also been made (40). More research is needed to clarify the role of vortex generation in swimming propulsion.

Stroke Technique

Just as running speed is the product of stride length and stride rate, swimming speed is the product of stroke length (SL) and stroke rate (SR). Of the two, SL is more directly related to swimming speed among competitive freestyle swimmers (10). Comparison of male and female swimmers performing at the same competitive distances reveals nearly identical SRs, but longer SLs resulting in higher velocities for the males (29). At slower speeds, skilled freestyle swimmers are able to maintain constant, high levels of SL, with a progressive reduction in SL as exercise intensity increases due to local muscle fatigue (20). The same phenomenon has been observed over the course of distance events, with a general decrease in SL and swimming speed over the course of the race (11). Research suggests that recreational freestyle swimmers seeking to improve swimming performance should concentrate on applying more force to the water during each stroke to increase SL, as opposed to stroking faster. Among backstrokers, although the ability to achieve a high swimming speed is related to SL at submaximal levels, increased speed is achieved through increased SR and decreased SL (21). Better performance in the breaststroke is associated with minimal vertical motion of the total body CG during the stroke cycle (34).

Another technique variable of importance during freestyle swimming is body roll. In one study, competitive swimmers were found to roll an average of approximately 60° to the nonbreathing side (26). The contribution of body roll is important since it enables the swimmer to employ the large, powerful muscles of the trunk rather than relying solely on the muscles of the shoulder and arm. Body roll can influence the path of the hand through the water almost as much as the mediolateral motions of the hand relative to the trunk (26). In particular, an increase in body roll has been shown to increase the swimmer's hand speed in the plane perpendicular to the swimming direction, thereby increasing the potential for the hand to develop propulsive lift forces (28).

SUMMARY

The relative velocity of a body with respect to a fluid and the density, specific weight, and viscosity of the fluid affect the magnitudes of fluid forces. The fluid force that enables flotation is buoyancy. The buoyant force acts vertically upward, its point of application is the body's center of volume, and its magnitude is equal to the product of the volume of the displaced fluid and the specific gravity of the fluid. A body floats in a static position only when the magnitude of the buoyant force and body weight are equal and when the center of volume and the center of gravity are vertically aligned.

Drag is a fluid force that acts in the direction of the freestream fluid flow. Skin friction is a component of drag that is derived from the sliding contacts between successive layers of fluid close to the surface of a moving body. Form drag, another component of the total drag, is caused by a pressure differential between the lead and trailing edges of a body moving with respect to a fluid. Wave drag is created by the formation of waves at the interface between two different fluids, such as water and air.

Lift is a force that can be generated perpendicular to the freestream fluid flow by a foil-shaped object. Lift is created by a pressure differential in the fluid on opposite sides of a body that results from differences in the velocity of the fluid flow. The lift generated by spin is known as the Magnus force. Propulsion in swimming appears to result from a complex interplay of propulsive drag and lift.

INTRODUCTORY PROBLEMS

For all problems, assume that the specific weight of fresh water equals 9810 N/m^3 and the specific weight of sea (salt) water equals $10{,}070 \text{ N/m}^3$.

1. A boy is swimming with an absolute speed of 1.5 m/s in a river where the speed of the current is 0.5 m/s. What is the velocity of the swimmer with respect to the current when the boy swims directly upstream? Directly downstream? (Answer: 2 m/s in the upstream direction; 1 m/s in the downstream direction)

2. A cyclist is riding at a speed of 14 km/hr into a 16 km/hr headwind. What is the wind velocity relative to the cyclist? What is the cyclist's velocity with respect to the wind? (Answer: 30 km/hr in the direction of the wind; 30 km/hr in the direction of the cyclist)

3. A skier traveling at 5 m/s has a speed of 5.7 m/s relative to a headwind. What is the absolute wind speed? (Answer 0.7 m/s)

4. A 700 N man has a body volume of 0.08 m^3. If submerged in fresh water, will he float? Given his body volume, how much could he weigh and still float? (Answer: Yes; 784.8 N)

5. A racing shell has a volume of 0.38 m^3. When floating in fresh water, how many 700 N people can it support? (Answer: 5)

6. How much body volume must a 60 kg person have to float in fresh water? (Answer: 0.06 m^3)

7. Explain the implications for flotation due to the difference between the specific weight of fresh water and the specific weight of sea water.

8. What strategy can people use to improve their chances of floating in water? Explain your answer.

9. What types of individuals may have a difficult time floating in water? Explain your answer.

10. A beach ball weighing 1 N and with a volume of 0.03 m^3 is held submerged in sea water. How much force must be exerted vertically downward to hold the ball completely submerged? To hold the ball one-half submerged? (Answer: 301.1 N; 150.05 N)

ADDITIONAL PROBLEMS

1. A cyclist riding against a 12 km/hr headwind has a velocity of 28 km/hr with respect to the wind. What is the cyclist's absolute velocity (Answer: 16 km/hr)

2. A swimmer crossing a river proceeds at an absolute speed of 2.5 m/s on a course oriented at a 45° angle to the 1 m/s current. Given that the absolute velocity of the swimmer is equal to the vector sum of the velocity of the current and the velocity of the swimmer with respect to the current, what is the magnitude and direction of the velocity of the swimmer with respect to the current? (Answer: 3.3 m/s at an angle of 32.6° to the current)

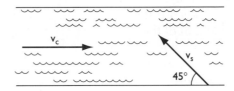

3. What maximum average density can a body possess if it is to float in fresh water? Sea water?

4. A scuba diver carries camera equipment in a cylindrical container that is 45 cm long, 20 cm in diameter, and 22 N in weight. For optimal maneuverability of the container under water, how much should its contents weigh? (Answer 120.36 N)

5. A 50 kg person with a body volume of 0.055 m³ floats in a motionless position. How much body volume is above the surface if the water is fresh? If the water is salt water? (Answer: 0.005 m³; 0.0063 m³)

6. A 670 N swimmer oriented horizontally in fresh water has a body volume of 0.07 m³ and a center of volume located 3 cm superior to the center of gravity.
   a. How much torque does the swimmer's weight generate?
   b. How much torque does the buoyant force acting on the swimmer generate?
   c. What can the swimmer do to counteract the torque and maintain a horizontal position?
   (Answer: 0; 20.6 N-m)

7. Based on your knowledge of the action of fluid forces, speculate as to why a properly thrown boomerang returns to the thrower.

8. Explain the aerodynamic benefits of drafting on a bicycle or in an automobile.

9. What is the practical effect of streamlining? How does streamlining alter the fluid forces acting on a moving body?

10. Explain why a curve ball curves. Include a discussion of the aerodynamic role of the seams on the ball.

1. Slice a hollow ball such as a table tennis ball or a racquet ball in half and float one-half of the ball (concave side up) in a container of water. Gradually add lead shot to the half ball until it floats with the cut edge at the surface of the water. Remove the half ball from the water, then measure its diameter and calculate its volume. Weigh the ball along with the lead shot that was placed in the ball. Using your measurements, calculate the specific weight of the water in the container. Repeat the experiment using water at different temperatures, or using different liquids.

2. Position a container of water on a scale and record its weight. Insert your hand, fingers first, into the water until the water line is at the wrist joint. Record the weight registered on the scale. Subtract the original weight of the container from the new weight, divide the difference in half, and add the result to the original weight of the container to arrive at the target weight. Slowly evaluate your hand from the water until the target weight is reached. Mark the water line on your hand. What does this line represent?

3. Determine how the procedure in Laboratory Exercise 2 might be modified to locate the plane passing through the center of volume of the forearm segment. Test your procedure to verify that it is correct.

4. Using a stopwatch, time yourself while riding up an escalator. Either measure or estimate the length of the escalator and calculate the escalator's speed. Again using a stopwatch, time yourself while carefully running up the moving escalator and calculate your speed. Calculate your speed relative to the speed of the escalator.

5. Use a variable-speed fan and a spring scale to construct a mock wind tunnel. Position the fan so that it blows vertically upward and suspend the spring scale from a rigid arm above the fan. This apparatus can be used to test the relative drag on different objects suspended from the scale. Notice that relative drag among different objects may change with fan speed.

REFERENCES

1. Allman WF: Pitching rainbows: The untold physics of the curve ball. In Schrier EW and Allman WF, eds: *Newton at the bat*, New York, 1981, MacMillan-Charles Scribner's Sons.

2. Bartlett R, Müller E, Lindinger S, Brunner F, and Morriss C: Three-dimensional evaluation of the kinematic release parameters for javelin throwers of different skill levels, J Appl Biomech 12:58, 1996.

3. Berger MA, deGroot G, and Hollander AP: Hydrodynamic drag and lift forces on human hand/arm models, J Biomech 28:125, 1995.

4. Berger MA, Hollander AP, and deGroot G: Technique and energy losses in front crawl swimming, Med Sci Sports Exerc 29:1491, 1997.

5. Brown RM and Counsilman JE: The role of lift in propelling swimmers. In Cooper JM, ed: *Biomechanics*, Chicago, 1971, Athletic Institute.
6. Capelli C and diPrampero PE: Effects of altitude on top speeds during 1 h unaccompanied cycling, Eur J Apl Physiol 71:469, 1995.
7. Capelli C et al: Energy cost and efficiency of riding aerodynamic bicycles, Eur J Appl Physiol 67:144, 1993.
8. Cappaert JM, Pease DL, and Troup JP: Three-dimensional analysis of the men's 100-m freestyle during the 1992 Olympic games, J Appl Biomech 11:103, 1995.
9. Counsilman JE: *Science of swimming*, Englewood Cliffs, NJ, 1968, Prentice-Hall, Inc.
10. Craig AB, Jr. and Pendergast DR: Relationships of stroke rate, distance per stroke, and velocity in competitive swimming, Med Sci Sports Exerc 11:278, 1979.
11. Craig AB, Jr et al: Velocity, stroke rate and distance per stroke during elite swimming competition, Med Sci Sports Exerc 17:625, 1985.
12. Denoth J, Luethi SM, and Gasser HH: Methodological problems in optimization of the flight phase in ski jumping, Int J Sport Biomech 3:404, 1987.
13. Ellington CP: Unsteady aerodynamics of insect flight, Symp Soc Exp Biol 49:109, 1995.
14. Firby H: *Howard Firby on swimming*, London, 1975, Pelham Books, Ltd.
15. Ganslen RV: Aerodynamic factors which influence discus flight. Research report, University of Arkansas.
16. Gnehm P, Reichenback S, Alpeter E, Widmer H, and Hoppeler H: Influence of different racing positions on metabolic cost in elite cyclists, Med Sci Sports Exerc 29:818, 1997.
17. Hay JG and Yu B: Critical characteristics of technique in throwing the discus, J Sports Sci 13:125, 1995.
18. Jackson PS: Performance prediction for Olympic kayaks, J Sports Sci 13:239, 1995.
19. Jin H, Shimizu S, Watanuki T, Kubota H, and Kobayashi K: Desirable gliding styles and techniques in ski jumping, J Appl Biomech 11:460, 1995.
20. Keskinen KL and Komi PV: Stroking characteristics of front crawl swimming during exercise, J Appl Biomech 9:219, 1993.
21. Klentrou PP and Montpetit RR: Energetics of backstroke in swimming in males and females, Med Sci Sports Exerc 24:371, 1992.
22. Kolmogorov SV and Duplishcheva OA: Active drag, useful mechanical power output and hydrodynamic force coefficient in different swimming strokes at maximal velocity, J Biomech 25:311, 1992.
23. Kolmogorov SV, Rumyantseva OA, Gordon BJ, and Cappaert JM: Hydrodynamic characteristics of competitive swimmers of different genders and performance levels, J Appl Biomech 13:88, 1997.
24. Kyle CR and Burke E: Improving the racing bicycle, Mechanical Engineering, 106:34, 1984.
25. Kyle CR and Caiozzo VJ: The effect of athletic clothing aerodynamics upon running speed, Med Sci Sports Exerc 18:509, 1986.
26. Liu Q, Hay JG, and Andrews JG: Body roll and handpath in freestyle swimming: An experimental study, J Appl Biomech 9:238, 1993.
27. Maglischo E: *Swimming faster: A comprehensive guide to the science of swimming*, Palo Alto, Calif, 1982, Mayfield Publishing Co.

28. Payton CJ, Hay JG, and Mullineaux DR: The effect of body roll on hand speed and hand path in front crawl swimming—a simulation study, J Appl Biomech 13:300, 1997.

29. Pelayo P, Sidney M, Kherif T, Chollet D, and Tourny C: Stroking characteristics in freestyle swimming and relationships with anthropometric characteristics, J Appl Biomech 12:197, 1996.

30. Rayner JM: Dynamics of the vortex wakes of flying and swimming vertebrates, Symp Soc Exp Biol 49:131, 1995.

31. Remizov LP: Biomechanics of optimal flight in ski jumping, J Biomech 17:167, 1984.

32. Richardson RS and Johnson SC: The effect of aerodynamic handlebars on oxygen consumption while cycling at a constant speed, Ergonomics 37:859, 1994.

33. Roberson JA and Crowe CT: *Engineering fluid mechanics,* 2nd ed, Boston, 1980, Houghton Mifflin Co.

34. Sanders RH, Cappaert JM, and Pease DL: Wave characteristics of Olympic breaststroke swimmers, J Appl Biomech 14:40, 1998.

35. Schleihauf RE: A hydrodynamic analysis of swimming propulsion. In Terauds J and Bedingfield E, eds: *Swimming III,* Baltimore, 1979, University Park Press.

36. Schleihauf RE et al: Models of aquatic skill sprint front crawlstroke, N Z J Sports Med p 6, Mar 1986.

37. Takamoto M, Ohmichi H, and Miyashita M: Wave height in relation to swimming velocity and proficiency in front crawl stroke. In Winter D et al, eds: *Biomechanics IX-B,* Champaign Ill, 1985, Human Kinetics Publishers, Inc.

38. Toussaint HM et al: Effect of a triathlon wet suit on drag during swimming, Med Sci Sports Exerc 21:325, 1989.

39. Townend MS: *Mathematics in sport,* New York, 1984, John Wiley & Sons, Inc.

40. Ungerechts BE: On the relevance of rotating water flow for the propulsion in swimming. In Jonsson B, ed: *Biomechanics X-B,* Champaign, Ill, 1987, Human Kinetics Publishers, Inc.

41. van Ingen Schenau GJ: The influence of air friction in speed skating, J Biomech 15:449, 1982.

42. Ward-Smith AJ: A mathematical analysis of the influence of adverse and favourable winds on sprinting, J Biomech 18:351, 1985.

43. Ward-Smith AJ: The influence of aerodynamic and biomechanical factors on long jump performance, J Biomech 16:655, 1983.

ANNOTATED READINGS

Kyle CR: Athletic clothing, Sci Am 254:104, 1986.
Presents an overview of the aerodynamic improvements that have been made in clothing for several different sports based on wind tunnel research on drag.

Morriss C and Bartlett R: Biomechanical factors critical for performance in the men's javelin throw, Sports Med 21: 438, 1996.
Reviews biomechanics research on the men's javelin throw to promote understanding of how elite javelin throwers achieve success.

Scrier EW and Allman WF, eds: *Newton at the bat,* New York, 1981, MacMillan–Charles Scribner's Sons.
Chapters on baseball pitching, golf, boomerangs, and frisbee include interesting and entertaining discussions of the actions of fluid forces.

Townend MS: *Mathematics in sport*, New York, 1984, John Wiley & Sons, Inc.
The chapter on sailing provides a technical analysis of the aerodynamics and hydro-dynamics of sailing and windsurfing.

RELATED WEB SITES

Circulation and the Magnus Effect
http://www.phys.virginia.edu/classes/311/notes/aero/node2.html
Shows a diagram and includes discussion of the Magnus force with practical examples.

The Curve with Top Spin
http://math.gmu.edu/~vle3/proj2/top.html
Presents graphed trajectories of balls thrown with different angular velocities of top spin.

The Curve with Back Spin
http://math.gmu.edu/~vle3/proj2/back.html
Presents graphed trajectories of balls thrown with different angular velocities of back spin.

Drag Force in a Medium
http://rowlf.cc.wwu.edu:8080/~vawter/PhysicsNet/Pages/DragForce.html
Discusses the relationship between drag, the drag coefficient, and the Reynolds number, with formulas provided.

The Drag Force on a Sphere
http://www.ma.iup.edu/MathDept/Projects/CalcDEMma/drag/drag0.html
Provides links to pages on the graph of the drag coefficient versus Reynolds number and two models for drag force.

The Javelin
http://www-ct.sprintlink.co.za/~bellevue/javelin/index.html
Describes javelin aerodynamics, along with many other javelin-related topics.

NASA Education On Line
http://trc.dfrc.nasa.gov/trc/ntps/index.html
NASA site that provides links for students and instructors to sites on Newton's Laws, Lift, Drag, Thrust, and Aerodynamic Performance.

Physics in Action
http://www.blueneptune.com/~xmwang/physDemo.html
Provides links to a number of graphical animations of the effect of the Magnus force.

Relative Velocity
http://physics.bu.edu/py105/notes/RelativeV.html
Discusses relative velocity with practical quantitative examples for both one- and two-dimensional motion.

Relative Velocity
http://erebus.phys.cwru.edu/phys/courses/p123/vlabs/vectors/node39.html
Presents a problem and solution on relative velocity from two reference frames.

Relative Velocity Applet
http://www.math.gatech.edu/~carlen/2507/notes/classFiles/partOne/RelVel.html
Interactive application that enables control of two moving points and provides the ability to graph the absolute and relative motions of the points.

Sites of Science: Buoyancy
http://nyelabs.kcts.org/nyeverse/shows/s105.html
Provides links to a laboratory experiment illustrating buoyancy and to an Exploratorium page on buoyancy phenomena.

Appendix A

BASIC MATHEMATICS AND RELATED SKILLS

NEGATIVE NUMBERS

Negative numbers are preceded by a minus sign. Although the physical quantities used in biomechanics do not have values that are less than zero in magnitude, the minus sign is often used to indicate the direction opposite the direction regarded as positive. Therefore, it is important to recall the following rules regarding arithmetic operations involving negative numbers:

1. Addition of a negative number yields the same results as subtraction of a positive number of the same magnitude:

$$6 + (-4) = 2$$
$$10 + (-3) = 7$$
$$6 + (-8) = -2$$
$$10 + (-23) = -13$$
$$(-6) + (-3) = -9$$
$$(-10) + (-7) = -17$$

2. Subtraction of a negative number yields the same result as addition of a positive number of the same magnitude:

$$5 - (-7) = 12$$
$$8 - (-6) = 14$$
$$-5 - (-3) = -2$$
$$-8 - (-4) = -4$$
$$-5 - (-12) = 7$$
$$-8 - (-10) = 2$$

3. Multiplication or division of a number by a number of the opposite sign yields a negative result:

$$2 \times (-3) = -6$$
$$(-4) \times 5 = -20$$
$$9 \div (-3) = -3$$
$$(-10) \div 2 = -5$$

4. Multiplication or division of a number by a number of the same sign (positive or negative) yields a positive result:

$$3 \times 4 = 12$$
$$(-3) \times (-2) = 6$$
$$10 \div 5 = 2$$
$$(-15) \div (-3) = 5$$

EXPONENTS

Exponents are superscripted numbers that immediately follow a base number, indicating the number of times that number is to be self-multiplied to yield the result:

$$5^2 = 5 \times 5$$
$$= 25$$
$$3^2 = 3 \times 3$$
$$= 9$$
$$5^3 = 5 \times 5 \times 5$$
$$= 125$$
$$3^3 = 3 \times 3 \times 3$$
$$= 27$$

SQUARE ROOTS

Taking the square root of a number is the inverse operation of squaring a number (multiplying a number by itself). The square root of a number is the number that yields the original number when multiplied by itself. The square root of 25 is 5, and the square root of 9 is 3. Using mathematics notation these relationships are expressed as the following:

$$\sqrt{25} = 5$$
$$\sqrt{9} = 3$$

Because -5 multiplied by itself also equals 25, -5 is also a square root of 25. The following notation is sometimes used to indicate that square roots may be either positive or negative:

$$\sqrt{25} = \pm 5$$
$$\sqrt{9} = \pm 3$$

ORDER OF OPERATIONS

When a computation involves more than a single operation, a set of rules must be used to arrive at the correct result. These rules may be summarized as follows:

1. Addition and subtraction are of equal precedence; these operations are carried out from left to right as they occur in an equation:

$$7 - 3 + 5 = 4 + 5$$
$$= 9$$
$$5 + 2 - 1 + 10 = 7 - 1 + 10$$
$$= 6 + 10$$
$$= 16$$

2. Multiplication and division are of equal precedence; these operations are carried out from left to right as they occur in an equation:

$$10 \div 5 \times 4 = 2 \times 4$$
$$= 8$$
$$20 \div 4 \times 3 \div 5 = 5 \times 3 \div 5$$
$$= 15 \div 5$$
$$= 3$$

3. Multiplication and division take precedence over addition and subtraction. In computations involving some combination of operations not of the same level of precedence, multiplication and division are carried out before addition and subtraction are carried out:

$$3 + 18 \div 6 = 3 + 3$$
$$= 6$$
$$9 - 2 \times 3 + 7 = 9 - 6 + 7$$
$$= 3 + 7$$
$$= 10$$
$$8 \div 4 + 5 - 2 \times 2 = 2 + 5 - 2 \times 2$$
$$= 2 + 5 - 4$$
$$= 7 - 4$$
$$= 3$$

4. When parentheses ( ), brackets [ ], or braces { } are used, the operations enclosed are performed first, before the other rules of precedence are applied:

$$2 \times 7 + (10 - 5) = 2 \times 7 + 5$$
$$= 14 + 5$$
$$= 19$$

$$
\begin{aligned}
20 \div (2 + 2) - 3 \times 4 &= 20 \div 4 - 3 \times 4 \\
&= 5 - 3 \times 4 \\
&= 5 - 12 \\
&= -7
\end{aligned}
$$

USE OF A CALCULATOR

Simple computations in biomechanics problems are often performed quickly and easily with a hand-held calculator. However, the correct result can be obtained on a calculator only when the computation is set up properly and the rules for ordering of operations are followed. Most calculators come with an instruction manual that contains sample calculations. It is worthwhile to completely familiarize yourself with your calculator's capabilities, particularly use of the memory, before using it for solving problems.

PERCENTAGES

A percentage is a part of 100. Thus, 37% represents 37 parts of 100. To find 37% of 80, multiply the number 80 by 0.37:

$$80 \times 0.37 = 29.6$$

The number 29.6 is 37% of 80. If you want to determine the percentage of the number 55 that equals 42, multiply the fraction by 100%:

$$\frac{42}{55} \times 100\% = 76.4\%$$

The number 42 is 76.4% of 55.

SIMPLE ALGEBRA

The solution of many problems involves setting up an equation containing one or more unknown quantities represented as variables such as x. An equation is a statement of equality implying that the quantities expressed on the left side of the equals sign are equal to the quantities expressed on the right side of the equals sign. Solving a problem typically requires calculation of the unknown quantity or quantities contained in the equation.

The general procedure for calculating the value of a variable in an equation is to isolate the variable on one side of the equals sign and then to carry out the operations among the numbers expressed on the other side of the equals sign. The process of isolating a variable usually involves performing a series of operations on both sides of the equals sign. As long as the same operation is carried out on both sides of the equals sign, equality is preserved and the equation remains valid:

$$x + 7 = 10$$

Subtract 7 from both sides of the equation:

$$x + 7 - 7 = 10 - 7$$
$$x + 0 = 10 - 7$$
$$x = 3$$
$$y - 3 = 12$$

Add 3 to both sides of the equation:

$$y - 3 + 3 = 12 + 3$$
$$y - 0 = 12 + 3$$
$$y = 15$$
$$z \times 3 = 18$$

Divide both sides of the equation by 3:

$$z \times 3 \div 3 = \frac{18}{3}$$
$$z \times 1 = \frac{18}{3}$$
$$z = 6$$
$$q \div 4 = 2$$

Multiply both sides of the equation by 4:

$$q \div 4 \times 4 = 2 \times 4$$
$$q = 2 \times 4$$
$$q = 8$$
$$x \div 3 + 5 = 8$$

Subtract 5 from both sides of the equation:

$$x \div 3 + 5 - 5 = 8 - 5$$
$$x \div 3 = 3$$

Multiply both sides of the equation by 3:

$$x \div 3 \times 3 = 3 \times 3$$
$$x = 9$$
$$y \div 4 - 7 = -2$$

Add 7 to both sides of the equation:

$$y \div 4 - 7 + 7 = -2 + 7$$
$$y \div 4 = 5$$

Multiply both sides of the equation by 4:

$$y \div 4 \times 4 = 5 \times 4$$
$$y = 20$$
$$z^2 = 36$$

Take the square root of both sides of the equation:

$$z = 6$$

MEASURING ANGLES

The following procedure is used for measuring an angle with a pro-tractor:

1. Place the center of the protractor on the vertex of the angle.
2. Align the zero line on the protractor with one of the sides of the angle.
3. The size of the angle is indicated on the protractor scale where the other side of the angle intersects the scale. (Be sure to read from the correct scale on the protractor. Is the angle greater or less than 90 degrees?)

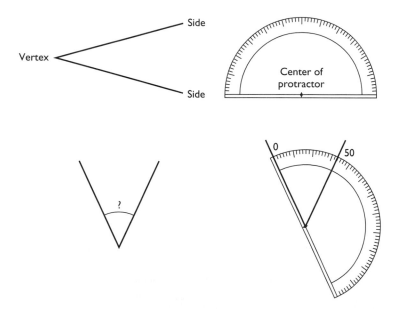

If you are unfamiliar with the use of a protractor, check yourself by ver-ifying the sizes of the following three angles:

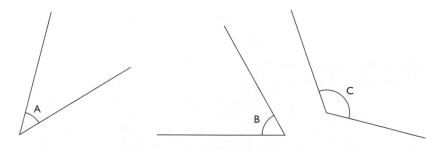

(Answer: A = 45°, B = 60°, C = 123°)

TRIGONOMETRIC FUNCTIONS

Trigonometric functions are based on relationships present between the sides and angles of triangles. Many functions are derived from a right triangle—a triangle containing a right (90°) angle. Consider the right triangle below with sides A, B, and C, and angles α, β, and γ:

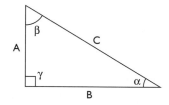

Side C, which is the longest side and the side opposite the right angle, is known as the *hypotenuse of the triangle.*

A commonly used trigonometric relationship for right triangles is the *Pythagorean theorem.* The Pythagorean theorem is an expression of the relationship between the hypotenuse and the other two sides of a right triangle:

> The sum of the squares of the lengths of the two sides of a right triangle is equal to the square of the length of the hypotenuse.

Using the sides of the labeled triangle yields the following:

$$A^2 + B^2 = C^2$$

Suppose that sides A and B are 3 and 4 units long, respectively. The Pythagorean theorem can be used to solve for the length of side C:

$$C^2 = A^2 + B^2$$
$$= 3^2 + 4^2$$
$$= 9 + 16$$
$$C^2 = 25$$
$$C = 5$$

Three trigonometric relationships are based on the ratios of the lengths of the sides of a right triangle. The sine (abbreviated *sin*) of an angle is defined as the ratio of the length of the side of the triangle opposite the angle to the length of the hypotenuse. Using the labeled triangle yields the following:

$$\sin \alpha = \frac{\text{opposite}}{\text{hypotenuse}} = \frac{A}{C}$$

$$\sin \beta = \frac{\text{opposite}}{\text{hypotenuse}} = \frac{B}{C}$$

With A = 3, B = 4, and C = 5:

$$\sin \alpha = \frac{A}{C} = \frac{3}{5} = 0.6$$

$$\sin \beta = \frac{B}{C} = \frac{4}{5} = 0.8$$

The cosine (abbreviated *cos*) of an angle is defined as the ratio of the length of the side of the triangle adjacent to the angle to the length of the hypotenuse. Using the labeled triangle yields the following:

$$\cos \alpha = \frac{\text{adjacent}}{\text{hypotenuse}} = \frac{B}{C}$$

$$\cos \beta = \frac{\text{adjacent}}{\text{hypotenuse}} = \frac{A}{C}$$

With A = 3, B = 4, and C = 5:

$$\cos \alpha = \frac{B}{C} = \frac{4}{5} = 0.8$$

$$\cos \beta = \frac{A}{C} = \frac{3}{5} = 0.6$$

The third function, the tangent (abbreviated *tan*) of an angle, is defined as the ratio of the length of the side of the triangle opposite the angle to that of the side adjacent to the angle. Using the labeled triangle yields the following:

$$\tan \alpha = \frac{\text{opposite}}{\text{adjacent}} = \frac{A}{B}$$

$$\tan \beta = \frac{\text{opposite}}{\text{adjacent}} = \frac{B}{A}$$

With A = 3, B = 4, and C = 5:

$$\tan \alpha = \frac{A}{B} = \frac{3}{4} = 0.75$$

$$\tan \beta = \frac{B}{A} = \frac{4}{3} = 1.33$$

Two useful trigonometric relationships are applicable to *all* triangles. The first is known as the Law of Sines:

> The ratio between the length of any side of a triangle and the angle opposite that side is equal to the ratio between the length of any other side of the triangle and the angle opposite that side.

With respect to the labeled triangle, this may be stated as the following:

$$\frac{A}{\sin \alpha} = \frac{B}{\sin \beta} = \frac{C}{\sin \gamma}$$

A second trigonometric relationship applicable to *all* triangles is the Law of Cosines:

> The square of the length of any side of a triangle is equal to the sum of the squares of the lengths of the other two sides of the triangle minus two times the product of the lengths of the other two sides and the cosine of the angle opposite the original side.

This relationship yields the following for each of the sides of the labeled triangle:

$$A^2 = B^2 + C^2 - 2BC \cos \alpha$$
$$B^2 = A^2 + C^2 - 2AC \cos \beta$$
$$C^2 = A^2 + B^2 - 2AB \cos \gamma$$

COMMON UNITS OF MEASUREMENT

This appendix contains factors for converting between metric units commonly used in biomechanics and their English system equivalents. In each case a value expressed in a metric unit can be divided by the conversion factor given to yield the approximate equivalent in an English unit, or a value expressed in an English unit can be multiplied by the conversion factor to find the metric unit equivalent. For example, to convert 100 Newtons to pounds, do the following:

$$\frac{100 \text{ N}}{4.45 \text{ N/lb}} = 22.5 \text{ lb}$$

To convert 100 pounds to Newtons, do the following:

$$(100 \text{ lb})(4.45 \text{ N/lb}) = 445 \text{ N}$$

VARIABLE	METRIC UNIT	← MULTIPLY BY DIVIDE BY →	ENGLISH UNIT
Distance	Centimeters	2.54	Inches
	Meters	0.3048	Feet
	Kilometers	1.609	Miles
Speed	Meters/second	0.447	Miles/hour
Mass	Kilograms	14.59	Slugs
Force	Newtons	4.448	Pounds
Work	Joules	1.355	Foot-pounds
Power	Watts	745.63	Horsepower
Energy	Joules	1.355	Foot-pounds
Linear momentum	Kilogram-meters/second	4.448	Slug-feet/second
Impulse	Newton-seconds	4.448	Pound-seconds
Angular momentum	Kilogram-meters^2/second	1.355	Slug-feet^2/second
Moment of inertia	Kilogram-meters^2	1.355	Slug-feet^2
Torque	Newton-meters	1.355	Foot-pounds

ANTHROPOMETRIC PARAMETERS FOR THE HUMAN BODY*

SEGMENT LENGTHS		
SEGMENT	MALES	FEMALES
Head and neck	10.75	10.75
Trunk	30.00	29.00
Upper arm	17.20	17.30
Forearm	15.70	16.00
Hand	5.75	5.75
Thigh	23.20	24.90
Lower leg	24.70	25.70
Foot	4.25	4.25

Segment lengths expressed in percentages of total body weight.

SEGMENT WEIGHTS		
SEGMENT	MALES	FEMALES
Head	8.26	8.20
Trunk	46.84	45.00
Upper arm	3.25	2.90
Forearm	1.87	1.57
Hand	0.65	0.50
Thigh	10.50	11.75
Lower leg	4.75	5.35
Foot	1.43	1.33

Segment weights expressed in percentages of total body weight

From Plagenhoef S, Evans FG, and Abdelnour T: Anatomical data for analyzing human motion, Res Q Exerc Sport 54:169, 1983.

*The values reported in these tables represent mean values for limited numbers of individuals as reported in the scientific literature.

SEGMENTAL CENTER OF GRAVITY LOCATIONS

SEGMENT	MALES	FEMALES
Head and neck	55.0	55.0
Trunk	63.0	56.9
Upper arm	43.6	45.8
Forearm	43.0	43.4
Hand	46.8	46.8
Thigh	43.3	42.8
Lower leg	43.4	41.9
Foot	50.0	50.0

Segmental center of gravity locations expressed in percentages of segment lengths; measured from the proximal ends of segments.

SEGMENTAL RADII OF GYRATION MEASURED FROM PROXIMAL AND DISTAL SEGMENT ENDS

SEGMENT	Males		Females	
	PROXIMAL	DISTAL	PROXIMAL	DISTAL
Upper arm	54.2	64.5	56.4	62.3
Forearm	52.6	54.7	53.0	64.3
Hand	54.9	54.9	54.9	54.9
Thigh	54.0	65.3	53.5	65.8
Lower leg	52.9	64.2	51.4	65.7
Foot	69.0	69.0	69.0	69.0

Segmental radii of gyration expressed in percentages of segment lengths

Glossary

absolute angle angular orientation of a body segment with respect to a fixed line of reference

acceleration the rate of change in velocity

acromioclavicular joint irregular joint between the acromion process of the scapula and the distal clavicle

active stretching the stretching of muscles, tendons, and ligaments produced by active development of tension in the antagonist muscles

acute loading application of a single force of sufficient magnitude to cause injury to a biological tissue

agonist role played by a muscle acting to cause a movement

amenorrheic having cessation of menses

anatomical reference position erect standing position with all body parts, including the palms of the hands, facing forward; considered the starting position for body segment movements

angle of attack angle between the longitudinal axis of a body and the direction of the fluid flow

angle of projection the direction at which a body is projected with respect to the horizontal

angular motion involving rotation around a central line or point

angular acceleration rate of change in angular velocity

angular displacement change in angular position

angular impulse change in angular momentum equal to the product of torque and the time interval over which the torque acts

angular momentum quantity of angular motion possessed by a body that is equal to the product of moment of inertia and angular velocity

angular velocity rate of change in angular position

anisotropic exhibiting different mechanical properties in response to loads from different directions

annulus fibrosus thick, fibrocartilaginous ring that forms the exterior of the intervertebral disc

antagonist role played by a muscle acting to slow or stop a movement

anthropometric related to the dimensions and weights of body segments

apex the highest point in the trajectory of a projectile

Archimedes principle physical law stating the buoyant force acting on a body is equal to the weight of the fluid displaced by the body

articular cartilage protective layer of dense white connective tissue covering the articulating bone surfaces at diarthrodial joints

articular capsule	double layered membrane that surrounds every synovial joint
articular fibrocartilage	soft tissue discs or menisci that intervene between articulating bones
average	occurring over a designated time interval
axial	directed along the longitudinal axis of a body
axis of rotation	imaginary line perpendicular to the plane of rotation and passing through the center of rotation
balance	ability to control equilibrium
ballistic stretching	a series of quick, bouncing-type stretches
base of support	area bound by the outermost regions of contact between a body and support surface or surfaces
bending	asymmetric loading that produces tension on one side of a body's longitudinal axis and compression on the other side
Bernoulli principle	an expression of the inverse relationship between relative velocity and relative pressure in a fluid flow
biomechanics	application of mechanical principles in the study of living organisms
bone atrophy	decrease in bone mass resulting from a predominance of osteoclast activity
bone hypertrophy	increase in bone mass resulting from a predominance of osteoblast activity
boundary layer	layer of fluid immediately adjacent to a body
bursae	sacs secreting synovial fluid internally that lessen friction between soft tissues around joints
cardinal planes	three imaginary perpendicular reference planes that divide the body in half by mass
center of mass (mass centroid, center of gravity)	point around which a body's weight and mass are equally balanced in all directions
center of volume	point around which a body's volume is equally distributed and at which the buoyant force acts
central skeleton	the skull, vertebrae, sternum, and ribs
centripetal force	force directed toward the center of rotation of any rotating body
close-packed position	joint orientation for which the contact between the articulating bone surfaces is maximum
coefficient of drag	unitless number that is an index of a body's ability to generate fluid resistance
coefficient of friction	number that serves as an index of the interaction between two surfaces in contact
coefficient of lift	unitless number that is an index of a body's ability to generate lift
coefficient of restitution	number that serves as an index of elasticity for colliding bodies
collateral ligaments	major ligaments that cross the medial and lateral aspects of the knee
combined loading	simultaneous action of more than one of the pure forms of loading
compression	pressing or squeezing force directed axially through a body
compressive strength	ability to resist pressing or squeezing force
concentric	contraction involving shortening of a muscle
contractile component	muscle property enabling tension development by stimulated muscle fibers
coracoclavicular joint	syndesmosis with the coracoid process of the scapula bound to the inferior clavicle by the coracoclavicular ligament

cortical bone compact mineralized connective tissue with low porosity that is found in the shafts of long bones

couple pair of equal, oppositely directed forces that act on opposite sides of an axis of rotation to produce torque

cruciate ligaments major ligaments that cross each other in connecting the anterior and posterior aspects of the knee

curvilinear motion along a curved line

deformation change in original shape

density mass per unit of volume

dislocating component the component of muscle force directed away from the center of the joint crossed

displacement change in position

dynamic equilibrium (D'Alembert's principle) concept indicating a balance between applied forces and inertial forces for a body in motion

dynamics branch of mechanics dealing with systems subject to acceleration

eccentric contraction involving lengthening of a muscle

electromechanical delay time between arrival of neural stimulus and tension development by the muscle

English system system of weights and measures originally developed in England and used in the United States today

epicondylitis inflammation and sometimes microrupturing of the collagenous tissues on either the lateral or medial side of the distal humerus; believed to be an overuse injury

epiphysis growth center of a bone that produces new bone tissue as part of the normal growth process until it closes during adolescence or early adulthood

extrinsic muscles muscles with proximal attachments located proximal to the wrist or ankle and distal attachments located distal to the wrist or ankle

failure loss of mechanical continuity

fast twitch fiber a fiber that reaches peak tension relatively quickly

first class lever lever positioned with the applied force and the resistance on opposite sides of the axis of rotation

flat bones skeletal structures that are largely flat in shape, for example, the scapula

flexion relaxation phenomenon when the spine is in full flexion, the spinal extensor muscles relax and the flexion torque is supported by the spinal ligaments

fluid substance that flows when subjected to a shear stress

foil shape capable of generating lift in the presence of a fluid flow

force push or pull; the product of mass and acceleration

form drag (profile drag, pressure drag) resistance created by a pressure differential between the lead and rear sides of a body moving through a fluid

fracture disruption in the continuity of a bone

free body diagram sketch that shows a defined system in isolation with all of the force vectors acting on the system

friction force acting at the area of contact between two surfaces in the direction opposite that of motion or motion tendency

frontal axis imaginary line around which sagittal plane rotations occur

frontal plane plane in which lateral movements of the body and body segments occur

fulcrum	point of support or axis about which a lever may be made to rotate
general motion	involving translation and rotation simultaneously
glenohumeral joint	ball and socket joint in which the head of the humerus articulates with the glenoid fossa of the scapula
glenoid labrum	rim of soft tissue located on the periphery of the glenoid fossa that adds stability to the glenohumeral joint
Golgi tendon organ	sensory receptor that inhibits tension development in a muscle and initiates tension development in antagonist muscles
hamstrings	the biceps femoris, semimembranosus, and semitendinosus
humeroradial joint	gliding joint in which the capitellum of the humerus articulates with the proximal end of the radius
humeroulnar joint	hinge joint in which the humeral trochlea articulates with the trochlear fossa of the ulna
iliopsoas	the psoas major and iliacus muscles with a common insertion on the lesser trochanter of the femur
iliotibial band	thick, strong band of tissue connecting the tensor fascia lata to the lateral condyle of the femur and the lateral tuberosity of the tibia
impact	collision characterized by the exchange of a large force during a small time interval
impacted	pressed together by a compressive load
impulse	product of force and the time over which the force acts
inertia	tendency of a body to resist a change in its state of motion
inference	the process of forming deductions from available information
initial velocity	vector quantity incorporating both angle and speed of projection
instant center	precisely located center of rotation at a joint at a given instant in time
instantaneous	occurring during a small interval of time
intraabdominal pressure	pressure inside the abdominal cavity; believed to help stiffen the lumbar spine against buckling
intrinsic muscles	muscles with both attachments distal to the wrist or ankle
irregular bones	skeletal structures of irregular shapes, for example, the sacrum
isometric	contraction involving no change in muscle length
joint flexibility	a term indicating the relative ranges of motion allowed at a joint
joint stability	the ability of a joint to resist abnormal displacement of the articulating bones
kinematics	the form, pattern, or sequencing of movement with respect to time
kinesiology	study of human movement
kinetic energy	capacity to do work by virtue of a body's motion
kinetic friction	constant magnitude friction generated between two surfaces in contact during motion
kinetics	study of the action of forces
kyphosis	extreme thoracic curvature
laminar flow	flow characterized by smooth, parallel layers of fluid
laws of constant acceleration	three formulas relating displacement, velocity, acceleration, and time when acceleration is unchanging
lever	simple machine consisting of a relatively rigid barlike body that may be made to rotate about an axis

lift	force acting on a body in a fluid in a direction perpendicular to the fluid flow
lift/drag ratio	the magnitude of the lift force divided by the magnitude of the total drag force acting on a body at a given time
ligamentum flavum	yellow ligament that connects the laminae of adjacent vertebrae; distinguished by its elasticity
linear	motion along a line that may be straight or curved, with all parts of the body moving in the same direction at the same speed
long bones	skeletal structures consisting of a long shaft with bulbous ends, for example, the femur
longitudinal axis	imaginary line around which transverse plane rotations occur
loose-packed position	any joint orientation other than the close-packed position
lordosis	extreme lumbar curvature
Magnus effect	deviation in the trajectory of a spinning object toward the direction of spin resulting from the Magnus force
Magnus force	lift force created by spin
mass	quantity of matter contained in an object
maximum static friction	maximum amount of friction that can be generated between two static surfaces
mechanical advantage	ratio of force arm/resistance arm for a given lever
mechanics	branch of physics that analyzes the actions of forces on particles and mechanical systems
menisci	cartilaginous discs located between the tibial and femoral condyles
meter	most common international unit of length on which the metric system is based
metric system	system of weights and measures used internationally in scientific applications and adopted for daily use by every major country except the United States
moment arm	shortest (perpendicular) distance between a force's line of action and an axis of rotation
moment of inertia	inertial property for rotating bodies that increases with both mass and the distance the mass is distributed from the axis of rotation
momentum	product of a body's mass and its velocity
motion segment	two adjacent vertebrae and the associated soft tissues; the functional unit of the spine
motor unit	a single motor neuron and all fibers it innervates
muscle spindle	sensory receptor that provokes reflex contraction in a stretched muscle and inhibits tension development in antagonist muscles
myoelectric activity	electric current or voltage produced by a muscle
net force	resultant force derived from the composition of two or more forces
neutralizer	role played by a muscle acting to eliminate an unwanted action produced by an agonist
normal reaction force	force acting perpendicular to two surfaces in contact
nucleus pulposus	colloidal gel with a high fluid content, located inside the annulus fibrosus of the intervertebral disc
osteoblasts	specialized bone cells that build new bone tissue
osteoclasts	specialized bone cells that resorb bone tissue
osteopenia	condition of reduced bone mineral density that predisposes the individual to fractures

osteoporosis	a disorder involving decreased bone mass and strength with one or more resulting fractures
parallel elastic component	passive elastic property of muscle derived from the muscle membranes
parallel fiber arrangement	pattern of fibers within a muscle in which the fibers are roughly parallel to the longitudinal axis of the muscle
passive stretching	the stretching of muscles, tendons, and ligaments produced by a stretching force other than tension in the antagonist muscles
patellofemoral joints	articulation between the patella and the femur
pelvic girdle	the two hip bones plus the sacrum, which can be rotated forward, backward, and laterally to optimize positioning of the hip joint
pennate fiber arrangement	pattern of fibers within a muscle with short fibers attaching to one or more tendons
perfectly elastic impact	impact during which the velocity of the system is conserved
perfectly plastic impact	impact resulting in the total loss of system velocity
periosteum	double-layered membrane covering bone; muscle tendons attach to the outside layer, and the internal layer is a site of osteoblast activity
peripheral skeleton	bones composing the body appendages
plantar fascia	thick bands of fascia that cover the plantar aspect of the foot
popliteus	muscle known as the unlocker of the knee because its action is lateral rotation of the femur with respect to the tibia
porous	containing pores or cavities
potential energy	capacity to do work by virtue of a body's position, calculated as the product of weight and height
power	rate of work production that is calculated as work divided by the time during which the work was done
pressure	force per unit of area
prestress	stress on the spine created by tension in the resting ligaments
primary spinal curves	curves that are present at birth
principal axes	three mutually perpendicular axes passing through the center of gravity that are referred to respectively as the transverse, anteroposterior, and longitudinal axes
principal moment of inertia	total body moment of inertia relative to one of the principal axes
projectile	a body in free fall that is subject only to the forces of gravity and air resistance
projection speed	the magnitude of projection velocity
pronation	combined conditions of dorsiflexion, eversion, and abduction
proprioceptive neuromuscular facilitation	a group of stretching procedures involving alternating contraction and relaxation of the muscles being stretched
propulsive drag	force acting in the direction of a body's motion
propulsive drag theory	theory attributing propulsion in swimming to propulsive drag on the swimmer
propulsive lift theory	theory attributing propulsion in swimming at least partially to lift acting on the swimmer
quadriceps	the rectus femoris, vastus lateralis, vastus medialis, and vastus intermedius
qualitative	involving nonnumerical description of quality

quantitative   involving the use of numbers

radial acceleration   component of angular acceleration directed toward the center of curvature that indicates change in direction

radian   unit of angular measure used in angular-linear kinematic quantity conversions; equal to 57.3 degrees

radiocarpal joints   condyloid articulations between the radius and three carpal bones

radioulnar joints   the proximal and distal radioulnar joints are pivot joints; the middle radioulnar joint is a syndesmosis

radius of gyration   distance from the axis of rotation to a point where the body's mass could be concentrated without altering its rotational characteristics

radius of rotation   distance from the axis of rotation to a point of interest on a rotating body

range   the horizontal displacement of a projectile at landing

range of motion   the angle through which a joint moves from anatomical position to the extreme limit of segment motion in a particular direction

reaction board   specially constructed board for determining the center of gravity location of a body positioned on top of it

reciprocal inhibition   the inhibition of the antagonist muscles resulting from activation of muscle spindles

rectilinear   motion along a straight line

relative angle   angle at a joint formed between the longitudinal axes of adjacent body segments

relative projection height   the difference between projection height and landing height

relative velocity   velocity of a body with respect to the velocity of something else, such as the surrounding fluid

repetitive loading   repeated application of a subacute load that is usually of relatively low magnitude

resultant   single vector that results from vector composition

right hand rule   procedure for identifying the direction of an angular motion vector

rotary component   the component of muscle force acting perpendicular to the attached bone

rotator cuff   band of tendons of subscapularis, supraspinatus, infraspinatus, and teres minor, which attach to the humeral head

sagittal axis   imaginary line around which frontal plane rotations occur

sagittal plane   plane in which forward and backward movements of the body and body segments occur

scalar   physical quantity that is completely described by its magnitude

scapulohumeral rhythm   a regular pattern of scapular rotation that accompanies and facilitates humeral abduction

scoliosis   lateral spinal curvature

second class lever   lever positioned with the resistance between the applied force and the fulcrum

secondary spinal curves   curves that do not develop until the weight of the body begins to be supported in sitting and standing positions

segmental method   procedure for determining total body center of mass location based on the masses and center of mass locations of the individual body segments

series elastic component   passive elastic property of muscle derived from the tendons

shear   force directed parallel to a surface

short bones   small, cubical skeletal structures, including the carpals and tarsals

skin friction (surface drag, viscous drag) resistance derived from friction between adjacent layers of fluid near a body moving through the fluid

slow twitch fiber a fiber that reaches peak tension relatively slowly

specific weight weight per unit of volume

spondylolisthesis complete bilateral fracture of the pars interarticularis, resulting in anterior slippage of the vertebra

spondylolysis presence of a fracture in the pars interarticularis of the vertebral neural arch

sports medicine clinical and scientific aspects of sports and exercise

stability resistance to disruption of equilibrium

stabilizer role played by a muscle acting to stabilize a body part against some other force

stabilizing component the component of muscle force directed toward the center of the joint crossed

static equilibrium motionless state characterized by $\Sigma F_v = 0$, $\Sigma F_h = 0$, and $\Sigma T = 0$

static stretching maintaining a slow, controlled, sustained stretch over time, usually about 30 seconds

statics branch of mechanics dealing with systems in a constant state of motion

sternoclavicular joint modified ball and socket joint between the proximal clavicle and the manubrium of the sternum

stiffness the ratio of stress to strain in a loaded material; that is, the stress divided by the relative amount of change in the structure's shape

strain amount of deformation divided by the original length of the structure or by the original angular orientation of the structure

strain energy capacity to do work by virtue of a deformed body returning to its original shape

stress distribution of force within a body, quantified as force divided by the area over which the force acts

stress fracture fracture resulting from repeated loading of relatively low magnitude

stretch reflex a monosynaptic reflex initiated by stretching of muscle spindles and resulting in immediate development of muscle tension

stretch-shortening cycle eccentric contraction followed immediately by concentric contraction

summation building in an additive fashion

supination combined conditions of plantar flexion, inversion, and adduction

synovial fluid clear, slightly yellow liquid that provides lubrication inside the articular capsule at synovial joints

system mechanical system chosen by the analyst for study

tangential acceleration component of angular acceleration directed along a tangent to the path of motion that indicates change in linear speed

tensile strength ability to resist pulling or stretching force

tension pulling or stretching force directed axially through a body

tetanus state of muscle producing sustained maximal tension resulting from repetitive stimulation

theoretical square law drag increases approximately with the square of velocity when relative velocity is low

third class lever lever positioned with the applied force between the fulcrum and the resistance

tibiofemoral joint dual condyloid articulations between the medial and lateral condyles of the tibia and the femur composing the main hinge joint of the knee

torque rotary effect of a force

torsion   load producing twisting of a body around its longitudinal axis

trabecular bone   less compact mineralized connective tissue with high porosity that is found in the ends of long bones and in the vertebrae

trajectory   the flight path of a projectile

transducers   devices that detect a signal

translation   linear motion

transverse plane   plane in which horizontal body and body segment movements occur when the body is in an erect standing position

turbulent flow   flow characterized by mixing of adjacent fluid layers

valgus   condition of outward deviation in alignment from the proximal to the distal end of a body segment

varus   condition of inward deviation in alignment from the proximal to the distal end of a body segment

vector   physical quantity that possesses both magnitude and direction

vector composition   process of determining a single vector from two or more vectors by vector addition

vector resolution   operation that replaces a single vector with two perpendicular vectors such that the vector composition of the two perpendicular vectors yields the original vector

velocity   change in position with respect to time

viscoelastic   having the ability to stretch or shorten over time

volume   space occupied by a body

wave drag   resistance created by the generation of waves at the interface between two different fluids, such as air and water

weight   attractive force that the earth exerts on a body

work   expression of mechanical energy that is calculated as force multiplied by the displacement of the resistance in the direction of the force

yield point (elastic limit)   point on the load-deformation curve past which deformation is permanent

Index

A

Abduction, 38, 39, 43
Absolute angles, 372
Acceleration
    angular, measurement of,
        380–382
    equation of constant
        acceleration, 352–360
    law of constant acceleration, 353
    linear and angular relationship,
        387–388
    measurement of, 335–340
    Newton's law, 397–398, 495
    radial acceleration, 388
    tangential acceleration, 387–388
Accelerometer, 56
Achilles tendinitis, 271
Acromioclavicular joint, 187
Active stretching, 134
Acute loading, 78–79
Adduction, 38, 39, 43
Agonists, skeletal muscle, 164–166
Air resistance, and projectile motion,
    343–344
Algebra
    algebraic calculations, types of,
        545–546
    vector algebra, 81–84
Amenorrhea, and female athletes,
    104–105
Amphiarthoses joints, 119
Anatomical levers, 450–451
Anatomical reference axes, 12
Anatomical reference planes, 30–32
Anatomical reference position,
    meaning of, 29
Angle of attack, 526, 528
Angle of projection, and projectile
    motion, 345–347, 349
Angles, 369–374
    absolute angles, 372
    instant center of rotation,
        373–374
    measurement of, 547

    relative angles, 370–372
    tools for measurement of,
        372–373
Angular acceleration
    measurement of, 380–382
    relationship to linear
        acceleration, 387–388
Angular displacement
    measurement of, 375–378
    relationship to linear
        displacement, 382, 384
Angular distance, measurement of,
    375–378
Angular impulse, 491–494
Angular kinematics
    angles, 369–374
    angular acceleration, 380–382
    angular displacement, 375–378
    angular distance, 375–378
    angular motion vectors, 382
    angular speed, 378–380
    angular velocity, 378–380
    average/instantaneous
        quantities, 382
    human movement studies,
        368–369
Angular kinetics
    angular momentum, 482–494
    centripetal force, 497–498
    inertia, 476–482
    Newton's laws, 494–496
Angular momentum, 482–494
    affecting factors, 483–485
    change in, 490–494
    conservation of, 485–487
    meaning of, 482–483
    transfer of, 488–490
Angular motion, 34, 36
Angular speed, measurement of,
    378–380
Angular velocity
    measurement of, 378–380
    relationship to linear velocity,
        384–385

Anisotropic, bone, 93
Ankle, 260–264
    injuries of, 270–272
    movement of, 261–264
    structure of, 262–264
Ankle bracing, 271
Annulus fibrosus, 287
Anorexia nervosa, signs of, 104
Antagonists, skeletal muscle, 164–165
Anthropometric factors, 3
Apex, 343
Appendicular skeleton, 94
Archimedes' principle, 511
Arthritis, 139
    osteoarthritis, 139
    rheumatoid arthritis, 139
Artibular fibrocartilage, 122–123
Articular capsule, 119–120
Articular cartilage, 94, 119, 122
Articular fibrocartilage, 122–123
Atrophy, bones, 100–102
Average quantities, 340–341
Axial forces, 75
Axial skeleton, 94
Axis of rotation, 34

B
Back injuries
    disc herniations, 317
    fractures, 315–317
    low back pain, 313–315
    soft tissue injuries, 315
    spondylolisthesis, 316–317
    spondylolysis, 316–317
Balance, and stability, 465
Ball-and-socket joint, 121, 122
Ballistic stretching, 136
Base of support, and stability,
    467–469
Bending, 75
Bernoulli principle, 524
Biomechanics
    meaning of, 3, 4
    problems in study of, 5, 7–12
    rationale for study of, 12
    scope of study of, 3–5
Body position terms
    anatomical reference axes, 12
    anatomical reference planes,
        30–32

anatomical reference position, 29
directional terms, 29–30
for motion, 32–36
for movements, 37–44
Body roll, 534
Bone growth
    adult bone development,
        102–103
    circumferential growth, 97
    longitudinal growth, 97
Bone problems
    epiphyseal injuries, 110
    fractures, 107–109
    osteoporosis, 102–107
Bones
    atrophy, 100–102
    hypertrophy, 99–100
    modeling/remodeling, 98–99
    structure and function, 92
    types of, 94
Bone tissue
    building blocks of, 91
    structure of, 91–92
Boundary layer, 516, 519–521
Boxer's fractures, 226
Breaststroke, knee injuries, 259
Bulimia, signs of, 104
Buoyancy, 510–514
    Archimedes' principle, 511
    center of volume, 511
    characteristics of, 510–511
    flotation, 512–514
Bursae, 120
    hip, 235
    knee, 250
    shoulder, 191–192
Bursitis, 138–139, 205

C
Cardinal planes, nature of, 30
Carpal tunnel syndrome, 9, 226
Carpometacarpal joints, 220, 221
Cat rotation, 489–490
Center of gravity, 65, 458–465
    location for human body,
        461–464, 555
    location for object, 459–461
    meaning of, 458–459
Center of mass, 458–459
Center of volume, 511

Centripetal force, 497–498
  meaning of, 497
  measurement of, 497
Chondromalacia, knee, 259
Circumduction, 44
Circumferential growth, bone, 97
Close-packed position, joints,
    124–125
Coefficient of drag, 515
Coefficient of friction, 404–408
Coefficient of lift, 523–524, 525
Coefficient of restitution,
    417–420
Collateral ligaments, 249
Combined loading, 77
Comminuted fracture, 108
Compression, 73
Compressive strength, bone, 91
Concentric contraction, muscle, 162
Condyloid joint, 121, 122
Conservation of mechanical
    energy, 414
Contractile component, muscle,
    147, 148
Contusions, hip, 245
Coracoclavicular joint, 189
Cortical bone, 91–92
Cosine, 549
Couple, 439
Cruciate ligaments, 249, 256
Curvilinear motion, 33–34, 35, 36

D
D'Alembert's principle, 454–458
Dancers injuries, of foot, 272
Deceleration, 337–338, 340
Deformation, 77–78
Density, 70–71
Depression, 38, 40
De Quervains disease, 226
Diarthroses joints, 119
Directional terms, body position
    terms, 29–30
Disc herniations, 317
Dislocations
    elbow, 213–215
    joints, 138
    shoulder, 203–204
Displacement

angular, measurement of,
    375–378
linear, measurement of, 328–330
linear and angular relationship,
    382, 384
Distance
    angular, measurement of,
        375–378
    linear, measurement of, 328–330
Dorsiflexion, 38
Dowager's hump, 104
Drag, 514–523
    coefficient of drag, 515
    form drag, 518–522
    meaning of, 514–515
    propulsive drag, 531–532
    skin friction drag, 516–517
    theoretical square law, 515–516
    wave drag, 522–523
Drop finger deformity, 226
Dynamic equilibrium, 454–458
Dynamics, 3
Dynamography, 80

E
Eating disorders, and female athlete
    triad, 104–105
Eccentric contraction, muscle, 164
Elastic limit, 78
Elbow, 206–216
    humeroradial joint, 206
    humeroulnar joint, 206
    ligaments of, 207
    loading of, 211, 213
    movements of, 207–211
    muscles, listing of, 209
    proximal radioulnar joint, 206
    supination, 210–211, 212
Elbow injuries
    dislocations, 213–215
    epicondylitis, 215–216
    overuse injuries, 215–216
    sprains, 213
Electrogoniometer, 55–56, 129, 373
Electromechanical delay
    muscular force, 170
    skeletal muscle, 170
Electromyography, 79–80
Elevation, 38, 40

Endurance, muscular, 176
Energy
    conservation of mechanical
        energy, 414
    kinetic energy, 422–423, 425
    measurement of, 422–423
    potential energy, 422–424
    strain energy, 423
    work/energy relationship,
        425–428
English system
    conversion of metric/English
        units, 552
    measurement units, 17
Epicondylitis, elbow, 215–216
Epiphyseal injuries, bone, 110
Epiphysis, 97
Equation of constant acceleration,
    352–360
Equilibrium, 436–458
    dynamic equilibrium, 454–458
    levers, 445–451
    static equilibrium, 452–454
    torque, 436–445
Eversion, 39, 41
Exercise
    and bone density, 99–100
    osteoporosis prevention,
        105–107
Exponents, 543
Extension, 37–38
Extrinsic muscles, hand, 223–224

F

Failure, mechanical, 78
Fascia, 126–127
Fast twitch, muscle fibers,
    154–158, 162
Fatigue, muscle fatigue, 176–177
Female athlete triad, 104–105
First class lever, 445–447
Fissured fracture, 108
Flat bones, 94
Flavum, 292
Flexibility. *See* Joint flexibility
Flexion, 37–38
Flexion relaxation phenomenon,
    308, 310
Flotation, 512–514

of human body, 512–514
of objects, 512
Fluids
    buoyancy, 510–514
    drag, 514–523
    laminar flow, 508
    lift force, 523–530
    properties of, 509–510
    propulsion in, 530–534
    and relative velocity, 507
    turbulent flow, 508
Foil shape, 524–529
Foot, 265–274
    alignment anomalies of,
        272–274
    injuries of, 271–272
    loading of, 269–270
    movement of, 266–269
    muscles of, 268
    structure of, 265–266
Force, 63–65
    couple, 439
    lift force, 523–530
    moment arm, 437–440
    muscular, 167–169
    rotary force, 436–440
    torque, 436–445
Form drag, 518–522
Fractures, 107–109
    hip, 245
    stress fractures, 109
    types of, 108
    vertebral, 315–317
Free body diagram, 64–65
Friction, 403–410
    coefficient of friction, 404–408
    example, in daily activities,
        408–410
    kinetic friction, 403, 405
    maximum static friction,
        403, 405
Frontal axis, 32
Frontal plane, nature of, 30, 31
Frontal plane movements, 38–40
Fulcrum, 445

G

General motion, 32
Glenohumeral joint, 189–190, 197

Gliding joint, 121
Golfer's injuries
    golfer's epicondylitis, 216
    wrist overuse, 226
Golgi tendon organs, 132, 135
Goniometer, 55, 129, 372–373
Gravity
    center of gravity, 458–465
    Newton's law, 402
    and projectile motion, 342–343
Greenstick fracture, 108
Ground reaction force (GRF),
        398–401

H

Hamstrings, 237, 241
Hand, 219–226
    injuries of, 226
    movements of, 221–223, 226
    muscles, listing of, 224–226
    structure of, 219–221
Hinge joint, 121, 122
Hip, 234–246
    loading of, 243–245
    movement of, 235, 237, 239,
        242–243
    muscles of, 237–243
    structure of, 234–237
Hip injuries
    contusions, 245
    fractures, 245
    strains, 245–246
Hula movement, 488–489
Human body
    anthropometric parameters,
        554–555
    body position terms, 29–44
Humeroradial joint, 206
Humeroulnar joint, 206
Humerus, and shoulder
        movements, 199
Hyperextension, 37–38
Hypertrophy, bones, 99–100
Hypotenuse of the triangle,
        548–549

I

Iliopsoas, 237
Iliotibial bands, 127, 250

Illiotibial band friction syndrome,
        258–259, 274
Impact, 417–420
    coefficient of restitution,
        417–420
    perfectly elastic impact, 417
    perfectly plastic impact, 417
Impacted, bone, 108
Impulse, 72, 412–416
    angular impulse, 491–494
    manipulation of, 414–416
    relationship to momentum,
        412, 414
Inclinometers, 373
Inertia, 63, 476–482
    momentum of, 476–482
    Newton's law, 397, 495
Inference, and problem solving, 17
Initial velocity, 352
Instantaneous quantities, 340
Instant center of rotation, 373–374
Intermetacarpal joint, 220
Intermetatarsal joint, 265
Interphalangeal joint, 221, 223, 265
Intervertebral discs, 286–290
Intraabdominal pressure, 311–312
Intrinsic muscles, hand, 223, 225
Inversion, 39, 41
Irregular bones, 94
Isometric contraction, muscle, 164

J

Joint flexibility, 127–138
    influencing factors, 130
    and injury, 130–132
    range of motion, 127–129
    static and dynamic flexibility, 127
    stretching, 132–138
Joint injuries, 138–139
    arthritis, 139
    bursitis, 138–139
    dislocations, 138
    osteoarthritis, 139
    sprains, 138
Joints
    articular cartilage, 122
    articular fibrocartilage, 122–123
    classification of, 119–122
    elbow, 206

foot, 265–266
hand, 219–221
hip, 234–235
knee, 250
ligaments, 123–124, 125–126
and muscles, 125–126
resultant joint torques,
441–445
shoulder, 187–190
stability of, 124–127
tendons, 123–124, 126
two-joint/multijoint muscle,
166–167
wrist, 216
Joule (J), 421

K

Kinematic quantities
acceleration, 335–340
average quantities, 340–341
displacement, 328–330
distance, 328–330
instantaneous quantities, 340
speed, 330, 334
velocity, 330–335
Kinematics
meaning of, 3, 326–327
See also Angular kinematics;
Linear kinematics
Kinesiology, meaning of, 3
Kinetic energy, 422–423, 425
Kinetic friction, 403, 405
Kinetic quantities, units for, 72
Kinetics
basic concepts related to, 63–67,
70–72
meaning of, 3
mechanical loads, 73–79
vector algebra, 81–84
See also Angular kinetics
Kinetics measurement tools, 79–80
dynamography, 80
electromyography, 79–80
Knee, 247–260
bursae, 250
ligaments, 248–250
loading of, 253–256
menisci, 248
movements of, 251, 253

muscles of, 251–253
patellofemoral joint, 250
tibiofemoral joint, 247
Knee braces, 11, 258
Knee injuries
breaststrokers, 259
chondromalacia, 259
illiotibial band friction
syndrome, 258–259
ligament injuries, 256, 258
shin splints, 259–260
Kyphosis, 292, 293

L

Laminar flow, fluids, 508
Lateral flexion, 38, 40
Law of acceleration, 397–398
angular analogue of, 495
Law of cosines, 550
Law of gravitation, 402
Law of inertia, 397
angular analogue of, 495
Law of reaction, 398–401
angular analogue of, 496
Law of sines, 550
Laws of constant acceleration, 353
Leighton flexometer, 129
Levers, 445–451
anatomical levers, 450–451
first/second/third class levers,
445–447
mechanical advantage,
448–449
operation of, 445
purposes of, 447
Lift force, 523–530
angle of attack, 526, 528
Bernoulli principle, 524
coefficient of lift,
523–524, 525
foil shape, 524–529
lift/drag ratio, 526
magnus effect, 529–530
meaning of, 523
Ligament injuries
hip, 235
knee, 256, 258
Ligaments
ankle, 262

Ligaments—*Cont.*
  elbow, 207
  hip, 235, 236
  and joints, 123–124, 125–126
  knee, 248–250
  spine, 290, 292
Linear acceleration
  measurement of, 336
  relationship to angular
    acceleration, 387–388
Linear displacement
  measurement of, 328–330
  relationship to angular
    displacement, 382, 384
Linear distance, measurement of,
  328–330
Linear kinematics
  friction, 403–410
  human movement studies, 328
  impact, 417–420
  impulse, 412–416
  measures in. *See* Kinematic
    quantities
  momentum, 411–412
  Newton's laws, 396–402
  projectile motion, 341–360
Linear momentum, 411
Linear motion, 33–34
Linear velocity
  measurement of, 330–335
  relationship to angular velocity,
    384–385
Little Leaguer's elbow, 216
Long bones, 94
Longitudinal axis, 32
Longitudinal growth, bone, 97
Loose-packed position, joints, 125
Lordosis, 292, 293
Low back pain, 313–315
Lower extremity
  ankle, 260–264
  foot, 265–274
  hip, 234–246
  knee, 247–260

M
Magnus effect, 529–530
Magnus force, 529, 530
Mass, 63
Mass centroid, 458–459

Mathematics
  algebra, 545–546
  angles, measurement of, 547
  calculator for computations, 545
  exponents, 543
  negative numbers, 542–543
  order of operations, 544–545
  percentages, 545
  square roots, 543
  trigonometry, 548–550
Maximum static friction, 403
Measurement units
  conversion of metric/English
    units, 552
  English system, 17
  metric system, 21
Mechanical advantage, 448–449
Mechanical loads, 73–79
  acute loading, 78–79
  axial forces, 75
  bending, 75
  combined loading, 77
  compression, 73
  effects of loading, 77–78
  repetitive loading, 78–79
  shear, 74
  stress, 74–75
  tension, 73–74
  torsion, 77
Mechanics, 3
Menisci, knee, 248, 258
Menstrual cycle, amenorrhea and
  female athletes, 104–105
Metacarpophalangeal joint, 220, 223
Metatarsophalangeal joint, 265
Meter, 328–329
Metric system, conversion of
  metric/English units, 552
Mobility impairment, 7–8
Moment arm, 437–440
Momentum, 411–412
  angular momentum, 482–494
  linear momentum, 411
  principle of conservation of
    momentum, 411–412
  relationship to impulse, 412, 414
Momentum of inertia, 476–482
  determination of, 479–481
  of human body, 481–482
  meaning of, 477–478

principal moment of inertia, 482
and radius of gyration, 480–481
Motion
angular motion, 34, 36
general motion, 32, 36
linear motion, 33–34
Motion segment, 284
Motor units, 154–155
activation of, 162
Movement analysis tools
accelerometer, 56
electrogoniometer, 54
electromagnetic magnet
tracking, 55
goniometer, 55
video analysis, 53–54
Movements
frontal plane movements, 38–40
qualitative analysis of, 44–52
sagittal plane movements, 37–38
transverse plane movements,
40–43
Muscle contraction
concentric contraction, 162
eccentric contraction, 164
isometric contraction, 164
Muscle fibers, 149–160
composition of, 150–153
fast twitch, 154–158
fast twitch fibers, 162
motor units, 154–155
and muscle contraction, 150
parallel arrangement, 158–161
pennate arrangement, 158–161
slow twitch, 154–158
slow twitch fibers, 162
Muscles
and joints, 125–126
See also Skeletal muscle
Muscle spindles, 133, 135
Muscular endurance, 176
Muscular force, 167–169
electromechanical delay, 170
force-velocity relationship,
167–169
length-tension relationship, 169
Muscular power, 175–176
Muscular strength, 171–174
Myoelectric activity, measurement of,
79–80

N
Negative numbers, 542–543
Net force, 65
Neutralizers, skeletal muscle,
165–166
Newton's laws, 396–402
angular analogues of, 494–496
law of acceleration, 397–398
law of gravitation, 402
law of inertia, 397
law of reaction, 398–401, 402
Normal reaction force, 404–408
Nucleus pulposus, 287

O
Oblique fracture, 108
Occupational biomechanics,
study of, 9
Osteoarthritis, 139
Osteoblasts, 97
Osteoclasts, 97
Osteopenia, 102
Osteoporosis, 7, 102–107
age-related, features of, 103–104
and menstrual cessation,
104–105
prevention of, 105–107
risk factors, 106
Overuse injuries
ankle and foot, 271–272
elbow, 215–216

P
Parallel arrangement, muscle fibers,
158–161
Parallel elastic component, muscle,
147, 148
Passive stretching, 134–135
Patellofemoral joint, 250, 255–256
Pelvic girdle, 235, 237
Pennate arrangement, muscle fibers,
158–161
Percentages, 545
Perfectly elastic impact, 417
Perfectly plastic impact, 417
Periosteum, 97
Pivot joint, 121, 122
Planes, anatomical reference planes,
30–32
Plantar arches, 265–266

Plantar fascia, 266
Plantar fascitis, 271–272
Popliteus muscle, 251
Porous, bone, 91
Potential energy, 422–424
Power
    meaning of, 421
    measurement of, 421
    muscular, 175–176
Pressure, 67
Pressure differential, 518
Pressure drag, 518–522
Prestress, 292
Primary curves, spine, 292
Principal axes, 482
Principal moment of inertia, 482
Problem solving, 13–18
    inference in, 17
    qualitative problems, 13–15
    quantitative problems,
        13–16
    steps in, 18
Profile drag, 518–522
Projectile motion, 341–360
    and air resistance, 343–344
    and angle of projection,
        345–347, 349
    equation of constant
        acceleration, 352–360
    and gravity, 342–343
    horizontal/vertical components
        of, 341–342, 354–355
    optimum conditions for,
        350–352
    and projection speed,
        347–348
    and relative projection height,
        348, 350
    trajectory, 344
Projection speed, and projectile
    motion, 347–348
Pronation, 39, 41, 42, 269
Proprioceptive neuromuscular
    facilitation, 136–138
Propulsion
    in fluid medium, 530–534
    propulsive drag, 531–532
    propulsive drag theory, 532
    propulsive lift theory, 533

    stroke technique, 533–534
    vortex generation, 533
Propulsive drag, 531–532
Propulsive drag theory, 532
Propulsive lift theory, 533
Pythagorean theorem, 548–549

Q
Quadriceps, 251
Qualitative analysis of movement,
    44–52
    planning for, 46–49
    sources of information for,
        45–46, 51
    steps in, 49–52
Qualitative problem solving, 13–15
Quantitative problem solving,
    13–16

R
Radial acceleration, 388
Radial deviation, 38
Radian, 376, 378
Radiocarpal joint, 216
Radius of gyration, 480–481
Radius of rotation, 384
Range of motion
    joints, 127–129
    measurement of, 128–129
Range of projectile, 347–348
Reaction, Newton's law, 398–401, 496
Reaction board, 462, 464
Reciprocal inhibition, 133
Rectilinear motion, 33, 35, 36
Relative angles, 370–372
Relative projection height, and pro-
    jectile motion, 348, 350
Relative velocity, and fluids, 507
Repetitive loading, 78–79
Restitution, coefficient of, 417–420
Resultant, 81, 83
Resultant joint torques, 441–445
Retinacula, 216
Rheumatoid arthritis, 139
Right hand rule, 382
Rotary force, 436–440
Rotation, 35, 36
Rotational injuries, shoulder, 205
Rotational movements, 40

Rotator cuff injuries, 204–205
   rotator cuff impingement
      syndrome, 204
   swimmer's shoulder, 204–205
   tears, 205
Rotator cuff muscles, 190, 192
Runner injuries
   of ankle/foot, 271–272
   runner's knee, 259

S

Saddle joint, 121, 122
Sagittal axis, 32
Sagittal plane, nature of, 30, 31
Sagittal plane movements, 37–38
Sarcomeres, 151
Scalar, 81
Scapular muscles, actions of,
   193–194
Scapulohumeral rhythm, 192
Scapulothoracic joint, 190–191
Scheuermann's disease, 292
Scoliosis, 293–294
Secondary curves, spine, 292
Second class lever, 445–447
Segmental method, 462, 464
Series elastic component, muscle,
   147, 148
Sernoclavicular joint, 187
Shear, 74
Shin splints, 259–260
Short bones, 94
Shoulder, 186–206
   acromioclavicular joint, 187
   bursae, 191–192
   coracoclavicular joint, 189
   glenohumeral joint, 189–190
   loading of, 199–202
   movements of, 192–198
   muscles, listing of, 196
   rotator cuff muscles, 190, 192
   scapulothoracic joint, 190–191
   sernoclavicular joint, 187
Shoulder injuries
   dislocations, 203–204
   rotational injuries, 205
   rotator cuff problems, 204–205
   subscapular neuropathy, 206
Sine, 549

Skeletal muscle
   agonists, 164–166
   antagonists, 164–165
   body temperature effects, 177–178
   elasticity of, 147–149
   electromechanical delay, 170
   extensibility of, 147–148
   irritability of, 149
   length changes, 162–164, 169
   muscle fatigue, 176–177
   muscle fibers, 149–160
   muscular endurance, 176
   muscular force generation,
      167–169
   muscular power, 175–176
   muscular strength, 171–174
   neutralizers and stabilizers,
      165–166
   tension development in, 149,
      162–164, 169
   two-joint/multijoint muscles,
      166–167
Skeletal system
   appendicular skeleton, 94
   axial skeleton, 94
   bone, 92–110
   diagram of, 95
   joints, 118–139
Skin friction drag, 516–517
Slow twitch, muscle fibers,
   154–158, 162
Specific weight, 70–71
Speed
   angular, measurement of, 334
   linear, measurement of, 330
Spine
   abnormal curves of, 292–294
   injuries of back/neck, 313–317
   intervertebral discs, 286–290
   ligaments of, 290, 292
   loading of, 304, 307–313
   movements of, 294–296, 298, 304
   muscles of, 296–297, 299–304,
      305–307
   vertebrae, 284–286
   vertebral column, 282–284
Spiral fracture, 108
Spondylolisthesis, 316–317
Spondylolysis, 316–317

Sports biomechanics, study of, 9–11
Sports medicine
    branches of, 6
    definition of, 5
Sports Medicine Division of the
        United States Olympic
        Committee (USOC), 11
Sprains
    ankle, 270–271
    elbow, 213
    nature of, 138
Square roots, 543
Stability, 465–469
    affecting factors, 465, 467–469
    meaning of, 465
Stabilizers, skeletal muscle,
        165–166
Static equilibrium, 452–454
Statics, 3
Static stretching, 136
Stiffness, bone, 91
Strain energy, 423
Strains
    hip, 245–246
    nature of, 92
Strength, muscular, 171–174
Stress, mechanical, 74–75
Stress fractures, 109
    ankle/foot, 272
    vertebrae, 316–317
Stress reaction, bone, 109
Stretching, 132–138
    active stretching, 134
    ballistic stretching, 136
    neuromuscular response to,
        132–134
    passive stretching, 134–135
    proprioceptive neuromuscular
        facilitation, 136–138
    static stretching, 136
Stretch reflex, 133, 134
Stretch-shortening cycle, muscle,
        147, 148
Stroke technique, propulsion, 533–534
Subscapular neuropathy,
        shoulder, 206
Subtalar joint, 265
Summation, motor units, 155
Supination, 39, 41, 42, 269
Surface drag, 516–517

Swimmer injuries
    breaststroker's knee, 260
    swimmer's back, 292–293
    swimmer's shoulder, 204–205
Synarthroses joints, 119
Synovial joints, 119–122
    types of, 121–122

T
Tailor's muscle, 237
Tangent, 549
Tangential acceleration, 387–388
Tarsometatarsal joints, 265
Tendons, and joints, 123–124, 126
Tendon sheaths, 120
Tennis elbow, 215–216
Tensile strength, bone, 91
Tension, 73–74
Tetanus, motor units, 155
Theoretical square law, 515–516
Third class lever, 445–447
Tibiofemoral joint, 247, 253–254
Torque, 71–72, 436–445
    meaning of, 436–441
    resultant joint torques, 441–445
Torsion, 77
Trabecular bone, 92
Trajectory, 344
Transducers, 80
Translation, 33
Transverse fracture, 108
Transverse plane, nature of, 30, 31
Transverse plane movements, 40–43
Trigonometry
    hypotenuse of the triangle,
        548–549
    law of cosines, 550
    law of sines, 550
    Pythagorean theorem, 548–549
    sine/cosine/tangent, 549
    for vector problems, 84
Turbulent flow, 508

U
Ulnar deviation, 38
Upper extremity
    elbow, 206–216
    hand, 219–226
    shoulder, 186–206
    wrist, 216–219

V

Valgus, 272–274
Varus, 272–274
Vector algebra, 81–84
    graphic solutions, 84
    trigonometric solutions, 84
    vector composition, 81–82, 83
    vector resolution, 82–84
Vector composition, 83
Velocity
    angular, measurement of,
        378–380
    linear and angular relationship,
        384–385
    linear, measurement of,
        330–335
    and muscular force, 167–169
Vertebrae
    degeneration/alterations to, 290
    structure of, 284–286
    vertebral column, 282–284
Video analysis, human movement,
    53–55
Viscous drag, 516–517

Visoelastic response, 149
Volume, 67
Vortex generation, 533
Vortex shedding, 533

W

Watts (W), 421
Wave drag, 522–523
Weight, as force, 66–67
Wolff's law, 98–99
Work, 420–421
    meaning of, 420
    measurement of, 421
    work/energy relationship,
        425–428
Wrist, 216–219
    bones of, 217
    injuries of, 226
    movements of, 217–219
    muscles, listing of, 218, 219
    structure of, 216

Y

Yield point, 78